Lecture Notes in Electrical Engineering

Volume 839

The book series *Lecture Notes in Electrical Engineering* (LNEE) publishes the latest developments in Electrical Engineering - quickly, informally and in high quality. While original research reported in proceedings and monographs has traditionally formed the core of LNEE, we also encourage authors to submit books devoted to supporting student education and professional training in the various fields and applications areas of electrical engineering. The series cover classical and emerging topics concerning:

- Communication Engineering, Information Theory and Networks
- Electronics Engineering and Microelectronics
- Signal, Image and Speech Processing
- Wireless and Mobile Communication
- Circuits and Systems
- Energy Systems, Power Electronics and Electrical Machines
- Electro-optical Engineering
- Instrumentation Engineering
- Avionics Engineering
- Control Systems
- Internet-of-Things and Cybersecurity
- Biomedical Devices, MEMS and NEMS

For general information about this book series, comments or suggestions, please contact leontina.dicecco@springer.com.

To submit a proposal or request further information, please contact the Publishing Editor in your country:

China

Jasmine Dou, Editor (jasmine.dou@springer.com)

India, Japan, Rest of Asia

Swati Meherishi, Editorial Director (Swati.Meherishi@springer.com)

Southeast Asia, Australia, New Zealand

Ramesh Nath Premnath, Editor (ramesh.premnath@springernature.com)

USA, Canada:

Michael Luby, Senior Editor (michael.luby@springer.com)

All other Countries:

Leontina Di Cecco, Senior Editor (leontina.dicecco@springer.com)

**** This series is indexed by EI Compendex and Scopus databases. ****

More information about this series at https://link.springer.com/bookseries/7818

P. Satish Rama Chowdary · Jaume Anguera ·
Suresh Chandra Satapathy · Vikrant Bhateja
Editors

Evolution in Signal Processing and Telecommunication Networks

Proceedings of Sixth International Conference on Microelectronics, Electromagnetics and Telecommunications (ICMEET 2021), Volume 2

Springer

Editors
P. Satish Rama Chowdary
Department of Electronics
and Communication Engineering
Raghu Institute of Technology
Visakhapatnam, Andhra Pradesh, India

Jaume Anguera
Department of Electronics
and Telecommunication
Ramon Llull University
Barcelona, Spain

Suresh Chandra Satapathy
School of Computer Engineering
Kalinga Institute Industrial Technology
University
Bhubaneswar, Odisha, India

Vikrant Bhateja
Department of Electronics
and Communication Engineering
Shri Ramswaroop Memorial College
of Engineering and Management
(SRMCEM)
Lucknow, Uttar Pradesh, India

Dr. A.P.J. Abdul Kalam Technical
University (AKTU)
Lucknow, Uttar Pradesh, India

ISSN 1876-1100 ISSN 1876-1119 (electronic)
Lecture Notes in Electrical Engineering
ISBN 978-981-16-8553-8 ISBN 978-981-16-8554-5 (eBook)
https://doi.org/10.1007/978-981-16-8554-5

This Springer imprint is published by the registered company Springer Nature Singapore Pte Ltd.
The registered company address is: 152 Beach Road, #21-01/04 Gateway East, Singapore 189721, Singapore

Organizing Committee

Chief Patron

Er. Prabhat Ranjan Mallick, Chairman KGI and Founder of BEC

Co-chief Patron

Er. Alok Ranjan Mallick, Honorable Chairman, BEC

Convener

Prof. (Dr.) B. N. Biswal, Director (A&A), BEC, BBSR

General Chairs

Prof. Jaume Anguera, Universitat Ramon Llull, Spain
Dr. Suresh Chandra Satapathy, School of Computer Engineering, Kalinga Institute of Industrial Technology (Deemed to be University) Bhubaneswar, Odisha, India
Department of Electronics and Communication Engineering, Shri Ramswaroop Memorial College of Engineering and Management (SRMCEM), Lucknow.

Program Chairs

Prof. P. Satish Rama Chowdary, Department of Electronics and Communication Engineering, Raghu Institute of Technology, Visakhapatnam, Andhra Pradesh, India
Dr. V. V. S. S. S. Chakravarthy, Department of Electronics and Communication Engineering Raghu Institute of Technology, Visakhapatnam, Andhra Pradesh, India

Organizing Committee

Dr. S. Samal (Professor, Aeronautical Engineering, BEC)
Prof. B. P. Mishra (HOD, Mechanical Engineering, BEC)
Prof. P. M. Dash (HOD, EE, BEC, BBSR)
Prof. M. K. Swain (HOD, CSE, BEC, BBSR)
Prof. S. J. Kar (CSE Department), BEC
Prof. G. S. Behera (EE Department), BEC
Prof. D. Panda (H&Sc Department), BEC
Dr. N. Chatterjee (H&Sc Department), BEC
Dr. U. Panigrahi (Mechanical Department), BEC
Dr. S. Sahoo (EE Department), BEC

Technical Advisory Committee

Prof. Xin-She Yang, Middlesex University, London, UK
Prof. Jaume Anguera, Universitat Ramon Llull, Spain
Prof. Nihad I. Dib, Jordan University of Science and Technology, Irbid, Jordan
Prof. Ganapati Panda, IIT Bhubaneswar, Odisha, India
Dr. Steven L. Fernandes, Creighton University, USA
Col. Prof. Dr. G. S. N. Raju, Vice Chancellor, CUTMAP
Dr. Seyedali Mirjalili, Griffith University, Queensland, Australia
Dr. Aurora Andujar, Fractus, Spain
Prof. R. K. Mishra, Berhampur University, Odisha, India
Dr. T. Srinivas, IISC Bengaluru
Prof. N. Bheema Rao, NIT Waranga
Prof. G. Sasibhushana Rao, Andhra University, Visakhapatnam
Dr. Babita Majhi, GG Central University, Bilaspur
Mr. V. Balaji, Head-Wireless Deployment, Reliance Corporate Park, Navi Mumbai
Dr. Lakshminarayana Sadasivuni, Andhra University (Retd.), Visakhapatnam
Prof. P. Rajesh Kumar, Andhra University, Visakhapatnam
Prof. A. Mallikarjuna Prasad, JNTUK
Prof. P. Mallikarjuna Rao, Andhra University, Visakhapatnam

Dr. Ch. Srinivasa Rao, JNTUK, Vizianagaram
Dr. Tummala Surya Narayana Murthy, UCEV, JNTUK, Vizianagaram
Dr. Tumati Venkateswara Rao, Sir C. R. Reddy College of Engineering, Eluru
Dr. Sudheer Kumar Terlapu, Shri Vishnu Engineering College for Women (A), Bhimavaram
Dr. Sonali Dash, Raghu Institute of Technology
Dr. M. Vamsi Krishna, Professor, Dhanekula Institute of Engineering and Technology
Prof. Somya Goyal, Manipal University Jaipur, Jaipur, Rajasthan

Preface

This book is a compilation of research papers presented in the Sixth International Conference on Microelectronics, Electromagnetics, and Telecommunications (ICMEET 2021) which was held at Bhubaneswar Engineering College, Bhubaneswar, Odisha, India, during August 27–28, 2021. Prior to this, five versions of the conference were organized consecutively from 2015 to 2019. Every time, the proceedings are published in Lecture Notes in Electrical Engineering Series of Springer.

The 6th ICMEET is an international colloquium, which aims to bring together academic scientists, researchers, and scholars to discuss the recent developments and future trends in the fields of microelectronics, electromagnetics, and telecommunication. Microelectronics research investigates semiconductor materials and device physics for developing electronic devices and integrated circuits with data-/energy-efficient performance in terms of speed, power consumption, and functionality. The books published so far with the title ICMEET in LNEE discussed various topics like analog, digital and mixed signal circuits, bio-medical circuits and systems, RF circuit design, microwave and millimeter wave circuits, green circuits and systems, analog and digital signal processing, nanoelectronics and gigascale systems, VLSI circuits and systems, SoC and NoC, MEMS and NEMS, VLSI digital signal processing, wireless communications, cognitive radio, and data communication.

ICMEET received 321 submissions. Each paper was peer-reviewed by at least two well-qualified reviewers and the members of the Program Committee. Finally, 101 papers were accepted for publication out of which 50 papers are included in this volume titled *Evolution in Signal Processing and Telecommunication Networks*—which is a collection of manuscripts with theme focusing on signal and image processing, electromagnetics, antenna, microwave circuits, telecommunication networks, optimization, and soft computing.

Several special sessions were offered by eminent professors on many cutting-edge technologies. The conference also featured two keynotes by Prof. (Dr.) Ran Cheng, Southern University of Science and Technology, China, with title "When Evolutionary Computation Meets Big Data: Challenges and Opportunities" and by

Dr. Kamlesh Verma, Scientist, Instruments Research and Development Establishment (IRDE), DRDO, Ministry of Defense, Government of India, with title "Visual Tracking in Un-stabilized Videos." The inaugural of the conference was graced by eminent people like Prof. (Dr.) Ran Cheng, Southern University of Science and Technology, and Prof. (Dr.) Naeem Hanoon, University of Teknologi, Malaysia, as Guest of Honor along with Prof. (Dr.) A. K. Nayak, Advisor—IIBM Group of Institutions, Former President and Fellow CSI as Chief Guest.

We would like to thank Chief Patron Er. Prabhat Ranjan Mallick, Chairman KGI and Founder of BEC and Co-chief Patron, Er. Alok Ranjan Mallick, Honorable Chairman, BEC, for their commitment and dedication to host the conference in their organization. Also, we wish to extend gratitude to Convener, Prof. (Dr.) B. N. Biswal, Director (A&A), BEC, BBSR, for his coordination through the success of the conference. We congratulate all the heads of the departments and the staff members of Bhubaneswar Engineering College, Bhubaneswar, for their efforts in coordinating the proceedings of the program.

Special thanks to all session chairs, track managers, and reviewers for their excellent support. Last but certainly not least, our sincere thanks go to all the authors who submitted papers and all the presenters for their contributions and fruitful discussions that made this conference a great success.

Visakhapatnam, India — P. Satish Rama Chowdary
Barcelona, Spain — Jaume Anguera
Bhubaneswar, India — Suresh Chandra Satapathy
Lucknow, India — Vikrant Bhateja

Contents

About the Editors

P. Satish Rama Chowdary is currently serving as Vice Principal and Head, Department of Electronics and Communication Engineering, Raghu Institute of Technology, Visakhapatnam. He obtained his M.E. from Andhra University and Ph.D. from Department of ECE, JNTUK Kakinada, in the field of Fractal Antennas. He currently holds the position of Executive Committee Member of IEEE Vizag Bay Section. He has 20 years of professional experience. His research interests are computational electromagnetics, fractal antennas and image processing. He has published two patents in the field of IoT and ML. He had published 40 journal and conference papers indexed by SCOPUS and SCIE. He co-edited proceedings of third and fifth ICMEET Conferences. He has also chaired several technical and special sessions in international conferences like FICTA, Soft Computing Informatics, ICMEET-2017, IC3T. He is Senior Member of IEEE, Life Member of Instrument Society of India, ICSES, and Soft Computing Research Society.

Jaume Anguera is IEEE Fellow, Co-founder and CTO of the technology company Fractus Antennas (Barcelona), Associate Professor at Ramon LLull University and Member of the GRITS research group. He is Inventor of more than 130 patents, and most of them licensed to telecommunication companies. He in Inventor of Virtual Antenna™ technology and Author of more than 230 widely cited papers, 6 books and international conferences (h-index 48). He has taught more than 20 antenna courses around the world. With over 20 years of R&D experience and provided solutions to Korean companies such as Samsung and LG. He has directed the master/doctorate thesis to more than 110 students. His biography appears in Who's Who in the World and in Science and Engineering. He is Associate Editor of the *IEEE Open Journal on Antennas and Propagation, Electronics Letters, International Journal of Electronics and Communications* and Reviewer in several IEEE and other scientific journals.

Suresh Chandra Satapathy holds a Ph.D. in Computer Science Engineering, is currently working as Professor of School of Computer Engineering and Dean—Research at KIIT (Deemed to be University), Bhubaneshwar, Odisha, India. He served as National Chairman (Educational and Research) of Computer Society of

India and am also Senior Member of IEEE. He played a key role in organizing more than 60 International Conferences and edited more than 80 Book Volumes from Springer LNCS, AISC, LNEE and SIST Series as Corresponding Editor. His research areas are swarm intelligence, machine learning, data mining and cognitive sciences. He developed two new optimization algorithms known as social group optimization (SGO), published in Springer Journal and Social Evolution and Learning Algorithm (SELO) published in Elsevier. He has more than 150 publications in reputed journals and conference proceedings. He is Part of the editorial board of *IGI Global*, Inderscience, Growing Science journals and also Guest Editor for *Arabian Journal of Science and Engineering* published by Springer. He is Editor-in-Chief of *IJIDSS* from Inderscience and Associate Editor of *KES Journal* from IOS press. I was awarded the Leadership in Academia Award in India by ASSOCHEM for the year 2017.

Vikrant Bhateja is Associate Professor, Department of Electronics and Communication Engineering, Shri Ramswaroop Memorial Group of Professional Colleges (SRMGPC), Lucknow, and also Head (Academics and Quality Control) in the same college. His area of research includes digital image and video processing, computer vision, medical imaging, machine learning, pattern analysis and recognition, neural networks, soft computing and bio-inspired computing techniques. He has more than 90 quality publications in various international journals and conference proceedings. Professor Bhateja has been on TPC and chaired various sessions from the above domain in international conferences of IEEE and Springer. He has been Track Chair and served in the core-technical/editorial teams for international conferences: FICTA 2014, CSI 2014 and INDIA 2015 under Springer-AISC Series and INDIACom-2015, ICACCI-2015 under IEEE. He is Associate Editor in *International Journal of Convergence Computing* (IJConvC) and also an Editorial Board Member of *International Journal of Image Mining* (IJIM) under Inderscience Publishers. At present, he is Guest Editor for two special issues floated in *International Journal of Rough Sets and Data Analysis* (IJRSDA) and *International Journal of System Dynamics Applications* (IJSDA) under IGI Global publications.

Analysis of Throughput and Spectral Efficiency of the CR Users with Channel Allocation

Sadhana Mishra, Ranjeet Singh Tomar, and Mayank Sharma

Abstract In cognitive radio network (CRN), users are intelligent radios which remain aware of the radio channel parameters such as frequency band, number of available channels, power requirement, and modulation schemes to sense and adapt parameters and reconfigure their radios accordingly. Earlier, the effect of channel allocation on spectral efficiency with varying IFQP cycle length is analyzed by the authors. It is studied that spectral efficiency with channel allocation is not only affected by IFQP cycle length but there are other parameters which also affects the performance parameters with channel allocation which is not studied till now as per authors knowledge. In this work, the effect of channel allocation is carried out on the performance parameters such as throughput and spectral efficiency of the CR users. For this purpose, a simulation scenario is designed in NetSim Version 9.0 simulator, and analysis is performed for CRN which consists of primary users and CR users. Further, these parameters are evaluated for varying length of channel bandwidth. Network delay time and payload are also demonstrated for varying number of channels.

Keywords Cognitive radio networks · Channel allocation · Throughput · Spectral efficiency · Payload transmitted · NetSim

1 Introduction

Spectrum is a very expensive resource in wireless communication due to its scarcity. It is found through research that this resource is not precisely utilized which is resulting

S. Mishra (✉) · R. S. Tomar · M. Sharma
ITM University Gwalior, Gwalior, Madhya Pradesh, India
e-mail: sadhanamishra.ec@itmuniversity.ac.in

R. S. Tomar
e-mail: ranjeetsingh@itmuniversity.ac.in

M. Sharma
e-mail: mayanksharma.ec@itmuniversity.ac.in

P. S. R. Chowdary et al. (eds.), *Evolution in Signal Processing and Telecommunication Networks*, Lecture Notes in Electrical Engineering 839,
https://doi.org/10.1007/978-981-16-8554-5_1

in scarcity of spectrum. For proper utilization of this resource, new technology has been evolved which is cognitive radio (CR), and it is considered as the most promising technology for next generation wireless networks to improve spectrum utilization. A CR can change its transmission parameters such as frequency, transmission power, modulation, and bandwidth according to the information obtained from the environment [1]. A spectrum hole has been defined as a band of frequencies assigned to a primary user, but at a particular time and specific geographic location, the band is not being utilized by that user. There is a problem of spectrum scarcity in wireless communication which could be resolved by proper utilization of spectrum. A lot of research studies on CR have been done to resolve the problem of spectrum scarcity [1]. With the rapid growth of wireless services, spectrum resource has become insufficient so it should be utilized more efficiently. Cognitive radio networks can increase the spectrum utilization by permitting cognitive radio users to obtain the spectrum bands which are not utilized by primary users in such a way so it will not create any interference to primary users. Authors in [2] proposed a relay selection approach with channel allocation for CR relay channels. NetSim CR module works on the IEEE 802.22 CR standard, and the standard parameters are set from a given set of parameters. Besides, it can connect a CR with internetwork devices and run all the protocols supported in internetworks [3].

Figure 1 represents a generalized cognitive radio network which is consisting number of cognitive radio users, primary users, base stations, etc. Each cognitive radio user is allowed to access the licensed channels of primary users, which are not utilized by the primary users. Primary users are also known as licensed users/incumbent, whereas cognitive radio users are called as unlicensed users/secondary users.

Cognitive radio was proposed as the means to promote the efficient use of spectrum by exploiting the existence of spectrum holes. In this paper, primary users and cognitive radio users are operating in the 54–144 MHz frequency band. This frequency band is divided into 15 channels, bandwidth of each channel is considered, 6 MHz (54–60 MHz) is utilized by primary user for 10 s, and then cognitive radio users access these bands for their communication in such a way not to cause any

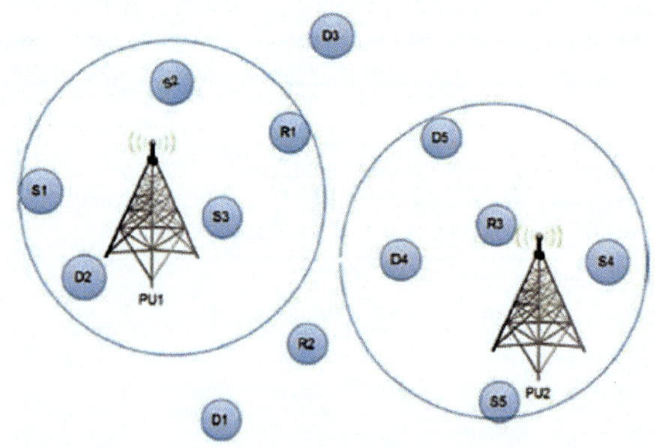

Fig. 1 Generalized cognitive radio network consisting of CR users, base stations, and primary users [8]

interference to primary users. With some interval, cognitive radio users access the band of primary users (Incumbent) in an opportunistic manner. This is also termed as dynamic access of spectrum band, and when primary user is not utilizing this band, a spectrum hole is created. It is also shown that the spectral efficiency is better when there is a greater number of channels available [4]. Further, throughput is also increasing with the increase in number of channels for a considered network scenario.

An important component of the cognitive radio concept is the ability to measure, sense, learn, and be aware of the parameters related to the radio channel characteristics, availability of spectrum and power, radio's operating environment, user requirements and applications, available networks (infrastructures) and nodes, local policies, and other operating restrictions. A spectrum hole has been defined as a band of frequencies assigned to a primary user, but at a particular time and specific geographic location, the band is not being utilized by that user [5]. Cognitive radio was proposed as the means to promote the efficient use of spectrum by exploiting the existence of spectrum holes. In cognitive radio terminology, primary users (incumbent) can be defined as the users who have higher priority or legacy rights on the usage of a specific part of the spectrum. On the other hand, secondary users (CR CPE), which have lower priority, exploit this spectrum in such a way that they do not cause interference to primary users. Therefore, secondary users need to have cognitive radio capabilities, such as sensing the spectrum reliably to check whether it is being used by a primary user and to change the radio parameters to exploit the unused part of the spectrum [6].

As illustrated in Fig. 2, IEEE 802.22 wireless regional area networks (WRANs) are designed to provide broadband access to data networks. According to Federal Communication (FCC) report, temporal and geographical variations in the utilization of the assigned spectrum range from 15 to 85% only. The WRAN systems will use vacant channels in the VHF and UHF bands allocated to the television broadcasting service in the frequency range between 54 and 862 MHz while avoiding interference to the broadcast incumbents in these bands. The wireless communication is known as opportunistic spectrum access and is a feature of cognitive radio. Cognitive radio refers to a smart radio that has the ability to sense the external environment and make intelligent decisions to adjust its transmission parameters

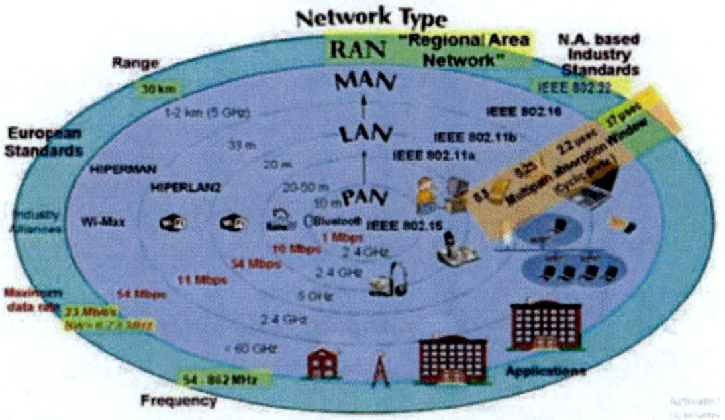

Fig. 2 Different IEEE standards for different networks with frequency bands [3]

according to the current state of the environment. Spectrum sharing is an important task of cognitive radio systems. Cognitive radio technology is predicted to change dynamic spectrum of networks. The spectrum is divided into licensed and unlicensed frequencies. The licensed spectrum is for the exclusive use of designated for primary users(PU) that utilizes UHF/VHF TV bands between 54 and 862 MHz. The unlicensed spectrum can be freely accessed by any secondary users(SU). SUs are equipped channel with cognitive radio capabilities to opportunistically access the spectrum. Cognitive radio capability allows SUs to temporarily access the PU's underutilized licensed channels. To improve spectrum usage efficiency, PUs can access the wireless network resources according to their license, while cognitive radio must combine with intelligent management methods.

A spectrum hole has been defined as a band of frequencies assigned to a primary user, but at a particular time and specific geographic location, the band is not being utilized by that user. Cognitive radio was proposed as the means to promote the efficient use of spectrum by exploiting the existence of spectrum holes. In cognitive radio terminology, primary users (Incumbent) can be defined as the users who have higher priority or legacy rights on the usage of a specific part of the spectrum. On the other hand, secondary users (CR CPE), which have lower priority, exploit this spectrum in such a way that they do not cause interference to primary users. Therefore, secondary users need to have cognitive radio capabilities, such as sensing the spectrum reliably to check whether it is being used by a primary user and to change the radio parameters to exploit the unused part of the spectrum. The effect of channel allocation on spectral efficiency with varying IFQP cycle length is analyzed in [7].

The rest sections of this paper are organized as follows: In Section II, performance parameters are explained. Section III presents modeling and configuration of a network which consists of CR users, base station and primary users. Section IV demonstrates the simulation results of channel allocation and discussion. Finally, Section V concludes this work.

2 Performance Parameters

The different parameters used to analyze the performance of the considered scenario are defined as follows:

Spectral Efficiency: It is the measure of the information rate that can be transmitted over a given bandwidth in a specific communication system. It is also a measure of how efficiently a limited frequency spectrum is utilized by the physical layer protocols and by the media access control (MAC) protocols.

Delay: It is the average amount of time taken calculated for all the packets to reach the destination from the source.

This delay is related to the distance between the user and base station.

Throughput: It is the rate of successfully transmitted data packets in unit time in the network during the simulation. The unit for throughput is usually bits per second.

The maximum data rates of the TX and RX depend on the bandwidth of the channel. The objective of the paper is it represents improving the communications quality of the radio. Maximizing the throughput deals with the data throughput rate of the system. Total user data or payload is delivered to their respective destination every second.

Throughput (in Mbps) = Total payload delivered to destination (in bytes)*8/ Simulation time (in micro seconds).

Interference time: It represents the interference time of the primary users (incumbents) with base station or cognitive radio CPE. The efficiency of spectrum sharing can be improved by minimizing the interference.

Spectrum sharing: It refers to providing the fair spectrum scheduling method, and one of the major challenges in open spectrum usage is the spectrum sharing.

Link Throughput: It refers to the total bytes transmitted over the link. The calculation of total bytes is required to figure out these parameters.

$$\text{Link Throughput (in Mbps)} = \text{Total bytes transmitted over the link} \\ * \text{ 8/Simulation Time (in micro seconds)}$$

3 System Model of a Network with CR Users and Primary Users

In NetSim, CR module works on the IEEE 802.22 CR standard as demonstrated in Fig. 2 different IEEE standards with different frequency bands. As depicted in Fig. 2, IEEE standard for cognitive radio is 802.22, and frequency band allocated is 54–862 MHz band which is licensed UHF/ VHF TV bands. It is found that most of the time and in most of the regions, only some of the bands of this licensed spectrum are utilized so there is no scarcity of availability of spectrum bands, but there is a problem of proper utilization of bands so with proper utilization of unused spectrum bands can increase spectrum efficiency. For all the scenarios, the standard parameters are set from a given set of parameters in NETSIM. In NETSIM, a cognitive radio network can be connected with internetwork devices and supports all the protocols supported in internetworks [3].

A generalized network scenario as depicted in Fig. 1 can be modeled in NetSim by using different end devices, networking devices, and connecting interfaces/links, etc., along with minimum one application. In creating scenario, device placement plays a very important role because basically it determines distance parameter among different nodes in the network. After this, configuring of devices and links are done through which property of each is set accordingly for investigation of a new technology, algorithm evaluation, developing prototype, to measure system performance, etc. It requires to develop a model which is its first step of the simulation. This developed model represents the characteristic parameters of the selected system such as

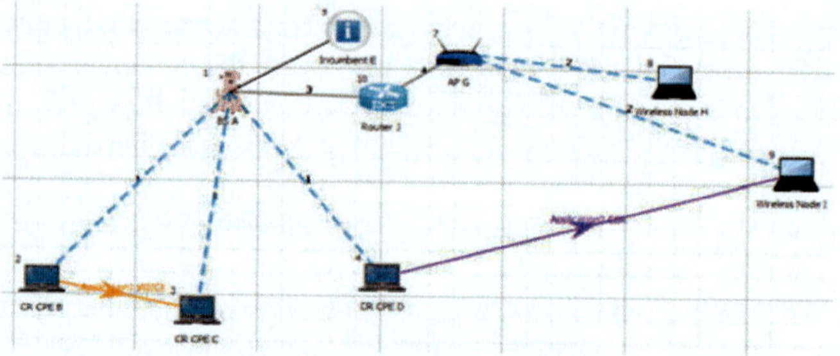

Fig. 3 Cognitive radio network scenario using NetSim

application parameter and network parameter, and operation of the system over time is represented by simulation.

In this section, a cognitive radio network scenario is created using NetSim as shown in Fig. 3. In Fig. 3, one base station, three CR users, one primary user, one router, and two wired nodes are interconnected by using wired/wireless links as shown in figure. As depicted in scenario, two applications are taken into consideration: One application is CR user–CR user communication, and other is CR user with wireless node which is in another network. After creating a scenario or modeling a network, next step is to configure the devices and links as per requirements. Inter-network and intra-network applications are demonstrated. Application 1 is created between CR user 1 and CR user 2. Similarly, Application 2 is created between CR user 3 and wireless node which is in different network.

First turn on the event trace, different parameters of network at different levels can be traced and demonstrated graphically in different figures for different number of channels.

4 Simulation Results and Discussion

This section presents simulation results for considered network scenario. Simulation parameters are assumed as shown in Table 1.

Table 1 Simulation parameters

Type of traffic	CBR
Operating frequency of PU and SU	54–144 MHz
Operational time	10 s
Distance between PU and SU	100 m
Simulation time	100 s
Channel bandwidth	6 MHz

Table 2 Cognitive radio network metrics for for 10 and 15_channel allocations

Network metrics	10 channels	15 channels
Simulation time (ms)	100,000.00	
Packets transmitted	256,578	281,484
Packets errored	3949	5090
Packets collided	386	559
Bytes transmitted (bytes)	12,360,571.0	14,748,816.00
Payload transmitted (bytes)	4,985,540.00	5,941,680.00
Overhead transmitted (bytes)	7,375,031.00	8,807,136.00

Simulation results are obtained by using NETSIM V9.0 network and traffic simulator [3]. In this work, a scenario is created which is represented in Fig. 3. An application throughput for Application 1 which is CR user 1 to CR user 2 and an application throughput 2 which is CR user 3 to wireless node 8 is plotted for varying number of channels. Here, node 1 and node 2 are CR users in Application 1 acting as CR source and CR destination, respectively, which corresponds to intra-network. Similarly, in Application 2, CR user 3 and wireless node 8 are acting as CR source and CR destination, respectively, which corresponds to internetwork.

As shown in Table 1, the operating frequency band for primary and secondary users is 54–174 MHz. This whole range is divided as 6 MHz channels for example, with one channel this operating frequency of primary and secondary user frequency band 54–60 MHz. Similarly, for five channels, range is 54–84 MHz, for ten channels, range is 54–114 MHz, and for 15 channels, range is 54–144 MHz. These different cases are created and evaluated over simulation time and demonstrated through plots in this section.

In Table 2, the values of different network metrics such as simulation time, transmitted packets, number of error packets, number of collided packets, transmitted bytes, transmitted payload, and transmitted overheads have been presented after performing the simulation for 10 and 15 number of channel allocations, respectively. In the same manner, network metrics with varying number of channel allocations such as 1, 5, 10, and 15 are calculated but not shown here due to page limitation, and these results are utilized for calculating performance metrics such as spectral efficiency and throughput of the system.

Figures 4, 5, 6, and 7 are representing link 1 and link 2 throughput with simulation time for 15 and 10 number of channels, respectively. These figures are plotted to analyze the link 1 throughput for different number of channel allocations, i.e., for number of channels of 1, 5, 10, and 15, respectively.

It is noticed from these figures that the link 1 maximum throughput is 2.049737 Mbps and minimum is zero for each case. Threre is another important observation is that the nature of the curve with simulation time for each channel which is showing that its mean and standard deviation are also different in each case which can be demonstrated with Figs. 4, 5 and 6. So we are getting different mean and standard deviation for different number of channels.

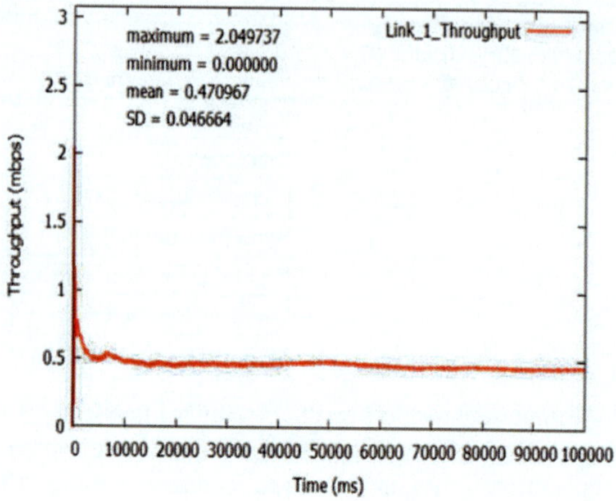

Fig. 4 Link 1 throughput for 15 number of channels

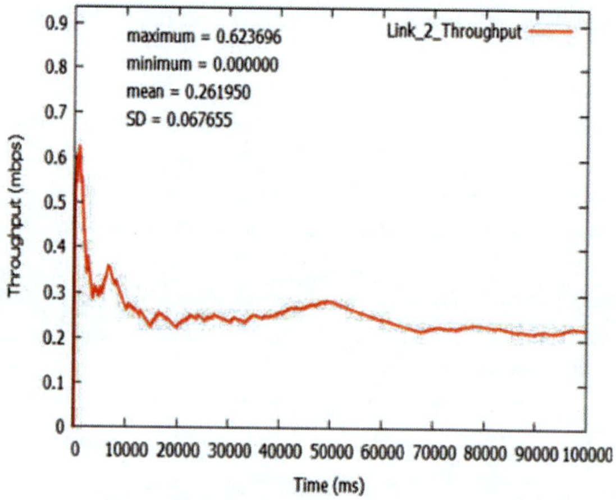

Fig. 5 Link 2 throughput for 15 number of channels

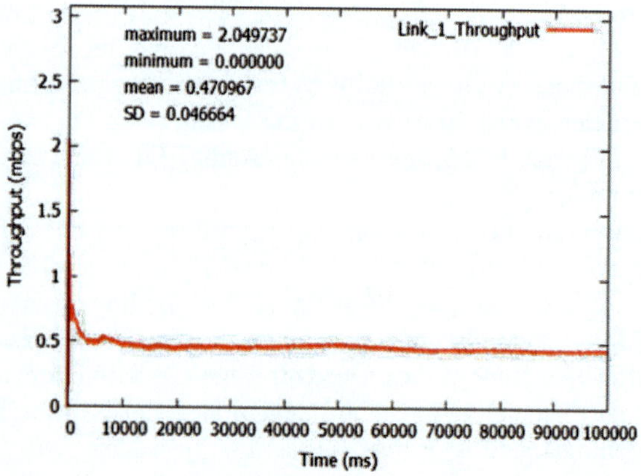

Fig. 6 Link 1 throughput for 10 number of channels

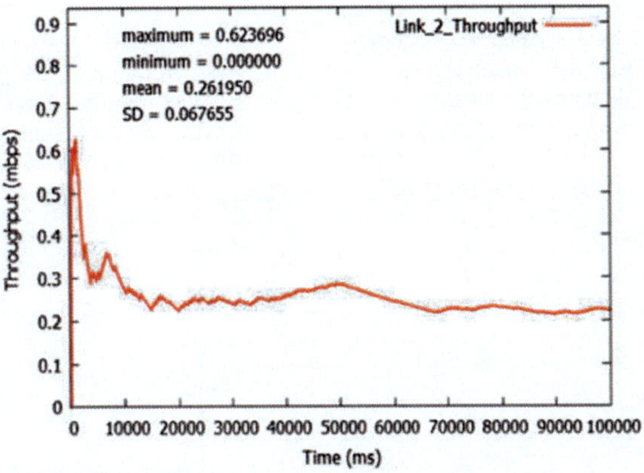

Fig. 7 Link 2 throughput for 10 number of channels

Some of the results are depicted in Figs. 6 and 7 for link 1 and link 2 throughput with simulation time. These results are plotted to analyze the link 1 and link 2 throughput for different number of channels.

It is noticed from these figures that the link 1 maximum throughput is 2.049737 Mbps and minimum is 0.00 for each case same as above. Further, Fig. 8 shows that the maximum throughput of link 2 is 0.6236 Mbps and minimum is zero.

Figure 8a, b demonstrates APP 1 throughput with simulation time, and Fig. 9a, b represents APP 2 throughput with simulation time. Here, only some of the results are demonstrated, and all results are utilized for analyzing the network performance parameters.

In Fig. 10, payload transmitted is demonstrated with different number of channels. It is noticed that operational time parameter is decreasing with increase in operational interval. As cognitive radio users will get more time to utilize the frequency band which is allocated to primary user, operational time of primary user will decrease.

In Fig. 11, throughput is shown with different number of channels. It is observed that as the number of channels increases, the throughput increases because users are getting more time to utilize more number of channels, as cognitive users are getting more benefit of spectrum access and primary user goes for idle.

Figure 12 demonstrates a very important parameter of cognitive radio network which is spectral efficiency. It is noticed in Fig. 12 that spectral efficiency performance metric is increasing with increase in number of channels, and it is because cognitive radio users are fully utilizing the spectrum of primary users.

It is noticed in Fig. 12 that the maximum spectrum efficiency is 0.08927 Mbps/Hz for 15 number of available channels.

In Fig. 13, interference time of primary user is demonstrated with different number of channels. It is noticed that this parameter is having some random nature. Basically, this interference time represents the interference of primary user with cognitive radio users or base station.

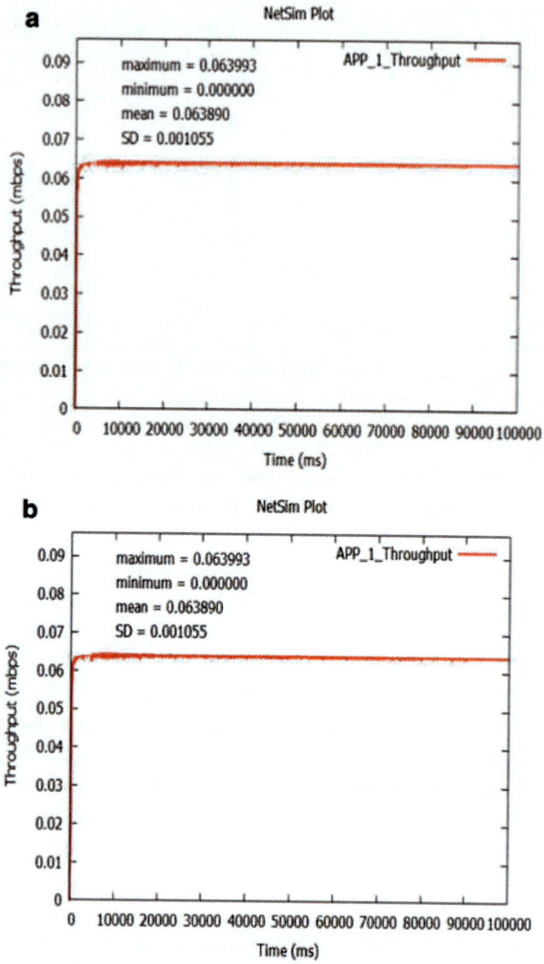

Fig. 8 **a** APP_1 throughput for 15_channel allocation. **b** APP_1 throughput for 10_channel allocation

5 Conclusion

In this work, performance of secondary users is analyzed with varying number of channel allocation in a scenario where CR users, base station, and primary users are present with varying channels. These performance parameters are analyzed using NETSIM network simulator. Performance parameters such as spectral efficiency, throughput, interference time, payload transmitted, and packet transmitted are analyzed and plotted graphically in NETSIM. It is noticed that with the increase in number of available channels, secondary users get more opportunity to send their data and utilize the available spectrum band in an opportunistic manner. As a result, both the performance parameters, i.e., spectral efficiency and throughput, increase which is also a practical case because at this time secondary users are getting more

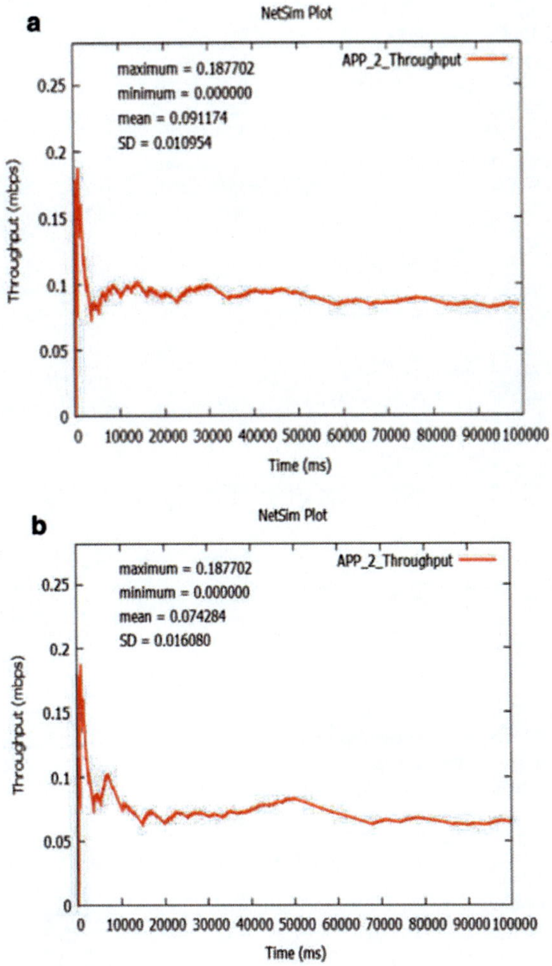

Fig. 9 **a** APP_2 throughput for 15_channel allocation. **b** APP_2 throughput for 10_channel allocation

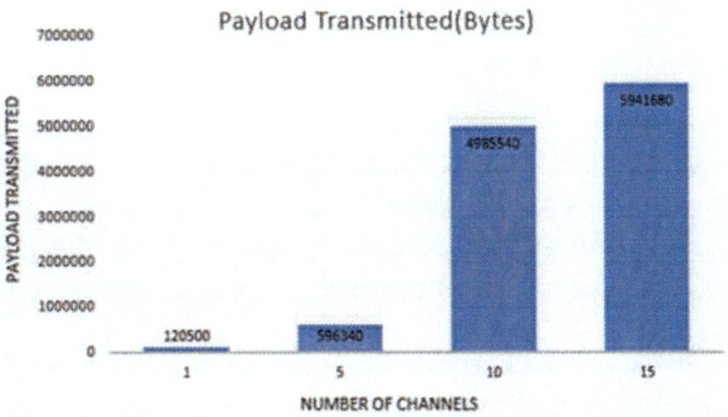

Fig. 10 Payload transmitted verses number of channels

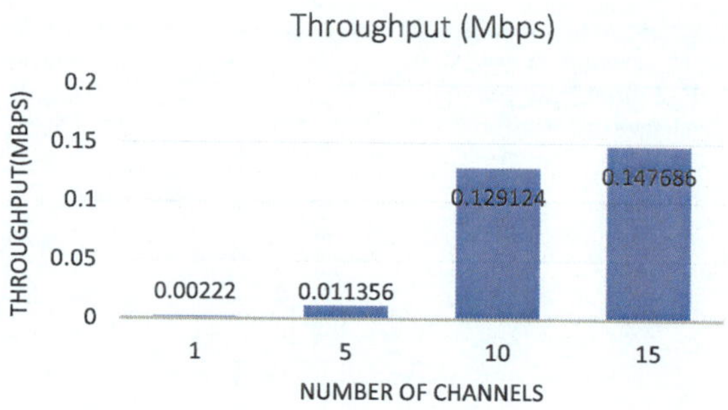

Fig. 11 Throughput (in Mbps) verses number of channels

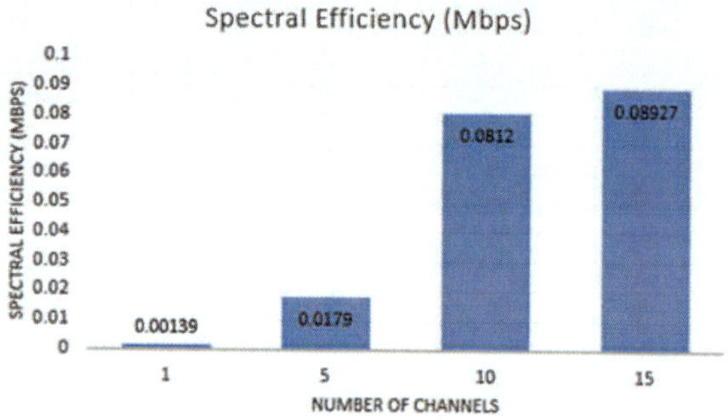

Fig. 12 Spectral efficiency (in Mbps) verses number of channels

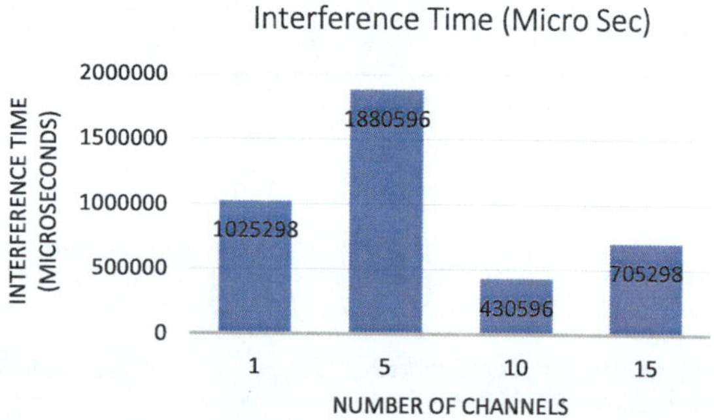

Fig. 13 Interference time (in microseconds) verses number of channels

opportunities for their operation with increase in channel allocation. Further, network parameters are also demonstration and discussed.

In the future, this work can be extended with the varying number of Primary User or Secondary Users and performance can be analyzed.

References

1. Haykin S (2005) Cognitive radio: brain-empowered wireless communications. IEEE J Sel Areas Commun 23(2):201–220
2. Mishra S, Trivedi A (2014, September) Relay selection with channel allocation for cognitive radio relay channels in CRN. In: IEEE eleventh international conference on wireless and optical communications networks (WOCN), pp 1–4
3. Network-Simulator-NetSim-www.tetcos.com
4. Mishra S, Trivedi A (2013, August) Exploiting opportunistic decode-and-forward cooperation for cognitive radio relay channels in multi-antenna cognitive radio networks. In: IEEE international conference on advances in computing, communications and informatics (ICACCI), pp 155–158
5. Mishra S, Trivedi A (2017) Performance study of MIMO transmissions with joint channel allocation and relay assignment. Wireless Personal Communication (WPC). Springer, Berlin. https://doi.org/10.1007/s11277-017-4317-x
6. Srinath G, Mishra S, Trivedi A (2018) Hybrid cognitive Gaussian two-way relay channel: performance analysis and optimal resource allocation. Phys Commun 27:106–115
7. Saifuddin KM, Ahmed AS., Reza KF, Alam SS, Rahman S (2017) Performance analysis of cognitive radio: NETSIM viewpoint. In: International conference on electrical information and communication technology (EICT), pp 7–9
8. Mishra S, Trivedi A (2017) Performance study of an improved resource allocation scheme with full-duplex relaying. Wireless Personal Communication (WPC), Springer.https://doi.org/10.1007/s11277-017-5002-9
9. Mitola J, Maguire GQ (1999) Cognitive radio: making software radios more personal. IEEE Pers Comm 6(4):13–18. https://doi.org/10.1109/98.788210.ISSN 1070-9916
10. Li L, Zhou X, Xu H, Li GY, Wang D, Soong A (2011) Simplified relay selection and power allocation in cooperative cognitive radio systems. IEEE Trans Wirel Commun 10(1):33–36
11. Mishra s, Trivedi A (2017) Performance study of MIMO transmissions with joint channel allocation and relay assignment, vol 96, issue No. 2. Wireless Personal Communication, Springer, Berlin, pp 2651–2665
12. Mishra S, Tomar RS, Ojha SS (2020, July–September) To analyse the performance parameters of CRN in the presence of the primary users. Int J Interdisc Res Innov 8(3):76–88

On Performance Improvement of Wireless Push Systems Via Smart Antennas

Ramesh Gottipati and L. Balaji

Abstract A typical network unit of a communication system consists of clients. A closed loop path is adopted, while information is pushed to the clients. The clients take the information and act according to the instructions. Usually, directive radiating systems are used for communicating between the users. However, in the due course of time, multi-directional radiating systems are employed by the broadcast servers. Unfortunately, the servers often fail to make use of the potential features of the system due to the geographical distribution of the clients bounded by the coverage area. In this paper, an adaptive smart antenna-based wireless push system is proposed. The smart antenna system can alter its beamwidth as the distribution of the clients. The technique also invites modifications to the broadcast schedules. A significant enhancement in the performance is witnessed finally.

Keywords Smart antenna · Beamwidth · Wireless push systems · Client distribution

1 Introduction

In a typical push system, the server acts as the broadcast unit from where the information is disseminated to the clients [1–10]. The push-based system (PS) helps all the clients to receive common information. In this case, they are supposed to blindly follow the instructions. When compared with the pull system, the push system has prominent applications mostly in telecommunication field. Moreover, the initialization expenditure is also quite low. Several low-cost broadcasting algorithms are incorporated for an efficient and reliable push system implementation. These algorithms are used to instill the adaptiveness which directly deals with the demands of client (Learning automation is one such technique of inducting the adaptiveness) [11–17]. In specific, this is useful in the push systems for underwater wireless networks.

R. Gottipati (✉) · L. Balaji
Vel Tech Rangarajan Dr. Sagunthala R & D Institute of Science and Technology, Avadi, Chennai, TamilNadu, India

P. S. R. Chowdary et al. (eds.), *Evolution in Signal Processing and Telecommunication Networks*, Lecture Notes in Electrical Engineering 839,
https://doi.org/10.1007/978-981-16-8554-5_2

In this work, the initial remarks are dedicated to the wireless data broadcasting techniques and systems. The major contribution of the system is to provide necessary information which is to be conveyed to the clients through schedules as per the existing or prevalent demands. This typically makes use of several algorithms which has been the focus point, while the core objective is to minimize the response time. This strategically effects the performance of push-based system and the reliability of the data broadcasted.

The PS cannot serve the application which has no prior information of the client demand. In [1], an adaptive PS is presented which has automated its learning, thereby adaptively taking decisions and serving the clients. This subsequently diminishes the complex computational steps. The technique incorporates a feedback system which can instill the adaptiveness in the structure through a decision-making scenario. A ten-element planar two-dimensional array is proposed for the purpose which is neat curated along the chassis of the radio gadget. Every element in the array refers to the loop antenna which is shorted at one end and operating in the λ/2 mode. However, the choice of the radiating element is according to the requirement in terms of radiation characteristics and other related features. The loop antenna also provides the ease of miniaturization and can easily blend with the decorum of the system and operates in the spectrum of 3.5 GHz.

A technique of sharing the same radiating system between the uplink and downlink has been proposed and presented as shown in Fig. 1. This refers to a typical antenna architecture with the capability of beamforming features along with excellent spectrum efficiency. The optimization concept makes use of the block coordinate descent technique with NP-hard in order to propose appropriate solution to beamforming and

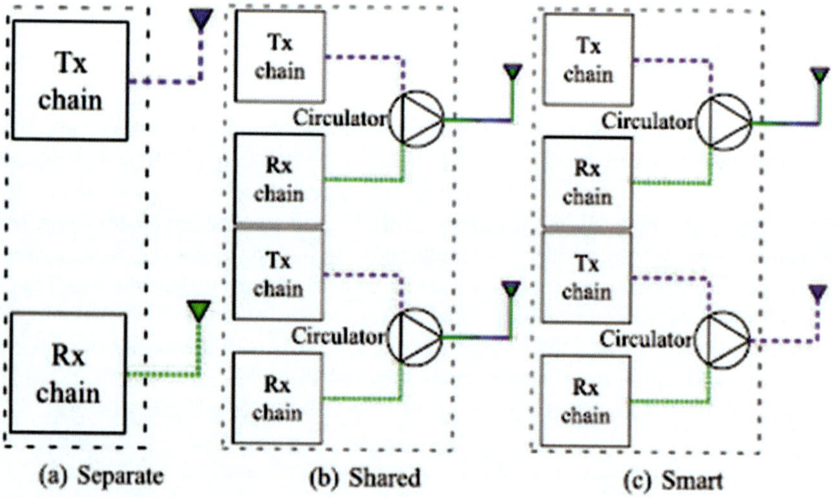

Fig. 1 Architecture representation of the full-duplex proposed smart system

power allocation problem. This ensures a very low level of self-interference power which is not a possible task in the half-duplex models.

Further, the manuscript is divided into five sections in which the resource allocation framework is discussed in Sect. 2 and the proposed model in Sect. 3. Following, the results are presented in Sect. 4, and overall conclusions are mentioned in Sect. 5.

2 Resource Allocation Technique

The performance of a push-based system relies heavily on the proper scheduling of the broadcast data. And it is possible to achieve the steer the beam by manipulating the modulated echo eliminates and minimizes the effects of component instability [5]. This typically involves in handling the phase characteristics in the frequency regime. The radio over fiber [6] is a technique by which the radio signals are communicated over the optical fiber system. It finally contributes to the broadbanding in an efficient scenario. This also provides a very low-loss transmission with minimal interference as it uses the fiber-optic means. The corresponding BER and eye diagram reveals the fact that it is feasible.

Reconfigurable intelligent surfaces (RISs) proposed in [8] belong to the class of smart environment in radio engineering. This includes the application of metamaterials and the array configuration to handle the beam steering and reconfigurability in antennas.

In [10], a rigorous insight into the 5G has been reported. And it's supported by not only smartphones but also different IoT devices to provide different services like smart building, smart city, and many more which will require a 5G antenna with low latency, low path loss, and stable radiation pattern clearly indicates the evolution of new technology with high-speed communications and advanced systems involving the Internet of Thing (IoT), multimodal antenna systems with selective operating conditions which features multiple objectives and multiple constraints are proposed in [10]. Cascaded radiating system is another technique proposed in [11] which can efficiently extend the coverage in the mm Wave 5G systems. A typical system model is given in Fig. 2 [14], where several mobile users are treated as the subscribers in which the content is directly pushed to them. This content is derived from the content provider and transferred to the user equipment in one hop.

3 Proposed System

From the above model presented in Fig. 2, it is evident that there are certain noticeable issues which have a direct impact on the performance of the system. These are termed as disadvantages and need to overcome through a new model which is consistent. The proposed model is based on the smart antenna configuration with capability to reschedule and adaptiveness inherently possessed. The proposed system makes

Fig. 2 System model

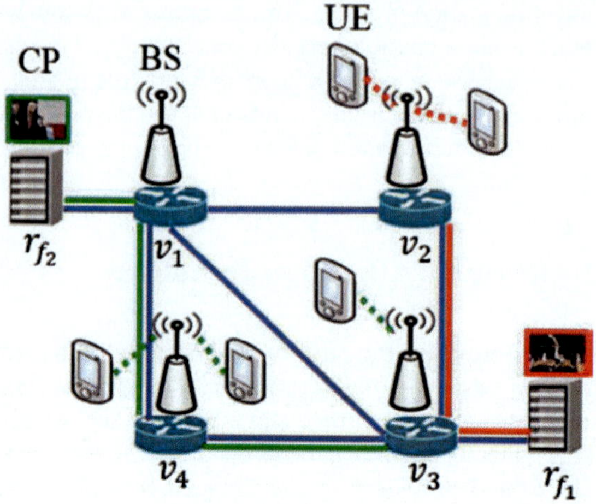

use of multiple antennas with multi-directional capabilities. This means that they are capable of changing the beam adaptively to any direction with a minor modifications to its electrical parameter distribution. This helps to frame out the broadcast server to increase performance. In spite of this, it is evident that the model of broadcasting system often fails to utilize the complete potential expressed by the multiple antenna system. However, this is not under consideration with the hindrance of the corresponding geographical distribution of clients. These clients are randomly distributed through the area known as the antenna coverage area[15]. Considering all the above, the proposed model is curated with adaptive features as given in Fig. 3. The adaptive smart antenna-based wireless push system has the features of controlling the

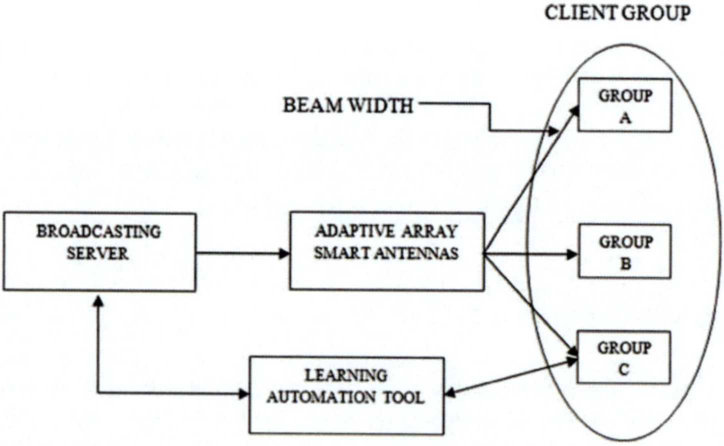

Fig. 3 Proposed system

beamwidth using the inherent characteristics of the smart antenna. It is possible to control the beam, sidelobe energy distribution as well as the null position in the radiation pattern using the corresponding design parameters which are both electrical and spatial with respect to the array configuration [18–20]. For instance, the current locations of the clients can be the radiation pattern of the smart array within the system area. This feature is blended with modifications to the broadcast schedules. This has to be achieved using the learning automation tool. This is typically performed at the broadcasting server side. This proposed model essentially enhances the performance as considered observed by from the perspective of the clients Demand.

The learning automation tool is the essential difference between the conventional and the smart system. It accomplishes adaptive decision-making robust to random environment. It learns and adjusts itself to an optimal action with the aid of iterative interactions with its environment [16]. The concerned probability distribution is used to update and derive new environment response accomplishing to automation. This finally leads to a particular action which is entirely based on the previous environmental responses. The learning automation with probability distribution can be used to determine the significance of the client in the given service area for a particular antenna. This is demonstrated in the proposed model structure given in Fig. 3. Using this learning automation tool to gets the information excludes the items that never demanded by the clients within the coverage area of the antenna system.

The computation of the signal-to-noise ratio is given as

$$\text{SNR} = \frac{1}{\sigma_N^2} \cdot P \quad (1)$$

Here, P refers to power transmitted, and the noise power is represented by σ_N^2.

4 Simulation Results

In this section, results pertaining to the study of performance of the proposed system are presented in terms of various plots. These plots provide a statistical and graphical reference to the insight of the performance. In the line, the mean response time is computed for several values of N1 which refers to the fixed antenna count. These fixed antennas performance is compared with that of corresponding number of the smart antenna network as shown in Fig. 4. It is evident from the plot that the smart antenna system has better response time and almost unaltered with increasing number of antennas in the smart network.

Similarly, in Fig. 5, the performance pertaining to five networks referred as N5 is evaluated between the smart and fixed antenna systems. The mean response time for various number of antennas in a five-network scenario is evaluated and plotted against the skew coefficients. Further, in Fig. 6, the same kind of study has been extended with the six networks this time, and the response time is computed accordingly and

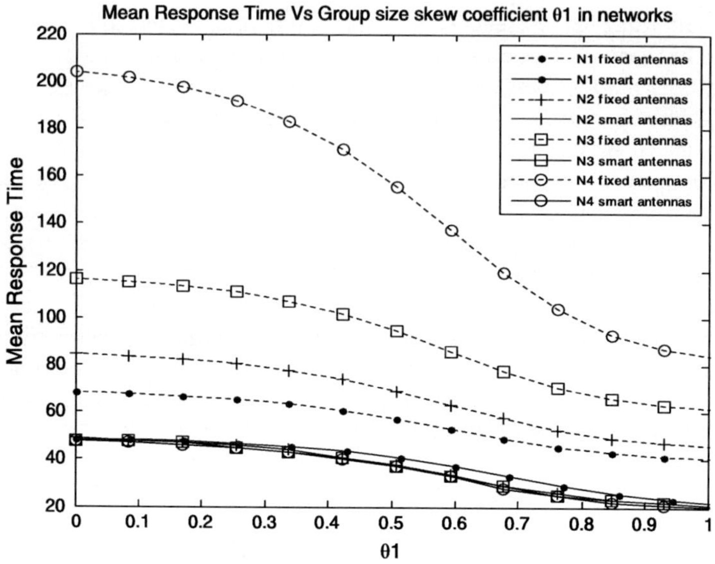

Fig. 4 Performance of smart antenna system

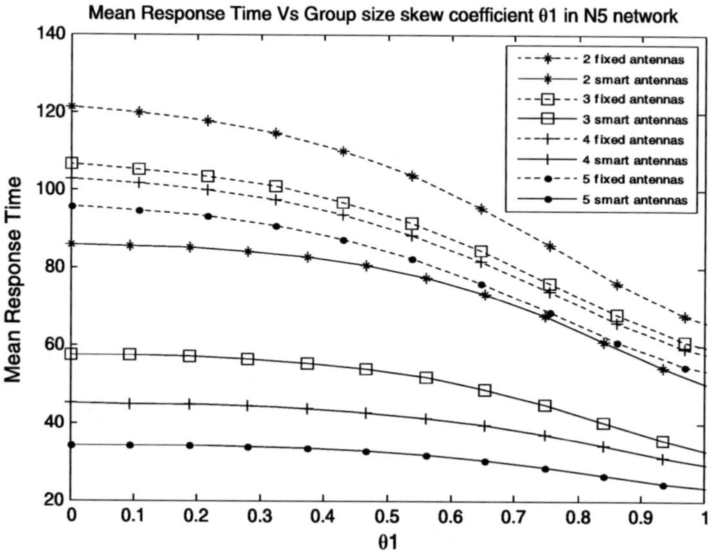

Fig. 5 Performance evaluation within five-network scenario

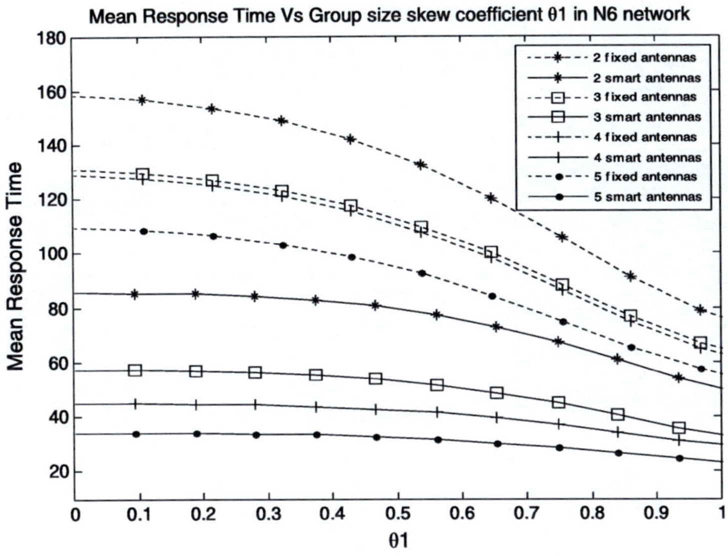

Fig. 6 Performance over six-network scenario

plotted. The performance of the smart antenna system is consistent with number of antenna elements and the networks and clearly shows its dominance.

The evaluation of the smart antenna system with respect to fixed antenna systems can be performed over the probability distribution characteristics. This is graphically presented in Fig. 7 where several probability distributions are considered and applied to both the fixed and smart antenna systems. It is possible to elevate the characteristics of the smart system over the conventional radiating system. The mean response time is certainly much less in the case of the smart systems.

The similar kind of response is witnessed in the case where the smart system is evaluated for several mobile client configurations. The number of antennae in every step of the study is varied from N1 to N5, and the corresponding mean response time with respect to the considered mobile client is evaluated and plotted in Fig. 8. The performance of the smart system is once again elevated as the reported mean response time is always less in every scenario simulated.

5 Conclusion

The smart antenna system is successfully simulated and elevated its performance in this work. The demonstration of varying the beamwidth in order to appreciate the smart and adaptive nature of the antenna is presented with the help of the simulated results. Once the antenna is assigned, it starts excluding the regions of no interest or technical no coverage from its schedules and starts broadcasting schedule the

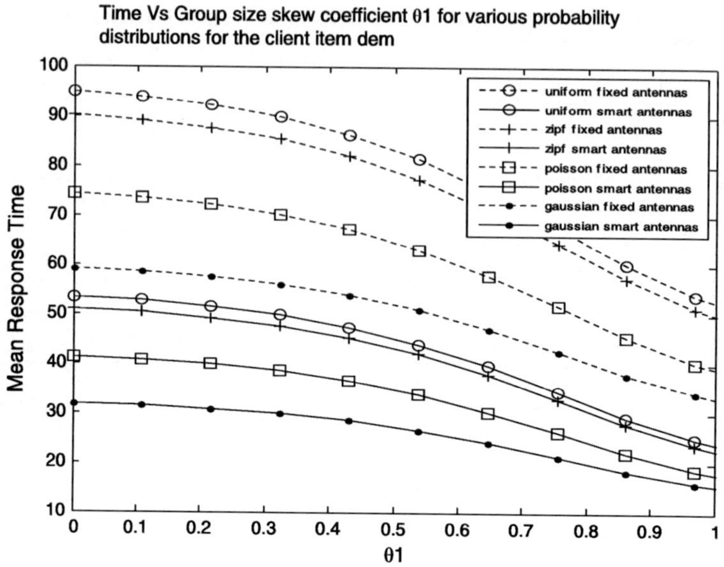

Fig. 7 Performance of several distribution scenarios

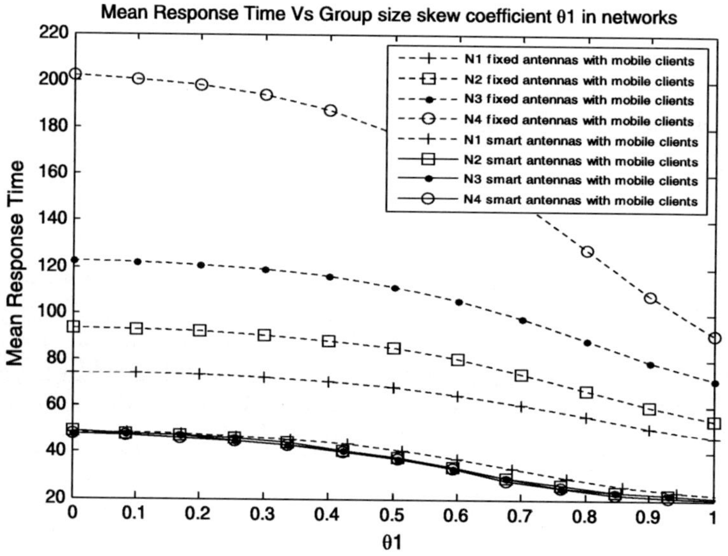

Fig. 8 Performance of several mobile clients

information items that refer to geographic areas that are out of its coverage. From the simulation results, it is evident that the antenna with smart features is dominant in every study when compared with the conventional fixed antenna system.

References

1. Alja'afreh SS, Altarawneh B, Alshamaileh MH, Almajali ER, Hussain R, Sharawi MS, Xing L, Xu Q (2021, February 23) Ten antenna array using a small footprint capacitive-coupled-shorted loop antenna for 3.5 GHz 5G smartphone applications. IEEE Access J
2. da Silva JMB, Ghauch H, Fodor G, Skoglund M, Fischione C (2021) Smart antenna assignments essential in full-duplex communications. IEEE Trans Commun 69(5)
3. Xu S, Song X, Gao Y, Li S, Cao J, Xie Z (2021) Secrecy-enhancing design for two-way energy harvesting cooperative networks with full-duplex relay jamming. IEEE Trans China Commun
4. Zhang X, Qi H, Zhang X, Han L (2020) Energy-efficient resource allocation and data transmission of cell-free internet of things.IEEE Internet Things J. 10.1109/ JIOT.2020.3030675
5. Siddiqui TA, Holopainen J, Viikari V (2021) Beam-steerable transponder based on antenna array and phased modulators. IEEE Antennas Wirel Propag Lett 20(3)
6. Li G, Deng J, Xin S, Huang X () A Radio over fiber system compatible with 3G/4G/5G for full spectrum access and handover with multi-scenarios. IEEE J Lightwave Technol. https://doi.org/10.1109/JLT.2021.3065568
7. Jaglan N, Gupta SD, Sharawi MS (2021) 18 element massive MIMO/diversity 5G smartphones antenna design for Sub-6 GHz LTE bands 42/43 applications. IEEE Open J Antennas Propag. Date of Publication 20 Apr 2021
8. Di Renzo M et al (2020, June 16) Reconfigurable intelligent surfaces vs. relaying: differences, similarities, and performance comparison. IEEE Open J Commun Soc
9. Pal A et al (2020, April 20) Novel interference suppression null steering antenna system for high precision positioning. IEEE Open Access J
10. Kumar S, Dixit AD, Malekar RR, Raut HD, Shevada LK (2020, September 1) fifth generation antennas: a comprehensive review of design and performance enhancement techniques. IEEE Open Access J.
11. Dong J, Wang S, Mo J (2020, May 12) Design of a twelve-port MIMO antenna system for multi-mode 4G/5G smartphone applications based on characteristic mode analysis. IEEE Access J
12. Kim E-S, Sung M, Lee JH, Lee JK, Cho S-H, Kim J (2020) Coverage extension of indoor 5G network using RoF-based distributed antenna system IEEE Access J. Date of Publication 26 Oct 2020
13. Blázquez-García R et al Smart relay architecture for over-the-horizon high quality communications with unmanned aerial vehicles
14. Hu J, Zhao Y, Yang K (2019) Modulation and coding design for simultaneous wireless information and power transfer. IEEE Commun Mag
15. Chen Z, Tao X et al (2019) Joint minimization of wired and wireless traffic for content delivery by multicast. IEEE Trans Wirel Commun 18(5)
16. Nicopolitidis P, Kakali V, Papadimitriou G, Pomportsis A (2012) On performance improvement of wireless push systems via smart antennas. IEEE Trans Commun 60(2)
17. Liaskos C, Petridou S, Papadimitriou G (2011) Towards realizable, low cost broadcast systems for dynamic environments. IEEE Trans Netw 19(2):383–392
18. Ahammed MJ, Swathi A, Sanku D, Chakravarthy VVSSS, Ramesh H (2018) Performance of firefly algorithm for null positioning in linear arrays. In: Proceedings of 2nd international conference on micro-electronics, electromagnetics and telecommunications. Springer, Singapore, pp 383–391

19. Chakravarthy VVSSS., Rao PM (2015) Circular array antenna optimization with scanned and unscanned beams using novel particle swarm optimization. Indian J Appl Res 5(4)
20. Chakravarthy VS, Rao PM (2015, March) Amplitude-only null positioning in circular arrays using genetic algorithm. In: 2015 IEEE international conference on electrical, computer and communication technologies (ICECCT). IEEE, pp 1–5

Hybridization of RF Switch and Related Aspects

Nirupam Sharma, Suma S. Lonkadi, B. H. M. Darukesha, Kamaljeet Singh, and A. V. Nirmal

Abstract Single pole dual throw switch is an important component in RF circuits. Traditional switch on the PCB has lot of criticalities such as repeatability, bigger footprint and reliability. Hybridization of the circuit results in overcoming these issues. It also needs development of various new process, selection of components, package, layouting etc., which play a very important role in reliable switch development. Thin film technology is employed for the realization of the switch. The realized circuit tested at ~400 MHz and shows insertion loss <1.5 dB and isolation of >45 dB.

Keywords Thin films · Photolithography · Hybrid microcircuits · Wire bonding

1 Introduction

RF Switch is a single pole dual throw (SPDT) switch used for switching between coherent carrier generation from receiver and non-coherent carrier generation from TCXO. It is designed to operate at ~400 MHz and has a typical specification of insertion loss <1.5 dB and isolation of >45 dB. The proposed development work was carried out in an effort to indeginize the RF Switch HMC used in S Band data transmitter. The pin configuration is presented in Fig. 1. Thin film hybrid microcircuit (HMC) technology is an electronic packaging technology realized on a ceramic substrate with metal pads and interconnection etched out using photolithography and wet etching processes. Component attachment to metal pads are carried out using conductive adhesives or solder, interconnections between active device dice and substrate are carried out using gold wire bonding and interconnection with package pins are carried out using gold ribbons. Wire bonding of 1.5 mil gold wire on thin film metallized pattern and component soldering for thin film HMC were the new processes qualified.

N. Sharma (✉) · S. S. Lonkadi · B. H. M. Darukesha · K. Singh · A. V. Nirmal
U R Rao Satellite Centre, Indian Space Research Organization, Bangalore 560017, India

P. S. R. Chowdary et al. (eds.), *Evolution in Signal Processing and Telecommunication Networks*, Lecture Notes in Electrical Engineering 839,
https://doi.org/10.1007/978-981-16-8554-5_3

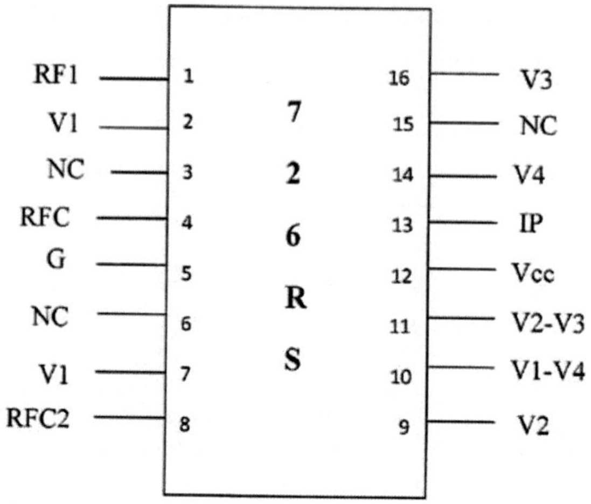

Fig. 1 Pin configuration of RF switch HMC

The development of RF Switch HMC for indegenization had a challenge of one to one replacement of imported module both in terms of package size and electrical specifications. The HMC development involved feasibility study for hybridization, component selection based on available substrate real estate, new process qualification, development of proto module, testing and qualification of the product for aerospace application.

2 Thin Film Technology for Hybridization

2.1 Feasibility Study

Thin films techniques are adopted at frequencies higher than 30 MHz, where grounding through shortest path, minimum track lengths, metallic enclosures translate to better performance. Feasibility of hybridization of a circuit encompasses the holistic effort towards the development of HMC. The available substrate footprint constraints dictated the component selection and termination required, SMT component and active devices availability in die form. Destined pin configuration of the HMC posed constraints on component placement, routing and providing shortest wire bond lengths to minimize losses. Development of HMC encompasses various steps such as substrate footprint selection, component selection, wire bond orientation, selection of ribbon bonds, package and pin selection. Typical flowchart for development of RF Switch HMC is presented in Fig. 2.

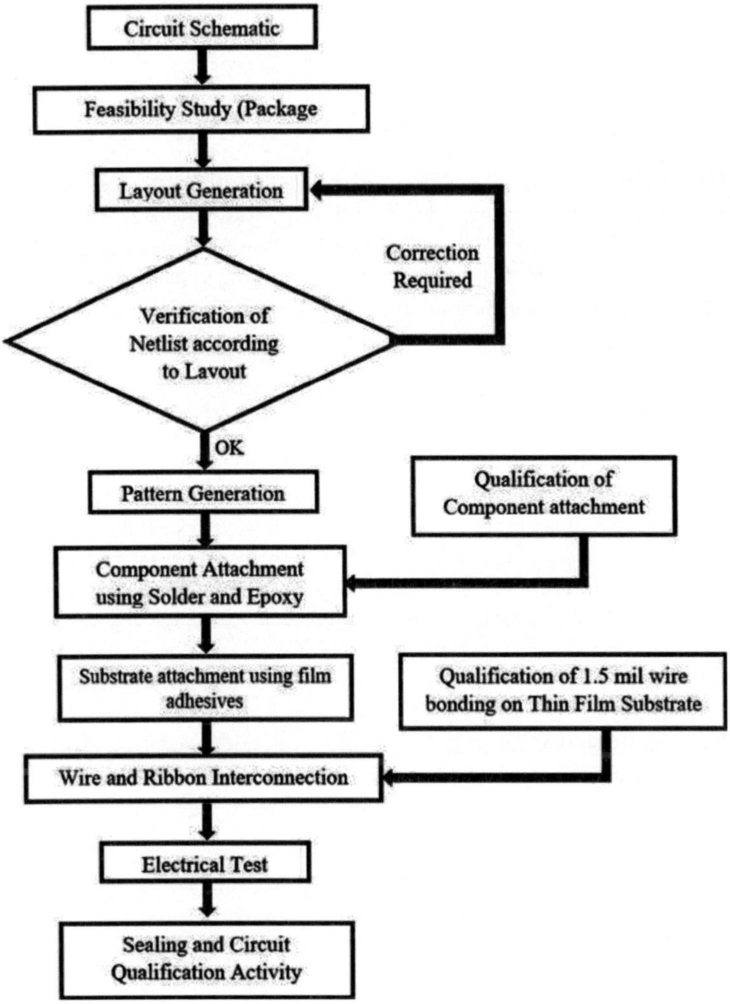

Fig. 2 Flowchart of development of RF switch HMC

2.2 *Component Selection*

All active components (ICs and Diodes) are used in bare form which confirms to MIL-PRF-Standards with Class-S element evaluation. Other passive components like chip capacitors, chip resistors and chip inductors are of space grade with failure Level-S. All these parts used are processable and mountable using conductive/ non-conductive adhesives or solder. Components are attached using epoxy and In:Pb::50:50 solder due to availability of various terminations.

2.3 Layout Generation

Single layer layout was generated following the schematic and layout design guidelines using CAD tool. Selection of solder terminated component reduced the interconnection pads size by 200 sq.mil. Pads requiring immediate ground connections were extended to the edges. Component placement, routing of interconnections and providing shortest wire bond lengths to minimize losses were carried out without deviating from the standards to accommodate in the package size of 0.975 in. × 0.775 in. Verification of netlist was carried out according to the schematic and layout was generated for pattern generation on 25 mil thick Alumina substrate.

2.4 Thin Film Technique

Photolithography mask plate was filmed using the layout design of the HMC. Two inch square substrate with Cr/Cu/Au metallization was spin coated with positive photoresist. UV exposure was carried out prior to development of patterns in the substrate. The pattern generation carried out using wet etching process followed by scribing using Nd:YAG laser to get the substrate size of 0.842 in. × 0.524 in. Wrap around ribbon connection were made between the pad and the ground plane. Device and components were attached to the substrate pads using electrically conductive epoxy using die bonding techniques. Few components were soldered to the pads using InPb(50:50) solder. Substrate to package base attachment was carried out using electrically conductive adhesive film. 1 mil wire connections were made from bare die to substrate pads and 1.5 mil wire bondings were required for post bonding. Ribbons were bonded between pads and package pins.

3 New Processes Qualified

3.1 Wire Bonding of 1.5 mil Gold Wire

Jumper connection between pads were carried out using 1.5 mil gold wire using ball and wedge bonding technique. Wire of dia 1.5 mil was chosen since the distance between the pads were long and 1 mil wire exhibited sagging which lead to shorting of pads. Dummy substrate pads were cleaned thoroughly using isopropyl alcohol followed by vapour decreasing for 50 s. Wire bondings were carried out and subjected to 300 °C for hour prior to destructive pull testing (DPT). Following the promising DPT results, wire bonds on Proto HMC were made and non-destructive pull test (NDPT) was carried out (Table 1).

Table 1 Optimized wire bonding parameters for RF switch HMC

Parameter	1 mil wire bonds	1.5 mil wire bonds
1st bond force (cN)	27	40
1st bond time (ms)	27	40
2nd bond force (cN)	50	70
2nd bond time (ms)	35	100
DPT strength (gF)	8.46	>20

3.2 *Soldering of SMT Components*

Fabrication of HMC usually favours attachment of components using Epoxy adhesives compared to solder attachment due to the demerit of flux residue presence after solder attachment. The challenges with RF switch HMC was attachment of 12 components using solder without leaving any flux residues. To achieve this, the component attachment process was initiated by soldering components to the respective pads followed by intensive cleaning using vapours of isopropyl alcohol and inspection. After visual inspection clearance, the substrate with soldered components were put for vacuum baking at 90 °C and 10 milli torr pressure for three days. Post vaccum baking, inspection for flux residues was carried out and cleared for epoxy attachment of rest of the components.

4 Realization of Prototype Module

Prototype module of RF Switch HMC was fabricated using the thin film technique discussed above. KOVAR package of size 0.975 in. × 0.775 in. with 16 pins was used for packaging the HMC. Package was hermetically sealed using seam sealing techniques and was tested. A prototype module of RF Switch HMC is presented in Fig. 3

Second proto module was fabricated following same process and tested electrically to ensure repeatability. Electrical test was conducted to verify the repeatability. The proto HMCs were tested for electrical performance at ambient as well as at temperatures extremes (−30 to +65 °C, dwell time 2 h. at each extremes). Derating was carried out for the design and cleared. Based on successful completion of proto evaluation, RF Switch HMC was cleared for qualification.

Fig. 3 Assembled RF switch HMC

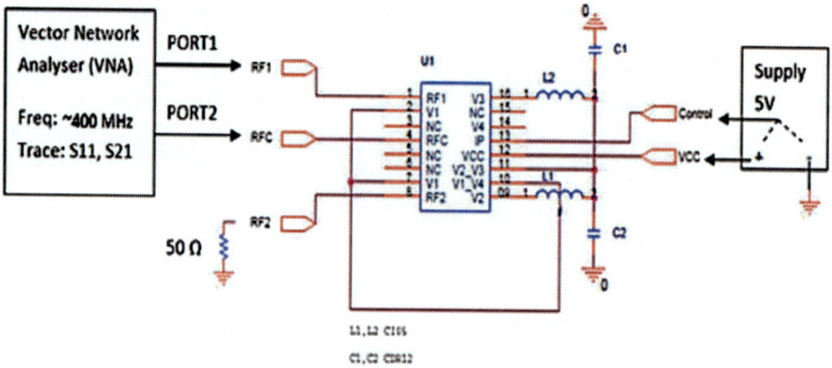

Fig. 4 Testing setup for RF switch HMC

5 Test Setup and Results

Testing of the switch require characterization of insertion loss, return loss at the frequency of ~400 MHz. Vector network analyzer, signal generator and supply voltage terminals are the major components. Testing procedure for RF Switch HMC is presented in Fig. 4. Test results are presented in Table 2.

Imported RF Switch HMC (Part No. DS 0075, Mfg: Daico Industries) is a SPDT switch with package dimension of 0.9 in. × 0.54 in. (14 pin package). The comparison of various specifications are presented in Table 3.

Table 2 Electrical test results of proto HMC

	Control (V)	S_{21} (dB)	S_{11} (dB)	Supply current (mA)
RF1-RFC	0	−1.43 (Insertion loss)	−1.43 (Return loss)	33
RF1-RFC	5	−47.01 (Isolation)	−21.3 (Return loss)	34
RF2-RFC	0	−45.16 (Isolation)	−23.5 (Return loss)	33
RF2-RFC	5	−1.36 (Insertion loss)	−1.43 (Return loss)	34

Table 3 Comparison between hybrid and imported RF switch

Parameter	Indeginized RF switch	Imported RF switch
Package dimension	0.97 in. × 0.77 in.	0.9 in. × 0.55 in.
Insertion loss (dB)	1.4	1.2
Isolation (dB)	47	50

6 Conclusion

RF Switch HMC was developed using thin film technology in an effort towards indegenization. Challenges towards realization of HMC led to the introduction of new processes into thin film HMC technology. Proto modules evaluation for electrical and environmental conditions were carried out. An analysis of performance between indigenous RF Switch and the imported module has been presented. RF Switch HMC is currently undergoing Circuit Type Qualification at URSC, Bangalore.

Acknowledgements The authors wish to thank Director, URSC for his constant guidance and encouragement. The authors would also like to thank Mr. Nagesha J., Mr. K. Ashoka and Mr. Poorna Nageshwara Rao of TnFD, URSC Bangalore for fabrication and assembly of the RF Switch HMC. The authors would like to appreciate the contributions of CSG and HQCD of URSC Bangalore towards development of RF Switch HMC.

References

1. Jones RD (1982) Hybrid circuit design and munufacture. Marcell Dekker, Inc.
2. Linden AE, Topfer ML, Cote RE, Mayer SE (1969) Thick-film microcircuit design and manufacture. NEPCON
3. Reimer DE (1968) Hybrid medium-scale integration. In: 1968 Hybrid electronics symposium
4. Finlayson MJ (1968) Ultrasonic wire bonding of microelectronics devices. NEPCON

5. Stanley WW (1982) Hybrid microelectronic interconnection. In: Electronic packaging and production
6. Ilgenfritz R, Mogey L (1970) The introduction and packaging of thick-film microcircuits. In: Proceedings of electronic components conference

Estimation of Gender Using Convolutional Neural Network

Kathi Mohan Goud and Shaik Jakeer Hussain

Abstract Gender prediction is one of the key challenges in computer vision. It is easier to decide the gender by finding the presence of beard and moustache. For the non-beard and moustache images, gender prediction is always a challenging task. One of the approaches that is helpful in this case is the prediction by convolutional neural networks. At first 1173 male and 1134 female images are used in the training set. 80% of the training set images are used for training and 20% for validation purposes. Different models were trained by choosing different epochs. In order to estimate the efficiency of the classifiers, a testing set of 531 male images and 405 female images (which are not present in the training set) are used. The best gender efficiency achieved is 92.4% at 200 epochs. It is found that either increase in epoch or blocks neither directly nor inversely affects the gender prediction efficiency. The combination of classifiers also failed to get better results than 92.4%.

Keywords Gender prediction · Convolutional neural network

1 Introduction

Gender classification through computer vision is required in various applications for assessing the performance. The usage of technology differs with gender [1]. Dissimilar genders in a team improve interactional behaviours [2]. Transgender health can be diagnosed with voice-based gender classification process [3]. Humans can find the gender by his/her speech or vision. There were several ways a machine can estimate the gender of a human. Some of the approaches are following:

(a) By processing the speech signals [4, 5].
(b) By iris texture[6] that is used for face recognition.
(c) By understanding or processing the writings of human [7].
(d) By analysing the gait [8–12].

K. M. Goud (✉) · S. J. Hussain
Department of Electronics and Communication Engineering, Vignan's Foundation for Science, Technology and Research, Guntur, India

P. S. R. Chowdary et al. (eds.), *Evolution in Signal Processing and Telecommunication Networks*, Lecture Notes in Electrical Engineering 839,
https://doi.org/10.1007/978-981-16-8554-5_4

(e) By extracting the facial features [13].
(f) By segmenting the face into different parts [14].
(g) By using 3D geometric features [15].
(h) By occular texture information [16].
(i) By skin colour segmentation and geometric features of face [17].
(j) By using 3D shapes of the human body [18].
(k) By fusion of spatial features [19].
(l) By analysing the smile on face [20].

Many classifiers were developed and developing for gender prediction. The combination of classifiers can result in efficient gender detection [21]. Gender classification requires feature extraction and classification. For example, LBP can be used as a feature extractor, and SVM can be used for the purpose of classification [22]. In addition to LBP, usage of LPQ as feature extractor can improve the accuracy [23]. The decision tree as a classification technique has an advantage of easily interpretable rules and logic statements over SVM [24]. The feature extractor may be a single classifier or the combination of classifiers [25]. The intensity, brightness, shape, etc., are also need to be considered for the usage of gender classification in real time [26]. The fusion method is best in considering these [27]. Based on writing styles and topics of interest, the gender of a person can be estimated [28]. In addition to gender, emotion also can be estimated with speech [29]. In some situations, gender differences are used to predict the performance [5]. Gaussian mixture models can be used in gender prediction using speech [3]. Convolutional neural network is one of the approaches to the prediction of gender [30]. It is used to predict the gender from occular images of smart phone [31]. The CNN can also use in human gait recognition [32], facial emotion recognition [33], etc. Instead of using any CNN, it is easier to find the male gender by finding the beard or moustache presence. But it is not possible to decide the face having no beard or no moustache is a female.

2 Implementation of Convolutional Neural Network

Figure 7 illustrates the convolutional neural network for gender classification. Blocks 1 to 6 select the features, and the flatten layer to sigmoid activation aids in classification. Fig. 1 illustrates Block 1 in which 32 filters are used to extract the features. Batch normalization improves the efficiency of training, and max pooling reduces the size and improves the speed of training. Dropout aids in avoiding overfitting the model. Similar to Block 1, Block 2 has 64 filters, Block 3 consists of 64 filters, and Block 4, Block 5 and Block 6 have 128, 128, and 256 filters, respectively, which all aid in extracting the features, and the block diagrams of each are illustrated from Figs. 1, 2, 3, 4, 5, and 6.

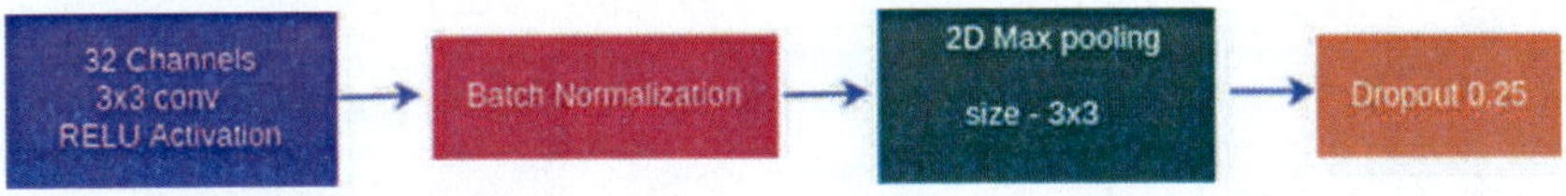

Fig. 1 Feature selection Block 1

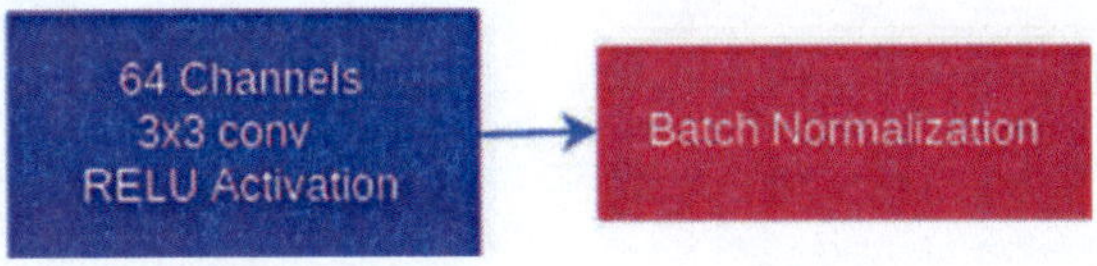

Fig. 2 Feature selection Block 2

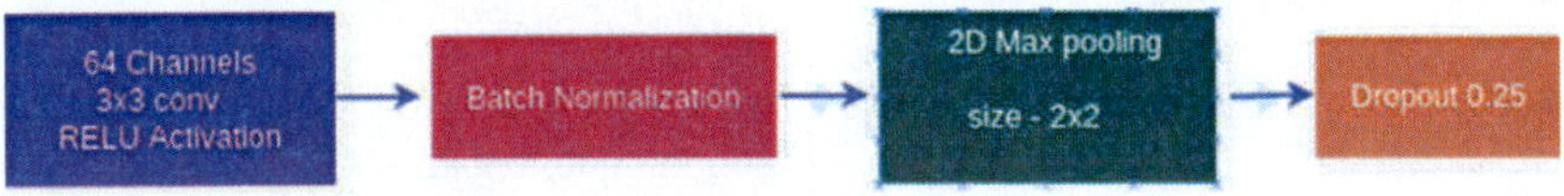

Fig. 3 Feature selection Block 3

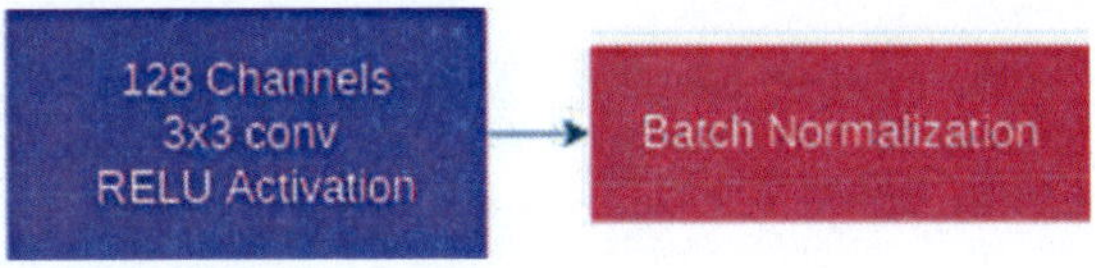

Fig. 4 Feature selection Block 4

3 Results and Discussion

$$\text{Accuracy} = (496 + 369)/936 = 0.924$$

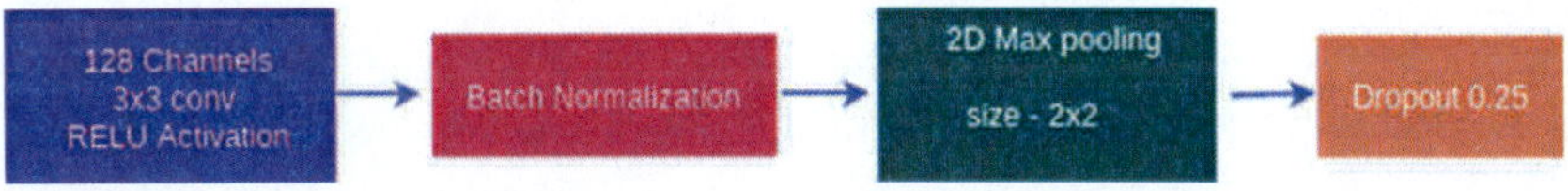

Fig. 5 Feature selection Block 5

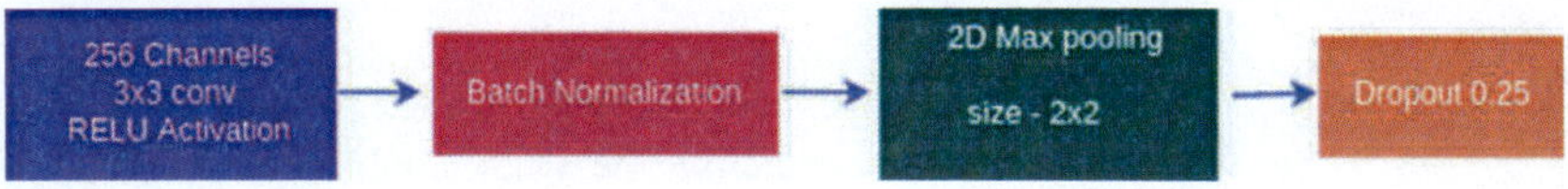

Fig. 6 Feature selection Block 6

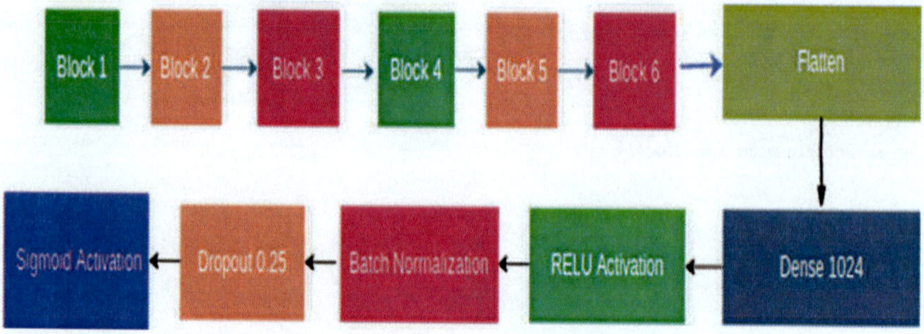

Fig. 7 Convolution neural network for gender detection

Table 1 True and predicted values of male and female

	Estimated by model		
True		Man	Woman
	Man	496	35
	Woman	36	369

Table 2 Performance metrices

	Male	Female
True positive	496	369
True negative	369	496
False positive	36	35
False negative	35	36
Precision	0.9323	0.9133
Recall	0.9376	0.9037
Macro F1	0.9349	0.908

Table 1 illustrates the comparison of predicted ones by the model with the actual values. The model truly predicts 496 males among 531 and 369 among 405 female images. The performance metrics are calculated from these and illustrated in Table 2. The overall accuracy is evaluated which is 92.4% and is appreciable accuracy.

4 Conclusion

Gender detection is one of the challenges that computer vision facing today. Neural Networks is one of the approach in predicting the gender. Different convolution neural networks were trained with images for gender prediction. The efficiency of the classifier is improved by altering the epochs and dropout. The different models with different blocks, epochs provide many classifiers and provide an option to choose the best one. The best accuracy attained is 92.4 %. It is required to improve this research by altering some more parameters or by using different methods.

References

1. Morris M, Venkatesh V, Ackerman P (2005) Gender and age differences in employee decisions about new technology: an extension to the theory of planned behavior. IEEE Trans 52:69–84. https://doi.org/10.1109/TEM.2004.839967
2. Ingram S, Parker A (2002) The influence of gender on collaborative projects in an engineering classroom. IEEE Trans Prof Commun 45(1):7–20. https://doi.org/10.1109/47.988359
3. Lagos D (2019) Hearing gender: voice-based gender classification processes and transgender health inequality. Am Sociol Rev 84:000312241987250. https://doi.org/10.1177/0003122419872504
4. Sedaaghi MH (2009) A comparative study of gender and age classification in speech signals. Iran J Electr Electron Eng [online] Available at http://ijeee.iust.ac.ir/article-1-108-en.html
5. Přibil J, Přibilová A, Matoušek J (2017) GMM-based speaker age and gender classification in Czech and Slovak. J Electr Eng 68(1). https://doi.org/10.1515/jee-2017-0001
6. Tapia JE, Perez CA, Bowyer KW (2015) Gender classification from the same iris code used for recognition. IEEE Trans Inf Forensics Secur 11(8). https://doi.org/10.1109/TIFS.2016.2550418
7. Alanazi SA (2019) Toward identifying features for automatic gender detection: a corpus creation and analysis. IEEE Access 7:111931–111943. https://doi.org/10.1109/ACCESS.2019.2932026
8. Yu S, Tan T, Huang K, Jia K, Wu X (2009) A study on gait-based gender classification. IEEE Trans Image Process 18(8):1905–1910. https://doi.org/10.1109/TIP.2009.2020535
9. Troje NF (2002) Decomposing biological motion: a framework for analysis and synthesis of human gait patterns. J Vis 2(5):371–387. https://doi.org/10.1167/2.5.2
10. Hu M, Wang Y, Zhang Z, Zhang D (2011) Gait-based gender classification using mixed conditional random field. IEEE Trans Syst Man Cybern Part B (Cybern) 41(5):1429–1439. https://doi.org/10.1109/TSMCB.2011.2149518
11. Lu J, Wang G, Moulin P (2014) Human identity and gender recognition from gait sequences with arbitrary walking directions. IEEE Trans Inf Forensics Secur 9(1):51–61. https://doi.org/10.1109/TIFS.2013.2291969
12. Bei S, Deng J, Zhen Z, Shaojing S (2019, October 15) Gender recognition via fused silhouette features based on visual sensors. IEEE Sens J 19(20):9496–9503. https://doi.org/10.1109/JSEN.2019.2916018
13. Bukar AM, Ugail H, Connah D (2016) Automatic age and gender classification using supervised appearance model. J Electron Imaging 25(6). https://doi.org/10.1117/1.JEI.25.6.061605
14. Khan K, Attique M, Syed I, Gul A (2019) Automatic gender classification through face segmentation. Symmetry 11(6):770. https://doi.org/10.3390/sym11060770
15. Ballihi L, Amor BB, Daoudi M, Srivastava A, Aboutajdine D (2012) Boosting 3-D-geometric features for efficient face recognition and gender classification. IEEE Trans Inf Forensics Secur 7(6):1766–1779. https://doi.org/10.1109/TIFS.2012.2209876
16. Eskandari M, Sharifi O (2019) Effect of face and ocular multimodal biometric systems on gender classification. IET Biometrics 8(4):243–248. https://doi.org/10.1049/iet-bmt.2018.5134
17. Mashagba E (2016) Real-time gender classification by face. Int J Adv Comput Sci Appl 7(3). https://doi.org/10.14569/IJACSA.2016.070347
18. Tang J, Liu X, Cheng H, Robinette KM (2011) Gender recognition using 3-D human body shapes. IEEE Trans Syst Man Cybern Part C (Appl Rev) 41(6):898–908. https://doi.org/10.1109/TSMCC.2011.2104950

19. Tapia JE, Perez CA (2013) Gender classification based on fusion of different spatial scale features selected by mutual information from histogram of LBP, intensity, and shape. IEEE Trans Inf Forensics Secur 8(3):488–499. https://doi.org/10.1109/TIFS.2013.2242063
20. Dantcheva A, Brémond F (2017) Gender estimation based on smile-dynamics. IEEE Trans Inf Forensics Secur 12(3):719–729. https://doi.org/10.1109/TIFS.2016.2632070
21. Khan A, Majid A, Mirza AM (2005, January 1) Combination and optimization of classifiers in gender classification using genetic programming. Int J Knowl Based Intell Eng Syst 11(1). https://doi.org/10.3233/KES-2005-9101
22. Sajja TK, Kalluri HK (2019) Gender classification based on face images of local binary pattern using support vector machine and back propagation neural networks. Adv Model Anal B 62(1):31–35, https://doi.org/10.18280/ama_b.620105
23. Zhang C, Ding H, Shang Y, Shao Z, Fu X (2018) Gender classification based on multiscale facial fusion feature. Hindawi Math Prob Eng 2018:6. https://doi.org/10.1155/2018/1924151(Article ID 1924151)
24. Khan MNA, Qureshi SA, Riaz N (2013) Gender classification with decision trees. Int J Signal Process Image Process Pattern Recog 6(1). http://article.nadiapub.com/IJSIP/vol6_no1/13.pdf
25. Khan SA, Ahmad M, Nazir M, Riaz N (2014) A comparative analysis of gender classification techniques. Middle East J Sci Res 20:1–13. https://doi.org/10.5829/idosi.mejsr.2014.20.01.11434
26. Geetha A, Sundaram M, Vijayakumari B (2018) Gender classification from face images by mixing the classifier outcome of prime, distinct descriptors. Soft Comput 23. https://doi.org/10.1007/s00500-018-03679-5
27. Perez C, Tapia J, Estévez P, Held C (2012) Gender classification from face images using mutual information and feature fusion. Int J Optomechatronics 6(1):92–119. https://doi.org/10.1080/15599612.2012.663463
28. Zhang Y, Dang Y, Chen H (2011) Gender classification for web forums. IEEE Trans Syst Man Cybern Part A Syst Humans 41(4):668–677. https://doi.org/10.1109/TSMCA.2010.2093886
29. Bisio I, Delfino A, Lavagetto F, Marchese M, Sciarrone A (2013) Gender-driven emotion recognition through speech signals for ambient intelligence applications. IEEE Trans Emerg Top Comput 1(2):244–257. https://doi.org/10.1109/TETC.2013.2274797
30. Liew SS, Hani MK, Radzi SA (2016) Gender classification: a convolutional neural network approach. Turkish J Electr Eng Comput Sci 24:1248–1264. https://doi.org/10.3906/elk-1311-58
31. Rattani A, Reddy N, Derakhshani R (2018) Convolutional neural networks for gender prediction from smartphone-based ocular images. IET Biometrics 7(5):423–430.
32. Arshad H, Khan MA, Sharif MI, Yasmin M, Tavares JMR, Zhang YD, Satapathy SC (2020) A multilevel paradigm for deep convolutional neural network features selection with an application to human gait recognition. Expert Syst e12541
33. Aamir M, Ali T, Shaf A, Irfan M, Saleem MQ (2020) ML-DCNNet: multi-level deep convolutional neural network for facial expression recognition and intensity estimation. Arab J Sci Eng 45(12):10605–10620

Human Abnormal Activity Detection in the ATM Surveillance Video

B. Prabha, P. Manivannan, and Puvvada Nagesh

Abstract Automatic teller machine (ATM) installation is increased in recent years, due to the financial needs and convenience of the customers. However, the ATM machine installation increases the criminal activities such as attacks on customer, snatching of customer money also increased which affects the bank customer's security. Detecting and localizing the human anomalous activities inside the ATM room ARE the most challenging area which received much research attention. Video surveillance system aimed at monitoring the human behaviours by learning patterns of normal and abnormal activities. Any event/activity which deviates from the regular activity is identified as abnormal/anomalous. We propose computer vision-based human abnormal activity recognition system which will efficiently detect the abnormal activities happening at the ATM room. We conducted the experiment to assess the proposed system performance with existing methods our system achieves better accuracy with less error rate when varying the bin size.

Keywords Abnormal activity detection · ATM surveillance video · Public safety

1 Introduction

Human abnormal activity detection [1] on automated video surveillance has received the more research attention in recent years due to the straight applicability in different applications such as public safety [2], health, military and human–computer interaction [3]. Public safety [4] involves safeguarding public from the crimes, and other potential threats are at high level. Installation of automated teller machine (ATM) [5] is increased due to the various banking transactions such as deposit, withdrawal of money is needed from one account to another. For this popularity of ATMs, recent days banking is more convenient. However, this convenience facilitates the bank customers the criminal activities such as attack the customers, money snatching also have rapidly been raised. Moreover, closed-circuit television (CCTV) camera is

B. Prabha (✉) · P. Manivannan · P. Nagesh
Department of CSE, Koneru Lakshmaiah Education Foundation, Vaddeswaram, Guntur, Andhra Pradesh, India

P. S. R. Chowdary et al. (eds.), *Evolution in Signal Processing and Telecommunication Networks*, Lecture Notes in Electrical Engineering 839,
https://doi.org/10.1007/978-981-16-8554-5_5

(a) Normal **(b) Abnormal**

Fig. 1 Normal and abnormal human posture inside the ATM room

installed inside/outside the ATM to provide security for the customers. But unfortunately, the CCTV camera systems are not sufficient and it does not efficiently detect suspicious/abnormal behaviours. Hence, we need automated surveillance system which can accurately monitor and detect the abnormal activities inside the ATM room. Furthermore, the system can automatically generate a report to the nearest control room to take immediate actions to avoid the crimes.

Figure 1 shows the normal as well as abnormal posture of customer inside the ATM room. The aim of developing the automated detection system is to acquire a model of the regular pattern from the training and detects action which deviates from the testing. The existing methods/approaches for detecting abnormality were used either handcrafted/deep features to describe the patterns of spatiotemporal features. Reddy et al. [6] proposed abnormal activity detection by extracting motion and appearance-based features extracted and abnormal activities are classified through support vector machine (SVM). Arpitha et al. [7, 8] extracted scale-invariant feature transform (SIFT) and Gabor features and abnormal activity detected through SVM. Kaltsa et al. [9] spatial information are extracted from spatiotemporal cubes and combined with the feature vector of HOG. Histogram of oriented swarm acceleration (HOSA) is used for extracting temporal dynamics. To obtain accurate abnormality pattern from the ATM surveillance video, we extracted histogram of gradients (HOG) and optical flow (OF) feature vectors and classified the normal/abnormal events through SVM classifier. The proposed system paper organization is as follows. In Sect. 2, closely related work of the proposed system is discussed. Section 3 presents the detailed description of ATM abnormal activity detection system. Section 4 presents experimental result, and comparison study of the proposed work with their performance is analysed.

2 Related Work

Due to the relative changes in behaviours of the abnormal actions [10], most of the existing methods approached the abnormality by modelling normal events which persist on training data. Chella et al. [11] developed the people tracking system for detecting abnormality in the human–computer interaction domain using posture estimation technique. In [7], a generative approach is proposed to capture the abnormal activities by extracting SIFT and Gabor feature vector by modelling the events through SVM classifier. Nithya Shree et al. [12] proposed the blob-based recognition with two segments such as framing and tracking of human. Rishabh [13] extracted 3D information that utilizes the grouping to characterize convexity quantity of the signals. [13] proposed two-stage algorithm through K-means clustering. One class support vector machine (OCSVM) is presented clustering-based approach to eliminate the outliers and grouping the spatiotemporal cubes for modelling normal activities. In [14], a generative method extracts temporal patterns and builds the model through fully connected layer of auto-encoder that learns from the spatiotemporal features. Bernard [15] proposed the real-time activity detection using posture through static camera.

Representation of silhouettes uses the 2D techniques for the observed person for real-time processing. Two methods are composed such as posture detection and temporal filtering. In the posture detection, the human posture is recognized by computing frames and in the posture temporal filtering method filter the human posture over information of the posture from the successive frame detection. Anomalous event detection is mostly based on temporal data with its structure of video. In [16], authors breaking this through spatiotemporal descriptor compute and compare with other classifiers. Gong et al. [17] proposed the augmented auto-encoder based on memory that utilized memory elements for representing the input data characterization. The proposed method depends on the learning the model through generalized approach and reconstructs the abnormal segments which make discrimination of the normal and abnormal events. Object of the proposed work is for detecting human abnormal activity from the surveillance video frame. HOG and OF feature descriptors are extracted to detect the human abnormality which happens inside the ATM by classifying normal/abnormal events through SVM classifier.

3 Human Abnormal Activity Detection in ATM Surveillance Videos

SVM classifier effectively classifies the normal and anomalous activities that are considered as an efficient for detecting abnormal activities in unstructured videos. Normal events of machine learning methods to learn from huge volume of data classify the events that will be the best fit of the data based on the prior knowledge. In generative models, SVM is a powerful and is widely used learning tool for activity

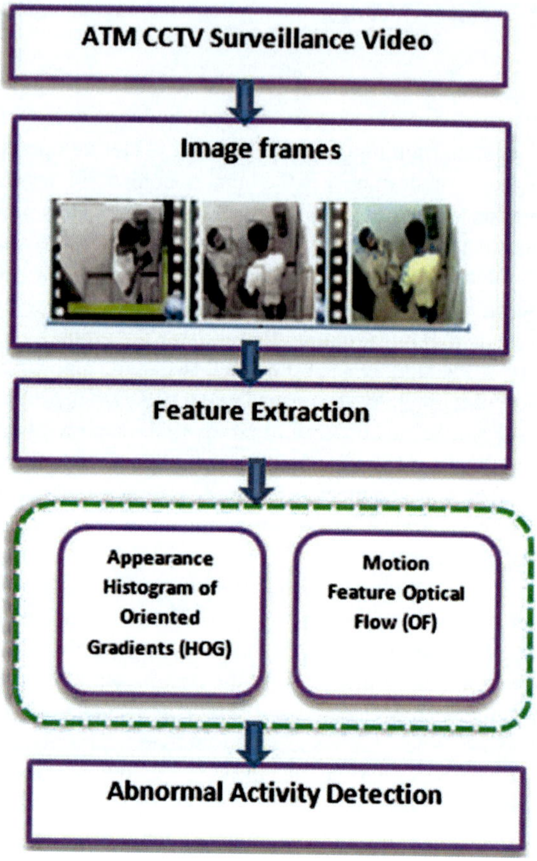

Fig. 2 Human abnormal event detection in ATM surveillance videos

recognition application due to its accuracy w.r.t high-dimensional data. Figure 2 shows architectural framework of the proposed work to detect abnormal events in ATM surveillance videos. Real-time CCTV ATM surveillance videos are considered as dataset for developing the proposed system. Videos are split into consecutive frames and converted the RGB frames into grayscale images. Furthermore, HOG and OF feature descriptors are extracted from grayscale images, and finally, the normal/abnormal activities are classified by using the SVM method. The proposed method contains two sections such as feature extraction and classification.

3.1 Feature Extraction

Feature extraction is the process of extracting descriptive appearance [15] and motion [18–21] features such as edges, blobs and trajectories/optical flow from consecutive

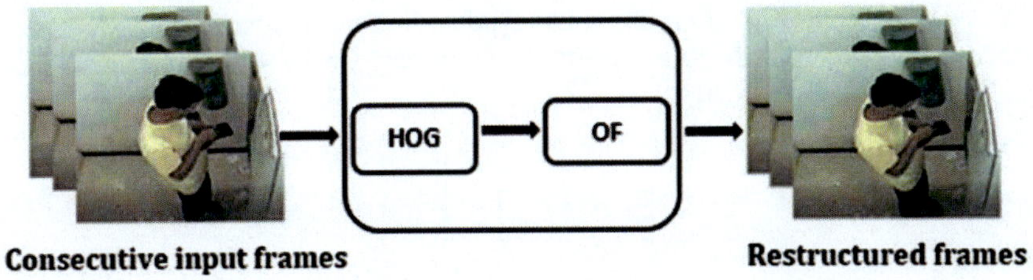

Fig. 3 Training phase of HOG and OF features for ATM video surveillance

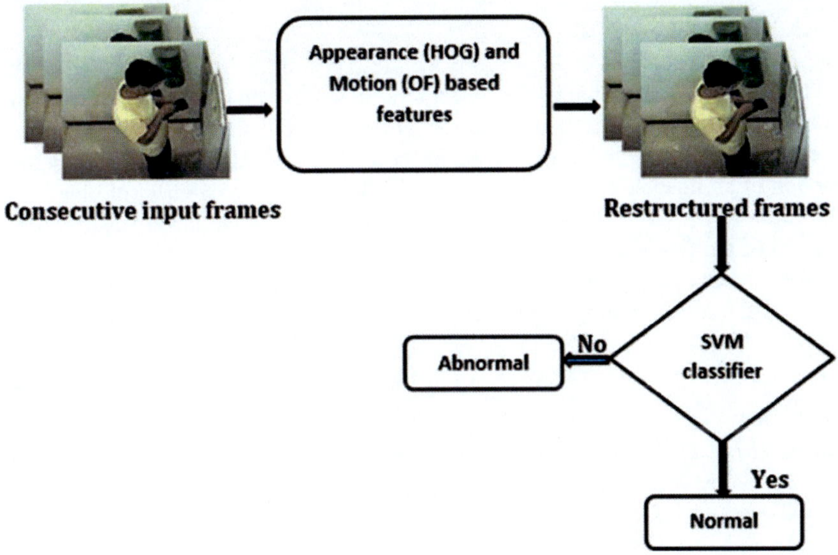

Fig. 4 Testing phase of HOG and OF features for ATM video surveillance

video frames. In the proposed method, the HOG [22–26] feature extracts local object shape and appearance and OF [27] captures the motion features from the ATM surveillance video. Figures 3 and 4 represent the feature extraction and abnormal activity classification using SVM classifier in detail.

3.1.1 Histogram of Oriented Gradients (HOG)

HOG is one of the popular feature descriptors to apply on video frames to extract unique patterns. This feature descriptor is mainly used on computer vision through different applications such as object detection and recognition. HOG features count the number of gradient occurrences in the region of interest of an image. In [24], HOG is generated by computing gradients in an image from Eq. (1) which capture silhouette and contour data. Gradient data is pooled to form 1-D to 2-D transformation which forms input to machine learning methods such as SVM [28], random forest and logistic regression [29].

$$\nabla f = \begin{bmatrix} gx \\ gy \end{bmatrix} = \begin{bmatrix} \frac{\delta f}{\delta x} \\ \frac{\delta f}{\delta y} \end{bmatrix} \tag{1}$$

where *x*, *y* represent pixel values, with linear sum of values.

3.1.2 Optical Flow (OF)

Optical flow [19, 30] captures the object motion, edges and surfaces on video clips generated by capturing the relative motion. This motion is defined as velocity distribution and brightness in an image. Estimation of motion is a basic method of computing motion of an image intensities, which can be described to movement of objects in the video scene. OF methods are depending on calculating motion estimation from Eq. (2) of the frame intensities over the time in a surveillance video. Then optical flow can be analysed to produce segmentations into segmented regions that might be associated through moving objects. OF is computed by using the following equation

$$I(a, b, t) = I(a + \delta a, b + \delta b, t + \delta t) \tag{2}$$

where $I(a, b, t)$ denotes image patch, time displacement is δt through distance $(\delta x, \delta y)$ with *x*- and *y*-directions.

3.2 *Abnormal Event Recognition: SVM Classifier*

Discriminative models are very effective in supervised learning rather than the generative models. SVM classification [31–34] has given better outcomes to different types of the data classification in discriminative models. Moreover, the extracted features such as HOG, OF from the ATM surveillance video, SVM used for classifying the events. In the proposed work, two-class SVM learns both normal/abnormal through training phase.

4 Experimental Result Analysis

The proposed method performance is evaluated by conducting the experiments on six real-time videos. The proposed system is developed with the MATLAB 2018 in laptop having Intel core i5, 1.70 GHz processor with 8 GB RAM with image frames. The dataset is divided into two classes normal and abnormal. To achieve the better performance experiments conducted in frame level, HOG [35] and OF features are used in the proposed work. Table 1 shows performance evaluation and accuracy of the proposed work through bin size variations. The proposed system detects the abnormal activities such as when the customer is attacked by the other person which is identified through bounding box is shown in Fig. 5.

Table 1 shows the proposed method performance which is well through varying the bin size in OF [36]. The proposed approach can detect abnormal activities (customer

Table 1 Error rate and accuracy comparison of ATM real-time surveillance video dataset

Bin value	Bin size	α value	Iteration	Error rate	Accuracy
Video frame level	4	1	3	33.5	66.5
	32	1	3	25.6	74.4
	42	1	3	45.2	54.8
	16	1	3	31.2	68.8
	50	1	3	40.8	59.2
	45	**1**	**3**	**14.5**	**85.5**

Bold represents the final parameters and values chosen for the experimentation of the proposed system.

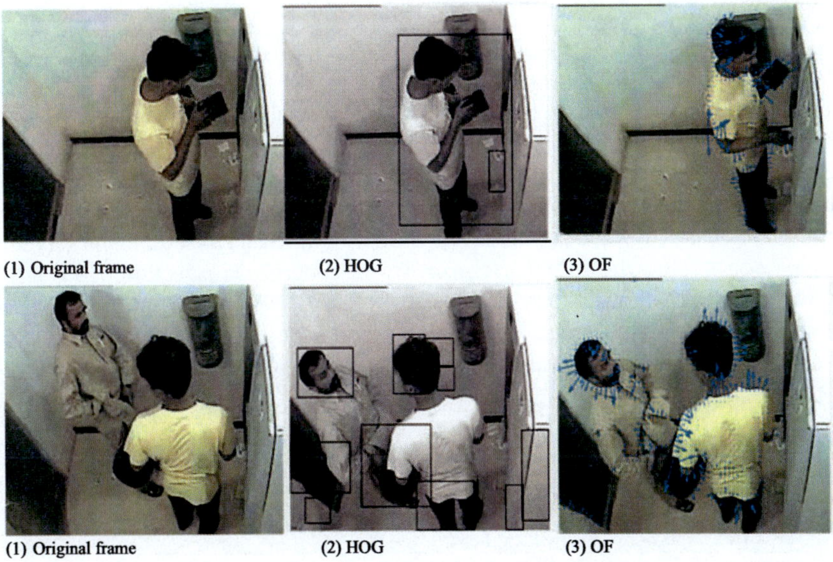

Fig. 5 Abnormal activity detection of frames on ATM surveillance video. Top row: single person: original frames (1), HOG detection (2), and optical flow detection (3). Bottom row: multiple persons: original frames (1), HOG detection (2), and optical flow detection (3)

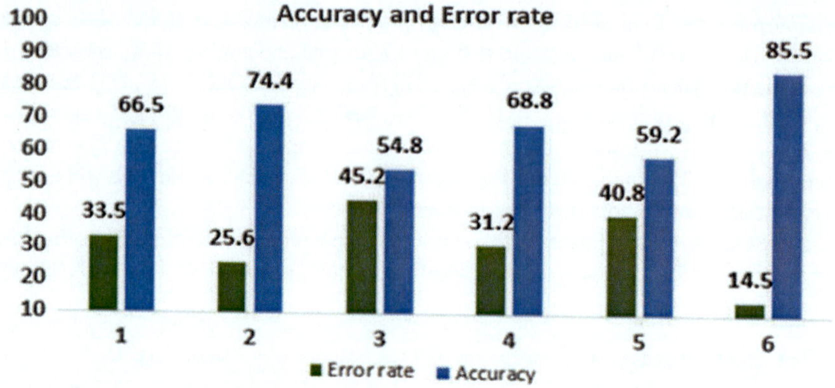

Fig. 6 Performance evaluation of error rate and accurate rate of the proposed system

attacked by the other person) in the frame. Since this ATM surveillance approach extracts discriminative shape and motion (HOG, OF) detection features, anomalies detection rate is very fast than the handcrafted method. Figure 6 depicts the proposed system error rate and accuracy comparison values in Table 1.

5 Conclusion and Future Work

Machine learning approaches are playing promising role in anomalous activity detection in the ATM room in recent years. We propose SVM classification-based method for anomalous activity detection in ATM video. Furthermore, extracted feature vectors are determining the abnormality detection in input video. We conducted experiments on the real-time ATM surveillance video that show the proposed method is able to detect the abnormality well through appearance (HOG) and motion (OF)-based feature descriptor. SVM classifier along with the feature vector in the surveillance video provides the better performance in the abnormality detection. The results are compared by varying bin size, and it depicts that the proposed system performance is improved (less error rate).

References

1. Popoola OP, Wan K (2012) Video-based abnormal human behavior recognition—a review. IEEE Trans Syst Man Cybern 42(6):865
2. Borges PVK, Conci N, Cavallaro A (2013) Video-based human behavior understanding: a survey. IEEE Trans Circ Syst Video Technol 23(11):199
3. Liu H, Chen S, Kubota N (2013) Intelligent video systems and analytics: a survey. IEEE Trans Industr Inf 9(3):1222

4. Gupta M, Gao J, Aggarwal CC, Han J (2014) Outlier detection for temporal data: a survey. IEEE Trans Knowl Data Eng 25(1)
5. Yi M (2011) Abnormal event detection method for ATM video and its application. In: Lin S, Huang X (eds) Advanced research on computer education, simulation and modeling. CESM 2011. Communications in Computer and Information Science, vol 176. Springer, Berlin, Heidelberg.https://doi.org/10.1007/978-3-642-21802-6_30
6. Reddy V, Sanderson C, Lovell BC (2009) Improved anomaly detection in crowded scenes via cell-based analysis of foreground speed, size and texture. In: Proceedings of IEEE conference on computer vision and pattern recognition, p 1446
7. Arpitha K, Honnaraju B (2018) Vision based anomaly detection system for ATM. Int Res J Eng Technol 5(4)
8. Inthiyaz S, Madhav BTP, Kishore PVV (2017) Flower segmentation with level sets evolution controlled by colour, texture and shape features. Cogent Eng 4(1)
9. Kaltsa V, Briassouli A, Kompatsiaris I, Strintzis MG (2014) Swarmbased motion features for anomaly detection in crowds. In: 2014 IEEE international conference on image processing (ICIP). IEEE, pp 2353–2357
10. Kishore PVV, Kumar KVV, Kumar EK, Sastry ASCS, Kiran MT, Kumar DA, Prasad MVD (2018) Indian classical dance action identification and classification with convolutional neural networks. Adv Multimedia 1(9)
11. Chella A, Dindo H, Infantino I (2006) People tracking and posture recognition for human-robot interaction. Vision Based Human-Robot Interaction, Palermo, Italy
12. Nithyashree RS, Sah R, Gowda SN (2016) Surveillance video based robust detection and notification of real time suspicious activities in indoor scenarios
13. Ionescu RT, Smeureanu S, Popescu M, Alexem B (2019) Detecting abnormal events in video using narrowed normality clusters. In: 2019 IEEE winter conference on applications of computer vision (WACV). IEEE, pp 1951–1960
14. Hasan M, Choi J, Neumann J, Roy-Chowdhury AK, Davis LS (2016) Learning temporal regularity in video sequences. In: Proceedings of the IEEE conference on computer vision and pattern recognition, pp 733–742
15. Agrawal R, Gupta N (2016) Real time hand gesture recognition for human computer interaction. In: IEEE 6th international conference on advanced computing (IACC), pp 470–475. https://doi.org/10.1109/IACC.2016.93
16. Del Giorno A, Bagnell JA, Hebert M (2016) A discriminative framework for anomaly detection in large videos. In: European conference on computer vision. Springer, Berlin, pp 334–349
17. Gong D, Liu L, Le V, Saha B, Mansour MR, Venkatesh S, Hengel A (2019) Memorizing normality to detect anomaly: memory augmented deep autoencoder for unsupervised anomaly detection. arXiv preprint arXiv:1904.02639
18. Chandola V, Banerjee A, Kumar V (2009) Anomaly detection: a survey. ACM Comput Surv 41(3):15
19. Bashir F, Khokhar A, Schonfeld D (2005) Automatic object trajectory-based motion recognition using Gaussian mixture models. In: Proceedings of IEEE conference on multimedia and expo, p 1532
20. Bashir FI, Khokhar AA, Schonfeld D (2007) Object trajectory-based activity classification and recognition using hidden markov models. IEEE Trans Image Process 16(7):1912
21. Duan-YuChen P-C (2011) Motion-based unusual event detection in human crowds. J Vis Commun Image Representn 22(2):178–186
22. Hu W, Tan T, Wang L, Maybank S (2004) A survey on visual surveillance of object motion and behaviors. IEEE Trans Syst Man Cybern 34(3):334
23. Li W, Mahadevan V, Vasconcelos N (2014) Anomaly detection and localization in crowded scenes. IEEE Trans Pattern Anal Mach Intell 36(1):18
24. Dalal N, Triggs B (2005) Histograms of oriented gradients for human detection. In: Proceedings of IEEE conference on computer vision pattern recognition, pp 886–893
25. Razia S, Prathyusha PS, Krishna NV, Sumana NS (2018) A comparative study of machine learning algorithms on thyroid disease prediction. Int J Eng Technol 7(28):315

26. ElBedwehy MN, Behery GM, Elbarougy R (2020) Face recognition based on relative gradient magnitude strength. Arab J Sci Eng 45(12):9925
27. Hu W, Xie N, Li L, Zeng X, Maybank S (2011) A survey on visual content-based video indexing and retrieval. IEEE Trans Syst Man Cybern 41(6):797
28. Kumar BS, Naren J, Vithya G, Prahathish K (2019) A novel architecture based on deep learning for scene image recognition. Int J Psychosoc Rehabil 23(1):400
29. Reddy AN, Krishna D (2018) A survey on applications and performance of deep convolution neural network architecture for image segmentation. Int J Pure Appl Math 118(19):43
30. Horn BK, Schunck BG (1981) Determining optical flow. Artif Intell 17(1):185–203
31. Satti SK, Devi KS, Dhar P, Srinivasan P (2021) A machine learning approach for detecting and tracking road boundary lanes. ICT Express 7(1):99–103. ISSN 2405-9595.https://doi.org/10.1016/j.icte.2020.07.007
32. Chandra DV, Sastry JKR (2019) Developing autonomous vehicle systems using machine learning techniques and comparison of SVC and Naive Bayes algorithms. Int J Innov Technol Exploring Eng (IJITEE) 8(7)
33. Rao SNT, Vemuru S, Moturi S (2019) Optimized feature extraction and hybrid classification model for heart disease and breast cancer prediction. Int J Recent Technol Eng (IJRTE) 7(6)
34. Kotkar VA, Sucharita V (2019) Particle filtering based opticle flow computation model for crowd anomaly detection using gaussian mixture model. Int J Recent Technol Eng (IJRTE) 7(68)
35. Devi JY, Keerthi G, Priya KJ, Lakshmi BG (2018) Anomaly detection in health care. Int J Mech Eng Tech (IJMET) 9(1):89–94
36. Razia S, Narasingarao MR, Bojja P (2017). Development and analysis of support vector machine techniques for early prediction of breast cancer and thyroid. J Adv Res Dyn Control Sys 9:869–878

Moving Object Detection Using Optical Flow and HSV

Date Archana and Shah Sanjeevani

Abstract Detection of moving object in a video sequence has become a demanding task due to changing background, varying appearance of the moving object and when moving object share similar colour with the background. In this paper, combination of two methods is done to detect moving object. An optical flow method along with Hue Saturation and Value colour space is proposed to detect camouflage moving object. The algorithm is tested on various standard data sets. Both the qualitative and quantitative results were obtained. The experimental results show that the method adjust itself according to various challenges and gives best qualitative and quantitative results.

Keywords Moving · Camouflage · Optical · Video · Background

1 Introduction

Moving object detection in different challenging environment is a very difficult task in computer vision [1]. The challenging environment may be dynamic background like waving trees and water fountain, or changing illumination conditions, cluttered background, object occlusion and camouflage were moving object share similar colour with background. Basically, the methods used for detecting moving object are frame differencing [2, 3], optical flow [4], background subtraction [5, 6] and temporal differencing. Each method is having its own benefits and limitations. In case of frame difference method, the moving object with large uniform area is determined as background. Thus, the holes are seen in the detected moving object which is known as aperture problem. Detection accuracy depends upon the rate at which frame arrived and speed of moving object. Background subtraction is mostly suitable for static background. The optical flow technique is complex and needs extra hardware

D. Archana (✉)
Sinhgad College of Engineering, Pune, India

S. Sanjeevani
Smt. Kashibai Navale College of Engieering, Pune, India

P. S. R. Chowdary et al. (eds.), *Evolution in Signal Processing and Telecommunication Networks*, Lecture Notes in Electrical Engineering 839,
https://doi.org/10.1007/978-981-16-8554-5_6

for real-time applications. The prior knowledge of background is not required and gives best results in case of dynamic background and moving camera. HSV colour model represents the colour in a way the human eye can recognise. HSV background subtractions work well in dynamic background by continuously updating the background. Thus, this paper gives a combination of optical flow and HSV [7] for detection of moving object. More focussed is given on camouflage data set where moving object and background share similar colour.

2 Methodology

Optical flow methods use the flow vectors of moving object from frame to frame to detect object in a video [8–10]. Motion detection is done by estimating flow vectors by Farneback method. Optical flow technique can detect objects that are moving in video sequences even though the camera is moving. In this method, it is assumed that the intensity of moving object or pixel remains constant from frame to frame. It is computed by taking two images current and displaced.

$$I_{x,y,t} = I\left(x + \delta_x; y + \delta_y; z + \delta_z\right) \tag{1}$$

Using Taylor series, above equation is expanded to:

$$I\left(x + \delta_x; y + \delta_y; z + \delta_z\right) = I_{x,y,t} + \frac{\mathrm{d}I}{\mathrm{d}x}\delta x + \frac{\mathrm{d}I}{\mathrm{d}y}\delta y + \frac{\mathrm{d}I}{\mathrm{d}t}\delta t + H.O.T \tag{2}$$

Discard higher order terms (H.O.T.),

$$\frac{\mathrm{d}I}{\mathrm{d}x}\delta x + \frac{\mathrm{d}I}{\mathrm{d}y}\delta y + \frac{\mathrm{d}I}{\mathrm{d}t}\delta t = 0 \tag{3}$$

$$\frac{\mathrm{d}I}{\mathrm{d}x}\delta x/\delta t + \frac{\mathrm{d}I}{\mathrm{d}y}\delta y/\delta t + \frac{\mathrm{d}I}{\mathrm{d}t} = 0 \tag{4}$$

$$\frac{\mathrm{d}I}{\mathrm{d}x}V_x + \frac{\mathrm{d}I}{\mathrm{d}y}V_y + \frac{\mathrm{d}I}{\mathrm{d}t} = 0 \tag{5}$$

$$I_x V_x + I_y V_y = -1 I_t \tag{6}$$

Here, V_x, V_y are flow vectors, and I_x, I_y are the imitative of the intensities of the image at co-ordinate (x; y; t). By using magnitude of V_x, V_y, motion vectors are obtained for object detection by using thresholding technique. Values of motion vector are calculated as:

$$\mathrm{Th} = \sqrt{V_x^2 + V_y^2} \tag{7}$$

On this Th value, thresholding method is applied. At last, morphological operations are applied to detect the final moving object. Optical flow techniques are mathematically complicated, and to use in real time, it requires specially designed hardware. Optical flow along with brightness measure is used to detect the moving object constituting camouflage. It is assumed that brightness of the object remains constant from frame to frame.

Algorithm

1. Take video as an input.
2. Extract the frame.
3. Take the initial frame and convert it to grey.
4. Create the HSV colour image.
5. Save the previous frame data.
6. Get the next frame and convert it into greyscale.
7. Calculate the dense optical flow using farneback method.
8. Obtain the flow magnitude and direction angle
9. Update the colour image.
10. Display the resulting frame.

OpenCV provides algorithm to find the heavy optical flow. It computes the flow vectors for all the points in the frame. It uses Gunner Farneback's algorithm. An array with optical flow vectors, (u, v) is obtained. Next, find magnitude value, and direction is calculated. Direction is nothing but Hue estimates of the image, whereas magnitude gives information about value plane. Then, represent object in HSV colour space.

3 Results

The algorithm is implemented and tested on standard data set for qualitative and quantitative analysis. The developed algorithm is tested on standard databases containing camouflage video sequences. The database used is changedetection.net and labelled and annotated sequences for integral evaluation of segmentation algorithms (LASIESTA).

Changdetection.net provides 2 video databases 2012 data set and 2014 data set with different challenging situations. From this, sofa and highway sequence is selected for testing the algorithm. Sofa consists of 2750 frames with size 320 × 240, and highway contains 1700 frames with size 320 × 240.In this sequence, colour of cloth is similar to the colour of sofa. LASIESTA consists of many real indoor and outdoor sequences with varying challenging conditions in moving object detection. Two camouflage sequences ICA_01 and ICA_02 were selected from indoor sequence category to test our algorithm. In first sequence, A person moves in a closed area and remains static for a certain second in front of entry which is of same colour like his clothing. Total frames are 350. In second sequence, a person moves and pose in front of a surface with same colour to his clothes. A tree is continuously moving on the table. Total number of frames are 525. The results are compared qualitatively and quantitatively using above data sets. All experiments have been performed on greyscale frame sequences (Fig. 1).

As seen from result in case of I_CA_01 sequence, camouflage object is detected with some false positives. Moving plant in the background is falsely detected as foreground moving object in sequence I_CA_0 2. In case of sofa sequence along with the moving object, some intermittent objects and noise are detected. All moving cars are detected by merged case of outdoor sequence.

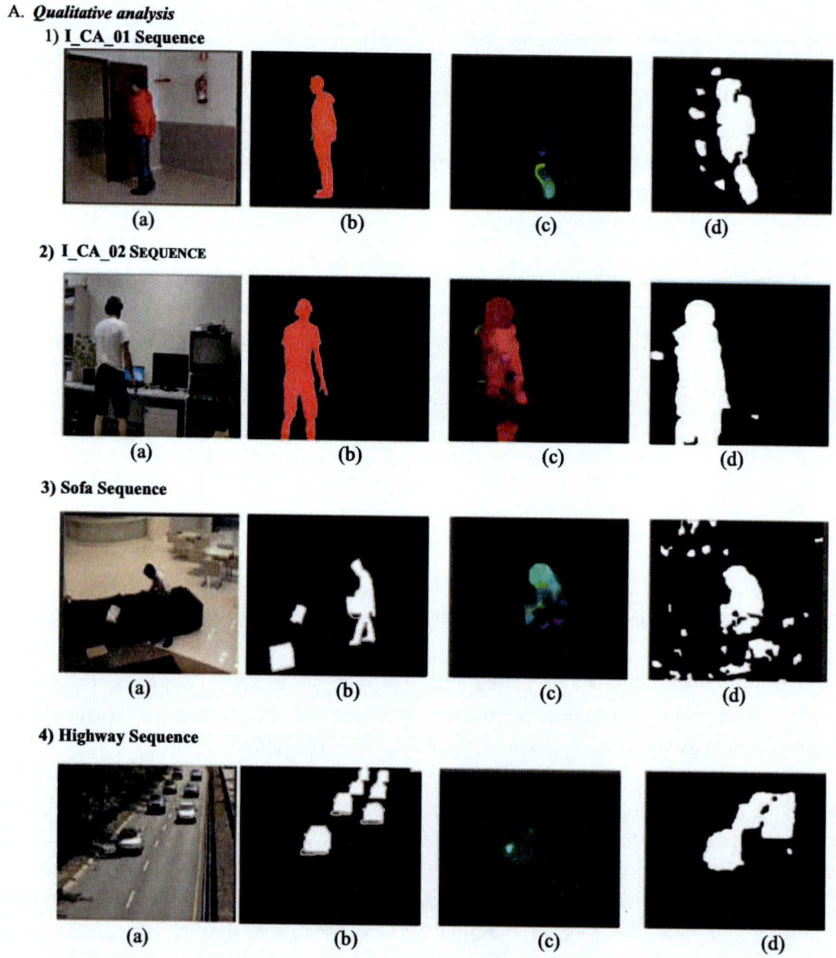

Fig. 1 Qualitative analysis on different database **a** input frame **b** ground truth image **c** HSV image **d** binary image

B. ***Quantitative Analysis***

The three parameters have been evaluated to analyse the performance of the developed algorithm. Performance evaluation tells how well the object is detected with a smaller number of misclassifications.

1. Recall compares the entire count of true positives with that of the entire count of true positives that are present in the ground truth:

$$\text{Recall} = \frac{\text{Number of correctly identifying pixel}}{\text{Number of foreground pixels in ground truth.}} \tag{8}$$

2. Precision compares the pixels that are truly positives to the complete number of true positive pixels that are detected by the method:

$$\text{Precision} = \frac{\text{Number of correctly identifying pixel}}{\text{Number of foreground pixels detected by algorithm}} \tag{9}$$

   An object is detected accurately when the detection rate is large without affecting the precision.
3. The F-measure is calculated as follows:

$$F \text{ measure} = \frac{2 * \text{recall} * \text{precision}}{\text{recall} + \text{precision}} \tag{10}$$

The F-measure gives the achievement of classification in precision-detection rate space. For this, F-measure must be closed to 1 (Table 1; Fig. 2).

Above table shows the performance evaluation parameter results of the implemented algorithm. In case of I_CA_01, 77% of object is detected. The implemented algorithm shows very good results for I_CA_02 sequence; recall is 92%, whereas F-measure is 86%. It indicates that most of the object is detected. Object is precisely detected in case of highway sequence is very good near about 80%. The algorithm performs very well in case of camouflage object detection.

Table 1 Quantitative analysis

Database	Recall	Precision	F-measure
I_CA_01	0.77	0.72	0.75
I_CA_02	0.92	0.65	0.86
Sofa	0.84	0.75	0.76
Highway	0.79	0.84	0.77

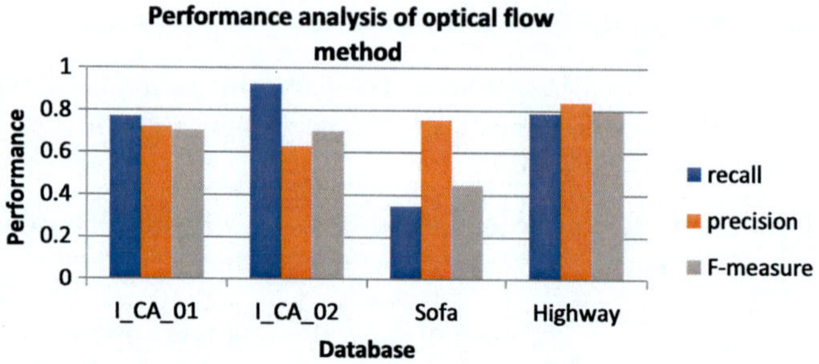

Fig. 2 Performance analysis of optical flow method

4 Conclusion

Camouflaged moving object is detected by implementing optical flow using HSV. The object is detected using optical flow and then represented in HSV colour space. The testing of algorithm is done in different environment which consists of indoor as well as outdoor camouflage sequences from standard data sets. Method detects moving object with less accuracy and noise. Not applicable for real-time applications due to less speed. In some cases, two moving objects were detected as one. Thus, from above, it is cleared that camouflaged moving object can be detected using optical flow representation in HSV colour space method.

References

1. Collins R, Lipton A, Kanade T (2000) Introduction to the special section on video surveillance. In: Proceeding of IEEE transactions on pattern analysis and machine intelligence, pp 745–746
2. Zhan C, Duan X, Xu S, Song Z, Luo M (2007) An improved moving object detection algorithm based on frame difference and edge detection. In: Fourth international conference on image and graphics, pp 519–523
3. Migliore D, Matteucci M, Naccari M (2006) A revaluation of frame difference in fast and robust motion detection. In: Proceedings of the 4th ACM international workshop on Video surveillance and sensor networks, pp 215–218
4. Brox T, Bruhn A, Papenberg N, Weickert J (2004) High accuracy optical flow estimation based on a theory for warping. In: Proceedings of 8th European conference on computer vision, Springer LNCS 3024, vol 4, pp 25–36
5. Wren C, Azarbayejani A, Darrel T, Pentland A (1997) Pfinder: realtime tracking of the human body". In: Proceedings of IEEE transactions on pattern analysis and machine intelligence, pp 780–785
6. Piccardi M (2004) Background subtraction techniques: a review. In: IEEE international conference on systems, man and cybernetics, vol 4, pp 3099–3104
7. Riza K, Gwangwon K, Youngbae K, Sungbum P, Jongan P (2008) Efficient image retrieval using adaptive segmentation of HSV color space. In: International conference on computational sciences and its applications, pp 491–496

8. Watanabe M, Takeda N, Onoguchi K (1996) A moving object recognition method by optical flow analysis. In: Proceedings of the 13th international conference on pattern recognition, vol 1, pp 528–533
9. Patel MP (2014) Moving object detection with moving background using optic flow. In: IEEE international conference on recent advances and innovations in engineering (ICRAIE-2014), pp 1–6
10. Zinbi Y, Chahir Y (2008) Moving object segmentation using optical flow with active contour model. In: IEEE conference on information and communication technologies: from theory to applications (ICTTA), pp 1–5

Propagation of Data Using Free Space Under Different Weather Conditions

K. Sony, Sarat K. Kotamraju, and K. Ch Sri Kavya

Abstract Free-space optical communication is a license-free spectrum which doesn't require any spectrum allocations. As the medium of communication is free space, it might undergo atmospheric turbulences such as rain, fog, haze, and snow. The link need to be designed with utmost care such that it should be able to overcome the turbulences and the signal need to be transmitted for long distances without loss of signal strength. Appropriate mitigation techniques are incorporated which will reduce the attenuation in the free space link. In this work, optical link which works for 1–5 Kms are studied, and parameters such as bit error rate and quality factor are observed for different wavelengths.

Keywords Free-space optical link (FSO) · Bit error rate (BER) · Quality factor (Q)

1 Introduction

Wireless optical communication systems can be classified into two categories: indoor system and outdoor systems (FSO). In the FSO, it is again broadly classified into terrestrial and space links. The subdivision can be briefed as inter orbital links, Inter satellite links, deep space links [1, 2].

The FSO links which are commercially available can generate data rates ranging from 10 Mbps–10 Gbps. There are many commercial providers which has all the equipment required for optical link setup.

A free-space optical link needs LOS connection between the transmitter and receiver for transfer of information from one system to the other. The process included multistep process, where the message signal needs to be modulated on the optical carrier such as free-space link or optical fiber, and then, the modulated signal undergoes different atmospheric turbulences. The components required in a free-space link are as follows:

K. Sony (✉) · S. K. Kotamraju · K. Ch Sri Kavya
Department of Electronics and Communication Engineering, Koneru Lakshmaiah Education Foundation, Vaddeswaram, India
e-mail: sonykarra@kluniversity.in

P. S. R. Chowdary et al. (eds.), *Evolution in Signal Processing and Telecommunication Networks*, Lecture Notes in Electrical Engineering 839,
https://doi.org/10.1007/978-981-16-8554-5_7

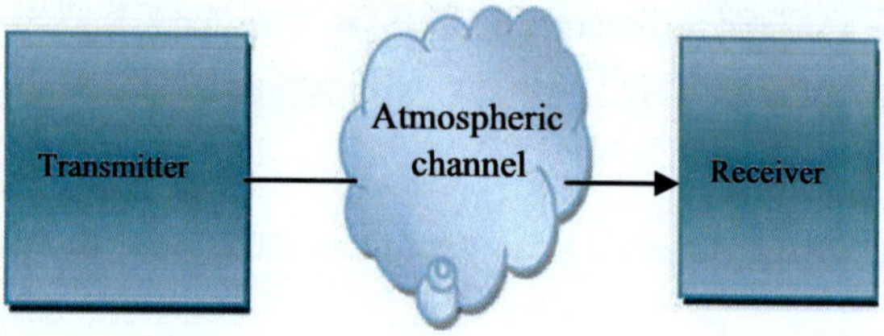

Fig. 1 Block diagram of FSO link

1. Transmitter (Optical source)
2. Atmospheric channel (Turbulence channel)
3. Receiver (Photo detector)

The general block diagram of FSO link includes all the three components which are interlinked with one another (Fig. 1).

1.1 Transmitter

The basic function is in general to modulate the information signal onto the carrier, and then, the signal is propagated through the sir and then will be received by the received [3, 4]. It required a modulator, driver circuit, and a telescope which can collimate the beam and collect the signal and direct them to the receiver. The modulation used in free space optical link is intensity modulation, where the driver current of the source is varied in accordance with the message signal.

1.2 Atmospheric Channel

The propagating medium in free space optical link is free space. Different atmospherics and environmental factors such as fog, rain, snow, or cloud create quite a influence on the free-space optical link [5, 6]. These factors are not fixed and varies for any instant of time. So, channel is one of the delimit of the free-space optical link.

1.3 Receiver

At the receiver, the transmitted data need to be recovered, and the components involved are optical filter, photo detector, and demodulator. The receiver at the other end collects the information and directs it to the photo detector [7, 8]. The filter will reduce any unwanted background noises and converts the optical signal into the electrical signal. Regenerator and spectrum analyzers are also connected to estimate the signal strength received by the receiver.

2 System Design and Analysis

Wireless in FSO communication system, the most important parameter for design is to choose the wavelength. As the gain of the system (Antenna) is directly proportional to the operational wavelength, as the wavelength increases the quality of the link improves, the selection of the wavelength depends purely on the transmitter and receiver components, safety of the eye, and cost-effectiveness.

The wavelength in generally is considered from the range [880–1550 nm] where the absorption of molecules is high at 850 nm [9–11].

In free space, the loss factor mainly depends on the wavelength and the link range R. R and L are inversely proportional to each other.

In this paper, at different wavelengths such as 880, 1310, 1530 nm, the propagation link is studied, and various parameters such as quality factor and bit error rate are observed.

Initially, the link is transmitted at 1 km distance, and good quality factor and less bit error rate are observed. As the distance is increasing, the quality factor reduces, and the bit error rate is increasing. In the given system layout, the link which is designed is successfully working till 5 km [12, 13].

The modulator which is used in the design is Mach Zender modulator, and a PIN photodetector is used to detect the signal at the output. At the input end, any kind of modulation formats can be considered such a RZ or NRZ; either of them will yield us better results [14, 15].

The system is illustrated in Figure 2 which includes all the components required for the design of the simulation setup.

A tabular column is plotted to showcase the different distances which are used in propagation along with the relation between the quality factor and the bit error rate (Fig. 3).

It is clearly observed from the table that for 1 km the quality factor is 860, and as the distance is increasing, it has come down to 15.3716 [18, 19].

For increased distance, the eye height also reduces which is a major drawback in any communication system [20, 21, 22].

3 Results

See Figures. 4, 5, 6, 7, 8, 9, 10, 11, 12, and 13.

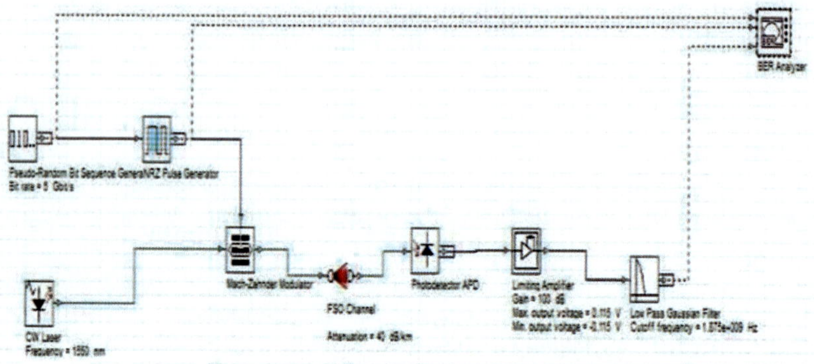

Fig. 2 System layout

4 Conclusion

The work in this paper demonstrates that at 1–5 km the link will work effectively, but with reduced quality factor and BER.

To improve the working efficiency, new modeling such as mitigation techniques need to be incorporated into the work which will reduce the bit error rate of the system.

In future, the link design can be modeled till 10 km range with adaptive optics.

Fig. 3 Tabular representation of various distances versus the quality factor bit error rate, eye diagram, and the threshold values [16, 17]

DISTANCE	MAX Q FACTOR	MIN BER	EYE HEIGHT	THRESHOLD	DECISION INST
1KM	860.718	0	0.00597297	0.000742555	0.53125
2KM	256.736	0	0.000957549	0.00012695	0.53125
3KM	91.6699	0	0.00026508	5.90E-05	0.53125
4KM	36.0261	1.61E-284	9.01E-05	4.37E-05	4.69E-01
5KM	15.3716	2.91E-53	3.20E-05	1.97E+05	5.00E-01

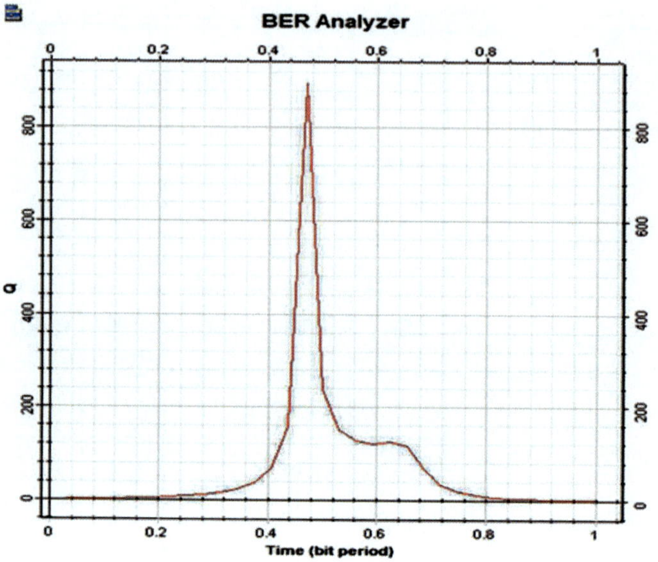

Fig. 4 At 1 km

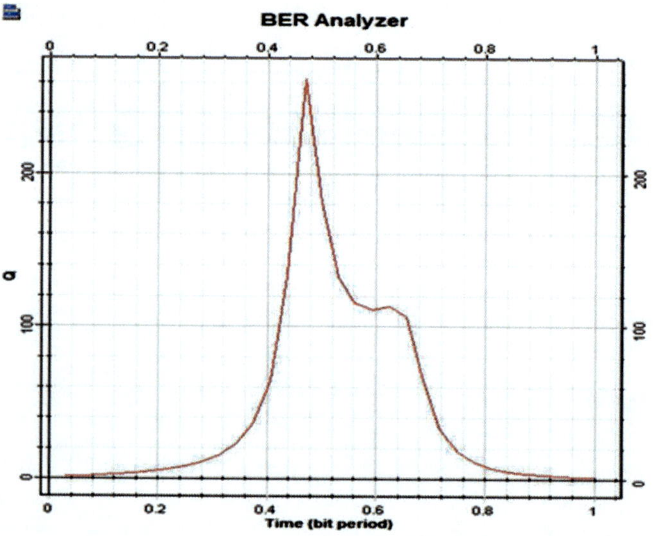

Fig. 5 At 2 km

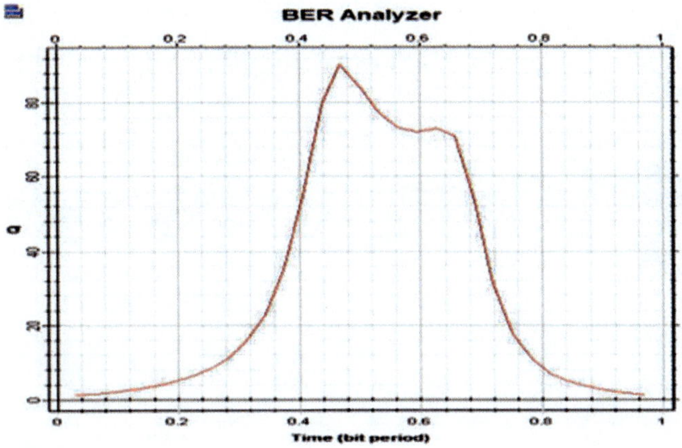

Fig. 6 At 3 km

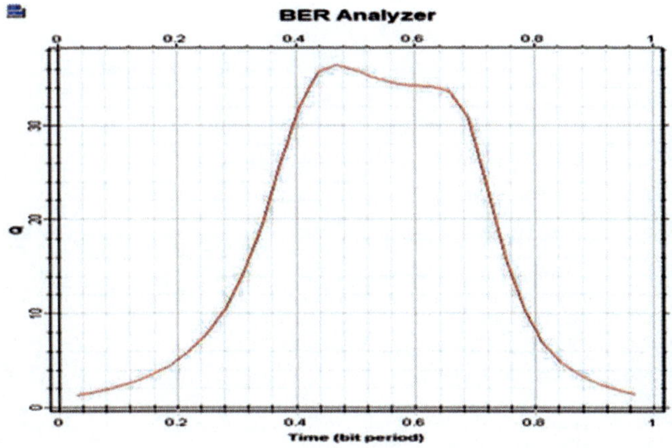

Fig. 7 At 4 km

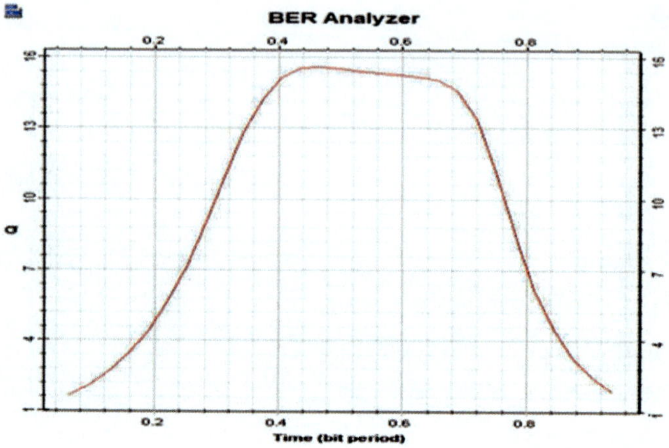

Fig. 8 At 5 km

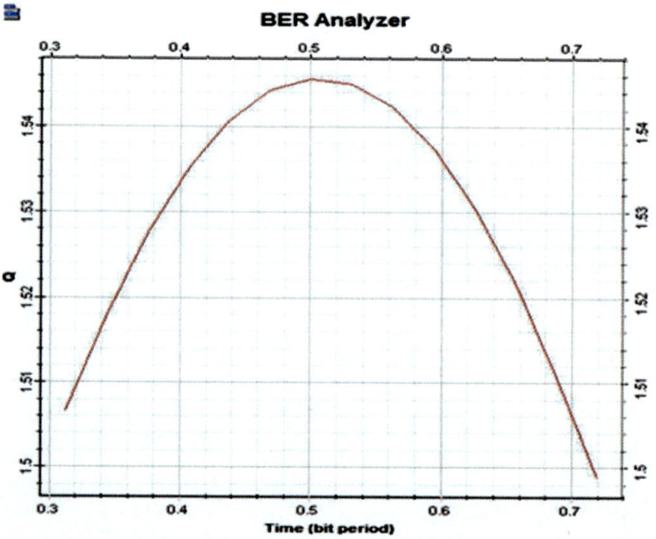

Fig. 9 Power at 1 W

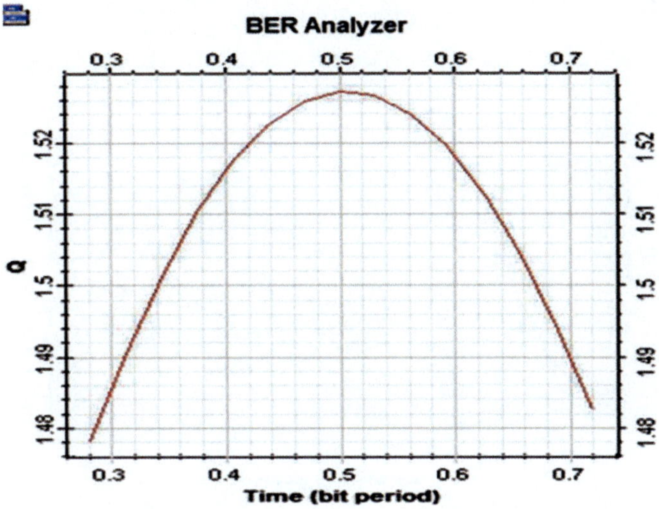

Fig. 10 Power at 2 W

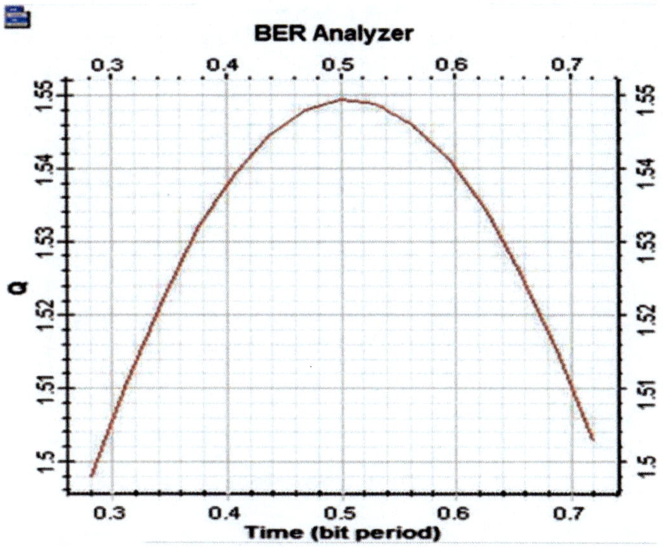

Fig. 11 Power at 3 W

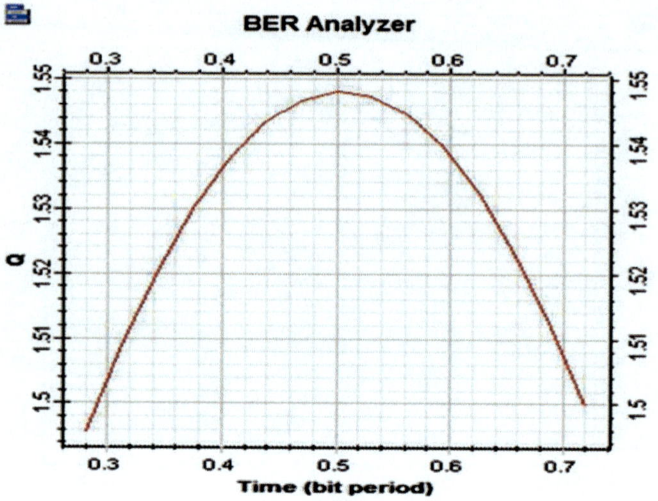

Fig. 12 Power at 4 W

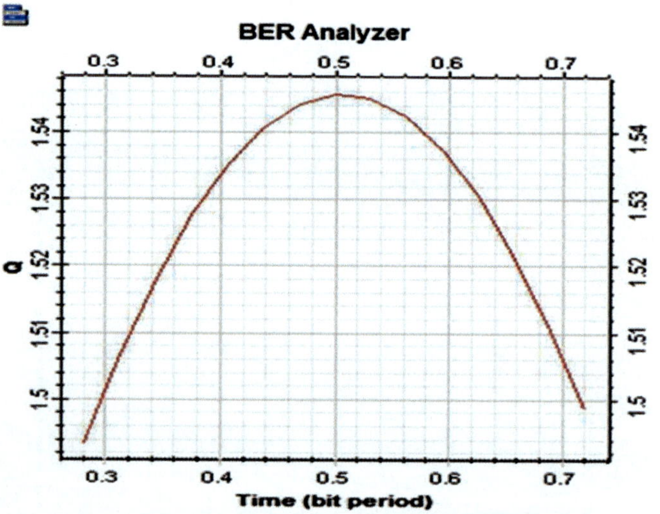

Fig. 13 Power at 5 W

References

1. Sridara K (2012) free space optical communication. Int J Latest Res Sci Technol 1(3):202–205
2. Indoor optical wireless communication: potential and state of art. IEEE Communications Magazine. September 2011
3. Killinger D (2002) Free space optics for laser communication through air. Opt Photon News 13(3):36–42

4. Ghassemlooy Z, Popoola WO (2010). Terrestrial freespace optical communications, mobile and wireless communications network layer and circuit level design. In: Fares SA, Adachi F (eds). ISBN: 978-953-307-042-1. InTech, Available from: http://www.intechopen.com/articles/show/title/terrestrial - free-spaceoptical-communications
5. Iniguez RR, Idrus SM, Sun Z (2008) Atmospheric transmission limitations. In: Optical wireless communications -IR for wireless connectivity. Taylor & Francis Group, LLC, London, p 40
6. Jia L, Liu JQ, Taylor DP (2007) Optical communication using subcarrier PSK intensity modulation through atmospheric turbulence channels. IEEE Trans Commun 55:1598–1606
7. Popoola WO, Ghassemlooy Z (2009) BPSK subcarrier intensity modulated free space optical communication in atmospheric turbulence. J Light Wave Technol 27:967–973
8. Fried DL (1967) Optical heterodyne detection of an atmospherically distortion wave front. Proc IEEE 55:57–77
9. Zhu X, Kahn JM (2002) Free space optical communication through atmospheric turbulence channels. IEEE Trans Commun 50:1293–13000
10. Kim I, Mcarthur B, Korevaar E (2001) Comparison of laser beam propagation at 785 and 1550 Nm in fog and haze for optical wireless communications. In: Proceedings Spie, vol 4214, pp 26–37
11. Akiba M, Ogawa K, Walkamori K, Kodate K, Ito S (2008) Measurement and simulation of the effect of snow fall on free space optical propagation. Appl Opt 47(31):5736–5743
12. Achour M (2002) Simulating free space optical communication; Part I, rain fall attenuation. In: Proceedings of Spie, vol 3635
13. Sony K (2017) Modelling the systems for improvised performance in FSO networks. J Adv Res Dyn Control Syst 9 Issue Special Issue 14:2627–2636
14. Sony K (2018) Propagation studies in free space using RZ and NRZ optical links under atmospheric turbulence. Int J Eng Technol 7(3.27):627–630
15. Immadi G et al (2018) Estimating the performance of free space optical link under adverse weather conditions by using various models. Wirel Pers Commun 103(2):1603–1613
16. Immadi G et al (2014) Estimation of Ku band satellite signal propagation impairment due to rain in tropical environment using ITU-R. Int J Appl Eng Res 20:7149–7168
17. Sree Madhuri A, Venkata Narayana M, Immadi G (2018) Performance analysis of single and multiple channel FSO system under turbulent conditions using various models. Int J Eng Technol 7(12):14–18
18. Sony K (2019) Evaluating the performance of FSO links using various attenuation models. Int J Innov Technol Exploring Eng 8(6):223–228
19. Durga Indira N (2019) Analysis of optical link using 4x1 WDM. Int J Innov Technol Exploring Eng 8(6):1682–1685
20. Immadi G, Venkata Narayana M, Kotamraju SK, Sree Madhuri A (2018) Estimating the performance of free space optical link under adverse weather conditions by using various models. Wirel Person Commun 103:1603–1613
21. Sony K (2021)Analysis of FSO links using various modulation formats and estimating the link performance for commercial and medical applications. Turk J Physiotherapy Rehabil 32(2):781–788
22. Sony K (2019) Analysis of FSO system under clear and rain conditions. Int J Recent Technol Eng 7(6):924–928

BEP Analysis of Filter Bank Multicarrier Under IQ Imbalance

R. Suraj, M. Venkatesh, C. Charumathi, Alekhya Kapavarapu, K. Pradeep Raj, R. Gandhiraj, and G. A. Shanmugha Sundaram

Abstract IQ imbalance occurs when there is a mismatch between the I and Q branch's digital to analog converter (DAC), filter or mixer. It has a considerable impact on signal detection and increases the bit error ratio (BER), which is undesired. This work mainly focuses on the impact of IQ imbalance in multi-carrier waveforms, especially conventional FBMC and pruned DFT-s FBMC that are potential candidates for 5G-NR. These waveforms are studied followed by subjecting them to IQ imbalances, and their performance is analysed through simulations, considering bit error probability (BEP) as the performance metric. It is also shown that pruned DFT-s FBMC is less sensitive to IQ imbalances compared to other waveform candidates.

Keywords FBMC · Pruned DFT-s FBMC · IQ imbalance · BEP analysis

1 Introduction

The emergence of fifth generation mobile communication system (5G) paved way to countless possibilities in deploying services such as ultra reliable low latency communication (URLLC), enhanced mobile broadband (eMBB) and massive machine type communication (mMTC). Various technologies, such as beamforming, massive MIMO and improved coding schemes [1] are utilized to achieve this purpose. Additionally, different potential waveform candidates are introduced that meets 5G requirements [2–4]. The current LTE standards make use of cyclic prefix OFDM modulation (CP-OFDM), which brought huge improvements than the previous generation, in terms of spectral efficiency, robustness to frequency selective channel and higher throughput. Some more applications of CP-OFDM include application in RADAR and satellite communication (SATCOM) [5, 6].

R. Suraj · M. Venkatesh · C. Charumathi · A. Kapavarapu · K. Pradeep Raj · R. Gandhiraj (✉) · G. A. Shanmugha Sundaram
SIERS Research Laboratory, Department of Electronics and Communication Engineering, Amrita School of Engineering, Amrita Vishwa Vidyapeetham, Coimbatore, India
e-mail: r_gandhiraj@cb.amrita.edu

P. S. R. Chowdary et al. (eds.), *Evolution in Signal Processing and Telecommunication Networks*, Lecture Notes in Electrical Engineering 839,
https://doi.org/10.1007/978-981-16-8554-5_8

Filter bank multi-carrier (FBMC) overcomes some of the drawbacks CP-OFDM has. Spectral efficiency of FBMC waveform is higher than CP-OFDM because it makes no use of CP. Moreover, since a bank of filters are used both at the transmitter and the receiver, the out of band emission (OOBe) is significantly reduced, which is very useful to mitigate the interference with the adjacent frequency bands. It also has very good resilience to carrier frequency offset (CFO) [7].

Regardless, these waveforms are susceptible to RF front end impairments such as IQ imbalance, amplifier distortions [8] and phase noise (PN). This has a huge impact on the received signal-to-noise ratio (SNR), which distorts the constellation and affects the detection. Moreover, this gets worse in higher carrier frequency, which is the case for 5G that is expected to operate in mmWave band. Therefore, it is very essential to take into account these RF impairments when one goes for real-time implementation.

Authors in [9] have proposed an advanced FBMC modulation scheme, called the pruned DFT-s FBMC, which has several advantages over the conventional FBMC modulation like lower PAPR and OOBe, lower complexity, better MIMO compatibility etc. To the best of our knowledge, the impact of IQ imbalance on FBMC and pruned DFT-s FBMC waveforms haven't been studied yet, which is the primary motivation of this work. BEP analysis is carried out to examine which waveform is the least sensitive to IQ imbalances.

2 Contributions

1. Initially, FBMC and pruned DFT-s FBMC waveforms, along with their corresponding mathematical model, are presented. This model also includes the effect of IQ imbalance on these waveforms, which is then utilized for simulations.
2. It is shown that pruned DFT-s FBMC not only has lower PAPR, OOBE and complexity, as mentioned in [9], but also is more robust to IQ imbalances than the other waveforms considered. This is proved by using bit error probability (BEP) as the metric to compare the waveforms. Moreover, a BEP gain curve is defined and provided, so as to understand the gain one gets at different SNR regimes when pruned DFT-s FBMC is used.
3. Finally, we have briefed the benefits of this analysis, i.e., uplink transmission in 5G communication systems under IQ imbalance, and shown that pruned DFT-s FBMC can be used instead of the currently used CP-OFDM and DFT-s OFDM-based waveforms.

3 IQ Imbalance

One of the major impairments one has to take into account while deploying any communication system is in-phase's and quadrature-phase's amplitude and phase imbalances, collectively known as IQ imbalance. This arises when there is a differ-

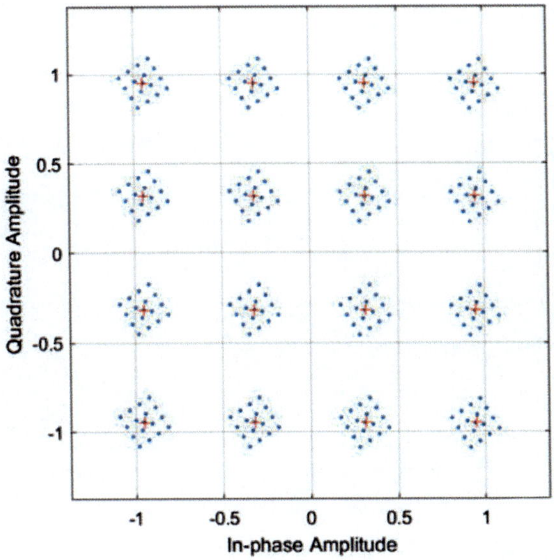

Fig. 1 Impact of IQ imbalance in 16 QAM OFDM systems

ence in the amplitude of the I and Q signals, or when the phase difference between them is not exactly 90°. IQ imbalance can lead to severe degradation while demodulating the received RF signal, which can in turn affect the symbol detection and increases the bit error ratio (BER) and BEP. The effect of IQ imbalance in a 16 QAM OFDM system is illustrated in Fig. 1. This work focuses on the IQ imbalance that arises from the Superheterodyne receivers, as shown in Fig. 2 [10].

3.1 Mathematical Modelling

Let the received signal be $\tilde{x}(t)$. This signal can be the OFDM signal after CP is added, or the FBMC signal, after it is passed through the synthesis filter bank (SFB), both of which are converted to passband at the transmitter and transmitted over the air interface. The signal after it is corrupted by IQ imbalance can be represented as

$$x(t) = \mu\, \tilde{x}(t) + \nu\, \tilde{x}^*(t) \tag{1}$$

$$\mu = \cos(\delta/2) + j\epsilon\, \sin(\delta/2)$$

and

$$\nu = \epsilon\, \cos(\delta/2) - j\sin(\delta/2)$$

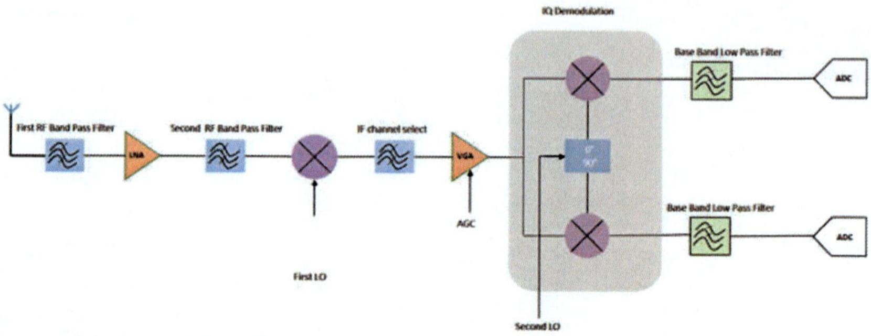

Fig. 2 Superheterodyne receiver

where ϵ and δ represent the amplitude gain and the phase imbalance between the I and Q branch respectively. These together contribute for the IQ imbalance. The amplitude imbalance ϵ can be written as [10] :

$$\epsilon = \frac{\epsilon_I - \epsilon_Q}{\epsilon_I + \epsilon_Q}$$

4 FBMC

Filter bank multi-carrier modulation overcomes some of the problems OFDM and DFT-s OFDM modulation has, such as very low OOBe, higher spectral efficiency, since it does not use CP, and can be used in low-latency use cases. Moreover, PAPR of FBMC is 15.82% better than that of OFDM [11]. The FBMC system model that is used in this paper is based on the system model from [12].

The continuous FBMC baseband signal can be expressed as :

$$s(t) = \sum_{n=1}^{N} \sum_{m=1}^{M} x_{m,n} \; p_{m,n}(t) \tag{2}$$

where $x_{m,n}$ are the constellation data points (typically pulse amplitude modulated (PAM)) at the mth subcarrier and the nth time index. M and N denote the total number of subcarriers and FBMC symbols, respectively. $p_{m,n}(t)$ is the basis function given as :

$$p_{m,n}(t) = \tilde{p}(T - nT) \; \mathrm{e}^{j2\pi m \Delta f \; (t-nT)} \mathrm{e}^{j(m+n)\frac{\pi}{2}} \tag{3}$$

where $\tilde{p}(t)$ is the prototype filter, Δf is the subcarrier spacing and T is the time spacing. To make it more easier to understand, the matrix-based system model from [12] has been used in this entire work. For that, initially, the basis pulse is sampled at a sampling rate $f_{\text{sampling}} = \Delta f.N$, so the number of samples is represented as N_{samples}, and stacked as a vector, to get $\mathbf{p}_{m,n} \in \mathbb{C}^{N_{\text{samples}} \times 1}$.

At the nth time index, the basis function is defined as

$$\mathbf{P}_n = [\mathbf{p}_{1,n}, \mathbf{p}_{2,n} \cdots \mathbf{p}_{M,n}\ ] \in \mathbb{C}^{N_{\text{samples}} \times M}$$

Therefore, its overall matrix notation is given as

$$\mathbf{P} = \ [\ \mathbf{P}_1, \mathbf{P}_2 \ldots \mathbf{P}_N\ ] \in \mathbb{C}^{N_{\text{samples}} \times MN}.$$

and the transmitted signal will now be denoted as :

$$\mathbf{s} = \mathbf{P}\,\mathbf{x} \tag{4}$$

where $x_n = [x_{1,n}, x_{2,n} \ldots x_{M,n}]^{\text{T}} \in \mathbb{C}^{M \times 1}$ is the nth transmitted symbols and $\mathbf{x} = vec\{x_1, x_1 \ldots x_N\} \in \mathbb{C}^{MN \times 1}$ is all such transmitted symbols stacked in a vector. The doubly selective channel is modelled as a convolution matrix $\mathbf{H} \in \mathbb{C}^{N_{\text{samples}} \times N_{\text{samples}}}$. Therefore, the received signal will be represented as:

$$\mathbf{r} = \mathbf{H}\,\mathbf{s} + \mathbf{n}, \tag{5}$$

where n is the additive white Gaussian noise (AWGN) with P_n as the noise power.

After receiving, the signal is demodulated with $\text{Q} = \text{P}^H$, which becomes,

$$\begin{aligned} \mathbf{y} &= \mathbf{Qr} \\ &= \mathbf{Q}(\mathbf{H}\,\mathbf{s} + \mathbf{n}) \\ &= \mathbf{QHs} + \mathbf{Qn} \\ &= \mathbf{QHPx} + \mathbf{Qn} \end{aligned} \tag{6}$$

Since the interference introduced by the channel's off-diagonal element is relatively lower [13], it can be omitted, thus the equation (6) changes to

$$\mathbf{r} \approx \mathbf{diag\{h\}QPx + Qn} \tag{7}$$

where $\mathbf{h}$ is the one-tap channel of size $MN \times 1$.

5 Pruned DFT-s FBMC

The basic idea of pruned FBMC, as described in [9] is illustrated as follows:

- It is known that OFDM has poor PAPR. In order to resolve this, it is precoded with a DFT matrix $\mathbf{W}_{\text{L}}$, such that the entire system looks like single carrier transmission.

- But the problem with DFT-s OFDM is that, the OOBe is still high. This is dealt by setting the base pulses that are close to and at the edge to zero so that the adjacent band interference can be reduced.
- Instead of setting only the edge base pulses to zero, it is made zero for half of the base pulses. In other words, it can be looked as half the number of input symbols are set to zero, in order to reduce the OOBe.
- This system is made robust to multipath, by adding zero-tail overheads, similar to CP-OFDM. But this can lead to loss of spectral efficiency.
- Therefore, to make it more spectrally efficient, this entire OFDM system is transformed to FBMC system, by spectrally shaping each of the subcarrier by a prototype filter $\tilde{p}(t)$.
- Finally, the basis pulses are scaled up so that the power is constant over the transmission period.

Mathematical Details Numerous prototype filters are available for FBMC [14], out of which the most important ones are the PHYDYAS and the Hermite filter. Hermite filter has better time-frequency localization than PHYDYAS filter, so Hermite filter is used here. Because of this property, it makes it more robust to doubly selective channels, which is the channel of our interest. More specifically, a truncated Hermite filter is used [9], which is given as

$$\tilde{p}_{\text{Herm.Trunc.}}(t) = \begin{cases} \tilde{p}_{\text{Herm.}}(t) & \text{if } \frac{-1.56}{2\Delta f} \leq t \leq \frac{-1.56}{2\Delta f} \\ 0 & \text{otherwise} \end{cases} \tag{8}$$

where

$$\tilde{p}_{\text{Herm.}}(t) = \sqrt{\Delta f} \mathrm{e}^{-2\pi(t\Delta f)^2} \sum_{i=\{0,4,8,12,16,20\}} \alpha_i H_i \left\{ 2\sqrt{(\pi)} t \Delta f \right\} \tag{9}$$

The block diagram of pruned FBMC modulation, based on [12], is given in Fig. 3, except it is now modelled along with the IQ imbalance. $S_{1,n}, S_{2,n} \ldots S_{M/2,n}$ are the symbols that needs to be transmitted at the nth instance. $\mathrm{b}_1, b_2 \ldots b_{M/2}$ are the one tap

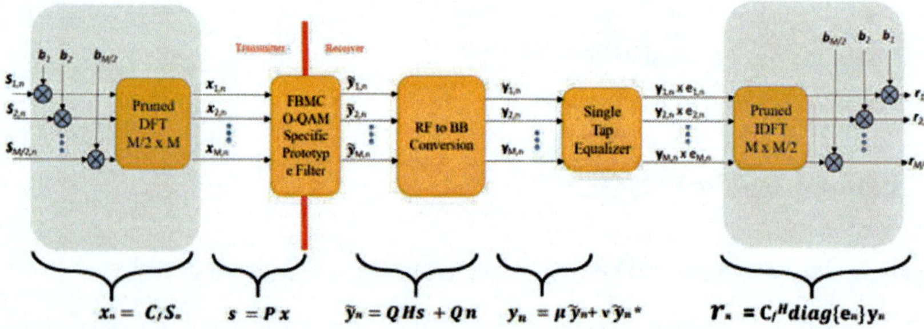

Fig. 3 Pruned FBMC block diagram under IQ imbalance

scaling factors, and C_f is the precoding matrix. Therefore, $x_k = C_f \times S_n$ is FBMC-OQAM modulated to get $s = Px$.

This signal is now affected by doubly selective channel, and after receiving, it is demodulated with $Q = P^{\mathrm{H}}$ to get

$$\tilde{y}_n = \mathrm{QHs} + \mathrm{Qn}$$

Upon reception, it is first converted from passband to baseband, which introduces IQ imbalance, which is written as

$$y_n = \mu\tilde{y}_n + \nu\tilde{y}_n^*$$

. This is then passed on to the single-tap equalizer with equalizing factors e_n and then to the pruned IDFT block, whose outputs are the detected symbols r_n, given as:

$$r_n = C_\mathrm{f}^{\mathrm{H}}\mathrm{diag}\{e_n\}y_n$$

6 Simulation Results

We considered doubly-selective channel, specified by the 3GPP (38.900 channel model [40, Sect. 7.7.3]), with tapped delay line (TDL)-A as the power delay profile that has 300ns long delay spread. The number of subcarriers is chosen to be $M = 256$, with 16 QAM modulation and $\Delta f = 15$ kHz. The carrier frequency is set to be 2.5 GHz.

The bit error probability (BEP) is calculated and plotted for various values of IQ imbalances for all the waveforms. This can be calculated by

$$\mathrm{BEP} = \mathbb{E}\left\{\frac{1}{N}\sum_{n=1}^{N}\frac{1}{M/2}\sum_{\tilde{m}=1}^{M/2}\mathrm{BEP}_{\mathrm{AWGN}}\{\mathrm{SINR}_{\tilde{m},n}(H)\}\right\} \tag{10}$$

where $\mathrm{BEP}_{\mathrm{AWGN}}\{.\}$ denotes the bit error probability of an AWGN channel.

Figures 4 and 5 show the BEP performance of FBMC and pruned FBMC waveforms subjected to various IQ imbalances. The amplitude imbalance is varied as $\epsilon = 0.1$ dB and $\epsilon = 1$ dB and the phase imbalance is varied as $\delta = 0°$ and $\delta = 10°$. With these values, at a higher velocity of 200 km/h, Figs. 4 and 5 are obtained.

To analyse which waveform is better, we set some parameters with velocity = 0 km/h, and the amplitude and the phase imbalances are kept the maximum (the worst case), i.e. $\epsilon = 1$ dB and $\delta = 10°$, and calculated the BEP for all the waveforms in one graph, given in Fig. 6. It can be seen that, pruned FBMC is less sensitive to IQ imbalance and outperforms the other waveforms. Performance of conventional FBMC is almost similar to that of DFT-s CP-OFDM under IQ imbalance. Specifically, FBMC is more sensitive to IQ imbalance in higher SNR regime (from 25 dB). This

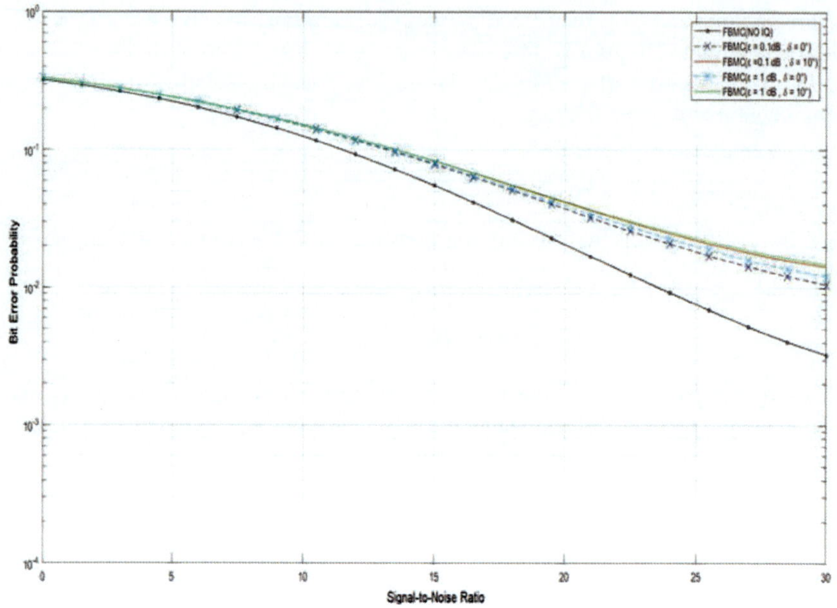

Fig. 4 FBMC under IQ imbalance, with ϵ = [0.1 dB, 1 dB] and δ = [0°, 10°]

is also reflected in the BEP gain curve, given in Fig. 7. It is defined as the amount of BEP improvement one gets by using pruned DFT-s FBMC, which is the difference between the BEP of the waveform considered and the BEP of pruned DFT-s FBMC at all the SNR values.

$$\begin{aligned}\text{BEP Gain} = |\ [\text{BEP}_x(1)\ \ \text{BEP}_x(2)\ \ldots\ \text{BEP}_x(k)] \\ -[\text{BEP}_y(1)\ \ \text{BEP}_y(2)\ \ldots\ \text{BEP}_y(k)]\ |\end{aligned}$$

where x denotes pruned DFT-s FBMC and y can be CP-OFDM, DFT-s CP-OFDM or the conventional FBMC and k denotes the SNR index. More is the difference between the BEPs between pruned DFT-s FBMC and the waveform considered, more is the gain. Therefore, out of all the waveforms, the gain one gets by using pruned DFT-s FBMC over FBMC is more than that of DFT-s OFDM, followed by CP-OFDM.

7 Conclusion

IQ imbalances causes significant degradation of performance in multi-carrier transmission schemes. Especially, in a doubly selective channel, it can be worse. In this paper, we have studied FBMC and pruned DFT-s FBMC waveforms when they are

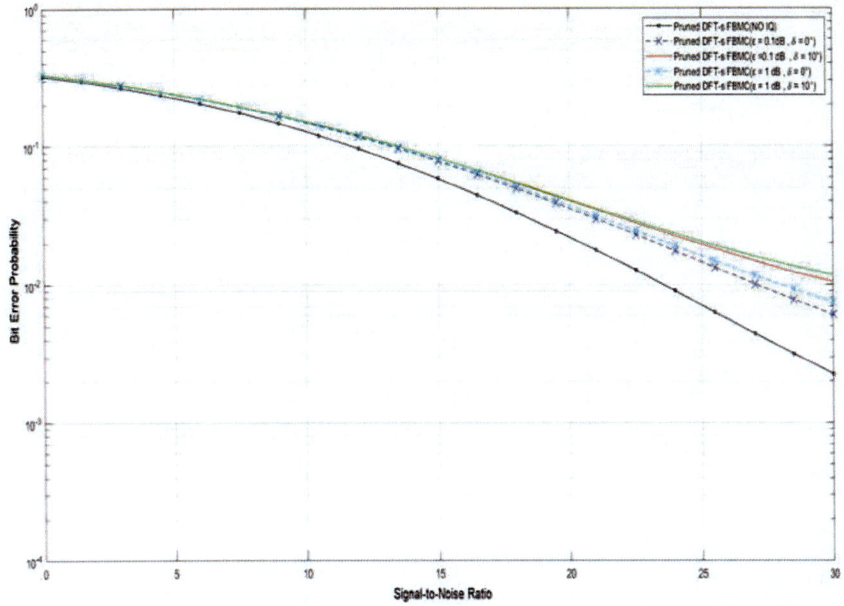

Fig. 5 Pruned DFT-s FBMC under IQ imbalance, with $\epsilon = [0.1 \text{ dB}, 1 \text{ dB}]$ and $\delta = [0°, 10°]$

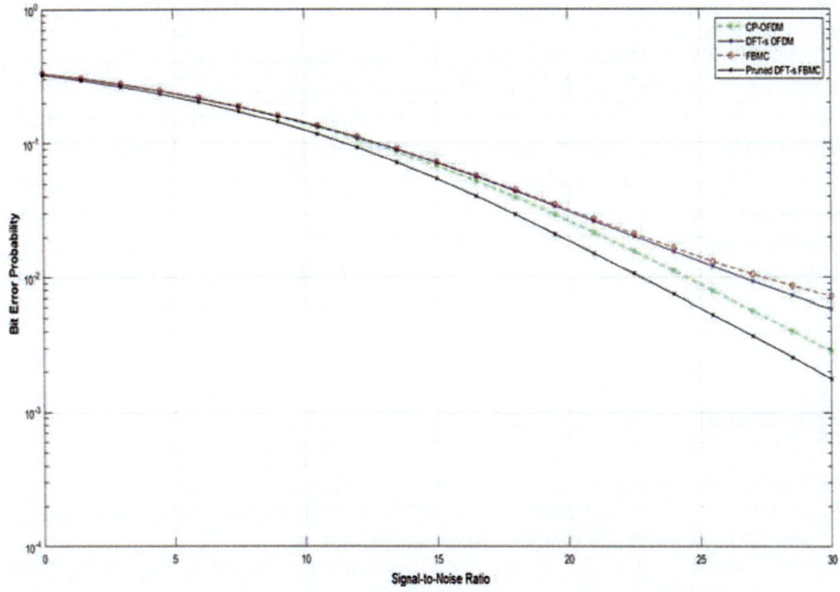

Fig. 6 Comparison of all the waveforms under IQ imbalance, with $\epsilon = 1$ dB and $\delta = 10°$

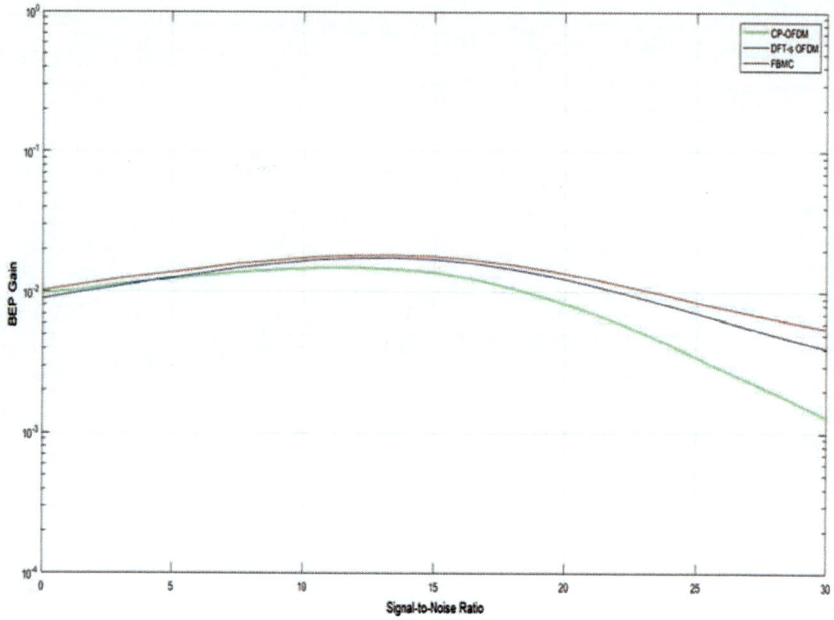

Fig. 7 Gain in terms of BEP

subjected to IQ imbalance, and it has been shown that pruned FBMC is more robust to IQ imbalances and outperforms other waveforms, in doubly selective channel. This can be useful, as the author in [9] has mentioned the potential applications of pruned DFT-s FBMC, which includes uplink transmissions in wireless communications. When the user density is high, which is envisaged for 5G, the necessity of waveform having lower OOBe, PAPR and is more resilient to IQ imbalance is magnified. In such a situation, pruned DFT-s FBMC can be used as an alternative to the currently used uplink waveforms such as CP-OFDM and SCFDMA, and can provide a significant performance gain in the presence of IQ imbalance.

References

1. Pechetti J, Hallingar B, Prasad PVND, Kumar N (2021) Performance analysis of polar codes for 5G wireless communication network. In: 2nd EAI international conference on big data innovation for sustainable cognitive computing. Springer, pp 361–378
2. Lin H, Siohan P (2016) Major 5G waveform candidates: overview and comparison. In: Signal processing for 5G: algorithms and implementations
3. Van Eeckhaute M, Bourdoux A, De Doncker P, Horlin F (2017) Performance of emerging multi-carrier waveforms for 5G asynchronous communications. EURASIP J Wirel Commun Netw 2017(1):1–15

4. Liu Y, Chen X, Zhong Z, Ai B, Miao D, Zhao Z, Sun J, Teng Y, Guan H (2017) Waveform design for 5G networks: analysis and comparison. IEEE Access 5:19282–19292
5. Suganya M, Gandhiraj R (2016) Applicability of MIMO and OFDM technology to SATCOM. In: 2016 International conference on communication and signal processing (ICCSP). IEEE, pp 0916–0920
6. Vineeth VS, Shanmugha Sundaram GA, Soman KP (2020) Influence of clutter on 5G waveform modulated radar signals. In: AIP conference proceedings, vol 2222. AIP Publishing LLC, p 030006
7. Bai Q, Nossek JA (2010) On the effects of carrier frequency offset on cyclic prefix based OFDM and filter bank based multicarrier systems. In: 2010 IEEE 11th international workshop on signal processing advances in wireless communications (SPAWC). IEEE, pp 1–5
8. Priyanga G, Shanmugha GA, Sundaram, Soman KP (2018) Combined influence of channel and amplifier on multicarrier modulated radar signals. Procedia Comput Sci 143:30–38
9. Nissel R, Rupp M (2018) Pruned DFT-spread FMBC: Low PAPR, low latency, high spectral efficiency. IEEE Trans Commun 66(10):4811–4825
10. Tarighat A, Bagheri R, Sayed AH (2005) Compensation schemes and performance analysis of IQ imbalances in OFDM receivers. IEEE Trans Signal Process 53(8):3257–3268
11. Joseph Jeffery JM, Masapalli L, Nookala VM, Dasari SP, Kirthiga S (2020) Peak to average power ratio and bit error rate analysis of multicarrier modulation techniques. In: 2020 International conference on communication and signal processing (ICCSP). IEEE, pp 1443–1446
12. Nissel R, Schwarz S, Rupp M (2017) Filter bank multicarrier modulation schemes for future mobile communications. IEEE J Sel Areas Commun 35(8):1768–1782
13. Nissel R, Rupp M (2017) OFDM and FBMC-OQAM in doubly-selective channels: calculating the bit error probability. IEEE Commun Lett 21(6):1297–1300
14. Sahin A, Guvenc I, Arslan H (2013) A survey on multicarrier communications: Prototype filters, lattice structures, and implementation aspects. IEEE Commun Surveys Tutor 16(3):1312–1338

Optical Letter Recognition for Roman-Text

Buddha Hari Kumar and P. Chitra

Abstract The reading abilities of computers are still way behind that of human beings. It is a too difficult task to build OCR for human use. Optical letter recognition is a scanned image which is converted into editable text. In this process, the image pictures are encoded with the help of Unicode. OCR system helps to add a volume or a journal product directly into an electronic PC document and alter the record utilizing a word processor. After removing the noise from the image and normalizing the image, the next step is to segment the image. Several segmentation techniques are used, such as line segmentation, word segmentation, and letter segmentation. It will be helpful to recognize the letter in the document image. After the letters are classified, the identified letters are grouped to reproduce the first symbol strings. The proposed method for developing high-performance OCR software for Roman-text includes the extraction of image, preprocessing highlights, and the creation of classification algorithms.

Keywords OCR (optical letter recognition) · Processing · VPP · HPP

1 Introduction

It consists of the transformation of transcribed, touch-typed, or electronic data into instrument-encrypt data using images scanned from handwritten, typewritten, or printed documents [1]. OCR process called scanning is used to convert paper files to electronic files. OCR can perform OCR not just for English but also for many other languages, making it a versatile tool. On the whole, OCR can be split between two systems: lattice coordinating and include extraction. In contrast to matrix matching, matrix matching is more common and is also more limited.

An "OCR" system uses a camera for text recognition and software for analyzing images. Typically, OCR systems leverage hardware (specialized circuit boards) and

B. Hari Kumar (✉) · P. Chitra
Department of Electronics & Communications, Sathyabama Institute of Science and Technology, Jeppiaar Nagar, Chennai 600119, Tamilnadu, India

P. S. R. Chowdary et al. (eds.), *Evolution in Signal Processing and Telecommunication Networks*, Lecture Notes in Electrical Engineering 839,
https://doi.org/10.1007/978-981-16-8554-5_9

software in their recognition process, but some inexpensive systems make their recognition solely through software. In the legal profession, advanced OCR systems have already made it possible in recent years to perform searches that once required a good deal of time and effort within minutes.

Assigning the information, a scanner observes as a letter to a bitmap pattern (also called pattern matching) storage system. This program assigns the equivalent plain text letter to a selected bitmap if it matches one of the chose bitmaps inside a given level of similitude

2 Ease of Use

2.1 Use of OCR

The backup and archival of thousands of books and documents are routinely performed in professional environments (such as libraries and offices). The scanning process produces PDF documents containing scanned images of original paper documents. It is difficult to look for a particular expression or name in such huge volumes of scanned documents. Please ensure that the paper size template you are maintaining is correct [2].

2.2 Working of OCR

OCR chips away at a truly rudimentary rule. It utilizes sensible techniques to look at and perceive letters in the document to be changed over. These templates provide a way to compare the letters in a file with predefined sets of letters [2]. The product recognizes the light and dull regions on the picture and duplicates the dim examples. These are then contrasted with the letters accessible in the layout. These processes all happen in milliseconds, which is amazing. All the letters are tested thusly, and the OCR attempts to arrive at 100% exactness. These processes all happen in milliseconds, which is amazing.

2.3 Need of OCR

OCR innovation rings in a paperless office climate. OCR makes looking, arranging, erasing, and summing up documents a lot simpler. The lot of flow is rescued in this strategy as physical impedance and person deviation do not become an integral factor straightforwardly. This method results in significant time savings since no manual interference or human deviations are present. Hundreds of reams of material may be transformed into searchable documents in hours by moving them to a computer system. OCR has indeed revolutionized the way data is processed and stored around the world, allowing companies to save time, money, and space. You can now digitize and convert handwritten documents, accounting ledgers, and other files into accessible PDF (versatile record design) documents. You would require OCR programming with the objective, so one sheet of PC saw and open content would be arranged above the separated nature; this sheet and the inquiry capacity of the per user bid be firmly cooperating to utilize the PDF document agreeable and useful. You need OCR programming to get this heap of things done.

3 OCR Procedure Details

See Fig. 1

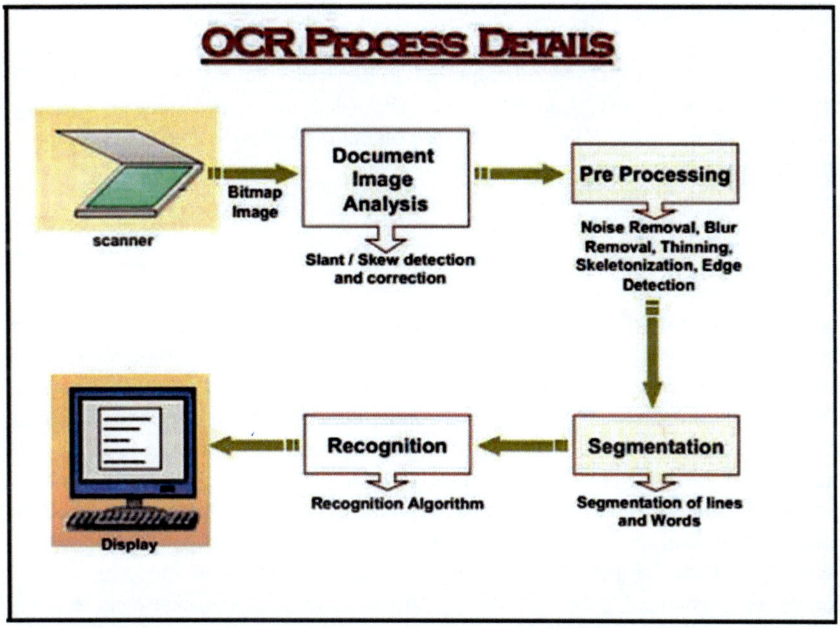

Fig. 1 Block diagram of OCR

3.1 Steps Associated with OCR Processor

3.1.1 Optical Letter Reading

A level bed reader is generally worn at three thundered dots per inch which changes the composed word on the sheet being shifted into an electronic picture.

3.1.2 Report Picture Scanning

The electronic picture of the content is analyzed for the existence of inclination or inclines, and in this manner, these are taken out. A considerable amount of runoff writing has mixes of content and chart, diagrams, and different types of outlines [3]. It is accordingly significant that the content region is distinguished individually out of different pictures and can be limited and separated.

3.1.3 Prehandling

In this stage, various cycles are applied to the text picture like upheaval and dark departure, bipartite, reducing, edge acknowledgment, and some morphological cycles, to set up an OCR picture of the content locale that is freed of commotion and dark.

3.1.4 Bipartite

Record picture bipartite is an invaluable procedure to change over a faint picture into two-tone. Overall bipartite and locally flexible bipartite are two notable kinds of bipartite procedures. There are not many classes of bipartite strategies, for example, circle graph-construct, collect-construct, decay-construct, thing property-construct, structural bipartite and endemic adaptable, etc. Circle graph-construct strategies use the things of the circle graph, similar to zeniths and pass or hole. Gathering construct procedures, use partitioning of the image picture element into two packs to find the cutoff regard. Entropy-construct techniques use entropy information. Article trademark construct techniques utilize nature of pictures like border planning or measure the likeness between the first and the bipartite picture. Structural-construct methods primarily use higher solicitation probability or the association among dot and aboriginal adaptable construct procedures figure close by limit subject to the information contained in the neighborhood of each picture element. We have used circle graph-construct properties to bipartite the records taken as an educational assortment. The digitalized text pictures are first changed over into two-tone pictures using a circle graph-construct peak way. Now we address thing dot by 1 and establishment picture

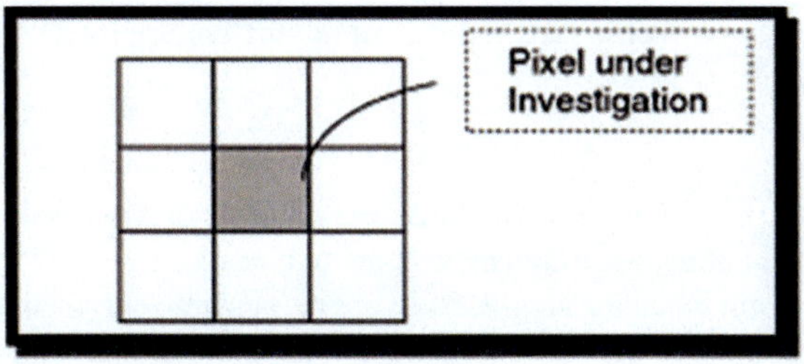

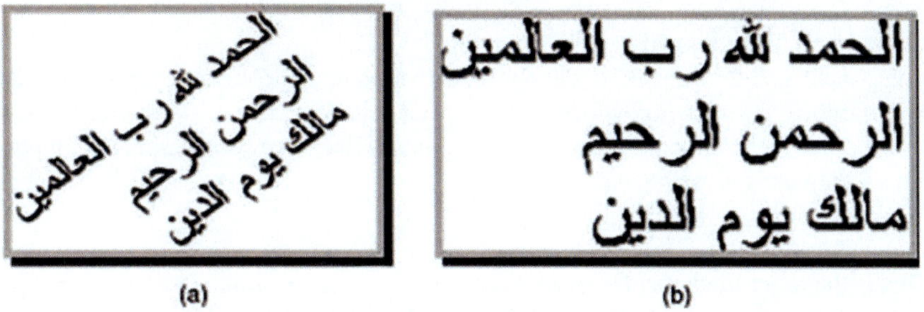

Fig. 2 Eight connected components. **a** Original image altered by 24; **b** picture with float rectification

element by 0. The two-tone picture generally shows distensions and gouges in the nature similarly as separated article dot above the establishments, whichever sponge by a shrewd, flatten motion.

3.1.5 Slant Identification/Adjustment

In pre-taking care phase for this task, we center on end of non-supportive data that can be contemplate as commotion and inclination area and modification. To take out the upheaval as bound picture element for some irregular picture element, we inspect for the presence of an abutting picture element in all viable eight different ways as below manifest in 1.1. If a picture element exists in any tolerable manners, then the picture element is everything except a confined picture element [4]. Any skewness induced during capturing the image is corrected. To handle the issue of slant location and remedy, we have used the going with glide revision framework; first we choose the upset mark of the text by enrolling the diversions of all the line divides that can be worked middle of any set of dull picture element in the image; then the looking at focuses is figured. To reduce the estimation cost, one could apply this association just to a picked window of text instead of the whole picture expecting that the entire sheet is altered correspondingly everywhere [5]. The point that has the greatest number of occasions is believed to be the place of turn of the picture. By choosing the mark of insurgency the check drift is rectified by turning the picture by a comparative point to the contrary side (Fig. 2).

Segmentation

In case the entire picture involves text just, the image is first divided into free groups of context. These contexts are then distributed into terms ultimately terms into single figures. At the point when the single letters are recognized, confined, and divided out in a text picture, it transforms into an issue of user choices of affirmative calculation to get the context in the image into a context analyzer.

Component Withdrawal

Component withdrawal is excessively seen as topological element examination. A sort of optical person acknowledgment does not depend on exact coordinating to set layouts. The program looks for normal components like open spaces, shut structures, lines—diagonals crossing, etc.

Image Recognition

This is the most fundamental stage, wherein affirmation estimation is registered to the pics present in the note pic which is divided at one each level. Due to associative letter codes for recognizing with its source and will be corrected by the P.C which is then applied to individual code by a processor to be displayed on the monitor, where it will be locally corrected, leveled, and saved into another suitable record for further analysis [6].

Post-processing

Post-handling is the last action to happen in a progression of OCR preparing stages. Fundamentally, the point of post-handling is to recognize and address semantic incorrect spellings in the OCR which yield text after the information picture has been checked and totally prepared. Essentially, there are two sorts of OCR blunders: non-word mistakes and genuine word mistakes. A no word blunder is a word that is perceived by the OCR framework; in any case, it does not identify with any section in the vocabulary. For example, at the point when "how are you Pradeep" is perceived by the OCR framework as "huw are yuu pradoop," then, at that point "How" is supposed to be a non-word mistake on the grounds that "huw" is not letterized in the English language. Conversely, a genuine word mistake is a word that is perceived by the OCR framework and relates to a passage in the dictionary, though it is linguistically erroneous regarding the sentence where it has happened. For example, when "how are you pradeep" is perceived by the OCR framework as "huw are yuu pradoop," then, at that point "you" is viewed as a genuine word blunder since "yuu" in spite of the fact that is linguistically right (accessible in the Latin lang.), its utilization in the group is syntactically mistaken. Commonly, non-term and certifiable blunders fall under three classes of mistakes: cancellation, addition, and replacement blunders. The erasure mistake happens when at least one letter is disposed of or taken out from inside the first word, for instance, misrecognizing "battle" as "bottle," "cattle," "eattle," or even "use" as "ose." The incorporation botch happens when somewhere

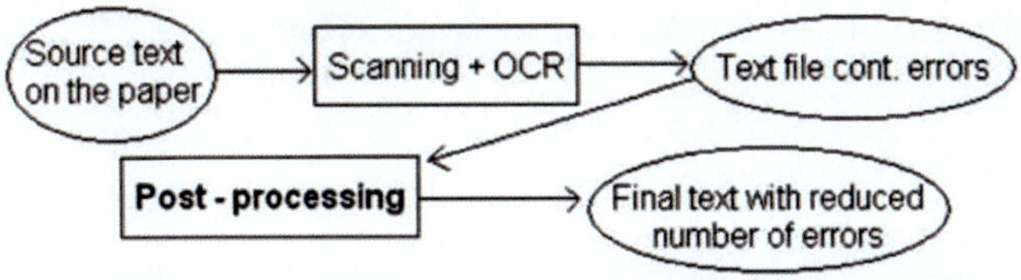

Fig. 3 Post-handling

around one extra letter is added or stiffed to the main word, for instance, mis-seeing "above" as "abuve" or even "abovo." The substitution botch happens when something like one letter is coincidentally changed in the main word, for instance, changing the individual "o in 'above' to 'u' or changing the individual 'e' in 'Above' to o" [7].

The unusual context of the article being solved by a long shot solution, i.e., the solitary guilty party for creating OCR blunders and therefore originating OCR frameworks either to work loosely step by step or to flop individually totally. Accordingly, innumerable post-handling approaches and calculations were proposed trying to identify and exact OCR mistakes (Fig. 3).

4 Outcomes and Analysis

In this paper, we endeavored diverse archive pictures like printed text records and composed reports as input pictures for testing. Also, we noticed the yield precision as displayed in the accompanying figures. On the off chance that the content is created by a sort machine certainly, the text dimension will be uniform all over the place. Betwixt two narrative groups, there is a restricted level belt with either empty picture element or not a lot of picture element. Therefore, applying horizontal projection profile (HPP) and recognizing the valleys in it, text line gatherings can be recuperated [8].

A model is displayed in Fig. 4. Particularly the modifiers (the two vowels and consonants) more often than not concur with the adjusting letters as displayed in Fig. 4. Such nontrivial blends of letters make the whole course of letter division very testing.

Term Dissection

From the detached narrative group of lines, terms get disunite. Regularly, applying vertical projection profile (VPP) is the best solution available, and while distinguishing some specific cutoff outperforming level spaces, best approach is terms that are separated from a narrative group. A model is shown in Fig. 5

PC made contents might contain distinctive text dimensions and various styles (e.g., strong, italic, and so forth) and antagonistically change the edge an incentive for recognizing secluded words. Henceforth, perceiving a viable edge esteem is undeniably challenging. It may change twice or even more even a lone text line [1].

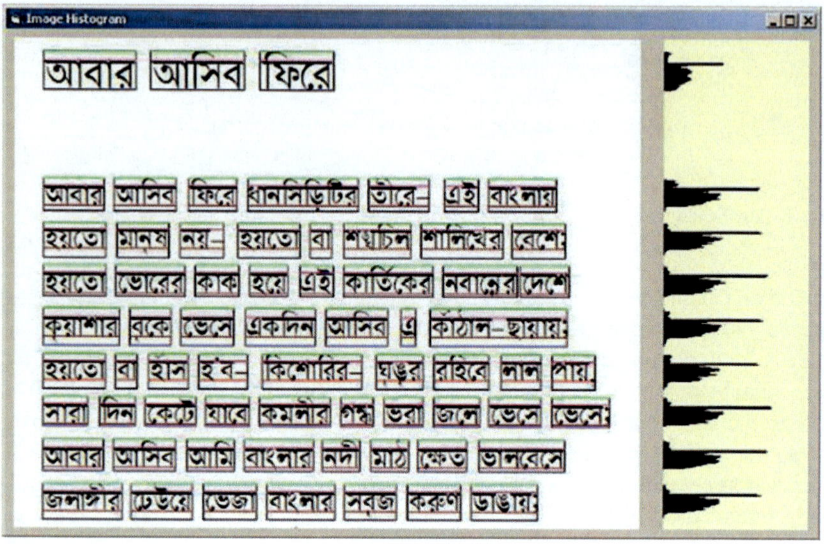

Fig. 4 Text line extraction from report utilizing HPP

Fig. 5 Word separation from text line using VPP

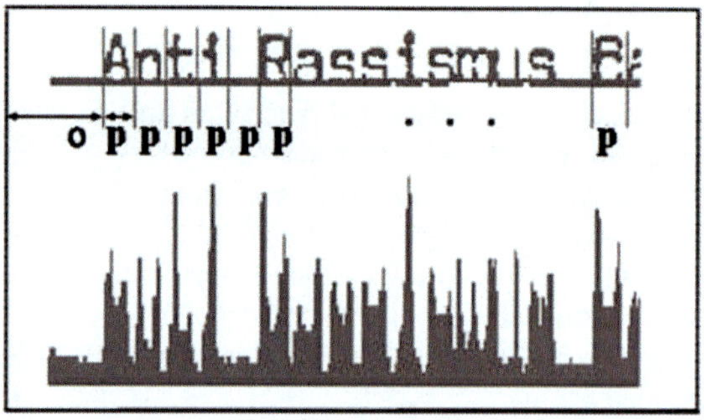

Of course, considering the practice of type-instrument made contents is certain, it is significantly more straightforward to work out the cutoff worth of the opening in average of two progressive terms in a band.

Letter Division

Division of letters from the secluded words is the toughest detailing of the content division stage. Since, in PC made writing a few letters in a compartment term may to some extent cross-over with each other, it turns out to be truly challenging to confine those letters appropriately. Particularly, the transformers (the two vowels and consonants) more often than not cross-over with the altering letters as shown in Fig. 6. Such non-trivial mixes of letters make the whole course of letter division incredibly testing. Plus, a few images, as Chandra-Bindu, regularly separate two

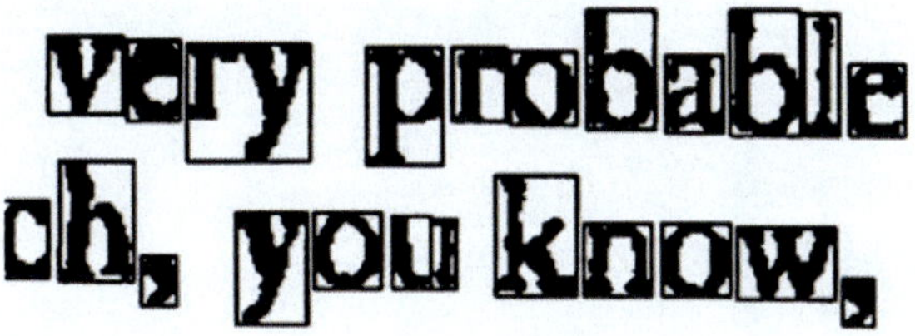

Fig. 6 Letter division of assorted instrument Latin language content

Fig. 7 Input image-1

successive letters in a word; then, at that point segregating those turns into a difficult situation. A model is displayed in Fig. 6.

This issue can be overwhelmed by applying shape following instrument or by executing eager quest procedure for letter division. In these strategies, syllables are separated as opposed to single person or letter from each word [9]. Be that as it may, the letters in a kind machine made Latin content rarely cross-over with different letters. This is the reason if there should be an occurrence of type-machine made contents letter division from separated words is a lot simpler and precise. Here, each letter can be segregated in a rectangular area. Subsequently, the acknowledgment cycle becomes less complex and quicker. A model is displayed in Fig. 6.

For informatics record, the precision is above 98% considering the way that it does not have any more fuss and has good essential. In next instance for informatics record with huge dimension and bold tones, we expect near 100% perfection considering the way that as the size grows precision in yield in like manner increases as shown in Figs. 7 and 8.

By handcrafted report, the exactness is near to 70% on the grounds that there might be commotion in the image and composing might be changed starting with one individual then onto the next individual. Indeed, even we can accomplish better exactness in transcribed report additionally by working on certain components in optical letter recognition like slant perception and change for dip portrayal, and varying more components, we can reduce the masque as viewed in Figs. 9 and 10.

5 Final Thoughts

The paper that has been planned here is about optical letter recognition for acknowledgment of Latin language print file images. We focused on directing the trials to gauge its exhibition in which we have great outcomes on sensible assorted better archives.

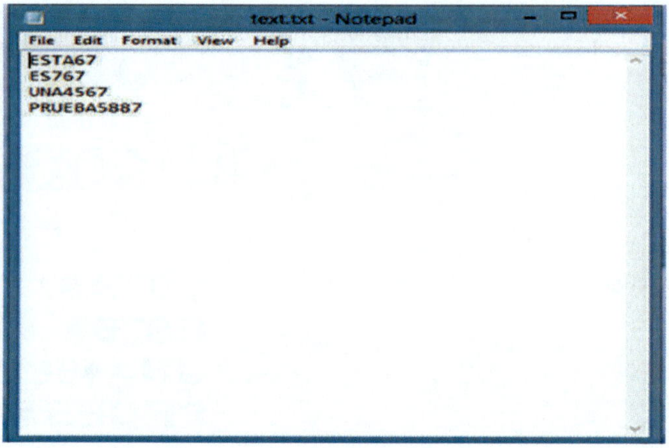

Fig. 8 Transmitting well-printed archive into changeable form

Fig. 9 Input image-2

Anyway exhibiting the optical letter recognition differs with wide variety of texts, dimensions, and manner. For Latin language content reported picture with adequate huge textual style, the acknowledgment exactness is over 90%, and for little text dimension, the acknowledgment precision decays and will be in the scope of 80–90% [10].

We likewise tried for various transcribed record pictures on this OCR framework for which the acknowledgment proficiency fluctuates from half to 70%.

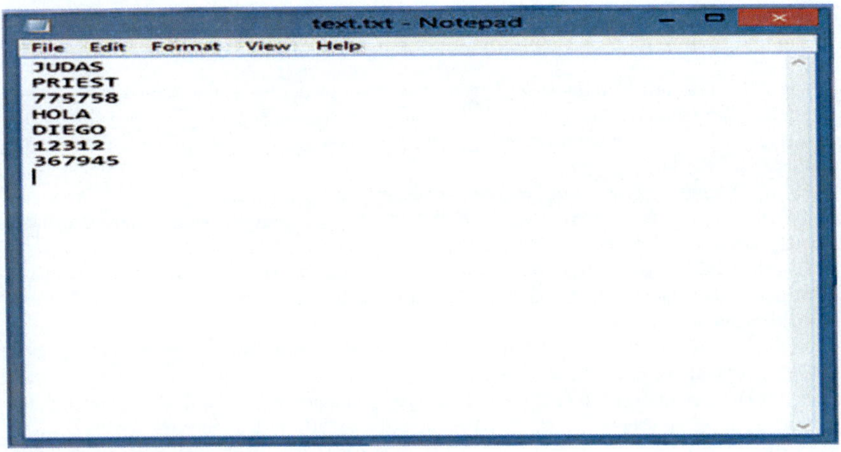

Fig. 10 Converting handwritten record into editable text

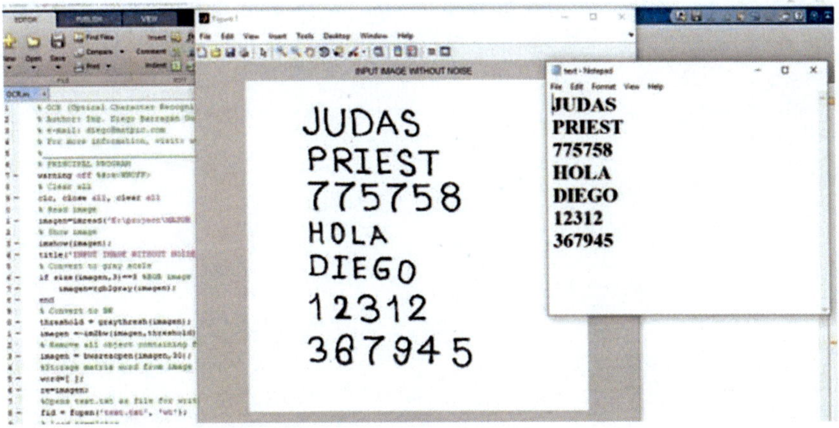

Fig. 11 Converting hand written record into editable text

6 Future Scope

In the event that we consider tremendous number of elements, for example, measurable and primary in the element direction easily expand the exactness of optical letter recognition to huge extension.

References

1. Ali MN, Tan G (2019) Bidirectional encoder–decoder model for Arabic named entity recognition. Arab J Sci Eng 44(11):9693–9701. Jiang X, Satapathy SC, Yang L, Wang SH, Zhang YD (2020) A survey on artificial intelligence in Chinese sign language recognition. Arab J Sci Eng 1–36
2. Pal U, Chaudhuri BB (n.d.). Automatic identification of English, Chinese, Arabic, Devnagari and Bangla script line. In: Proceedings of sixth international conference on document analysis and recognition. https://doi.org/10.1109/icdar.2001.953896
3. Al-zawqari A, Al-qahtani A, Al-zawqari A (2016) Ocr construct feature extraction and template matching algorithms for Qatari number plate. In: International conference on machine learning and cybernetics
4. Al Rabbani Alif M, Ahmed S, Hasan MA (2017) Isolated Bangla handwritten letter recognition with convolutional neural network. IEEE
5. Foong O-M, Hairuman IFB (2011) OCR signage recognition with skew & slant correction for visually impaired people
6. Garlapati BM, and Chalamala SR (2017) A system for handwritten and printed text classification. IEEE
7. Thakral B, Kumar M (2014) Devanagari handwritten text segmentation for overlapping and conjunct letters—A proficient technique. IEEE
8. Srivastav, Kumar J (2010) Text detection in scene images using stroke width and nearest-neighbor constraints
9. Kumar M, Kim YC, Lee GS (2010) Text detection using multilayer separation in real scene images
10. Rajeshbaba M, Anitha T (2013) Detect and separate localization text in various complicated—Colour image

Visual Words based Static Indian Sign Language Alphabet Recognition using KAZE Descriptors

Misaj Sharafudeen, Sujitha David, and Philomina Simon

Abstract Sign language is a powerful tool used by differently-abled people as a means of communication. It facilitates and simplifies the interaction of the differently abled population with the general public. Sign language (SL) recognition can narrow down the communication gap that occur due to the lack of knowledge of sign language. This paper proposes a static sign language alphabet recognition based on KAZE descriptors and the bag of visual words (BoVW) model. KAZE is a 2D feature recognition and description algorithm that works entirely in nonlinear scale space. An efficient skin segmentation method based on hybrid HSV-YCbCr color space is also proposed in the paper to obtain skin mask without noise. Linear discriminant analysis is used for dimensionality reduction of the features extracted and multilayer perceptron neural network is used as classifier in this method. Results demonstrate that the proposed method obtained an accuracy of 99.23% with KAZE descriptor and MLP classifier.

Keywords KAZE · Bag of visual words · Machine learning · Dimensionality reduction · Sign language recognition

1 Introduction

In this technologically advancing world, there is an increasing demand to empower the differently abled people of society. Sign language (SL) recognition has become a prominent research topic in image processing and computer vision in the last decade. Sign language is used by differently abled people for communication. But for effective communication between normal and differently abled people, both should have a better understanding of sign language. This is challenging, thus becomes a complex task. To address this need, sign language recognition systems have been introduced. Sign language uses expressions, hand gestures, and body movements to make com-

M. Sharafudeen · S. David · P. Simon (✉)
Department of Computer Science, University of Kerala, Thiruvananthapuram, Kerala, India
e-mail: philominasimon@keralauniversity.ac.in

P. S. R. Chowdary et al. (eds.), *Evolution in Signal Processing and Telecommunication Networks*, Lecture Notes in Electrical Engineering 839,
https://doi.org/10.1007/978-981-16-8554-5_10

munication easier for differently abled people. Based on the differences in regional, cultural, and native languages, there are about 138–300 disparate forms of sign languages all around the world [1], including American Sign Language (ASL), British Sign Language (BSL), and Indian Sign Language (ISL). Each of them contain a well-disciplined set of gesture codes where every sign has a distinct meaning. Converting these gestures into words or alphabets of existing languages is referred to as sign language recognition. Mainly, two different approaches have been explored for efficient sign language recognition. The glove-based approach and the vision-based approach [2]. In glove-based technique, finger motion is obtained by a network of sensors. Vision-based recognition can be achieved by either static recognition or dynamic recognition. Poses and configurations are included in static gesture recognition. Dynamic gestures consist of strokes, pre-strokes, postures, and phases used for the recognition of continuous sign language. We concentrate on static hand gesture recognition for identifying alphabets.

The aim of our work is in bridging the communication gap between normal people and the differently abled using Indian sign language (ISL) recognition. ISL is more complex as it uses both single- and double-hand gestures. Using two hands leads to the occlusion of features and prediction can go wrong. We also propose a method for skin masking for better recognition. The non-availability of benchmark datasets for ISL due to the variance in sign language with locality and usage of different symbols for the same alphabet pose serious challenges in recognizing alphabets. Figure 1 illustrates samples from the ISL dataset used for conducting the experiments.

The main contributions of this work can be summarized as follows:

- We explore the KAZE feature descriptor for sign language recognition.
- We propose a hybrid color space approach for skin segmentation that combines masks from the HSV and YCbCr color spaces.
- Compared with the state-of-the-art, this work achieved very good results using KAZE descriptors that are extracted in the nonlinear scale space.
- The proposed work achieved good results with no overfitting using the handcrafted machine-learned features.

The rest of the paper is organized as follows. Section 2 elaborates the related works in Sign Language Recognition. Section 3 illustrates the working of the proposed approach. Section 4 throws light into the working flow of the methods discussed. Section 5 elucidates the gist and highlights of the paper.

2 Literature Review

Sign language recognition is one of the key areas of vision-based research for the past two decades. In this section, we summarize the major works in this area. Revanth et al. suggested a method that uses the support vector machine to provide a foundation for the creation of a sign language recognition system, where they have used Oriented FAST and Rotated BRIEF (ORB) for feature extraction [3]. Using the depth image

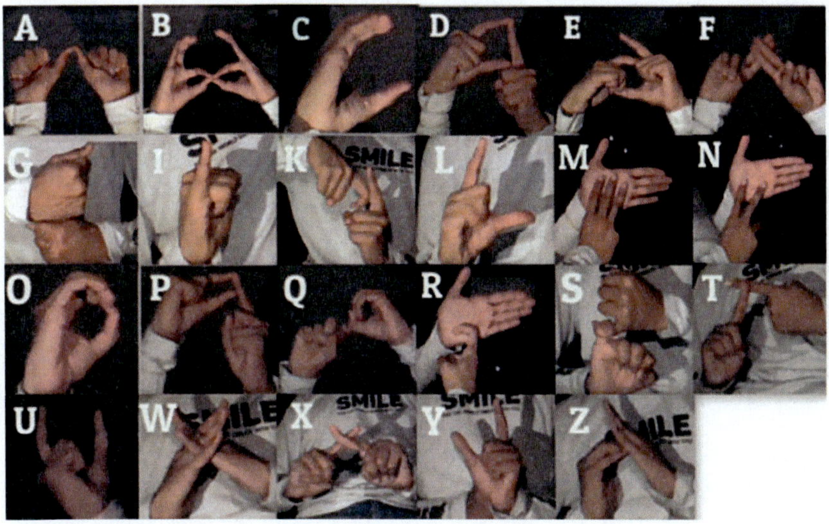

Fig. 1 ISL signs for alphabets A–Z (excluding H, J and V)

from the depth sensor, Gangrade et al. [4] investigated an alternative technique for more accurately recognizing hand movements. In this paper, the hand area is segmented from a depth picture in a cluttered environment using the Microsoft Kinect Sensor. The depth image obtained is then used to implement supervised machine learning by extracting and training the features of images.

In [5], Naren et al. generated an ISL fingerspelling recognition system. Initially, the Canny edge detection algorithm is applied over the preprocessed images followed by SURF feature extraction. The extracted features are then applied to K-means clustering to determine their BoWV. Finally, for classification, various machine learning algorithms such as SVM, naive Bayes, MLP, KNN, and random forest classifiers are used. A varied approach of using pseudo-inverse concept for classification is seen in [6]. In [7], Mali et al. proposed an SVM-based Indian Sign Language Recognition. Feature extraction is performed using principal component analysis (PCA), which removes redundant and unnecessary data. Using the SVM classifier, this recognizing system is capable of recognizing alphabetical and numerical signs with great accuracy.

Furthermore, Mariappan et al. [8] proposed a real-time recognition of Indian Sign Language. The skin segmentation function of OpenCV is used to identify and monitor regions of interest (ROI) for detecting real-time signs. The fuzzy C-means clustering machine learning algorithm is used for hand gesture training and prediction. Recognition of signs using hand gestures and facial expressions is proposed in [9]. The use of BRISK descriptors and bag of visual words (BoVW) for static sign language alphabet recognition has also given promising results with an average accuracy of 95.27% [10].

3 Indian Sign Language Recognition Based on KAZE Descriptors and Multilayer Perceptron Classifier

Figure 2 schematically represents the outline of the proposed system model for the recognition of static gestures, representing alphabets of Indian Sign Language. The task is accomplished in a series of training and testing procedures. The proposed five-step ISL system comprises image preprocessing, feature extraction, visual vocabulary definition, feature restructuring, and classifier training.

Preprocessing Preprocessing of images is employed to enhance input images by suppressing unnecessary distortions. Resizing images to smaller scales ensures faster processing and better feature extraction. The entire dataset is initially rescaled to 200 × 200 followed by skin segmentation and edge detection of the region of interest (ROI).

- Hand Region Segmentation: A hybrid color space based on HSV and YCbCr is introduced in this paper for skin masking to segment the skin region of the hand. In Fig. 3, an efficient skin masking method from two different color spaces is demonstrated. The sign gesture is transformed from RGB to HSV and YCbCr color spaces. HSV and YCbCr masks are generated from the corresponding image representation by thresholding. This procedure retains a range of each color component value and sets pixels to binary 1 (white) while all pixels lying out of range are discarded and set to binary 0 (black). Pixels falling within $0<H<15$, $15<S<255$, $0<V<255$ produce the HSV mask. Those within $0<Y<255$, $135<Cb<180$,

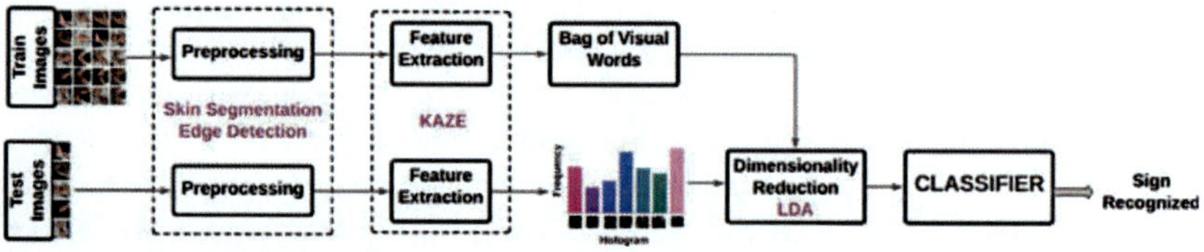

Fig. 2 Framework for Indian sign language alphabet recognition

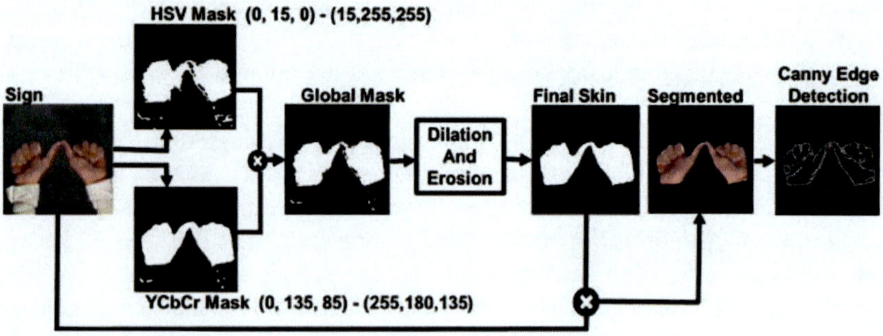

Fig. 3 Hybrid color space-based skin masking

85<Cr<135 produce the YCbCr mask. These range of values represent the wheatish colored Indian skin tone and was obtained through experimentation. A global mask is obtained by superimposing both masks. To remove remaining noises, a series of erosions and dilations are applied to the mask with an elliptical structural element of size (5,5). The final skin mask is then superimposed with the original image to segment the hand region.

- Edge detection: Canny edge detection is employed to generate smooth edge maps resulting from the noise-suppressing feature of the algorithm. The Canny method is the most often used for edge identification because of its non maximum suppression characteristic, which means that only edge points that are prominent in their neighborhood are regarded to be edges.

KAZE Feature Extraction In computer vision, feature matching algorithms are used for matching image feature points. These algorithms detect feature points (key points) in an image. In this work, we use KAZE descriptors for feature matching. Feature descriptors represent each key point using distinct patterns over neighboring pixels. Key points from the edge map are detected using the KAZE descriptor [11]. KAZE detects blob regions in the edge map from which sensitive features are extracted at several nonlinear scale levels. Moreover, the nonlinear diffusion filtering process manipulates nonlinear scale-spaces. The partial differential equations (PDE) filter the luminance of gesture images at increasing scale levels. Furthermore, PDEs using nonlinear functions are solved using the additive operative splitting (AOS) algorithm yielding a Hession matrix. The final descriptors of a key point are computed as the Hession matrix determinant. 64 descriptors are generated for each key point detected in the edge map using KAZE. Equation 1 describes the nonlinear diffusion of an image L.

$$\frac{\partial L}{\partial t} = \mathrm{div}(g|\Delta L_\sigma(x, y, t)| * \Delta L) \tag{1}$$

(x, y, t) represents the spatial coordinates at time t, div is the divergence operator, g the conductivity function, and Δ computes the gradient of image L that is smoothed using a Gaussian function with standard deviation σ.

Bag of Visual Words (BoVW) In the previous step, features (key points) acquired from an image is varied for each image. To equalize the number of features for classification, a bag of visual words (similar to the bag of words (BoW) model) is composed using the entire train set. A vocabulary of visual words is learned in the form of a codebook by converting key point vectors to codewords. BoVW follows two steps to standardize the number of features. Initially, the dictionary is constructed using k-means clustering where similar visual words in the dataset are bagged into k unique bags or clusters. These k codewords in the vocabulary represent the new feature set. The next step involves generating feature values for each training image by plotting histograms for the k codewords. The frequency with which the visual words of an image occur represents the feature values of the k numbered feature set. The k value is varied from 150 to 1000 for comparison.

Dimensionality Reduction The number of features completely depends on the selected value of k. Redundant and dependent features are removed using dimensionality reduction techniques by mapping features from higher to lower dimensional spaces. Any sign language dataset shows minimum between class variance. For this reason, linear discriminant analysis (LDA) is used.

Learning Feature vectors obtained from the training image dataset along with their labels are used to train different supervised learning classifiers and evaluated using the test feature vectors. Various classifier models used in our work are discussed in this section.

- Multiclass Support Vector Machine (SVM): SVM is a supervised learning model that finds a hyperplane maximally separating each class element as far as possible. Linear multiclass SVM regularized at $C = 0.001$ is trained and tested.
- K-Nearest Neighbor (KNN): In KNN algorithm, data points lying close together are related. The value of K represents the number of neighboring data points with which the data point to be classified is compared. Appropriate classes are assigned based on maximal voting. Further, the sample is labeled with the class that receives the most votes. $K = 5$ nearest neighbors are chosen for performance comparison.
- Logistic Regression: This supervised learning algorithm features binary relations in the form of probability values that vary in the range 0–1. The multi-class problem is treated as a one-versus-many scheme where each class is classified against the rest of the classes.
- Multilayer Perceptron (MLP): MLP is a feedforward network designed using multiple fully connected layers of basic perceptrons. MLP comprises input layers, hidden layers, and output layers. The network tries to capture patterns from input–output pairs by fine-tuning its hyperparameters. An MLP with five hidden layers iterated 1000 times is used in the experiment.

4 Results and Discussion

4.1 Experimental Setup and Dataset

The model is implemented in Python 3.7.3 running on Intel(R) Core(TM) i5-1021OU CPU @ 1.60 GHz. All experiments are conducted using machine learning classifiers available in the scikit.learn package. Results are processed by averaging individual results obtained over multiple executions of the entire procedure.

The dataset for ISL alphabet recognition system is taken from kaggle.com [12]. 23 distinct sign alphabets (dataset excludes letters H, J & V due to its dynamic nature) having 300 images each is chosen from the dataset. Each class is split in the ratio of 7:3. The accuracy measure is chosen as the evaluation metric since the entire dataset is balanced. The average accuracy of the model computed over N trials is given as

$$\text{Accuracy} = \frac{1}{N}\sum_{i=1}^{N} \frac{\text{Total number of True Positives (TP)}}{\text{Total number of images}} * 100 \qquad (2)$$

Table 1 Performance comparison of KAZE descriptor over varied k-values for classifiers

KAZE	MLP	Linear SVM	**Log_Reg**	KNN
$k = 150$	97.724	96.714	97.00	97.00
$k = 300$	98.164	97.874	97.681	97.922
$k = 400$	98.599	98.21	98.11	98.21
$k = 60$	98.792	98.55	98.88	98.84
$k = 800$	98.937	98.743	98.744	98.743
$k = 1000$	**99.323**	98.888	99.275	98.937

Table 2 Comparative performance analysis of keypoint descriptors

Descriptor	MLP	Linear SVC	Log_Reg	KNN
SIFT, $k = 150$	70.967	67.58	69.324	57.729
SIFT, $k = 1200$	86.23	85.12	84.63	85.07
ORB, $k = 150$	83.526	82.22	82.415	74.20
ORB, $k = 400$	86.714	84.978	85.024	82.947
AKAZE, $k = 150$	87.922	85.99	86.715	82.22
AKAZE, $k = 800$	93.574	92.75	92.608	92.995
KAZE, $k = 150$	97.487	96.52	96.28	95.94
KAZE, $k = 1000$	99.323	98.888	99.275	98.973

where TP represents the correctly classified instances.

In Table 1, the performance of different classifiers over different bags of clusters created from KAZE descriptors illustrates that classification accuracy increases in direct proportion to the value of k (number of vocabulary codewords). At $k = 1000$, almost perfect classification is noted with the MLP classifier.

A comparative analysis is performed with other feature matching algorithms. Table 2 depicts the performance analysis of SIFT, ORB, AKAZE, and KAZE over the minimum and maximum k values of their corresponding convergence. While ORB (Oriented FAST and Rotated BRIEF) and SIFT (Scale Invariant Feature Transform) detect corners from edge maps, KAZE and AKAZE (Accelerated KAZE) detect blobs as key points. ORB and SIFT perform on scale pyramid spaces. KAZE and AKAZE exploit their operation in the nonlinear scale space. However, the fast diffusion process in AKAZE has decelerated its performance in classification. It is inferred from both tables that learning using multilayer perceptron with KAZE descriptors is the most accurate and has outperformed all other models with an average accuracy of 99.323%, precision of 99.356%, and F1 score of 99.325%. Figure 4 shows the confusion matrix of the multi-class classification of the entire test set alphabets using KAZE at $k = 1000$ with the MLP classifier.

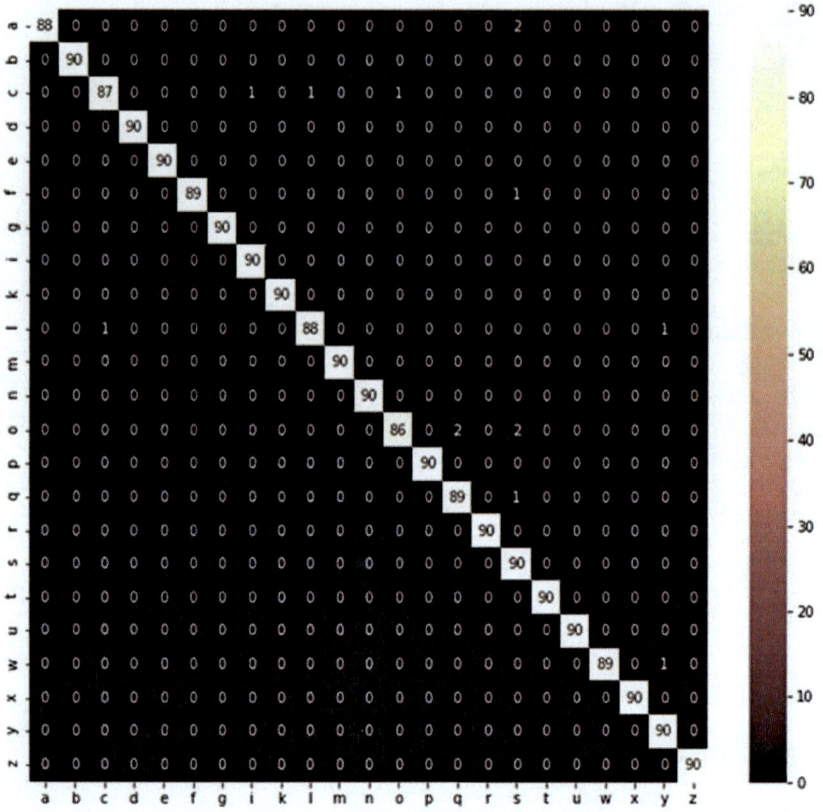

Fig. 4 Confusion matrix of model using KAZE at $k = 1000$ with MLP classifier

5 Conclusion

The static ISL alphabet recognition system is significant for enhancing informal communication between people with hearing disabilities and those who are not. The proposed static ISL alphabet recognition system is an effective approach for identifying ISL alphabets with the KAZE feature descriptor, the bag of visual words, and the multilayer perceptron classifier. An efficient skin segmentation method based on hybrid color space was also introduced in this paper. A comparative analysis is done with different classifiers. As future work, we intend to use CNN architectures for ISL classification. Furthermore, the research could be expanded by identifying common sign language action words.

References

1. Sign Language Alphabets From Around The World—Ai-Media, https://www.ai-media.tv/sign-language-alphabets-from-around-the-world/ Last Accessed 28 May 2021
2. Chanu OR, Pillai A, Sinha S, Das P (2017) Comparative study for vision based and data based hand gesture recognition technique. In: 2017 International conference on intelligent communication and computational techniques (ICCT). IEEE, pp 26–31
3. Revanth K, Raja NSM (2019) Comprehensive SVM based Indian sign language recognition. In: 2019 IEEE international conference on system, computation, automation and networking (ICSCAN). IEEE, pp 1–4
4. Gangrade J, Bharti J, Mulye A (2020) Recognition of Indian sign language using ORB with bag of visual words by Kinect sensor. IETE J Res, 1–15
5. Naren J, Venkatesan R, Rajendran P, Vasudha GS (2020) Indian sign language spelling finger recognition system. Smart systems and IoT: innovations in computing. Springer, Singapore, pp 845–855
6. Koc M et al (2020) Two pseudo-common vectors for pattern recognition. Arab J Sci Eng 45(12):10621–10635
7. Mali D, Limkar N, Mali S (2019) Indian sign language recognition using SVM classifier. In: Proceedings of international conference on communication and information processing (ICCIP)
8. Mariappan HM, Gomathi V (2019) Real-time recognition of Indian sign language. In: 2019 international conference on computational intelligence in data science (ICCIDS). IEEE, pp 1–6
9. Zhang C, Tian Y, Huenerfauth M (2016) Multi-modality American sign language recognition. In: 2016 IEEE International Conference on Image Processing (ICIP). IEEE, pp 2881–2885
10. Misaj S (2021) Indian sign language alphabet recognition using vocabulary of visual words. In: Proceedings of second international conference on artificial intelligence & society (ICAIS 2021), pp 56–59
11. Alcantarilla PF, Adrien B, Davison AJ (2012) KAZE features. European conference on computer vision (ICCV). Springer, Berlin, Heidelberg, pp 214–227
12. Indian Sign Language Translation Letters n Digits Dataset, https://www.kaggle.com/kartik2112/indian-sign-language-translation-letters-n-digits. Last Accessed 22 Nov 2020

Intensified Gray Wolf Optimization-based Extreme Learning Machine for Sentiment Analysis in Big Data

J. Rathika and M. Soranamageswari

Abstract Enhancement of the social network has boosted people's interest in researching product ratings and reviews before buying them. This has provided a way for doing research on domain adaptation, sentiment analysis, and natural language processing as separate areas of study. While a classifier learned from one domain may perform poorly on data from another domain, it is important to keep an open mind while exploring other classifiers. Reviews of dataset are currently being processed in big data. Algorithm developed for performing sentiment analysis in standalone machine or low quantity dataset will not give its best performance while performing sentiment analysis in big data. This paper proposes a classifier, namely intensified gray wolf optimization-based extreme learning machine (IGWO-ELM) for performing the sentiment analysis in big data. IGWO-ELM adopts the natural behavior of wolves to find the sentiments in big review dataset. Single layer hidden layer of ELM assists in performing better classification. Proposed classifier has been evaluated using accuracy and F-measure performance metrics. Results make an indication that the proposed classifier attains better classification accuracy than existing classifiers.

Keywords Classification · Sentiment analysis · Amazon · Optimization

1 Introduction

Recently, researchers started paying their attention to sentiment analysis for acquiring hidden information in massive volume of data (i.e., big data). There are many issues present concerning the technologies of the languages to achieve good results for what the people community needs [1, 2]. The automated text classification process

J. Rathika (✉)
Department of Computing, Coimbatore Institute of Technology, Coimbatore, India
e-mail: jradhika@cit.edu.in

M. Soranamageswari
Department of Information Technology, Government arts college, Coimbatore, India

P. S. R. Chowdary et al. (eds.), *Evolution in Signal Processing and Telecommunication Networks*, Lecture Notes in Electrical Engineering 839,
https://doi.org/10.1007/978-981-16-8554-5_11

is more intricate in a pessimistic or optimistic outlook for making decisions that vary from application to application and algorithm to algorithm. The task of making such decisions also fluctuates due to cultural or personal factors. The text should be in a precise manner, and if it is in poor format, then the task of classification will become worsen [3, 4]. These pitfalls can be commonly seen in some online Web sites like Flipkart or Amazon and social networks like Twitter or Facebook. Some of the often-faced issues are divided into two categories where one is based on a semantic approach and another is based on machine learning. Here, the semantic-based method detects the sentiments by featuring the words based on their usage from the dictionary. It also examines the preprocessed text and splits it into the number of multiple words. It also provides a choice for holding the words and validates their presence in the dictionary to make verification for the polarity sum value [5] (Fig. 1).

The approaches for learning will include training of classifiers with the supervised-based learning algorithms from the annotate text collection where every text will be commonly represented with word vectors, skip-grams, or n-grams. Opinion mining models the structure of the synaptic sentences, negation, subjectivity, irony, or intensification. Many different techniques like SVM, naïve Bayes, and KNN are commonly used for classification. Currently, deep learning and latent semantic analysis are most frequent used advanced methods by researchers [6, 7]. The technologies used are most commonly flexible and efficient for their volume to deduce and to comprehend the sentiments of the human along with their feelings. Opinions mining provides a way to understand the sentiment analysis. All the issues can be solved by manual training. The entire automated system does not involve any interventions of the human to measure the sentiments, and it is primarily brought to handle the issues in the domain area itself [8].

Fig. 1 Opinion mining structure

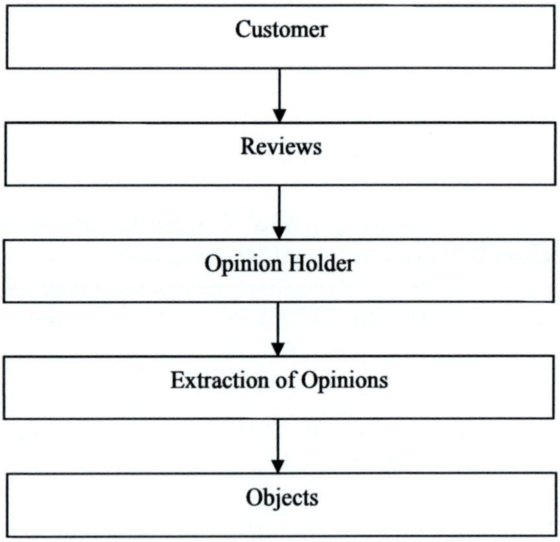

The main challenge which is in terms of opinion can be considered as either positive or negative depending upon the situation. All human does not disclose their views in the same way. The traditional processing of text is always based on the variations among two different sets of data in which the meaning of the text does not alter [9]. Classically, all the reviews will include both positive and negative states which are usually handled by examining the sentences of the text, one at a time.

1.1 Problem Statement

Making an analysis of service or product feedback will assist in improving the product quality or its service. Similarly, the reviews that are received from online shopping Web sites like Amazon, Flipkart not just assist the customer in purchasing a product, but it assists manufacturing company to understand the advantage and disadvantages of the product. Star rating options available online shopping Web sites are not enough to understand the product's worth [10]. Hence, it becomes necessary to go through the reviews given in text format to realize the product feature that lacks customer satisfaction. There exist multiple chances for the product to have thousands of reviews, and it becomes tough to read and analyze all the reviews. There arises a need for a novel methodology that provides a statistical-based report based on previous/former unsatisfied reviewers with the specific feature of the product. Also, domain independence is the major challenge faced in opinion mining [11, 12]. That is, the methodology that is performing well in one domain may not perform well in another domain.

1.2 Research Objective

The main objective of this research work is to analyze, design, and develop a bio-inspired optimization-based classifier for big data that predicts the opinions in cross-domains with increased accuracy.

2 Literature Review

Prospect theory [13] is proposed to analyze the relationship among the sentiments and its rating. The loss aversion and the sensitivity of diminishing samples were validated to calculate the sensitivity. The results implicate the utilization of absolute and relative measure for the cognitive bias service recovery. Freezing technique [14] is proposed for learning sentiment-based vector from LSTM and CNN. It integrates various deep learning methods from which it is observed where the clustered documents work better with ensemble technique. This method worked better for different datasets.

Learning method [15] is proposed to train the embedding aspects that are based on relationship between terms and categories. A cosine metric measure is introduced to study the alleviating limitations. It is used to initialize the existing models to solve the aspect-category sentiment analysis task. The result shows that the embedding aspect of the proposed technique was able to improve the analysis of sentiments efficiently. Deep study [6] is performed to analyze and explore the sentiments that are mostly utilized to detect opinions based on subject element. Various categories of techniques are employed for classifying the text, and it is based on the views which are either positive or negative. A two-step process was portrayed as the preliminary step. Public sentiment discovery [16] is a proposed technique to mine Twitter data to analyze the sentiments toward the prediction of stock movement. In this method, textual messages are clustered before it is processed. Apart from that, data mining algorithms are used to elaborate the word list to classify the sentiments.

Propagating sentiment signal [17] is proposed to analyze the sentiments in Twitter customer tweet. Estimation of polarity tweets is identified based on sentiments where supervised annotation feasibility and the tweet's polarity are also estimated. Trained data is identified as sufficient for handling the polarity reputation. Two-pass classifier [18] is proposed to predict the satisfaction level of a drug. A combination of SVM and ANN is used to review the customer comments that are collected in health care domain. The essential features were extracted from every review, and feature vectors are generated. The two-pass classifier is applied for predicting whether the review relies as positive or negative. Implicit aspect extraction [19] is proposed to extract the sentiment features at aspect-level. The aspects are specified with explicit words and retrieved through text formats. This method makes utilization of different techniques to identify the implicit aspects which were classified selected approaches. The issues and limitations of the aspect retrieval are also presented. Multimodal joint sentiment topic model [20] is proposed to analyze the sentiments present in weakly supervised blogs. Latent Dirichlet allocation is applied for analyzing the hidden topics and sentiment in messages which was based on emotion and personality of microblog users. The experiment shows effectiveness of unsupervised approaches for measuring the level of accuracy. Enhanced feature attention network [21] is proposed to increase the classification accuracy of target-dependent sentiments. The enhanced representation of features, position-based features of word, and speech features were studied. A multi-view network is developed to (i) study the target words modeling and (ii) increase target words based on sentiments and on context. Experimental studies were performed to validate the efficiency of the model, and the results indicate better accuracy than previous models. Optimization [22–29] can be applied in different domains to increase the expected results.

3 Intensified Gray Wolf Optimization-based Extreme Learning Machine

3.1 Intensified Gray Wolf Optimization (IGWO)

The main intention of optimization is to enhance the result to meet user expectations. By default, bio-inspired optimization algorithms are related to artificial intelligence toward taking the decision that ends with better results. In this paper, we have proposed an intensified gray wolf optimization (IGWO) for selecting the features in an effective way for performing the classification over sentiments. This sophisticated IGWO mimics natural gray wolf social hierarchies and hunting patterns and truly represents what wolves experience in the wild. When modeling gray wolf social hierarchy behavior, the group of wolves is divided into four equal halves, which are: alpha (al), beta (be), delta (de), and omega (om). IGWO considers al as the best fit solution, with be and de following close behind. Then, the other solutions are all found to belong to the om category. al, be, and de are the first three wolves to notice om and lead them toward possible prey locations. The wolves (i.e., om) make sure that they are encircling the target points specified by al, be, de when a new phase starts, and it is expressed as Eqs. (1) and (2).

$$\mathrm{Dst} = \left|H.C_{p(ci)} - C(ci)\right| \tag{1}$$

$$\mathrm{CP(ci + 1)} = C_{p(ci)} - F.\mathrm{Dst} \tag{2}$$

where ci represents the count of the current iteration, $C_{p(ci)}$ represents the prey's current position, and $C(\mathrm{ci})$ represents the wolf's current position where Dst represents the distance that is present between wolves and prey. F and H are expressed mathematically in Eqs. (3) and (4).

$$F = (2 \times \mathrm{le} \times rv_1) - \mathrm{le} \tag{3}$$

$$H = 2 \times rv_1 \tag{4}$$

where le represents a linear element decreasing from 3 to 0 during every iteration and rv_1 and rv_2 are random vectors where the values lie between [0, 1],

To establish whether al, be, and de are most likely located around the location of the prey, the IGWO algorithm assumes that these three numbers are at that location. The best solutions of al, be, and de are saved. The remaining wolves (i.e., *om*) will then be placed using the information from the saved solutions. Wolf locations are re-calculated using Eqs. (5)–(11).

$$\mathrm{Dst_{al}} = |(H_1.C_{\mathrm{al}}) - C| \tag{5}$$

$$\mathrm{Dst_{be}} = |(H_2.C_{\mathrm{be}}) - C| \tag{6}$$

$$\mathrm{Dst_{de}} = |(H_3.C_{\mathrm{de}}) - C| \tag{7}$$

$$H_1 = H_{\mathrm{al}} - (F_1.\mathrm{Dst_{al}}) \tag{8}$$

$$H_2 = H_{\mathrm{be}} - (F_2.\mathrm{Dst_{be}}) \tag{9}$$

$$H_3 = H_{\mathrm{de}} - (F_3.\mathrm{Dst_{de}}) \tag{10}$$

$$C(\mathrm{ci} + 1) = (H_1 + H_2 + H_3)/3 \tag{11}$$

where C_{al},C_{be}, C_{de} represent the location of alpha, beta, and delta wolves, C represents the solution of present iteration, and H_1, H_2, H_3 represent the randomly generated vectors. The exact distance between al, be, de and the present solution are estimated using Eqs. (5)–(7). Eqs. (8)–(11) are utilized in the calculation of the final location based on the present solution, i.e., after distance is defined. F_1, F_2, F_3 represent randomly generated vectors, and ci represents the current iteration number. The om step size is dependent on al, be, and de, and it is defined using Eqs. (5)–(11). The end locations of om wolves are computed using Eqs. (8)–(11).

3.2 *Extreme Learning Machine*

Extreme learning machine (ELM) is a novel learning methodology that focuses on different learning strategies. ELM has emerged from ELM, and modest empirical evidence points to its increased generality in several practical scenarios. The main intention of ELM is to resolve the issues that arise in single hidden layer feedforwarding neural network (SFFNN), and it is generalized by preprocessing the neural network. It contains a hidden layer with zero tuning parameter. To illustrate ELM's generalized SFFNN, Eq. (12) is used:

$$f_N(y) = \sum_{p=1}^{N} B_p u_p(y) = u(y)B \tag{12}$$

where B represents vectors output weight that lies between output node and hidden layer node N and B is defined as $\left[B_1, B_2, B_3, \ldots, B_{N-1}, B_N\right]$. y represents the hidden layer output vector, and it is defined as $u(y) = \left[u_1(y), u_2(y), u_3(y), \ldots, u_{N-1}(y), u_N(y)\right]$.

To map the dataset to the N-dimension hidden layer feature space from d-dimension input space, $u(y)$ must be used for feature mapping. Bartlett's hypothesis proposes that when weights are reduced, the effectiveness of feedforward neural networks increases. The system is better able to deal with the errors in training and the distribution of the model's output weights. It is mentioned as $minYB - U^2$ and B, where Y indicates the output matrix's hidden layer.

$$H = \begin{bmatrix} u(y_1) \\ \vdots \\ u(y_n) \end{bmatrix} = \begin{bmatrix} u_1(y_1) & \dots & u_N(y_1) \\ \vdots & \ddots & \vdots \\ u_1(y_n) & \dots & u_N(y_n) \end{bmatrix} \tag{13}$$

In order to optimize the distance of the separating margins of ELM feature space various classes, it is necessary to minimize output weights norm B. ELM also uses the minimal normal least square approach.

$$B = Y^+ U \tag{14}$$

where Y^+ represents the generalized inverse matrix of Moore–Penrose.

A kernel matrix that is known as kernel mapping function available for ELM is usually chosen when users have no idea about feature mapping or when a multiclass feature instance is present. It is mathematically illustrated as Eq. (15)

$$\varphi_{\text{ELM}} = YY^U : \varphi_{\text{ELM}_{i,j}} = u(y_p).u(y_q) = K(y_p, y_q) \tag{15}$$

where $u(y)$ represents the mapping function that maps the data to hidden layers feature space Y from input space. Using the orthogonal projection method, the generalized inverse of the matrix is calculated, i.e., $Y^+ = Y^U(YY^U)$. The output function of ELM is defined in Eq. (16).

$$F(y) = uB = u(y)Y^+\left(\frac{1}{C} + YY^+\right)^{-1} U = \begin{bmatrix} K(y, y_1) \\ \vdots \\ K(y, y_N) \end{bmatrix}^U \left(\frac{1}{C} + \varphi_{\text{ELM}}\right)^{-1} U \tag{16}$$

ELM's major characteristics are determined by two important parameters: p_1 and p_2 (which are locations). The result of this work involves a recuperative gray wolf optimization (IGWO), which is applied to the problem of identifying the two parameters in ELM. Evolutionary ELM (i.e., IGWO-ELM) was designed to adaptively identify the two essential parameters of ELM through the use of the IGWO technique. In IGWO-ELM, we have parameter optimization and classification property assessment methods; however, these two techniques are combined to form the whole IGWO-ELM methodology. The optimization technique is then ended, and the parameter pairs that meet the optima criterion are fed as input to ELM for sentiment

Table 1 Count of instances in dataset

Amazon product review dataset name	Total
Book	146,294
DVD	142,885
Electronics	88,127
Kitchen appliances	72,839

analysis. Accuracy of the classification is assessed using Eq. (17).

$$\text{AvgAccuracy} = \frac{\sum_{p=1}^{K} \text{testAccuracy}_p}{K} \tag{17}$$

4 Dataset and Performance Metrics

4.1 Dataset

This research work makes of Amazon product review datasets to evaluate the performance of proposed classifier against HES [30] and EBC [31]. In Amazon, different review datasets are available for different products. Among the different review datasets, this research work has chosen book, DVD, electronics, and kitchen appliances datasets. Count of instances available in the chosen datasets is provided in Table 1.

4.2 Performance Metrics

This research work makes use of accuracy and F-measure to measure the performance of proposed classifier against HES [30] and EBC [31]. Accuracy and F-measure are calculated using four variables which are true positive (TrPs), true negative (TrNg), false positive (FlPs), and false negative (FlNg).

TrPs: output of correctly predicted positive class.

TrNg: output of correctly predicted negative class.

FlPs: output of incorrectly predicted positive class.

FlNg: output of incorrectly predicted negative class.

4.2.1 Accuracy

It is the count of correctly predicted instances against the count of predictions made. It is mathematically expressed as Eq. (18)

$$\text{Accuracy} = \frac{\text{TrPs} + \text{TrNg}}{\text{TrPs} + \text{TrNg} + \text{FlP}s + \text{FlNg}} \tag{18}$$

4.2.2 F-measure

It is the measure of classification accuracy. It is mathematically expressed as Eq. (19)

$$F\text{ - measure} = \frac{2\text{TrPs}}{2\text{TrPs} + \text{FlP}s + \text{FlNg}} \tag{19}$$

5 Results and Discussion

5.1 Accuracy Analysis

In Fig. 2, the *x*-axis represents the number of datasets, while the *y*-axis represents the percentage of correctness (i.e., accuracy). The comparison in Fig. 2 shows that the power of IGWO-ELM (against HES [30] and EBC [31]) is readily apparent.

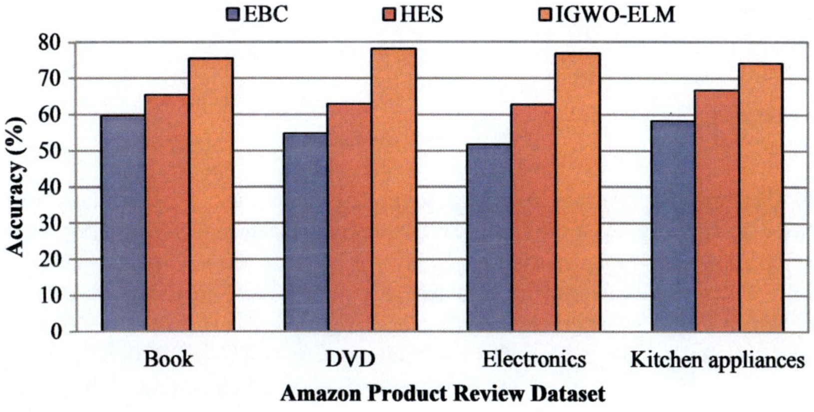

Fig. 2 IGWO-ELM versus accuracy

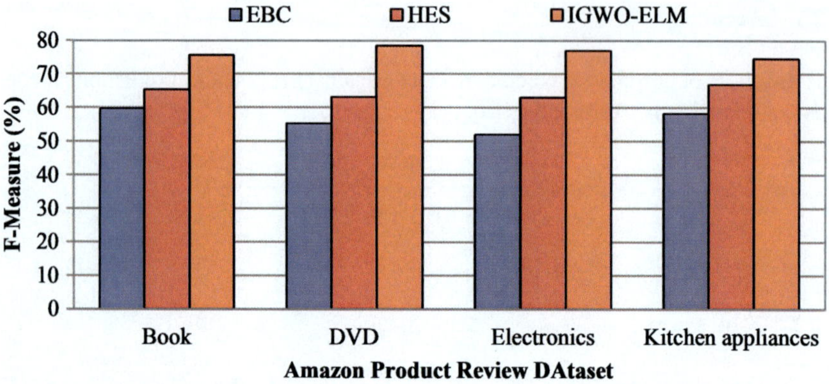

Fig. 3 IGWO-ELM versus F-measure

IGWO-ELM performs optimization during classification process and acquires superior outcomes, but HES [30] and EBC [31] perform classification alone. Neural network present in IGWO-ELM assists in improved accuracy.

5.2 *F-measure Analysis*

Figure 3 depicts the *x*-axis as the number of datasets, with the *y*-axis representing the proportion of F-measure. Effectiveness of feedforward neural network assists proposed classifier in attaining better F-measure than HES [30] and EBC [31]. Existing classifiers concentrate only on classifying the records where the optimization is not performed, and this makes them to attain lower accuracy than the proposed classifier.

6 Conclusion

This proposed classifier has to perform sentiment analysis in big data, namely intensified gray wolf optimization-based extreme learning machine (IGWO-ELM), where it adopts the natural characteristics of wolves to identify the sentiments. In IGWO-ELM, it detects the sentiments and fed as input to ELM to perform classification which results in better classification accuracy. Neural networks of ELM assist one step better accuracy. Classifier has been tested in MATLAB R2018b using the performance metrics accuracy and F-measure. IGWO-ELM has attained the average accuracy of 76.5523%, where entropy-based classifier and hybrid ensemble scheme have achieved 56.8198% and 64.5374%, respectively. Future enhancement of this research work can be focused with enhancing the classification accuracy even more by adopting enhanced machine learning techniques.

References

1. Mars A, Gouider MS (2017) Big data analysis to features opinions extraction of customer. Procedia Comput Sci 112:906–916. https://doi.org/10.1016/j.procs.2017.08.114
2. Abdar M et al (2020) Energy choices in Alaska: mMining people's perception and attitudes from geotagged tweets. Renew Sustain Energy Rev 124:109781. https://doi.org/10.1016/j.rser.2020.109781
3. Body T, Tao X, Li Y, Li L, Zhong N (2021) Using back-and-forth translation to create artificial augmented textual data for sentiment analysis models. Expert Syst Appl 178:115033. https://doi.org/10.1016/j.eswa.2021.115033
4. Sharma M, Kandasamy I, Vasantha WB (2021) Comparison of neutrosophic approach to various deep learning models for sentiment analysis. Knowl-Based Syst 223:107058. https://doi.org/10.1016/j.knosys.2021.107058
5. Huang F, Zhang X, Zhao Z, Xu J, Li Z (2019) Image–text sentiment analysis via deep multimodal attentive fusion. Knowl Based Syst 167:26–37. https://doi.org/10.1016/j.knosys.2019.01.019
6. Bhadane C, Dalal H, Doshi H (2015) Sentiment analysis: measuring opinions. In: Procedia computer science, vol 45©, pp 808–814. https://doi.org/10.1016/j.procs.2015.03.159
7. Bashir S et al (2021) Twitter chirps for Syrian people: sentiment analysis of tweets related to Syria chemical attack. Int J Disaster Risk Reduct 62:102397. https://doi.org/10.1016/j.ijdrr.2021.102397
8. Choi Y, Wiebe J, Mihalcea R (2017) Coarse-grained +/-effect word sense disambiguation for implicit sentiment analysis. IEEE Trans Affect Comput 8(4):471–479. https://doi.org/10.1109/TAFFC.2017.2734085
9. Kumar A, Narapareddy VT, Srikanth VA, Neti LBM, Malapati A (2020) Aspect-based sentiment classification using interactive gated convolutional network. IEEE Access 8:22445–22453. https://doi.org/10.1109/ACCESS.2020.2970030
10. Zhou J, Jin S, Huang X (2020) ADeCNN: an improved model for aspect-level sentiment analysis based on deformable CNN and attention. IEEE Access 8:132970–132979. https://doi.org/10.1109/ACCESS.2020.3010802
11. Bahri S, Bahri P, Lal S (2018) A novel approach of sentiment classification using emoticons. Procedia Comput Sci 132:669–678. https://doi.org/10.1016/j.procs.2018.05.067
12. Ma X, Zeng J, Peng L, Fortino G, Zhang Y (2019) Modeling multi-aspects within one opinionated sentence simultaneously for aspect-level sentiment analysis. Future Gener Comput Syst 93:304–311. https://doi.org/10.1016/j.future.2018.10.041
13. Sharma A, Park S, Nicolau JL (2020) Testing loss aversion and diminishing sensitivity in review sentiment. Tour Manag 77. https://doi.org/10.1016/j.tourman.2019.104020
14. Nguyen HT, Le Nguyen M (2019) An ensemble method with sentiment features and clustering support. Neurocomputing 370:155–165. https://doi.org/10.1016/j.neucom.2019.08.071
15. Tan X, Cai Y, Xu J, Leung H-FF, Chen W, Li Q (2020) Improving aspect-based sentiment analysis via aligning aspect embedding. Neurocomputing 383:336–347. Accessed: May 06, 2021. [Online]. Available: https://www.sciencedirect.com/science/article/pii/S0925231219317382
16. Li B, Chan KCC, Ou C, Ruifeng S (2017) Discovering public sentiment in social media for predicting stock movement of publicly listed companies. Inf Syst 69:81–92. https://doi.org/10.1016/j.is.2016.10.001
17. Giachanou A, Gonzalo J, Crestani F (2019) Propagating sentiment signals for estimating reputation polarity. Inf Process Manag 56(6):102079. https://doi.org/10.1016/j.ipm.2019.102079

18. Padmavathy P, Pakkir Mohideen S (2020) An efficient two-pass classifier system for patient opinion mining to analyze drugs satisfaction. Biomed Signal Process Control 57:101755. https://doi.org/10.1016/j.bspc.2019.101755
19. Ganganwar V, Rajalakshmi R (2019) Implicit aspect extraction for sentiment analysis: a survey of recent approaches, vol 165. Elsevier B.V, pp 485–491
20. Huang F, Zhang S, Zhang J, Yu G (2017) Multimodal learning for topic sentiment analysis in microblogging. Neurocomputing 253:144–153. Accessed: May 06, 2021. [Online]. Available: https://www.sciencedirect.com/science/article/pii/S0925231217304393
21. Yang M, Qu Q, Chen X, Guo C, Shen Y, Lei K (2018) Feature-enhanced attention network for target-dependent sentiment classification. Neurocomputing 307:91–97. https://doi.org/10.1016/j.neucom.2018.04.042
22. Ramkumar J, Vadivel R (2019) Performance modeling of bio-inspired routing protocols in cognitive radio ad hoc network to reduce end-to-end delay. Int J Intell Eng Syst 12(1):221–231. https://doi.org/10.22266/ijies2019.0228.22
23. Ramkumar J, Vadivel R (2021) Multi-adaptive routing protocol for internet of things based ad-hoc networks. Wirel Pers Commun, 1–23. https://doi.org/10.1007/s11277-021-08495-z
24. Ramkumar J, Vadivel R (2018) Improved frog leap inspired protocol (IFLIP)—for routing in cognitive radio ad hoc networks (CRAHN). World J Eng. 15(2):306–311. https://doi.org/10.1108/WJE-08-2017-0260
25. Vadivel R, Ramkumar J (2019) QoS-enabled improved cuckoo search-inspired protocol (ICSIP) for IoT-based healthcare applications, pp 109–121. https://doi.org/10.4018/978-1-7998-1090-2.ch006
26. Lingaraj M, Sugumar TN, Stanly Felix C, Ramkumar J (2021) Query aware routing protocol for mobility enabled wireless sensor network. Int J Comput Netw Appl 8(3):258. https://doi.org/10.22247/IJCNA/2021/209192
27. Ramkumar J, Vadivel R (2020) Improved Wolf prey inspired protocol for routing in cognitive radio Ad Hoc networks. Int J Comput Netw Appl 7(5):126–136. https://doi.org/10.22247/ijcna/2020/202977
28. Ramkumar J, Vadivel R (2020) Meticulous elephant herding optimization based protocol for detecting intrusions in cognitive radio ad hoc networks. Int J Emerg Trends Eng Res 8(8):4549–4554. https://doi.org/10.30534/ijeter/2020/82882020
29. Boopalan S, Jayasankari S (2021) Dolphin Swarm inspired protocol (DSIP) for routing in underwater wireless sensor networks. Int J Comput Netw Appl 8(1):44–53. https://doi.org/10.22247/IJCNA/2021/207981
30. Onan A, Korukoğlu S, Bulut H (2017) A hybrid ensemble pruning approach based on consensus clustering and multi-objective evolutionary algorithm for sentiment classification. Inf Process Manag 53(4):814–833. https://doi.org/10.1016/j.ipm.2017.02.008
31. Deshmukh JS, Tripathy AK (2018) Entropy based classifier for cross-domain opinion mining. Appl Comput Inform 14(1):55–64. https://doi.org/10.1016/j.aci.2017.03.001

Wideband Substrate Integrated Waveguide Based Dual-Polarized Antenna for Satellite Applications in Ku-Band

Meha Agrawal, Kapil Saraswat, and Trivesh Kumar

Abstract In this paper, a Substrate Integrated Waveguide based (SIWG) planar dual-polarized antenna is proposed for satellite communication operating in Ku-band. In this proposed work, dual polarization is excited by two orthogonal microstrip lines with enhanced isolation. In this work, three concentrated connected rings are placed in a circular slot of a circular SIWG cavity mimicking a split-ring resonator. The inductive coupling between the middle and innermost ring enhances the bandwidth as well as improves the isolation. The antenna structure exhibits (13.2–14.49 GHz) 9.25% impedance bandwidth with 24 dB isolation at 13.7 GHz. The realized broadside gain at the resonant frequency is 9.22 dBi. The highly efficient dual-polarized operation makes this design an attractive candidate for satellite applications in Ku-band and can be used for the CubeSat applications.

Keyword Dual polarization · Substrate integrated waveguide · Wideband antenna · Slot antenna

1 Introduction

In last decade, satellite communication technology has advanced dramatically with increased channel capacity and connection dependability. With respect to communication, diversification strategies such as polarization diversity have been developed to achieve these objectives [1]. Dual-polarized antennas (DPANs) provide polarization diversity by supporting two orthogonal modes by incorporating two orthogonal feeds

M. Agrawal (✉) · T. Kumar
Indian Institute of Information Technology, Design and Manufacturing, Jabalpur, India
e-mail: 1822604@iiitdmj.ac.in

T. Kumar
e-mail: trivesh@iiitdmj.ac.in

K. Saraswat
Central University of Rajasthan, Kishangarh, India
e-mail: kapils@curaj.ac.in

P. S. R. Chowdary et al. (eds.), *Evolution in Signal Processing and Telecommunication Networks*, Lecture Notes in Electrical Engineering 839,
https://doi.org/10.1007/978-981-16-8554-5_12

within the single radiating element. As compared to conventional single-polarized antennas, DPANs not only reduce the fading losses but also enhance frequency reuse [2, 3]. Additionally, DPAN offers wide bandwidth and high isolation despite the fact that it is hard to accomplish at the same time [4], and therefore, a trade-off between bandwidth and isolation is challenging.

The Substrate Integrated Waveguide (SIWG) technology provides certain benefits for designing an efficient DPAN in higher frequencies such as, low conductor losses, high gain, low cross-polarization levels, and high radiation efficiency. Also, SIWG offers low cost, easy fabrication, small size, and easy integration of the antenna with other planar RF-modules [5]. One of the major issues with the SIWG is the low bandwidth as the quality factor is significantly high [6]. However, by using the strategic mode analysis and by placing the metalized via holes at appropriate locations in the antenna structure the bandwidth and isolation between the multiple ports can be increased.

In this paper, substrate integrated waveguide-based DPAN is presented for satellite applications operating in the Ku-band which provide wide bandwidth and good port-to-port isolation. The circular SIWG cavity formed by metalized via holes offers enhanced isolation between the two orthogonal feed lines along with enhanced broadside gain. Wide bandwidth is achieved by loading the SIWG cavity with three nested metallic rings placed within a circular slot. These concentric metallic rings form slots. These slots are excited such that multiple modes are coupled which results in wide impedance bandwidth (IBW). An inductive coupling between the middle and innermost ring improves the port isolation. By following the framework mentioned above, the article is systematized as follows: Section 2 illustrates the geometrical design and implementation of the antenna along with the working of antenna. The simulated results of the antenna along with parametric analysis covering simulated radiation patterns and gain are presented in Section 3. Section 4 finally concludes the paper.

2 Wideband Dual-Polarized Antenna

2.1 Design and Working

The proposed SIWG-based DPAN antenna structure is shown in Figure 1. As shown in figure, a circular-shaped SIWG cavity is fed by two orthogonal ports. Two concentric rings with one circular patch are placed inside a SIWG cavity which forms three slots. Both the rings and the circular patch are connected through a metallic strip.

The design advancement of the proposed antenna is shown in Figure 2. As shown in Figure 2a, first, a circular SIWG cavity with radius ‘Rm’ is carefully designed with two orthogonal microstrip feed lines. The radius of the SIW circular cavity is calculated using predefined relation [7]:

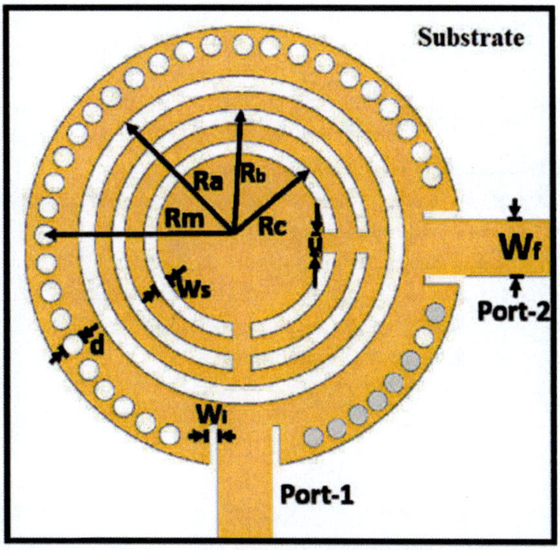

Fig. 1 Proposed antenna structure [where $R_m = 14$, $u = 0.4$, $w_s = 1$, $W_f = 4.7$, $d = 1$, $w_i = 0.3$, $R_a = 11.5$, $R_b = 7$, $R_c = 4.5$. (All the values are in mm)].

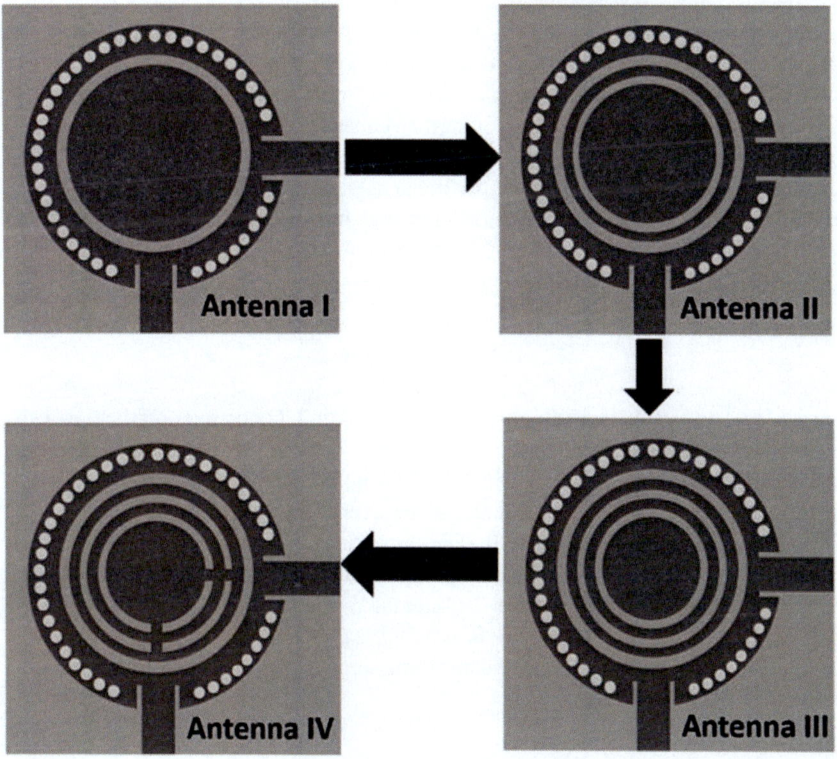

Fig. 2 Design advancement of the wideband SIWG-based DPAN

$$f_c = \frac{c}{2\pi\sqrt{\varepsilon_{\text{eff}}}}\frac{J_{nm}}{R_m} \tag{1}$$

where 'f_c' is the dominant frequency of operation, 'ε_{eff}' is the effective dielectric of the substrate material, and J_{nm} is the mth root of the nth-order Bessel function of the first kind. The values of 'J_{nm}' can be deduced from the listings in [8]. The top and the bottom layer of the low dielectric substrate is connected by embedding metallic via holes of diameter 1 mm (in this design total 38 vias are used to form a cavity). The SIWG cavity dimensions are chosen such that two orthogonal modes TE_{120} and TE_{210} appear in Ku-band. A circular ring slot (outer ring slot) etched inside the SIWG cavity acts as the main radiator (Antenna I). The radius of the outer ring slot 'Ra' is determined using the Equation [9]:

$$f_c = \frac{3.85 * c}{2\pi R_a\sqrt{\varepsilon_{\text{eff}}}} \tag{2}$$

where 'f_c' is the center frequency, and 'c' is the speed of light in free space. On excitation of the antenna structure, it radiates at 12.8 GHz exhibiting 4.154% impedance bandwidth along with 29 dB isolation. Then, in order to further enhance the bandwidth, another circular ring slot (middle ring slot) is etched inside the outer ring slot (Antenna II). This structure exhibits 6.1% bandwidth along with 29 dB isolation at 13.6 GHz frequency. The incorporation of one more circular ring slot (innermost ring slot) inside the middle ring slot further enhances the bandwidth (Antenna III) from 6.1 to 6.3% with 21 dB isolation at 13.7 GHz. The inclusion of multiple ring slots thereby decreases the overall aperture area leading to a shift in center frequency toward the higher value. A good impedance match is obtained when the diameter of the circular patch inside the innermost ring slot is [10]:

$$\text{Diameter}_{\text{patch}} = \frac{c}{\pi * f_c}\sqrt{\frac{1+\varepsilon_r}{2 * \varepsilon_{\text{eff}}}} \tag{3}$$

where 'ε_{eff}' is the dielectric constant. To maintain a good trade-off between the overall impedance bandwidth and isolation, two orthogonal patches are integrated connecting the middle and innermost slots (Antenna IV). These slots introduce an inductive coupling effect that enhances the overall performance of the proposed DPAN, and the antenna exhibits 9.25% impedance bandwidth along with 24 dB isolation between the ports. Figure 3 compares the S-parameter of the SIWG-based DPAN antennas shown in Figure 2 (Antenna I- IV) obtained through full wave EM simulation. Since the structure is symmetrical $S_{11} = S_{22}$, the proposed antenna structure is designed and simulated using commercially available full-EM simulation software CST Studio suite.

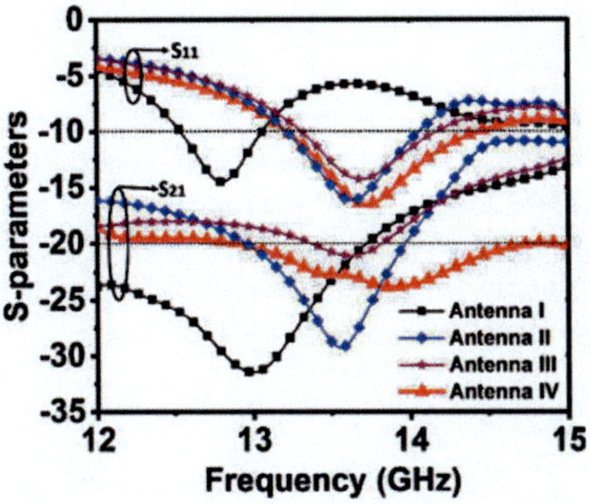

Fig. 3 Comparison of the S-parameters of the design advancement of the proposed antenna

2.2 *Surface Current Distribution*

The simulated surface current distribution of the proposed DPAN at 13.7 GHz is shown in Figure 4. When port-1 is excited at 13.7 GHz, the surface current is concentrated on the horizontal semi-circular portions of three ring slots (Fig. 4a), while when port-2 is excited, the surface current is concentrated at the vertical semi-circular edges of the three ring slots (Fig. 4b). However, high isolation is observed between the two orthogonal ports. Therefore, when port-1 is excited, negligible amount of radiation is leaked to port-2, and when port-2 is excited, negligible amount of radiation is leaked to port-1.

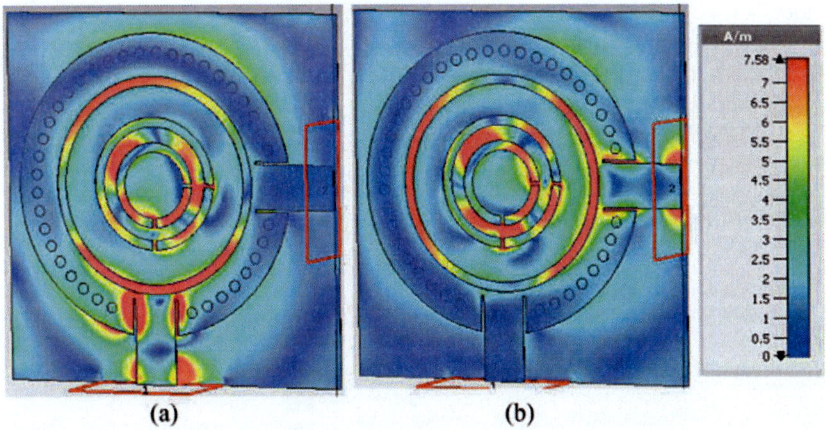

Fig. 4 Surface current distribution at 13.7 GHz on exciting port-1 and port-2, respectively

3 Simulated Results

The simulated S-parameters of the proposed antenna (shown in Fig. 1) are shown in Figure 5. The proposed structure exhibits a wide impedance bandwidth of 9.25% (13.21 GHz–14.49 GHz) at the center resonant frequency 13.7 GHz. It can also be observed that high isolation 24 dB is observed between the two feedlines. The minimum of 21 dB exists for the entire operating impedance bandwidth for S_{11}, S_{22} < -10 dB and S_{21}, S_{12} < -20 dB.

3.1 *Study of Parameter Variation*

The effect of variation of radius of the three ring slots and the width of inductive coupling inside the circular SIWG cavity is illustrated in Figure 6. It is observed that the variation of outer ring slot radius 'R_a' is proportional to the variation of the resonant frequency 13.7 GHz with a significant improvement in isolation (Fig. 6a). The variation of the radius of middle ring slot 'R_b' disturbs the overall current path thereby causing a variation in overall bandwidth and resonant frequency.

Figure 6b shows that maximum bandwidth along with maximum isolation is achieved when $R_b = 7$ mm. Figure 6c illustrates the corresponding change in reflection coefficient, bandwidth, and isolation on varying the radius of innermost ring slot 'R_c'. Although maximum bandwidth is observed at $R_c = 5.5$ mm, if bandwidth isolation trade-off is considered, then $R_c = 4.5$ mm is the most optimum radius. Figure 6d illustrates the variation of the width of inductive coupling patches. As the width increases, the overall current path is altered thereby improving the impedance matching of the antenna, however, the level of isolation is reduced. Therefore, at $u =$ 0.4 mm, a trade-off between wide bandwidth and high isolation can be considered.

The proposed wideband DPAN is designed on RT/duroid 5870 ($\varepsilon_r = 2.33$, $\tan\delta =$ 0.0012) substrate material with thickness 1.57 mm. The simulated and normalized

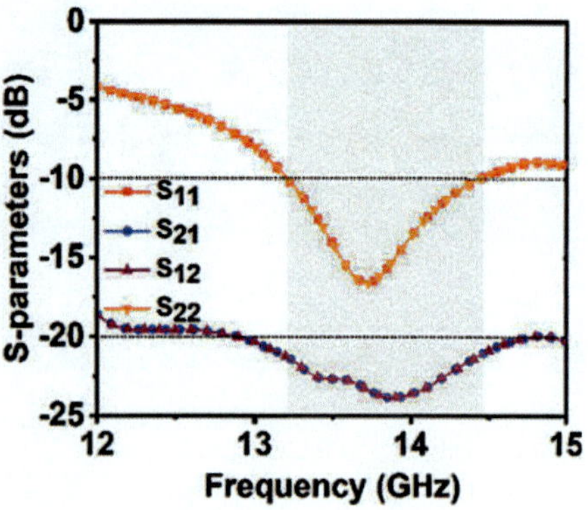

Fig. 5 S-parameters of the proposed SIWG-based DPAN

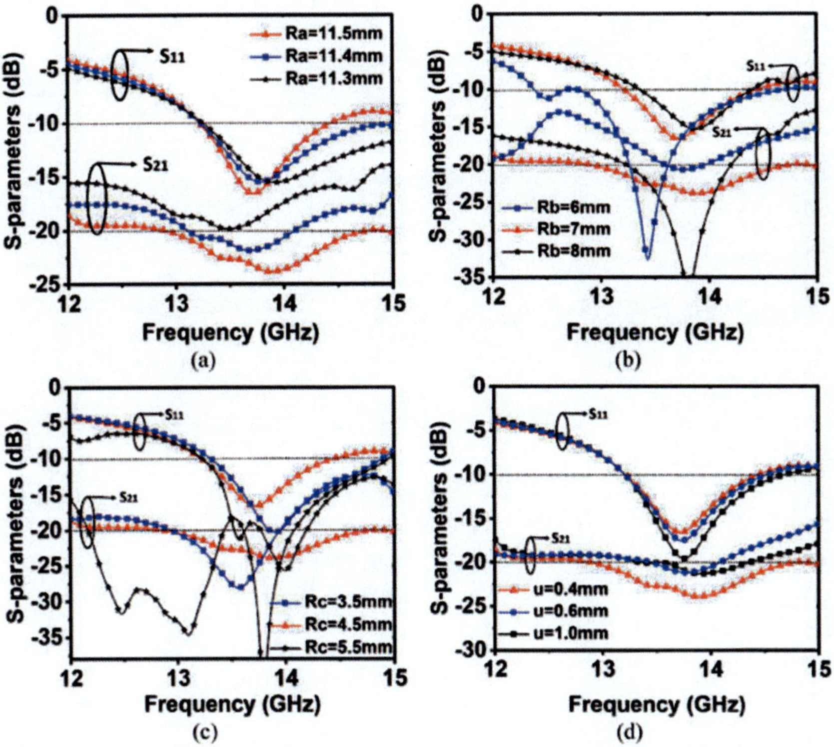

Fig. 6 Effect of variation of geometrical parameters of the proposed DPAN **a** radius of outer ring slot, **b** radius of middle ring slot, **c** radius of innermost ring slot, **d** width of the inductive coupling patches

radiation patterns in the *xz*-plane and *yz*-plane at the center frequency 13.7 GHz are shown in Figure 7. A unidirectional radiation pattern in broadside direction is observed as the SIWG is a cavity-backed structure. The maximum level of cross-polarization in the *xz*-plane and *yz*-plane is −18 dB. Figure 8 illustrates the peak realized gain with the frequency which shows 9.22 dBi at the center frequency. It can be observed that the proposed work exhibits maximum bandwidth isolation trade-off with the use of the incorporated nested ring slots along with inductive coupling between the two inner ring slots inside the SIWG cavity.

4 Conclusion

In this paper, a wideband SIWG-based dual-polarized antenna with nested ring slots has been presented for the satellite application. The two orthogonal patches provide inductive coupling between the middle ring slot and innermost ring slot

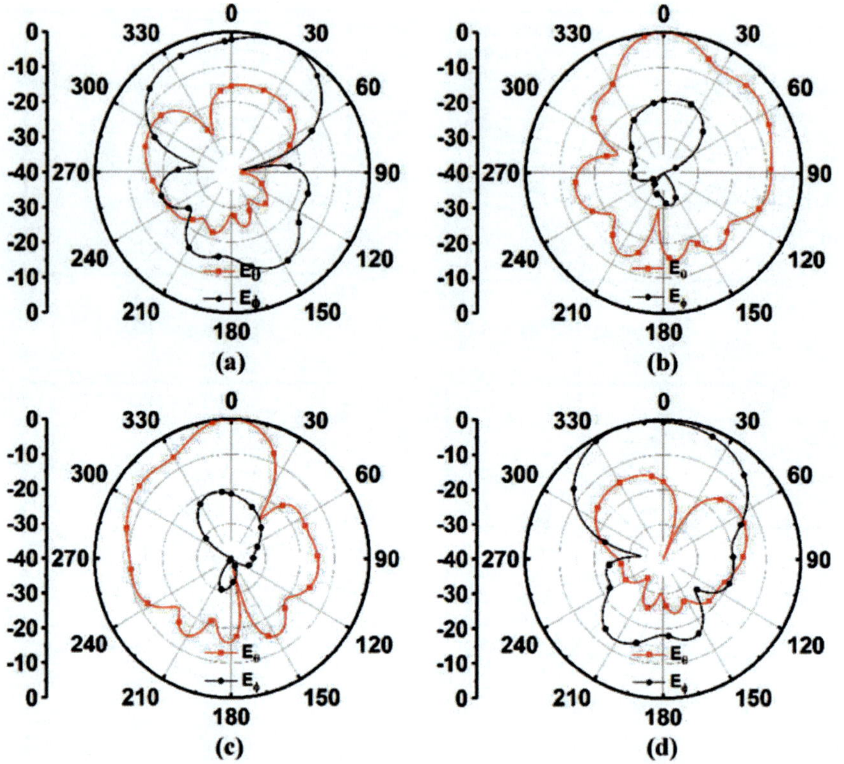

Fig. 7 Simulated radiation pattern at 13.7 GHz **a** Port-1 *xz*-plane, **b** Port-1 *yz*-plane, **c** Port-2 *xz*-plane, **d** Port-2 yz-plane

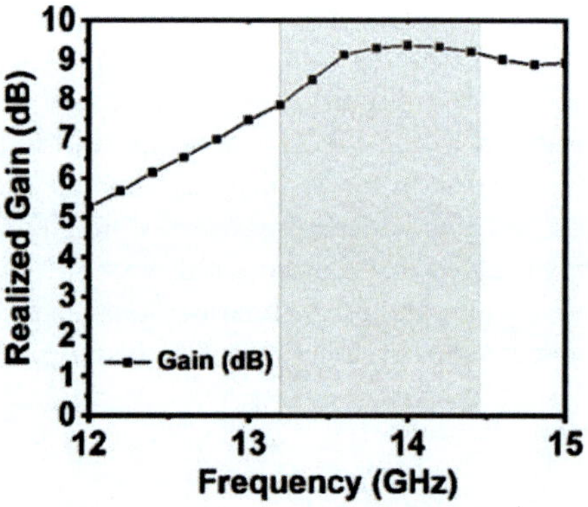

Fig. 8 Realized gain of the DPAN

thereby enhancing the bandwidth as well as isolation between the two ports. The diameter of the circular SIWG cavity is 28 mm. The overall size of the DPAN is 32 mm * 32 mm. The mutual coupling between the two inner ring slots improves the isolation and exhibits the overall impedance bandwidth of 9.25% at center frequency 13.7 GHz. The realized gain of 9.22 dBi at 13.7 GHz is obtained with a very low cross-polarization level at two orthogonal planes. The proposed antenna structure combines the benefits of SIWG antenna (enhanced gain, low cross-polarization levels, compactness) and those of traditional microstrip slot antenna. The inset microstrip feeding technique and compact structure make the design easily compatible with the various satellite applications in Ku-band.

References

1. Gosalia K, Lazzi G (2003) Reduced size, dual-polarized microstrip patch antenna for wireless communications. IEEE Trans Antennas Propag 51(9):2182–2186
2. Ren YJ, Hsu SH, Li MY, Chang K (2008) A dual-frequency dual-polarized planar airborne array antenna. In: IEEE antennas and propagation society international symposium, 1–4
3. Agrawal M, Kumar T (2020) A dual-band substrate integrated waveguide based dual linearly polarized antenna with high isolation. In: IEEE international conference electrical electronic and computer engineering (UPCON), 1–5
4. Mishra PK, Jahagirdar DR, Kumar G (2014) A review of broadband dual linearly polarized microstrip antenna designs with high isolation [education column]. IEEE Antennas Propag Mag 56(6):238–251
5. Kumar A, Raghavan S (2016) A review substrate integrated waveguide antennas and arrays. J Telecommun Electron Comput Eng (JTEC) 8:95–104
6. Lai Q, Fumeaux C, Hong W, Vahldieck R (2009) Characterization of the propagation properties of the half-mode substrate integrated waveguide. IEEE Trans Microw Theory Tech 57(8):1996–2004
7. Hong T, Zhao Z, Jiang W, Xia S, Liu Y, Gong S (2019) Dual-band SIW cavity-backed slot array using TM020 and TM120 modes for 5G applications. IEEE Trans Antennas Propag 67(5):3490–3495
8. Pozar DM (2009) Microwave engineering. Wiley, Hoboken
9. Wu Q, Wang H, Yu C, Hong W (2016) Low-profile circularly polarized cavity-backed antennas using SIW techniques. IEEE Trans Antennas Propag 64(7):2832–2839
10. Wong KL, Huang CC, Chen WS (2002) Printed ring slot antenna for circular polarization. IEEE Trans Antennas Propag 50(1):75–77

Detection of COVID-19 and Classification of Pneumonia Using Deep Neural Networks Algorithms

G. Bhavitha, A. Harsha Vardhan, K. Nikhil, Sk. Tasleema Farhan, and V. Saritha

Abstract COVID-19 which is a subclass of severe acute respiratory syndrome (SARS) is a viral disease which emerged from China in 2019. At first, there are shorthand of test kits available to diagnose the COVID-19 disease. The tests available to diagnose the COVID-19 are RT-PCR (real-time polymerase chain reaction), Rapid Antigen test and Antibody test. But in these, only RT-PCR has the high accuracy, and it is a time-taking process. It takes nearly from 4 to 48 h. Here, AI plays an important role in diagnosing the disease. In the recent years, AI becomes a part of medical field and is widely used in classification. The chest X-Rays are used to detect the COVID-19 using deep learning and the model used to detect the COVID-19 is ResNet18 which is a residual network containing 18 layers. In this work, we classified four types of classes to make sure that our model performance is better and classify accurately. The data set contain a total of 5365 images. In this, we used 80% of data for the training and 20% for validation. The accuracy obtained in classification of three classes is 96.67% and for four classes, the accuracy is 91%. We have also used another model for comparison which ResNet50 and achieved an accuracy of 75%.

Keywords Deep learning · Resnet18 · Resnet50 · COVID-19 · Python

1 Introduction

COVID-19(SARS COV2) is a form of pneumonia. First emerged at Wuhan Seafood Wholesale Market (Huanan Market) in Wuhan, China. The cases were identified from 2019. It is a positive-stranded RNA virus belongs to Coronaviridae, a subgroup B beta corona virus. It is also a 96.2% sequence of RATG13 virus. It infects endothelial cells living in blood vessels. The symptoms are fever, cough, short breathing, fatigue, headache, smell lose, sour throat and running nose. The total confirmed cases are 1165,772,430 and the total deaths are 3,437,545 as per 22 which is worldwide.

G. Bhavitha (✉) · A. Harsha Vardhan · K. Nikhil · Sk. Tasleema Farhan · V. Saritha
Department of Electronics and Communication Engineering, VR Siddhartha Engineering College (Autonomous), JNTU, Kakinada, India

P. S. R. Chowdary et al. (eds.), *Evolution in Signal Processing and Telecommunication Networks*, Lecture Notes in Electrical Engineering 839,
https://doi.org/10.1007/978-981-16-8554-5_13

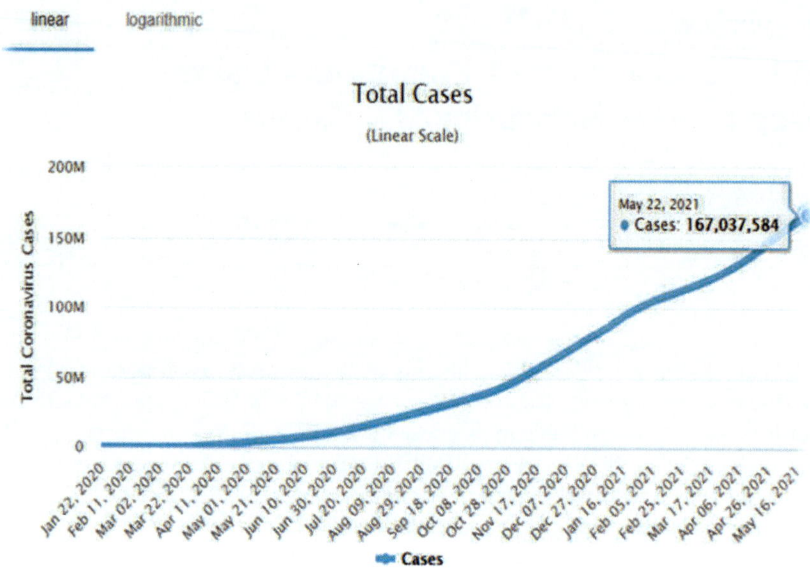

Fig. 1 The total corona cases worldwide

The most affected countries are the USA, India and Brazil respectively. The initial days, there are no test kits available readily to test the patients. The tests used to detect the corona virus are RT-PCR test, Antigen test and Antibody test. RT-PCR test is a gene detection method. It collects samples from nasal and throat. It requires high laboratory and takes 48 h to know whether a person is infected with COVID or not. The accuracy of this test is 97%. Antibody test takes blood samples to detect the antibodies against the COVID-19. It is fast but the accuracy is low and can be misleading. Rapid antigen test takes samples from nose to test the viral proteins. It takes 15 min and is not recommended for COVID detection. So as the RT-PCR test takes nearly 48 h which is time-consuming process, deep learning comes in place to support the doctors in detection of COVID-19 by performing convolution neural networks on x-rays. Nowadays, artificial intelligence produces a stable and accurate results, and it also plays an important role in the health industries. There are many techniques used to detect the COVID-19 such as using networks likeVGGNet16, Alex Net, InceptionV3 and MobileNetV2. In this work, we are using ResNet50 for classification of pneumonia and corona virus (Figs. 1 and 2).

2 Related Works

In paper [1], Amit Kumar has taken a total of 771 images. They have used various neural networks such as DenseNet201, ResNet50 and InceptionV3 and achieved an

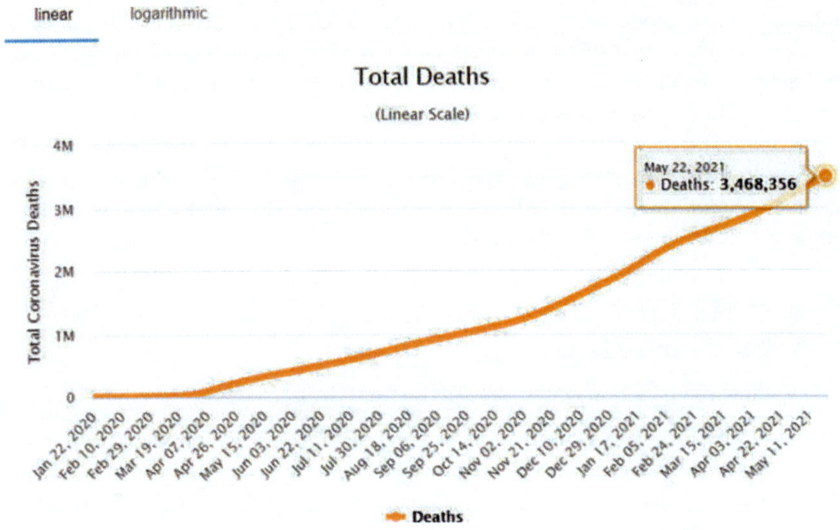

Fig. 2 The total deaths worldwide

accuracy of 93.6%, 95.3%, 94%, respectively, and sensitivity of 92%, 98% and 93%, respectively. They classified two types which are COVID +ve and COVID −ve. The total images for training are 771 images and for testing, the images are 235. They have used a sigmoid function in fully connected layer.

In paper [2], Linda Wang proposed a model of COVID-Net architecture. This is an open source available in [3]. The model was pretrained on the ImageNet model and achieved an accuracy of 93.3% and sensitivity of 95%. In this model, the classification is multi-classification and the classified types are normal, non-COVID-19 pneumonia and COVID-19. The data set contains a total of 1975 images. The data source is available here [3].

In paper [4], Afshar has proposed a model of COVID-caps architecture. This model has four convolutional layers and three capsule layers. This model classifies COVID +ve and COVID –ve with an accuracy of 98.3%. The fully connected layer uses sigmoid function.

In paper [5], Ahishali proposed a Chex Net architecture. This model has DenseNet-121 and a COVID-19Net architecture. It uses SoftMax function for multi-classification in output layer. This model classifies normal, viral pneumonia, bacterial pneumonia and COVID-19. The data set used is COVID Family data set. The accuracy obtained is 88.9%.

In paper [6], Qin used a CovNet model. This model is a 3D deep learning model. This model uses ResNet as a base model. The SoftMax function is used in fully connected layer. This model classifies a multi-classification. The classified types are COVID-19, CAP and non-COVID-19 pneumonia. The accuracy achieved is 90%.

In paper [7], Rachna made a comparison between different neural networks. The model used in this paper are Xception, InceptionV3 and ResNeXt. The accuracy achieved by this model are 97.97%, 96% and 93%, respectively. The total images used for classification is 6432 images. In these, 5467 images are used for training and 965 images for validation. The classification is multi-classification and the classified types are COVID-19, pneumonia and normal. The data set can be accessed here [8].

In paper [9], a deep learning convolution neural network with seven layers is used. The results of image processing and feature analysing COVID-19 gives 91.67% accuracy. Again, the data is further classified and trained using CNN for feature extraction. It achieves 100% accuracy.

3 Methods and Materials

3.1 Proposed Approach

The following model shows the proposed approach in this paper. The steps included in this model are Data Collection, Data Pre-processing, Training the model and results analysis (Fig. 3).

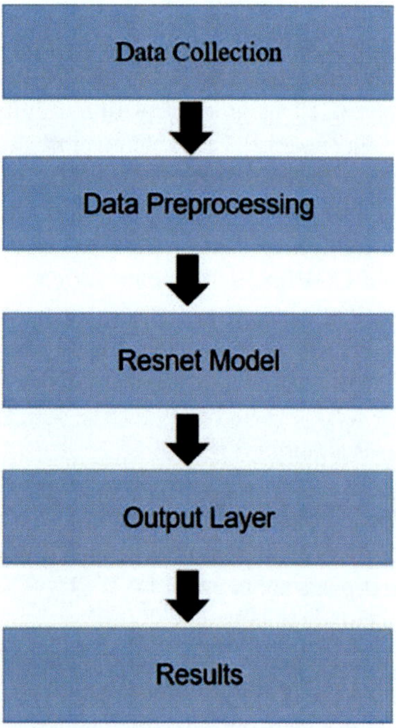

Fig. 3 Proposed approach

Table 1 Data set distribution

Label	Training samples	Testing samples
COVID-19	1200	268
Normal	1341	268
Viral pneumonia	1345	268
Bacterial pneumonia	1478	268

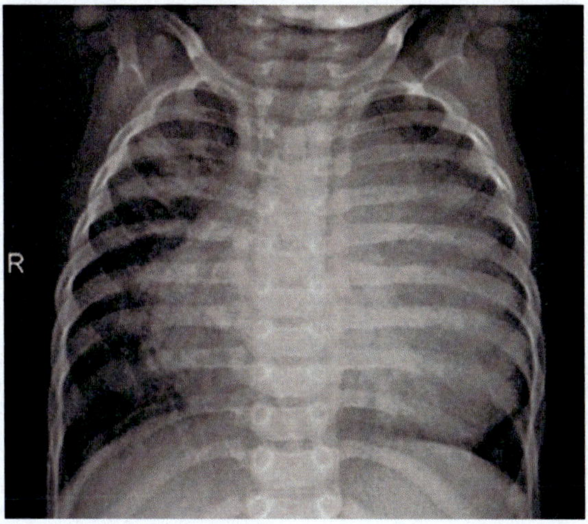

Fig. 4 Bacterial pneumonia

3.2 *Data sets*

The data sets used for this model are collected from Kaggle repository. The data sets are openly accessible to everyone and can be accessed here [10]. The bacterial data set can be accessed here [11]. A total of 5365 images have taken for model validation. The data set contains 1341 normal images, 1345 viral images, 1200 COVID images and 1478 bacterial images. The data set distribution is shown in Table 1.

Some of the examples of chest X-rays are: (Figs. 4, 5, 6 and 7)

3.3 *Data Pre-processing*

All the images in the data set have undergone by the following steps: (10 The images are reshaped into (224,224,3) (2) The flip used here is RandomHorizontalFlip. (3) The Zoom range is 0.5 and Sheer range is 0.5. The framework used here is Py torch.

Then, the images are applied to the trained model. By using Random_Split method, the data set is divided into 80% for training purpose and 20% for validation.

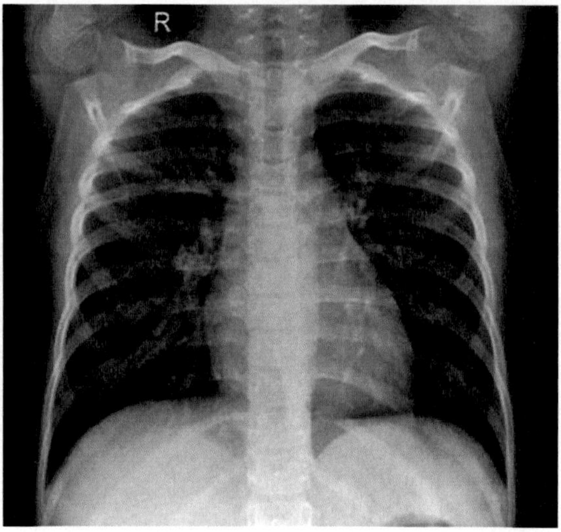

Fig. 5 Normal

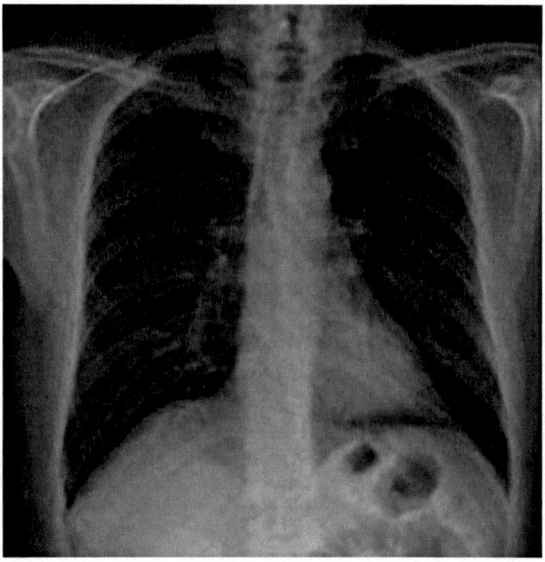

Fig. 6 COVID-19

3.4 Proposed Neural Networks

3.4.1 ResNet18

ResNet18 is a residual network having 18 layers. This model has identity connection block and convolution block. The convolution block will be applied if the input size

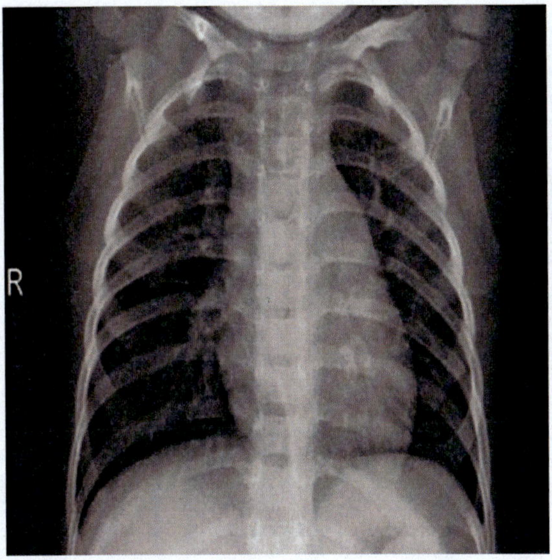

Fig. 7 Viral pneumonia

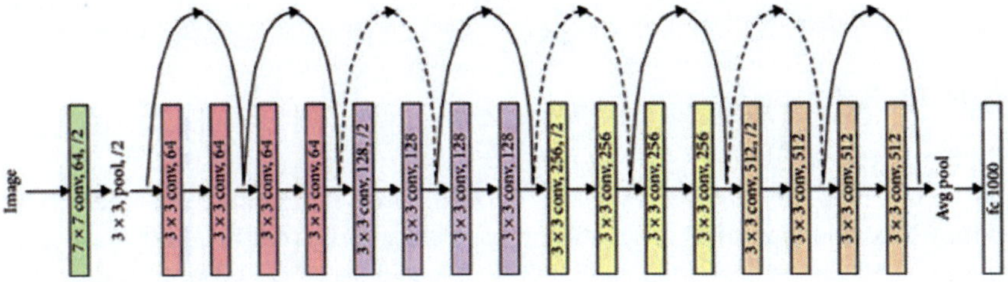

Fig. 8 Resnet-18 architecture

and output size are not matched and identity connection block is applied if the input and output sizes are matched. The loss function used in this model is categorical entropy loss, and the SoftMax function is used since it is a multi-classification. The total trainable parameters are 11,178,564. The model used here is pre-trained model. The model architecture is shown below.

To train this model Google Colab and Pytorch framework is used (Figs. 8 and 9).

3.4.2 ResNet50

ResNet50 is a residual Network having 50 layers. This model has identity connection block and convolution block. The convolution block will be applied if the input size and output size are not matched and identity connection block is applied if the input

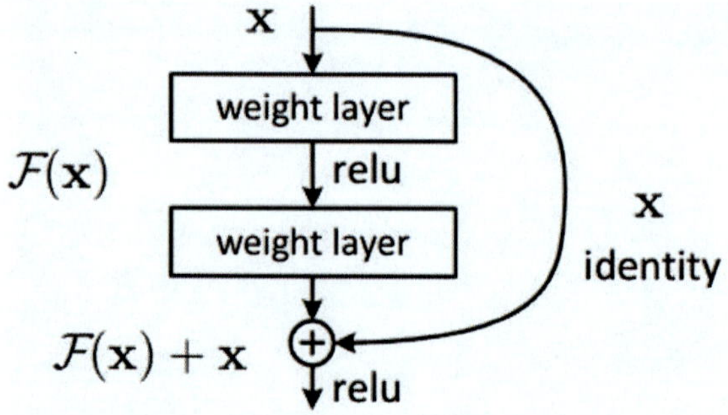

Fig. 9 Basic building block of residual network

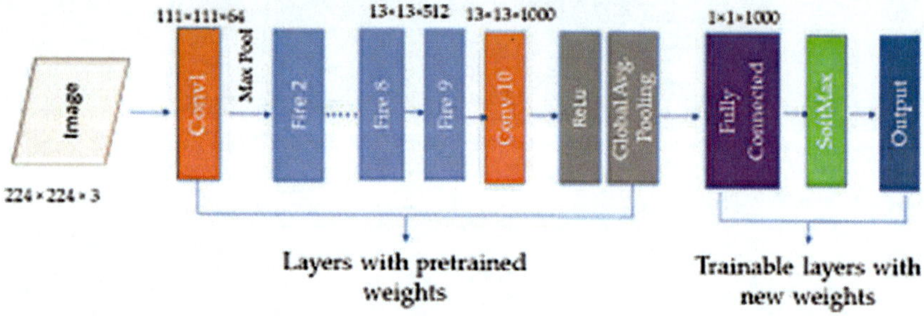

Fig. 10 Resnet50 architecture

and output sizes are matched. The loss function used in this model is categorical entropy loss, and the SoftMax function is used since it is a multi-classification. The total trainable parameters are 31,956,868. The model used here is pretrained model. The model architecture is shown in Fig. 10.

To train this model, Google Colab and Keras Framework is used.

4 Result Analysis

The proposed models are evaluated using confusion matrix and by using following parameters such as sensitivity, specificity, accuracy, $F1$ score and precision.

$$\text{Accuracy} = (\text{TP} + \text{TN}) * 100/(\text{TP} + \text{TN} + \text{FP} + \text{FN})$$

$$\text{Sensitivity or Recall} = \text{TP}/(\text{TP} + \text{FN})$$

$$\text{Precision} = \text{TP}/(\text{TP} + \text{FP})$$

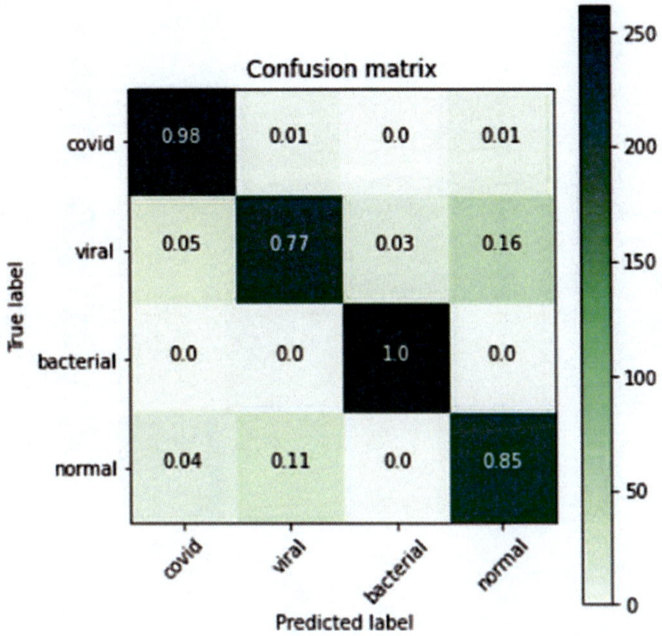

Fig. 11 Confusion matrix with normalization

$$\text{Specificity} = \text{TN}/(\text{TN} + \text{FP})$$

$$\text{F1 Score} = 2 * (\text{Recall} * \text{Precision})/(\text{Recall} + \text{Precision})$$

4.1 ResNet-18 Result Analysis—Confusion Matrix

To get the confusion matrix and classification report import confusion matrix from sklearn (Figs. 11 and 12).

Classification Report

See Table 2.

4.2 Resnet-50 Result Analysis—Confusion Matrix

See Figs. 13 and 14.

Classification Report

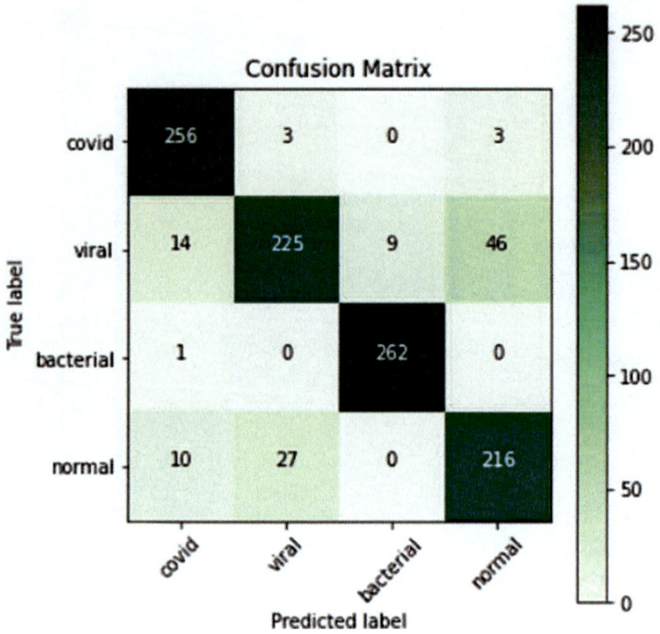

Fig. 12 Confusion matrix without normalization

Table 2 Testing data report

Label	Precision	Recall	F1-score	Support
COVID	0.95	0.98	0.96	266
Viral	0.85	0.82	0.83	273
Bacterial	1	0.97	0.98	256
Normal	0.84	0.86	0.85	277

See Table 3.

4.3 *Comparison Between Two Models*

See Table 4.

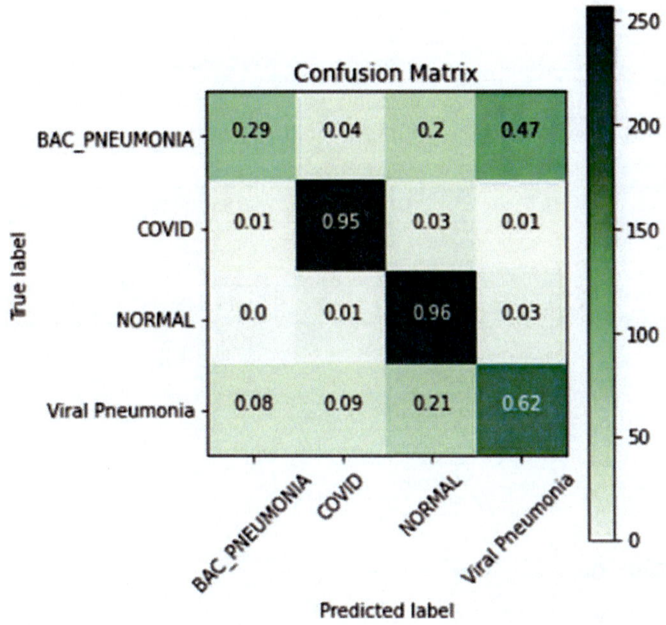

Fig. 13 Confusion matrix with normalization

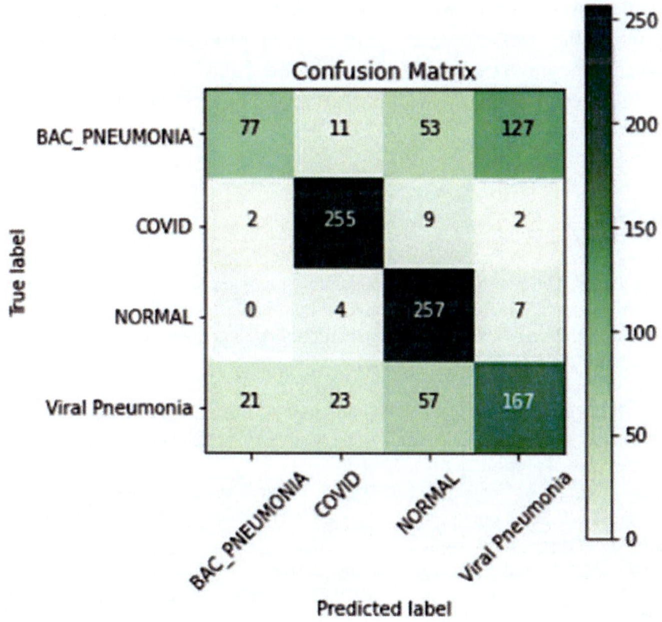

Fig. 14 Confusion matrix without normalization

Table 3 Testing data report

Label	Precision	Recall	F1 score	Support
COVID	0.98	0.88	0.93	268
Viral	0.57	0.72	0.64	268
Bacterial	0.72	0.57	0.63	268
Normal	0.83	0.87	0.85	268

Table 4 Comparison between ResNet-18 and ResNet-50

	ResNet-18	Resnet-50
Accuracy (%)	91	76
Sensitivity	0.9	0.7
Specificity	0.96	0.9
Precision	0.9	0.7
F1 score	0.9	0.7

5 Conclusion

In this paper, two models are proposed and compared for detection of COVID-19 along with different classes of pneumonia in order to have more accurate results. The proposed models are ResNet-18 and ResNet-50. The data sets are pre-processed in order to avoid the over fitting and under fitting problems. The ResNet-18 model achieved an accuracy of 91% and ResNet-50 model achieved an accuracy of 75% for same data set. By comparing these two, ResNet-18 model classifies more accurately.

References

1. Das AK, Gosh S, Thunder S, Sachin R, Agarwal automatic COVID-19 detection from X-Ray images using ensemble learning with convolutional neural network
2. Wang L, Lin ZQ, Wong A, COVID—net: a tailored deep convolutional neural network design for detection of COVID-19 cases from chest X-ray images
3. https://github.com/lindawangg/COVID-Net
4. Afshar P, Heidarian S, Naderkhani F, Oikonomou A, Plataniotis KN, Mohammadi A (2020) Covid- caps: a capsule network-based framework for identification of covid-19 cases from x-ray images. arXiv preprint arXiv:2004.02696
5. Ahishali M, Degerli A, Yamac M, Kiranyaz S, Chowdhury ME, Hameed K, Hamid, Mazhar R, Gabbouj M (2020) A comparative study on early detection of covid-19 from chest x-ray images. arXiv preprint arXiv:2006.05332
6. Li L, Qin L, Xu Z, Yin Y, Wang X, Kong B, Bai J, Lu Y, Fang Z, Song Y et al (2020) Artificial intelligence distinguishes covid-19 from community acquired pneumonia on chest CT. Radiology
7. Jain R, Gupta M, Taneja S, Hemanth DJ, Deep learning-based detection and analysis of COVID-19 on chest X-ray images.

8. Patel P, Chest X-ray (Covid-19& Pneumonia), Accessed at: https:/ww.kaggle.com/prashant268/chest-xray-covid19-pneumonia
9. Huang C, Wang Y, Li X, Ren L, Zhao J, Hu Y, Zhang L, Fan G, Xu J, Gu X, Cheng Z (2020) Clinical features of patients infected with 2019 novel coronavirus in Wuhan. China Lancet 395(10223):497–506
10. https://www.kaggle.com/tawsifurrahmancovid19-radiography-database
11. Kaggle datasets download -d Wilhelm Berghammer/chest-xrays-pneumonia-detection

Design of Arrayed Rectangular Probe Patch Antenna at 6.2 GHz for 5G Small Cell Applications

C. V. Sarang Mohan and P. Sudheesh

Abstract The current trend in 5G communication is to build a system relying on long-range communication transmissions. The issue with this topology is the obstacles that commonly block and attenuate the outgoing signals from their coverage. 5G small cells have been introduced here to achieve better coverage, especially in cities where obstacles are found at a high density. Patch antennas are convenient for the development of low-cost, lightweight, and low-profile antennas that can maintain high performance. This paper describes the design of an arrayed rectangular probe patch antenna operating at 6.2 GHz. The 4×4 non-conformal mesh antenna array structure improves directivity, gain, and radiation patterns from the regular patch antenna. This design will be useful for 5G small cell deployments. This antenna was designed and simulated using Ansys HFSS.

Keywords 5G · Small cells · Patch antenna · Rectangular probe · Arrayed antenna

1 Introduction

The evolution of 5G networks will be heterogeneous in terms of cell types, ranging from macrocell to femtocell. According to studies conducted in 2019, outdoor and indoor hot spot small cells are becoming increasingly important to meet capacity and user data rate demands [1]. The studies also emphasize the importance of operating such extremely dense small cell implementations in a dedicated spectrum to avoid issues with high-power macrocells. Beyond 3.4 GHz, frequency operation has recently received a lot of attention for improving the capacity of International Mobil Telecommunications (IMT) systems [2]. Small cells operate similarly to macrocells, but their performance can be improved by using transmission techniques such as multiple input multiple output, millimeter waves, and beamforming. Low-powered transmissions can be integrated through the small cell concept. They are classified

C. V. Sarang Mohan · P. Sudheesh (✉)
Department of Electronics and Communication Engineering, Amrita School of Engineering, Amrita Vishwa Vidyapeetham, Coimbatore, India
e-mail: p_sudheesh@cb.amrita.edu

P. S. R. Chowdary et al. (eds.), *Evolution in Signal Processing and Telecommunication Networks*, Lecture Notes in Electrical Engineering 839,
https://doi.org/10.1007/978-981-16-8554-5_14

into three types: femtocells, picocells, and microcells. All differ in terms of their coverage area and the number of users that they can support. Small cells can offload the macrocells which are overloaded. It will reduce the complexity of hardware, making the implementation easier. Mounting structures can be placed on indoor walls as well as on lamp posts or small towers outside. Femtocell is the smallest one, which is used in residential and commercial applications. It broadens the indoor coverage and increases throughput. Picocells are suitable for small enterprises that can handle more users with more coverage. It has recently been used primarily on airplanes. Microcells are large and can accommodate a large number of users and cover a small geographical area. All can be connected through backhaul using fiber, wired, and microwave connections. General features of small cells are.

- 10–500 m of coverage
- Backhaul options include wired, fiber, and microwave.
- 100 mW–5 W
- Low-to-medium price
- 8–200 users are supported.

5G has a larger spectrum compared to lower generations. But when the frequency of the spectrum increases, there is a high chance of blockage of signals. Small cells are deployed in 5G to get adequate coverage. It has a lower interaction with the macro cell tier [3]. Line of sight and non-line of sight transmissions are considered for a millimeter wave communication scenario [4, 5]. The main observation in these works is that when the base station density is reasonably high, the per-base station coverage probability performance begins to deteriorate. The author in [6] recommends using a directional antenna, which makes sure signal delivery of the transmitter and receiver are in direct line of sight. MIMO, OFDM, and beamforming help to improve signal strength, thereby improving multiplexing and beamforming gain [7, 8]. The presence of several multipath components inside the indoor channel requires the use of statistical parameters in channel modeling [9]. The small cells act as a communication bridge toward devices in 5G, and city planners might care about where to place these structures and at what height for optimal coverage. Above 6 GHz, it is commonly referred to as the Super Data Layer, and it supports high data rates. If a millimeter wave is used for small cells, then it can be made even smaller and fit into tiny spaces [10].

Patch antennas can be powered in a variety of ways. The two main types are contacting and no contacting. The RF current is delivered to the radiating patch via a contacting micro strip line and a probe. The no contacting type of coupling is used for electromagnetic field coupling [11]. Depending on the desired characteristics, a dielectric substrate is placed between two conductors in a patch antenna. The lower conductor is known as the ground plane, and the upper conductor is known as the patch. It can be designed in a variety of shapes, including rectangular, square, or circular [12]. This paper is a design of an arrayed rectangular probe feeding a patch antenna at 6.2 GHz. A 4×4 array structure is compact and can be used for small cell applications. It has a significant gain compared to a normal rectangular patch antenna. Today, the probe-fed technique is also widely utilized. One of the advantages of a probe-fed patch antenna is that the feed point determines its input resistance [13]. Patch antennas with Teflon dielectric, Duroid 5880 dielectric, and polyethylene dielectric material support frequencies ranging from 5 to 7.2 GHz [14]. In this paper, a comparison of the rectangular probe patch antenna and the arrayed rectangular probe patch antenna has been shown. The Ansys HFSS tool is used to simulate return loss, gain, and directivity.

2 Design of Arrayed Rectangular Probe Patch Antenna

An antenna array is a regular association of identical antennas used to generate radiation of a specific shape. Because the number of radiating elements is multiplied, the radiated power is increased. Radiation is produced by the simultaneous addition of fields from each element [15]. As a result, the possible combinations are numerous and entail a wide range of performance, which is required in various applications.

2.1 *Rectangular Probe Patch Antenna*

The probe-fed patch's design variables are patch length, patch width, substrate thickness, substrate width, substrate length, coax inner and outer radius, and feed length. In a patch antenna, the substrate is primarily used to provide mechanical support for the metallization of the antenna. To provide this support, a dielectric substrate is required, which has an impact on the antenna's electrical performance. Here, Duroid 5880 is used as a substrate which has a dielectric constant of 2.2. Figure 1 shows the rectangular patch antenna with probe feeding.

The width of the patch antenna is calculated using Eq. (1).

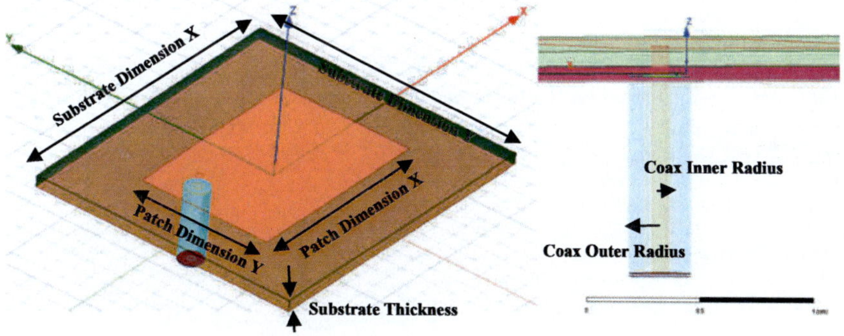

Fig. 1 Rectangular probe patch antenna

$$W = \frac{c}{2f_0\sqrt{\frac{1}{2} + \varepsilon_r}} \tag{1}$$

Here, c and f_0 represent the speed of light and resonant frequency, respectively. The length of the patch antenna depends on the resonant frequency and the effective dielectric constant. For calculating the effective dielectric constant, use Eq. (2) and then length by Eq. (3).

$$\varepsilon_{\text{reff}} = \frac{\varepsilon_r + 1}{2} + \frac{\varepsilon_r - 1}{2}\left[1 + 12\frac{h}{W}\right] \tag{2}$$

$$L_{\text{eff}} = \frac{c}{2f_0\sqrt{\varepsilon_{\text{eff}}}} \tag{3}$$

h denotes the thickness of the substrate. The patch has fringing effects. Because of these effects, the patch antenna's effective length seems to be longer than its actual length.

$$L = L_{\text{eff}} - 2\Delta L \tag{4}$$

In order to calculate the actual length in Eq. (4), first calculate ΔL, which is a function of the effective dielectric constant and the width to height ratio from Eq. (5).

$$\frac{\Delta L}{h} = 0.412\frac{(\varepsilon_{\text{eff}} + 0.3)\left(\frac{W}{h} + 0.264\right)}{(\varepsilon_{\text{eff}} - 0.258)\left(\frac{W}{h} + 0.8\right)} \tag{5}$$

Here, a probe feeding technique is being used which should match the characteristic impedance of 50 Ω.

$$Z_0 = \frac{138\left(\log_{10}\left(\frac{D_i}{D_o}\right)\right)}{\varepsilon_r} \tag{6}$$

Z_0 is the characteristic impedance shown in Eq. (6), D_i is the inner coaxial diameter and D_o is the outer coaxial diameter. It is followed by finding out the feed length in different X and Y axis using Eqs. (7) and (8). Inner coaxial diameter found as 0.04 cm and outer diameter as 0.13 cm.

$$X = \frac{L}{2\sqrt{\varepsilon_{\text{reff}}}} \tag{7}$$

$$Y = \frac{W}{2\sqrt{\varepsilon_{\text{reff}}}} \tag{8}$$

The patch dimensions obtained are 2.20 cm × 1.48 cm and the feed dimensions obtained are 0.65 cm × 0.40 cm. The substrate dimensions used are 3.5 cm × 3 cm × 0.11 cm.

2.2 4 x 4 Arrayed Rectangular Probe Patch Antenna

Sixteen elements of the regular rectangular probe patch antenna are arranged in a 4 × 4 square configuration. It can also be used to construct larger microstrip array antennas. HFSS meshes these 3D component arrays separately and stitches them together to form a non-conformal mesh to create a full field solution. A meshed patch antenna's gain and efficiency can be enhanced by decreasing the linewidth. If the linewidth is maintained exceedingly small, a meshed antenna is as effective as a solid patch. Some characterization was performed to obtain the optimal antenna design, such as changing the feed channel length, the patch dimensions, and the distance between the patch antennas. The patch antenna dimensions are 2.12 cm × 1.43 cm; the feed dimensions are 0.63 cm × 0.7 cm. The prototype of a rectangular patch antenna array was realized to be characterized experimentally as shown in Fig. 2.

The radiation pattern of each side of the single antenna is a coupled lattice pair. It means that one acts as the master and the other one is a slave. The size of the substrate has decreased a little to accommodate more antennas in the array. To obtain the same

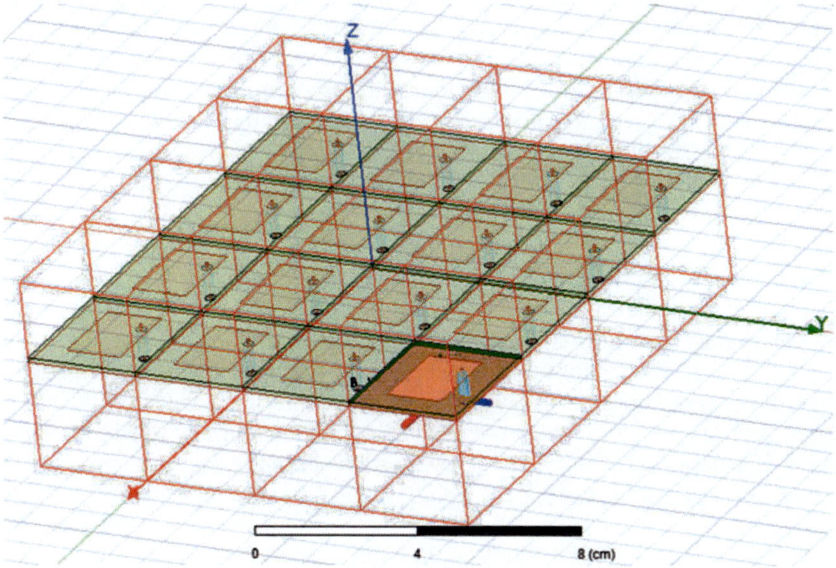

Fig. 2 4 × 4 arrayed rectangular probe patch antenna

frequency, other design parameters have been changed accordingly. It can be seen that each antenna is bonded together to form a non-conformal mesh. The radiation pattern would be a phased array, and it consists of main lobes and other side lobes. The main lobe has a significant increase in gain compared to the other lobes. So, the main lobe will be showing more directivity. In this case, the entire array is represented in HFSS, with each antenna element having its own excitation. Tuning can be done before and after the array creation. But a significant increase in gain is seen when optimization has been performed before the array is constructed.

The coaxial connector is linked to the backside of the ground plane. The coaxial center connector is soldered to the patch after passing through the substrate. Impedance bandwidth is one of the main parameters in antenna analysis. The dielectric constant has caused an increase in bandwidth. When it increases, the resonant frequency varies depending on the dielectric material used. Generally, dielectric constants range from 2.2 to 12, but for high efficiency and bandwidth, the dielectric substrate is thicker and the dielectric constant is lower. Because the size of the antenna is directly related to the wavelength at the resonant frequency, higher frequencies are possible.

The patch contains an inset cut that allows the of the feed line impedance to be matched to the patch. The inset position would be properly controlled to avoid the additional matching element. Overall, this model provides simple fabrication methods as well as simple modeling methods, particularly for impedance matching. Unlike other feed methods, this one allows the feed to be placed anywhere within the patch to match the input impedance. This enables easy fabrication and has spurious radiation at a low level [16]. The electric field lines in the antenna move along the substrate and also little to the outside air. The transverse electric magnetic mode is not possible in the transmission lines because of the variation in their phase velocities [17]. The antenna gain is a measure of the antenna's ability to concentrate energy in a certain direction in order to provide a more accurate picture of the radiation performance. In a sense, this is indicated in decibels and relates to the direction of the highest radiation [18]. The maximum gain of an antenna G is expressed in Eq. (9).

$$G = \eta D \tag{9}$$

$\eta =$ Antenna efficiency
$D =$ Directivity.

The antenna directivity is the ratio of radiation intensity to average radiation in one direction to all directions. It is also the ratio of the amount of energy coming in these directions to the amount of energy in a single directional antenna [18, 19]. It can be depicted from the radiation pattern; mainly it is from an isotropic point source that the measurement is performed. The effect of voltage to current ratio at a pair of terminals is called impedance Z_{in} is defined in Eq. (10) at the input terminal.

$$\mathrm{Z_{in}} = \mathrm{R_{in}} + \mathrm{jX_{in}} \tag{10}$$

$R_{in} =$ The real part, shows radiation losses.

X_{in} = The imaginary part, pair shows the antenna reactance and stored power.

The observation of the orientation of the electric field from source time shows polarization, which is a property of EM wave manifesting time varying direction as well as relative magnitude of the vector of electric field [19].

The orientation of the feed phase is angled at an angular orientation of 0, $\pi/2$, π, $3/2\pi$. The feeding network's feed lines of quarter-wavelength transformers and power dividers achieve phase arrangement and power distribution. When arranging the array elements, the mutual coupling losses should be minimized. The speed with which this method is simulated is an advantage. To provide a quick array simulation, the smallest amount of memory and time was used. A genetic algorithm is used for optimization. It is a random search procedure to find values close to the nominal position. Since there are more variables for optimization, this technique can be used to improve the return loss and get the resonating frequency to our desired frequency. Antenna elements were dispersed in a rectangular arrangement to create matching E-plane and H-plane array factor patterns. The user can specify the location of active, passive, and padding cells. This allows the user to create various lattice shapes.

3 Results

The antenna parameters are represented in Table 1. This paper compared the normal rectangular patch antenna with an arrayed rectangular patch antenna. The rectangular patch antenna array shows a gain of 19.67 dB, directivity of 19.48 dB, and return loss of −26.32 dB at the desired frequency and the regular rectangular patch antenna shows 7.2 dB, 7.1 dB, and −13.88 dB, respectively. The 3D radiation pattern is shown in Fig. 6. It can be seen that the red region of the radiation pattern in the lobes shows the maximum gain. The main lobe has the highest gain. The antenna concentrates the EM wave's energy along the main lobe. There are also side lobes with the lowest gain. Side lobes are unwanted radiation patterns that can never be completely removed. It also may pick up the unwanted signals. The gain vs frequency and directivity vs frequency are shown in Figs. 3 and 4. Directivity and gain are increasing with frequency. Figure 5 shows the return loss vs frequency plot. It can be seen that the antenna is resonating near the desired frequency.

Table 1 Antenna parameters

Parameters	Rectangular patch antenna	4 × 4 arrayed rectangular probe patch antenna
Gain (dB)	7.2	19.67
Directivity (dB)	7.1	19.48
Return loss (dB)	−13.88	−26.32
Permittivity	2.2	2.2
Frequency (GHz)	6.2	6.2

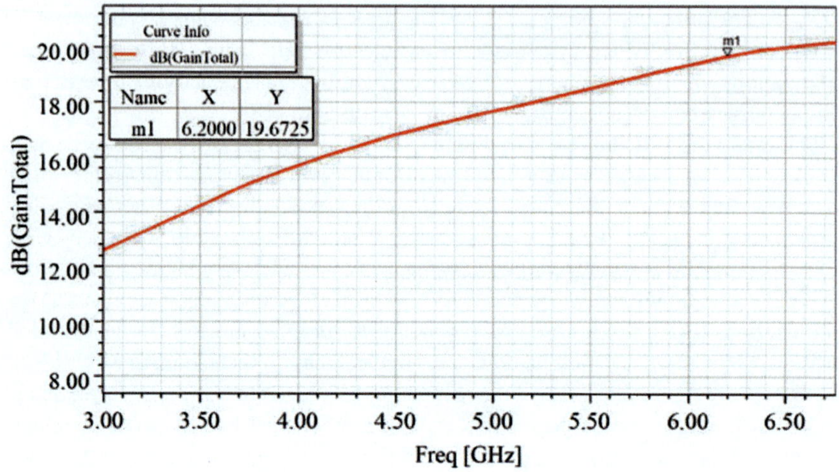

Fig. 3 The plot of gain versus frequency

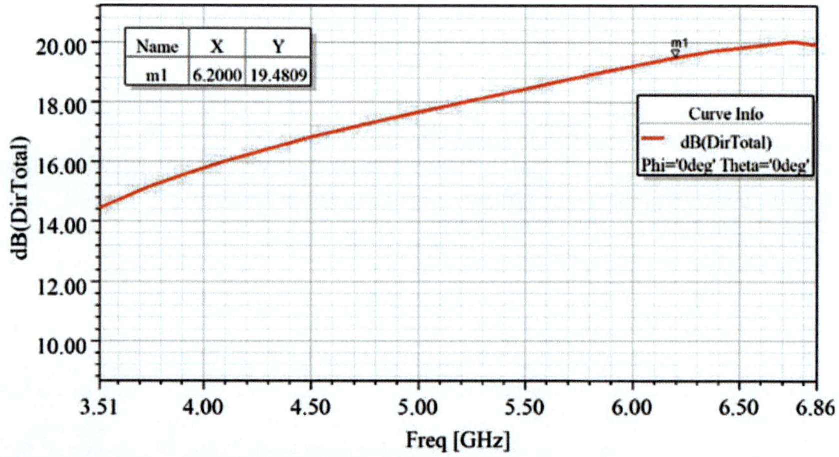

Fig. 4 The plot of directivity versus frequency

4 Conclusion

A return loss is the amount of dB at which the radiated waves are reflected at the desired frequency. The arrayed structure shows a significant decrease in return loss, which is 12.44 dB from a regular rectangular patch antenna. The gain of an antenna indicates how well it converts electrical energy to electromagnetic waves. The 3D radiation plot provides complete information about the gain of an antenna, since it shows the gain in every possible direction. It is desirable to decrease the side

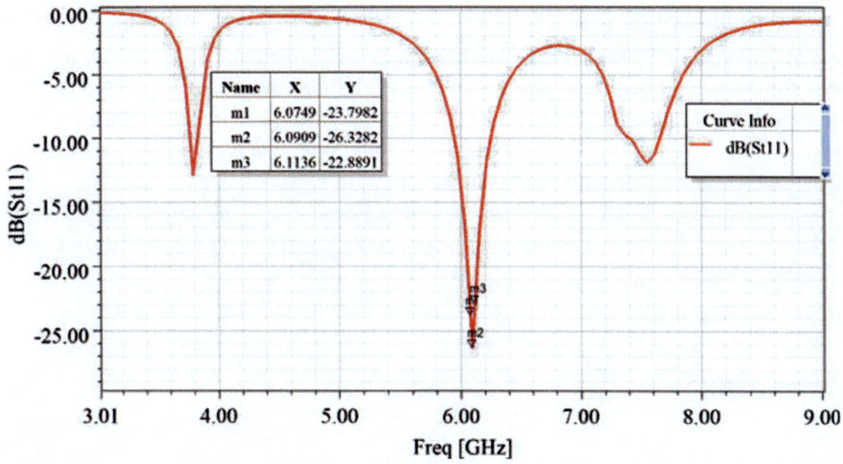

Fig. 5 The plot of return loss ($S11$) versus frequency

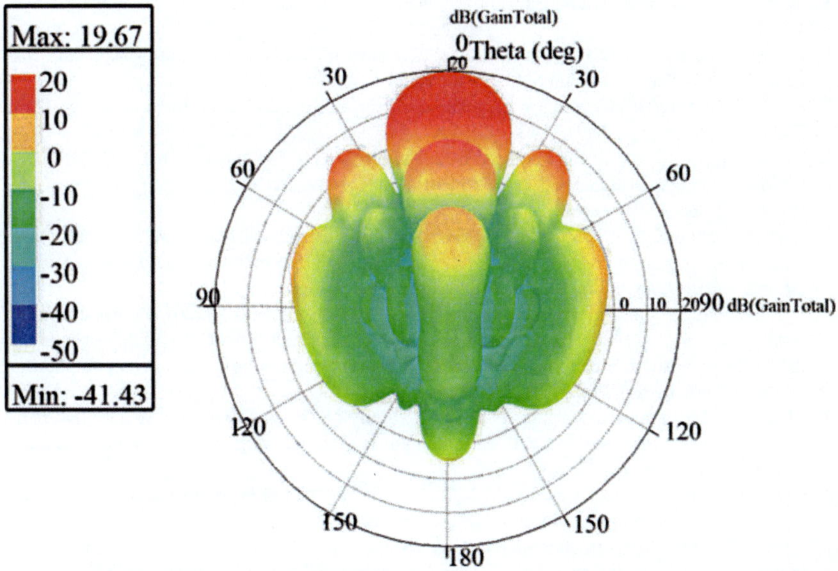

Fig. 6 Three-dimensional radiation pattern

lobe levels since excessive side lobe radiation wastes energy also can act as a noise source causing interference. The percentage of total energy radiated by an antenna that is contained inside the main lobe is referred to as beam efficiency. The main beam receives a significant magnitude of gain of 19.67 dB due to the use of sixteen elements as an array. This implies that the designed antenna has less power loss, a better radiation pattern, and better directivity. The array structure has a comparatively smaller dimension. It stitches together to form a non-conformal mesh to create a good field solution. It is relatively inexpensive and simple to manufacture. It also allows the mounting of electronic components on the backside. Most importantly, the patch antenna allows the design of compact and lightweight arrays. This antenna can be used for 5G small cell applications which require significant gain, as well as the structure should be small. This compact-sized antenna can be placed in a city square or denser place. To obtain adequate coverage, city planners also consider the position of the antenna.

References

1. Oughton EJ et al (2019) Assessing the capacity, coverage and cost of 5G infrastructure strategies: analysis of the Netherlands. Telematics Inform 37: 50–69
2. Hu C-N et al (2016) Millimeter-wave microstrip antenna array design and an adaptive algorithm for future 5G wireless communication systems. Int J Antennas Propag 2016
3. 3GPP (2013) TR 36.872: small cell enhancements for E-UTRA and EUTRAN—physical layer aspects
4. Zhang X, Andrews J (2015) Downlink cellular network analysis with multi-slope path loss models. IEEE Trans Commun 63(5):1881–1894
5. Bai T, Heath R (2015) Coverage and rate analysis for millimeter-wave cellular networks. IEEE Trans Wirel Commun 14(2):1100–1114
6. Kumar Goel L, Kottayil SK, Kirthiga S, Jayakumar M (2015) SMART GRID TECHNOLOGIES performance studies and review of millimeter wave MIMO beamforming at 60 GHz. Procedia Technol 21:658–666
7. Ignatius G, Murali KVU, Krishna NS, Sachin PV, Sudheesh P (2012) Extended Kalman filter based estimation for fast fading MIMO channels. In: 2012 international conference on devices, circuits and systems. ICDCS 2012, Coimbatore, pp 157–161
8. Narayanan L, Sellappan V, Kumar S, Kumar KV, VA MPK, Sudheesh P (2014) Basis expansion models for low complex parametric type channel estimation for MIMO-OFDM systems. In: Proceedings of the 5th national conference on recent trends in communication computation and signal processing RTCSP-2014, pp 35–37
9. Manojna DS, Kirthiga S, Jayakumar M (2011) Study of 2x2 spatial multiplexed system in 60 GHz indoor environment. In: 2011 international conference on process automation, control and computing (PACC), Coimbatore
10. Chataut R, Akl R (2020) Massive MIMO systems for 5G and beyond networks—overview, recent trends, challenges, and future research direction. Sensors 20(10):275
11. STT J, GB TR (2018) Rectangular microstrip patch antenna at ISM band. In: 2018 second international conference on computing methodologies and communication (ICCMC), pp 91–95. https://doi.org/10.1109/ICCMC.2018.8487877
12. Ansari, Namrata F, Ramapati M (2017) 1×1 rectangular patch antenna. In: 2017 IEEE international conference on power, control, signals and instrumentation engineering (ICPCSI). IEEE
13. Wang Z, Fang S, Fu S (2010) Wideband dual-layer patch antenna fed by a modified L-strip. J Microwaves, Optoelectron Electromagnet Appl 9(2):89–99
14. Ghosh S, Ghosh A, Sarkar I (2017) Design of probe feed patch antenna with different dielectric constants. 2017 devices for integrated circuit (DevIC). IEEE
15. Barrou O, El Amri A, Reha A, Microstrip patch antenna array and its applications: a survey
16. Pozar DM (1992) Microstrip antennas. Proc IEEE 80:79–91

17. MSc Thesis, ''Design of a patch Antenna'', chapter 3, Internet source, Florida State University, as on 2009/01/10.
18. Dafalla ZI et al (2004) Design of a rectangular microstrip patch antenna at 1 GHz. In: 2004 RF and microwave conference (IEEE Cat. No. 04EX924). IEEE
19. Nurulrodziah BT, Ghafar A (2005) Design of a compact microstrip antenna at 2.4 GHz. MSc thesis, Department of Electrical Electronics & Telecommunications, Faculty of Electrical Engineering, Universiti Teknologi, Malaysia

Design and Analysis of Fractal Loop Antenna Array for Rectenna Applications

Tirunagari Anilkumar, Meesala Mounika, Padala Rohini Lakshmi, Jana Ganga Raju, Maradana Sai Karthik Naidu, and B. Sridhar

Abstract In this article, a fractal-based planar printed loop antenna is designed for rectenna applications. The concept of Hilbert fractal is considered in this design to lower the operating frequency at compact dimensions. This planar printed loop antenna is designed on FR4 substrate with a formfactor of $45 \times 45 \times 1.6$ mm^3 and fed with coaxial feed. The characteristics of antenna after the insertion of fractal elements at peculiar symmetrical locations of the conducting square loop structure are investigated through simulation-based experiments. Further the radiation characteristics were analyzed with respect to single element, and the 1×2 array implementations (with different feed orientations) were experimented with same magnitude and phase (broadside case). The proposed array design is triple band in nature (1.42/1.92/2.42 GHz) and can accept the EM signals at the peak gain 2.77/4.12/6.16 dB, respectively.

Keywords Rectenna applications · Fractal loop antenna · Coaxial feed · Antenna array

1 Introduction

The use of various wireless electronic gadgets has been increasing day by day, and it can be simply said that we all are in the environment filled with electromagnetic waves. All the EM waves which are spread over are not utilized, and the energy associated with many such unused EM waves is getting wasted. The researchers are concentrating on the effective utilization of radiated EM energy through various sensing mechanisms [1, 2] including power conversion from harvested RF power to direct current. These are known as rectennas which can be used in wireless sensor networks and the latest IoT-based technologies as powering devices. These are the devices made in the combination of antenna and rectifier through an impedance matching network.

T. Anilkumar (✉) · M. Mounika · P. R. Lakshmi · J. G. Raju · M. S. K. Naidu · B. Sridhar
Department of ECE, Lendi Institute of Engineering and Technology, Vizianagaram, Andhra Pradesh, India

P. S. R. Chowdary et al. (eds.), *Evolution in Signal Processing and Telecommunication Networks*, Lecture Notes in Electrical Engineering 839,
https://doi.org/10.1007/978-981-16-8554-5_15

Many researchers reported various designs in this context and several antennas [3, 4] for wireless applications like Wi-Fi, WLAN, WiMAX, and X-band which are more ambient. Sometimes antennas with band notched and rejection band are required [5, 6] as per the need. Antenna arrays [7–9] play an important role owing to their enhanced radiation characteristics which are well suited at receiving mode of operation. In [10], a 1 × 4 array was designed for RF rectenna applications and utilized the beamwidth-enhancement method to obtain maximum efficiency in transmitting the power. The issues while connecting the antenna with rectifier circuit, particularly the complex nature of input impedance, were addressed in [11] and proposed a six-band dual CP rectenna with improved matching using an annular ring structure and a novel feeding. Progressive works were reported extending rectenna on paper electronics as discussed in [12] for self-sustainable devices aimed to convert voltage levels in the order of 1V. In [13], a rectenna was proposed with an antenna structure fed by CPW connected to a circular slot loop with stub matching. A fractal-based loop antenna structure with ILGP for proper impedance matching with the rectifier was proposed in [14]. Attempts were made in [15], to capture RF energy from nearly all directions by designing a dual-port L-probe antenna with stacked configuration (each individual port antenna covers half-sphere). Other configurations like compact flat dipole rectenna for IoT applications [16], circularly polarized rectenna with differential source feeding [17] were proposed. Fractal geometry-based rectenna [18] with multiple bending curves was reported with dual-tapered transmission line-based matching network to increase the rectification capacity. A dual-band (2.4 & 5.8 GHz) sickle-shaped antenna for dual-band applications was discussed in [19]. Developments are happening in the direction of waking up the wireless sensors using rectenna using packet injection algorithm [20]. Still there are challenges in designing the antenna arrays with suitable feeding systems for the desired beam pattern. It is desirable to have directional radiation beam and multidirectional to combat the variation of ambient energy over time and space.

In this article, design of an antenna array is proposed for 1x2 array configuration which was constructed based on Hilbert fractal loop geometry. Iteration level changes were investigated and presented in Sect. 2 with respect to their operating band performance. Section 3 emphasizes the array implementation with possible feed alignments and particularly emphasized the radiation characteristics with respect to the feed orientation at operating bands, and conclusions were drawn in Sect. 4.

2 Hilbert Fractal Loop Antenna Design

The proposed antenna is designed on FR4 substrate with an occupied volume of $45 \times 45 \times 1.6$ mm^3. At first, a square loop structure with a split at one of its edges is placed on the top of the substrate which serves as radiating patch.

The conventional ground plane is modified by reducing its dimensions to 25.12 mm in order to avoid reflections. A coaxial probe is given to the extended strip (on the right side) provided at the split region. As shown in Fig. 1a, a small

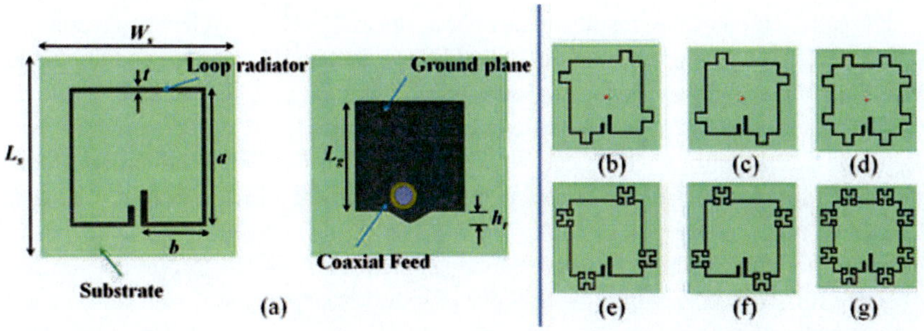

Fig. 1 Geometrical iterations of the antenna geometry **a** square loop antenna; Hilbert fractal iteration-1, **b** left-oriented, **c** right-oriented, **d** combination of **b** & **c**; Hilbert fractal iteration-2, **e** left-oriented, **f** right-oriented, **g** combination of **e** & **f**

isosceles triangular element is placed near the feed location in the ground plane to provide the impedance matching (Table 1).

The proposed antenna is the evolution of various iterations which is developed from the concept of Hilbert fractal. Two such iterations are considered to produce the antenna design. The fractalized elements appear like spikes on the edges of the square loop which are expected to lower the frequency of operation than earlier. The iterations from Fig. 1b–d depict the iteration-1-based evolution, and Fig. 1e–g represents the iteration-2. The operating band characteristics of these iterations are presented in Fig. 2a–b.

Table 1 Geometrical parameters of proposed antenna

Parameter	L_s	W_s	h	a	b	t	L_g	h_t	d
Size/mm	45	45	1.6	31.44	14.97	0.8	25.12	2.76	13.4

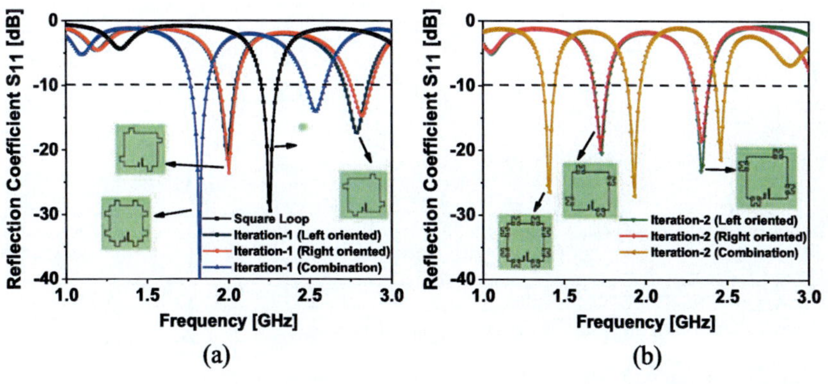

Fig. 2 Reflection coefficient vs. frequency characteristics of antenna iterations-1 (*figure on left*) and iterations-2 (*figure on the right*)

The square loop design results in the operating band 2.21–2.30 GHz and resonates at 2.26 GHz. When the first iteration was applied, the single resonant characteristic was disappeared and produced the dual-band behavior at 2 & 2.8 GHz approximately. There was not much significant difference observed for left-oriented and right-oriented iterations which can be seen in Fig. 2a, but the combination model of these two has shifted the resonant frequencies toward lower side at 1.82 and 2.54 GHz, respectively.

In the iteration-2, the square U-shaped segment was altered to produce second level of Hilbert fractal within the vicinity of that primary element, and this has led to three possibilities which were encountered in iteration-1. The resonant frequencies were shifted as there is an increased electrical length in iteration-2. Table 2 highlights the operating band and radiation characteristics of first level (iteration-1) of Hilbert fractal loop design and its second level (iteration-2). It is observed that the resonant frequencies of the combined fractal design (of both iterations) were found shifted toward lower frequency region from 2 to 1.82 GHz and 2.82–2.54 GHz for Iteration-1, and 1.73–1.41 GHz and the second band (at 2.34 GHz) has found created into two distinct resonant modes at 1.93 and 2.46 GHz, respectively. This was observed in the iteration-2, and due to the second-level fractal generation, the effective length has extended, thus causing these effects. Although the structure achieved lower frequencies of operation at compact dimensions, it is suffering

Table 2 Characteristics of the proposed fractal loop antenna iterations

Iteration	Operating bands (GHz)	% BW	f_r (GHz)	S_{11} (dB)	Peak gain (dB)	Radiation efficiency (%)
Iteration-1						
Left-oriented	1.94–2.04	5.03	1.99	−27.27	1.39	84.2
	2.71–2.85	5.04	2.79	−17.29	2.70	86.6
Right-oriented	1.94–2.05	5.51	2	−23.57	1.37	83.0
	2.75–2.88	4.62	2.82	−14.63	2.84	87.3
Combination	1.77–1.87	5.49	1.82	−46.42	0.705	84.9
	2.48–2.59	4.34	2.54	−13.94	2.39	87.7
Iteration-2						
Left-oriented	1.68–1.77	5.22	1.73	−20.44	0.32	81.8
	2.29–2.39	4.27	2.34	−23.14	1.39	80.4
Right-oriented	1.68–1.76	4.65	1.72	−19.83	0.23	82.6
	2.30–2.39	3.84	2.34	−18.47	1.47	81.7
Combination	1.37–1.44	4.98	1.41	−26.39	−1.13	83.3
	1.89–1.96	3.64	1.93	−26.99	0.26	81.7
	2.43–2.49	2.44	2.46	−21.30	−0.29	72.4

from sufficient far-field gain due to its reduced overall metallic area. Hence, array of antennas is considered as a possible solution of enhancing the far-field gain which is discussed in Sect. 3.

3 Development of Two-Element (1 x 2) Antenna Array and Its Performance Characteristics

The radiation characteristics of single element antenna observed in Table 2 are not sufficient for signal reception and will not effectively work for RF to DC power conversion. Hence, this has led to develop an array with two elements.

There aroused many combinations with respect to the orientation of the feed system. But among all combinations, only three were considered for the examination of array characteristics. The characteristics of these array combinations are discussed with respect to their operating band and the radiation behavior. During the analysis, two ports of the array antenna were provided with same excitation amplitude and same phase (Fig. 3).

3.1 Reflection Coefficient Versus Frequency Characteristics

As the antenna array is comprising two ports, there will have four scattering parameters in which two are self- (S_{11} & S_{22}) *also called reflection coefficients,* and the other two are mutual parameters (S_{12} & S_{21}) *also called mutual coupling coefficients.*

As it can be seen (in Fig. 4), all the three possible combinations of the antenna arrays operate at three bands. The coupling is below −10 dB level for the array with unidirectional feeds (array model-1) and vertically opposite-oriented feeds (Array model-2), whereas for array with the horizontally opposite-aligned feeds, the coupling is below −15 dB at the fundamental and second-order resonances. It is noticed that the operating bands are consistent for all possible combinations of the antenna array.

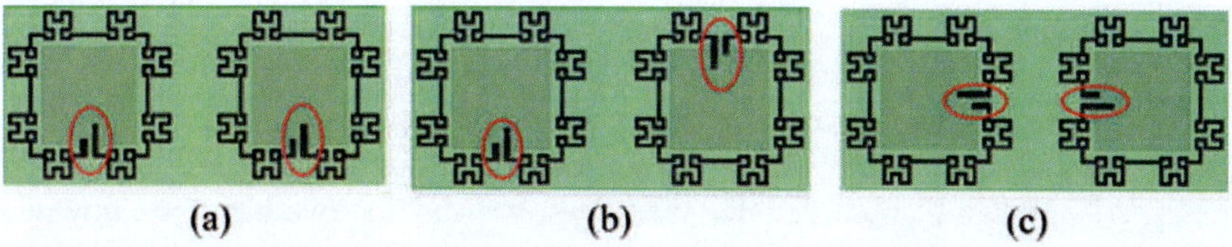

Fig. 3 Geometry of proposed two-element antenna array **a** feed facing along same direction, **b** vertical and opposite alignment, **c** horizontal and opposite alignment (*The ovals marked with red color represent the feed location in the array*)

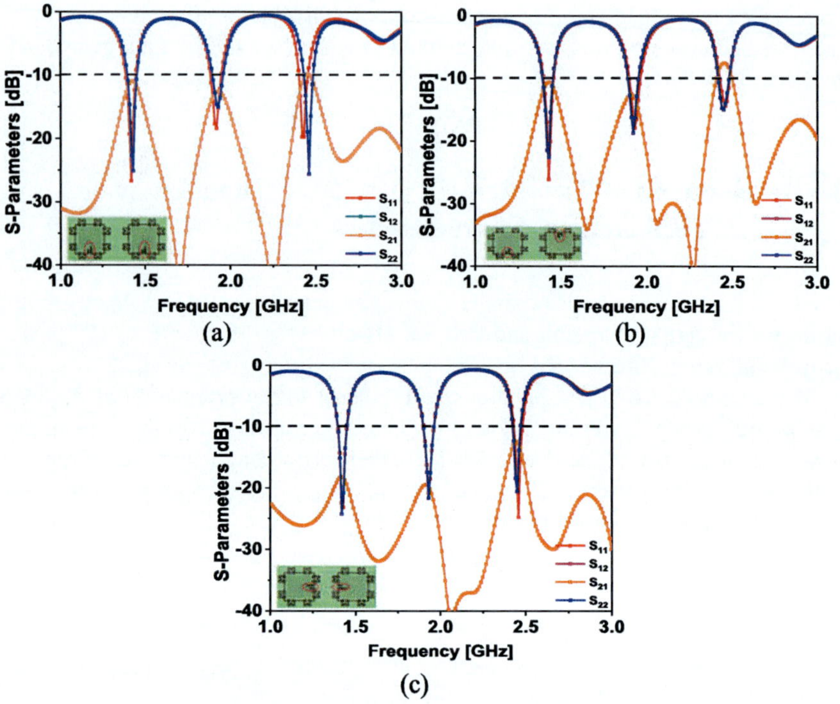

Fig. 4 S-parameters versus frequency characteristics of possible combinations of antenna arrays **a** feed facing along same direction, **b** vertical and opposite alignment, **c** horizontal and opposite alignment

3.2 Far-Field Radiation Characteristics

The far-field radiation patterns of the possible combinations of antenna arrays are stacked in Fig. 4. The patterns are obtained at three frequencies of their corresponding operating bands as mentioned in Table 2. For the first kind of arrangement, the combined pattern of two antenna elements produces the omnidirectional pattern in *XZ* plane and dumbbell-shaped in *XY* plane, direction of null along *y*-axis which is observed at first resonance. At mid-frequency, the direction of nulls is changed, and at third resonance, the pattern is bidirectional and looks like fan beam (Fig. 5).

For the vertical and opposite-oriented feed arrangement at first two operating bands, the null direction is parallel to the axis of feeding focusing two lobes diagonally. Also, it can be observed that the patterns of the first two bands are oriented like mirror images. At third band, for all possible cases, the pattern is oriented along *XZ* plane.

Table 3 emphasizes the operating bands and radiation performance of the antenna. However, all three models are producing almost closer operating bands and resonant frequencies, the peak gain values of the antenna at resonances were found as negative

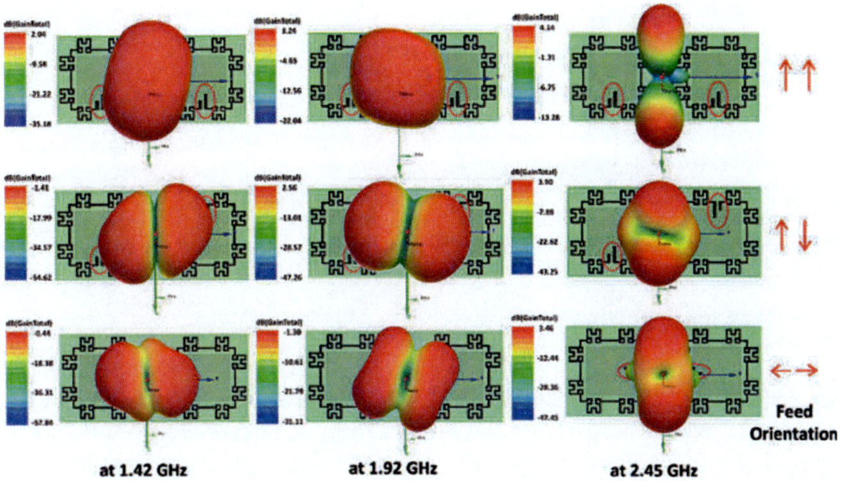

Fig. 5 3D far-field patterns of the proposed array antenna at three operating frequencies for array model-1 (*figure on the top*) array model-2 (*figure in the middle*) array model-3 (*figure in the bottom*)

Table 3 Characteristics of the possible combinations of 1 × 2 antenna array

Design	Operating bands (GHz)	% Bandwidth	f_r (GHz)	S_{11} (dB)]	Peak gain (dB)	Radiation efficiency (%)
Array model-1	1.38–1.44	4.25	1.41	−26.70	2.77	84.6
	1.87–1.94	3.67	1.91	−18.44	4.12	81.6
	2.39–2.44	2.07	2.42	−19.78	6.16	62.8
Array model-2	1.39–1.45	4.22	1.42	−26.12	−1.41	73.11
	1.89–1.95	3.12	1.92	−17.92	2.56	75.8
	2.42–2.48	2.44	2.45	−14.38	3.90	63.4
Array model-3	1.39–1.44	3.53	1.42	−23.08	−0.44	71.4
	1.89–1.95	3.12	1.92	−20.71	−1.30	66.3
	2.42–2.47	2.04	2.45	−24.72	3.46	54.3

which are not much appreciable for using this design as a rectenna element. This characteristic is observed for array model-2 and 3, and also, the reduction in radiation efficiency compared to model-1 was noticed.

Considering the array model-1 for its suitability toward rectenna applications, Fig. 6 portrays the illumination of EM waves and their reception by the aperture of proposed array antenna in horizontal and vertical orientations which are operating at three bands.

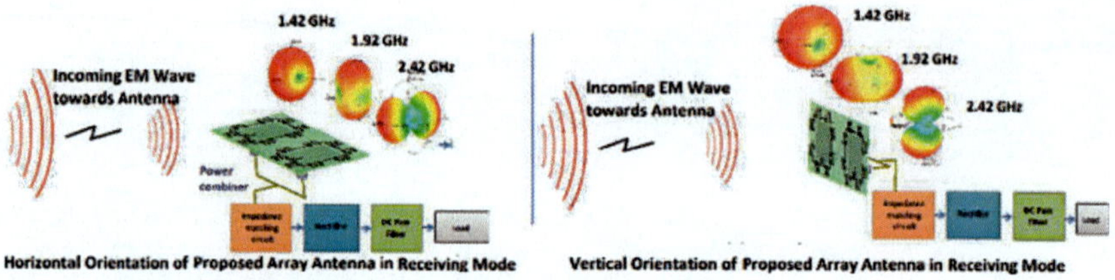

Fig. 6 Possible orientation of the array antenna in receiving mode

4 Conclusion

A modified square loop antenna incorporated with Hilbert fractal elements and fed with coaxial feed is presented in this article. The 1 × 2 array design of the same structure is proposed as a design with enhanced far-field gain compared to single element. The proposed design is intended to use for rectenna applications to effectively capture the ambient electromagnetic signals and convert them to DC for the possible utilization in the DC-powered IoT modules, smart devices, or gadgets.

Future scope: The proposed fractal loop antenna array design can be further frequency reconfigured, and RF switchable rectifier can be embedded within the design for obtaining the rectenna functionality. The multidirectional radiation can be obtained by frequency switching to combat the variation of ambient energy over time and space.

Acknowledgements Authors are thankful to the R&D Center, Department of ECE, Lendi Institute of Engineering & Technology, Vizianagaram, for providing the Ansys Electronics simulation facilities.

References

1. Allam V (2019) An IoT controlled octahedron frequency reconfigurable multiband antenna for microwave sensing applications. IEEE Sens Lett 3(10):1–4
2. Nadh BP, Munuswamy SK, Manikonda VR, Poluri VVK (2020) Windmill-shaped antenna with artificial magnetic conductor-backed structure for wearable medical applications. Int J Numer Modell Electron Netw Devices Fields 33(6):e2757
3. Monika M, Kumar MS, Prudhvinadh B (2019) Dual band reconfigurable compact circular slot antenna for WiMAX and X-band applications. Radioelec Comm Sys 62(9):474–485
4. Tilak GBG, Sarat KK, Korada CSK, Rao MV (2020) AMC backed circularly polarized dual band antenna for Wi-Fi and WLAN applications. J Electr Eng 71(5):298–307
5. Rao MV, Tirunagari A (2018) Conformal band notched circular monopole antenna loaded with split ring resonator. Wirel Pers Comm 103(3):1965–1976
6. Sowmithri P, Renuka M (2019) A triangular slot band notched MIMO antenna for UWB applications. J Adv Res Dyn Cont Sys 11(2):1002–1009
7. Madhav BTP, Rao TV, Tirunagari A (2018) Design of 4-element printed array antenna for ultra-wideband applications. Int J Microw Opt Technol 13(1):8–17

8. Anjaneyulu G, Varma JS (2019) Design and simulation of 8 element patch antenna array for X band satellite applications. In: International conference on smart systems and inventive technology (ICSSIT), pp 1176–1179
9. Paladuga Chowdary SR, Vedula CVSSS, Anguera J, Mishra RK, Andújar A (2018) Performance of beamwidth constrained linear array synthesis techniques using novel evolutionary computing tools. Appl Comp Electromag Soc J 33(3):273–278
10. Sun H, Geyi W (2016) A new rectenna using beamwidth-enhanced antenna array for RF power harvesting applications. IEEE Ant Wirel Propag Lett 16:1451–1454
11. Song C, Yi H, Carter P, Zhou J, Yuan S, Xu Q, Kod M (2016) A novel six-band dual CP rectenna using improved impedance matching technique for ambient RF energy harvesting. IEEE Trans Ant Propag 64(7):3160–3171
12. Kharrat I, Xavier P, Vuong T-P, Tourtollet GEP (2016) Compact rectenna design for lossy paper substrate at 2.45 GHz. Prog Electromag Res 62:61–70
13. Riviere J, Douyere A, Oree S, Luk J-D (2017) A 2.45 GHz ISM band CPW rectenna for low power levels. Prog Electromag Res C 77:101–110
14. Zeng M, Andrenko AS, Liu X, Li Z, Hong-ZT (2017) A compact fractal loop rectenna for RF energy harvesting. IEEE Ant Wirel Propag Lett 16:2424–2427
15. Shen S, Chiu C-Y, Murch RD (2017) A dual-port triple-band L-probe microstrip patch rectenna for ambient RF energy harvesting. IEEE Ant Wirel Propag Lett 16:3071–3074
16. Okba A, Takacs A, Aubert H (2018) Compact flat dipole rectenna for IoT applications. Prog Electromag Res 87:39–49
17. Kumar D, Chaudhary K (2018) Design of differential source fed circularly polarized rectenna with embedded slots for harmonics suppression. Prog Electromag Res 84:175–187
18. Shi Y, Jing J, Fan Y, Yang L, Wang M (2018) Design of a novel compact and efficient rectenna for wifi energy harvesting. Prog Electromag Res 83:57–70
19. Eid A, Hester JGD, Joseph C, Youssef T, Ali HR, Manos MT (2019) A compact source–load agnostic flexible rectenna topology for IoT devices. IEEE Trans Ant Propag 68(4):2621–2629
20. Bhatt K, Sandeep K, Pramod K, Chandra CT (2019) Highly efficient 2.4 and 5.8 GHz dual-band rectenna for energy harvesting applications. IEEE Ant Wirel Propag Lett 18(12):2637–2641

Design of High Isolation MIMO Antennas for Ultra-Wide Band Communication

T. Sathiyapriya, V. Gurunathan, J. Dhanasekar, and V. V. Teresa

Abstract Multiple input multiple output (MIMO) antenna performance could be downgraded due to poor isolation between the antenna elements. In this paper, a compact two-port MIMO antenna and four-port MIMO antenna are designed to improve isolation features of ultra-wide band (UWB) communications. The circular neutralizer is enabled as a decoupling structure to reduce mutual coupling between radiating elements. The total size of the proposed two-port and four-port UWB antennas are 21 mm × 31 mm and 48 mm × 31 mm, respectively. The substrate used for antenna design is FR4 with 1.6 mm thickness. The better isolation is achieved up to 29 dB in a frequency range of 3.1–10.6 GHz with proper impedance matching. The proposed UWB antenna has been desirable in terms of antenna size, geometric complexity, bandwidth, gain, and isolation level.

Keywords MIMO · Ultra-wide band (UWB) · Isolation · Neutralization line and Impedance matching

1 Introduction

Ultra-wide band (UWB) has risen in prominence in the field of wireless communications in recent years due to its fast data rate over a short distance, low power density, and wide frequency spectrum [1]. These are utilized in a number of applications, including GPS, remote sensing, and direct broadcast systems (DBS). Low profile, compactness, wider impedance matching, and high directivity properties are

T. Sathiyapriya (✉) · V. Gurunathan
Dr. Mahalingam College of Engineering and Technology, Pollachi, India

J. Dhanasekar · V. V. Teresa
Sri Eshwar College of Engineering, Coimbatore, India
e-mail: dhanasekar.j@sece.ac.in

V. V. Teresa
e-mail: teresa@sece.ac.in

P. S. R. Chowdary et al. (eds.), *Evolution in Signal Processing and Telecommunication Networks*, Lecture Notes in Electrical Engineering 839,
https://doi.org/10.1007/978-981-16-8554-5_16

major problems for UWB systems. Following the Federal Communication Commission's (FCC) clearance of the unlicensed frequency band, UWB technology becomes more appealing to all wireless researchers [2]. The FCC has assigned the frequency range 3.1–10.6 GHz to UWB systems, which transmit low-power signals over a wide bandwidth. The performance of UWB systems is influenced by a number of factors. Simultaneous fading is one of them, and it may be solved with MIMO technology, which uses multiple antennas at both the transmitter and receiver to enhance the dependability of various communication systems. One of the advantages of a MIMO system is that it allows numerous users to access various services at the same time, resulting in increased channel capacity.

If mutual coupling and correlation between antenna elements are limited, MIMO technology may be employed successfully in communication systems. When many antennas are used in a compactable device, the space between antenna parts is reduced, resulting in poor isolation. Furthermore, designing UWB matching or decoupling structures is time and space demanding. As a result, placing many UWB antennas with high isolation in space-constrained portable devices is a difficulty. Mutual coupling between the radiating components has been shown to impact the performance of MIMO antennas. In such a system, a distance between components must not be less than $\lambda/2$ of the lowest resonating frequency condition is to be met.

Different techniques and structures for improving isolation between radiating elements have been researched over several decades. With the addition of defected ground structure (DGS), orthogonally polarized antennas are proposed. Mutual coupling is reduced in both the E and H planes due to the orthogonal design and diamond-shaped DGS on the bottom side [1]. A double-sided circular monopole antenna with a dielectric resonator acts as a filter on the band that minimizes surface current. The decoupling structure links the EBG structure on both sides of the ground such that coupling decreases [2]. A double-side printed antenna with radiating components and electromagnetic band gap (EBG) structure on both the sides have been reported with isolation better than −20 dB [3]. The neutralization line technique involves connecting the patch with varying lengths of microstrip transmission line [4]. A fractal structure based on EBG is applied in four-element MIMO antennas. This system employs the stub structures to minimize the mutual coupling of the antenna by reducing surface current [5–9]. Changing the radiation elements structure is also a strategy for increasing isolation [10–14].

In this paper, a compact two-element and four-element MIMO antennas have been designed with improved isolation between elements over ultra-wide range of frequencies. The suggested antenna's geometry is made up of half-wavelength microstrip lines, rectangular patch, and semi-circular patch. As a decoupling network, a neutralization network with microstrip lines and a circular structure is used. For two port antennas, this approach improves isolation by up to −27 dB, while for four port antennas, it improves isolation by up to −29 dB. Both suggested MIMO antennas have a large bandwidth and resonate at frequencies between 3 and 10.6 GHz. To produce the compact proposed antenna, a typical UWB circular monopole antenna design is chosen and modified with the addition of microstrip lines.

2 Design of Two-Element Antenna System

2.1 *Proposed Antenna Geometry*

The proposed two-element antenna system geometry is depicted in Fig. 1. A semicircular shape of radius *R* is fixed on a microstrip line attached over the square patch. The tiny quantitude of copper is graved from the rectangular bottom corners to decrease the occupancy of the space. Both symmetrical antennas are configured to form a MIMO set-up at a distance of 4.9 mm. In order to decrease the mutual coupling, the circular neutralizer is located above the ground and between the MIMO antenna parts. The line of neutralization consists of a circular disc in the center of two thin microstrip lines. The circular disc permits a number of potential current decoupling paths. The size of the monopole ground plane is approximately 7.4 mm × 31 mm. To increase isolation, the split structure of 4.4 mm × 0.4 mm is inserted into the ground structure. The 2 × 2 MIMO antenna suggested is on a 1.6 mm thick FR4 substrate with a dielectric of 4.4 and loss tangent of 0.02. The feed line width is set at 3 mm to match the impedance of 50 Ω. Table 1 define the geometry parameters of the suggested two-element antenna system.

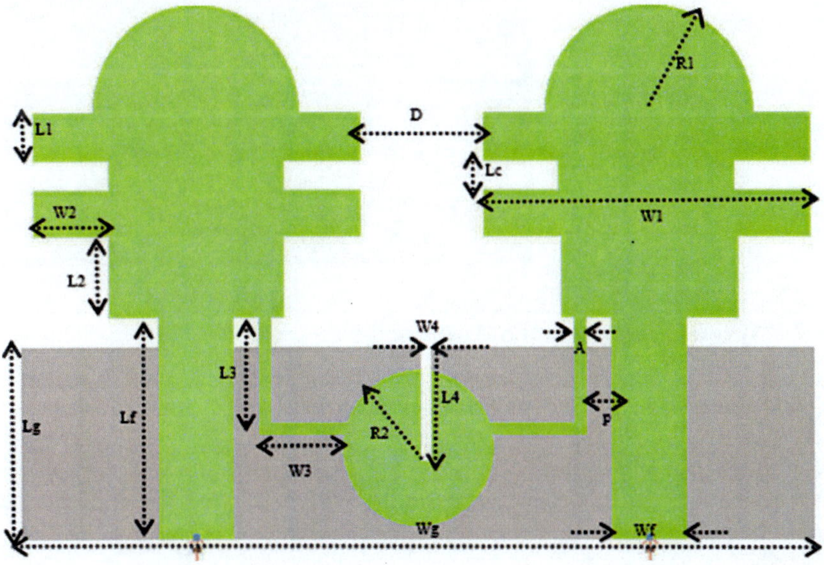

Fig. 1 Layout of two-port antenna system

Table 1 Geometry parameters of two-port system

Parameters	L_1	W_1	L_2	W_2	L_3	W_3	L_4	W_4	L_f
Values (millimeter)	2.0	13	3.2	3.0	4.5	3.3	4.4	0.4	8.5
Parameters	W_f	L_g	W_g	L_c	R_1	R_2	A	D	P
Values (millimeter)	3.0	7.4	31.0	1.1	4.2	3.0	0.5	4.9	1.0

Fig. 2 Single-element UWB antenna

2.2 *Single-Element UWB Antenna*

Figure 2 depicts the suggested single-element antenna design. The simulated scattering parameters of a two-port MIMO antenna is shown in Fig. 3. Here, the single-element antenna only resonates at 3.2 and 7.5–9.5 GHz, which are two distinct bands. As a result, using a single-element antenna, the UWB condition is not reached in the lower frequency band.

2.3 *Simulation and Analysis of Two-Element MIMO Antenna*

Figure 4 depicts how a single-element antenna is symmetrically replicated to produce a two-element array. Figure 5 illustrates the simulated S-parameters of a two port

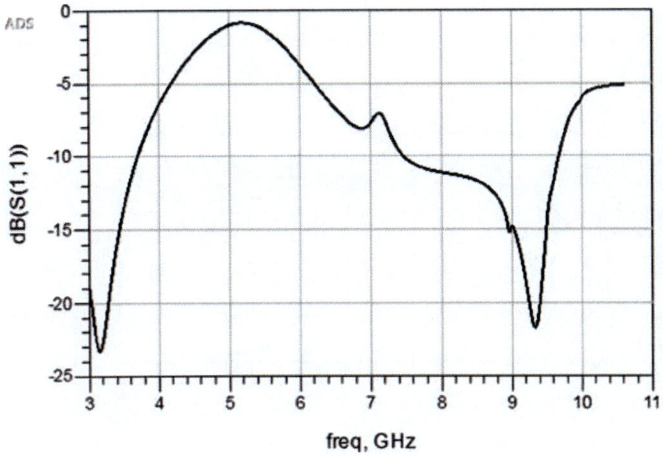

Fig. 3 S-Parameter (S_{11}) for single MIMO antenna

Fig. 4 Layout of 1 × 2 MIMO antenna

MIMO antenna with a circular neutralizer. The calculated frequency bandwidth of two port MIMO antenna system is 3.1–10.6 GHz, with a −27 dB isolation over the band.

This two-port MIMO neutralized line antenna has a suggested frequency bandwidth with excellent isolation even on two ends of the band (below and up). We found,

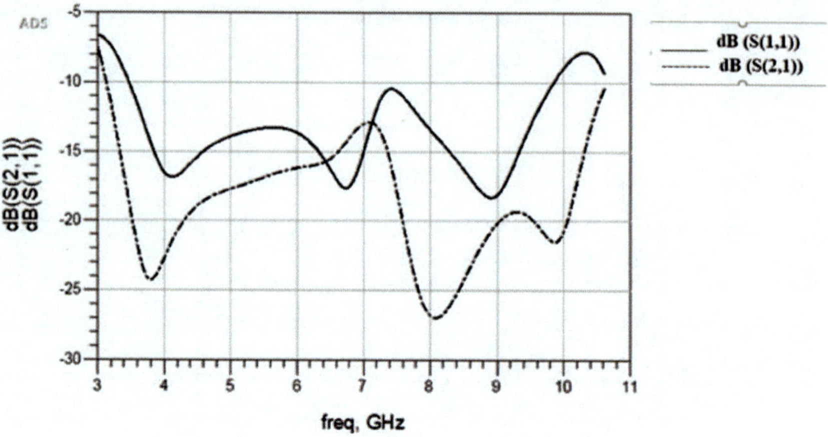

Fig. 5 S parameter results of two-element MIMO array antenna

from these results, that this approach was more effectively applied in different applications of wireless communication systems with increasing isolation between the two antenna components.

3 Design of Four-Element Antenna System

3.1 Layout of 4 Port (2 × 2) MIMO antenna

The two-element antenna is expanded in order to build the four-port ultra-wide band MIMO antenna. Figure 6 describes the geometry of the antenna. The antenna size is between 48 mm per 31 mm and the remaining characteristics are identical to the two port antenna. A vertical separation of 33.2 mm is maintained between the radiation components.

3.2 Simulation Results of Four-Element MIMO Array

Figure 7 illustrates the circular neutralization line approach to obtain proper simulated S-parameters for a 2 × 2 MIMO antenna. The antenna components have a simulated frequency bandwidth of 4–10.2 GHz, with a—29 dB isolation across the band. The neutralization decoupling structure is more successful in increasing isolation in both the lower and higher frequency bands.

Fig. 6 Layout of 2 × 2 MIMO antenna

3.3 Realized Gain and Directivity

The gain and direction of the proposed four-port MIMO antenna is shown in Figure 8. The simulated gain for the whole frequency range is 3 dB. The radiation efficiency simulated of the suggested design varied between 70.05 and 75.55% throughout the frequency spectrum. From the graph, it is obvious that the proposed decoupling structure is better than other isolation mitigation approaches.

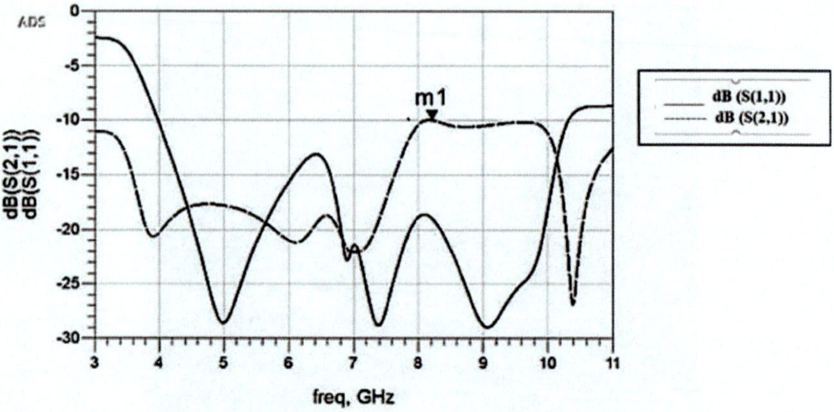

Fig. 7 S-parameter slot for 2 × 2 MIMO antenna

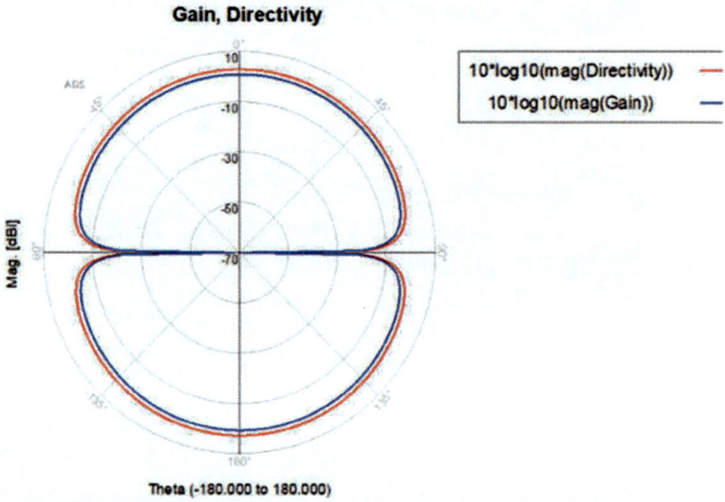

Fig. 8 Gain and directivity of 2 × 2 MIMO antenna

4 Radiation Characteristics of Four-Port Antenna System

The simulated radiation pattern of the four port antenna is shown at 4.5 GHz (Fig. 9a), 8.9 GHz (Fig. 9b), and 9.9 GHz (Fig. 9c), respectively. The Omni directional radiation pattern of H-plane and E-plane pattern is pattern is observed at the wide frequency spectrum.

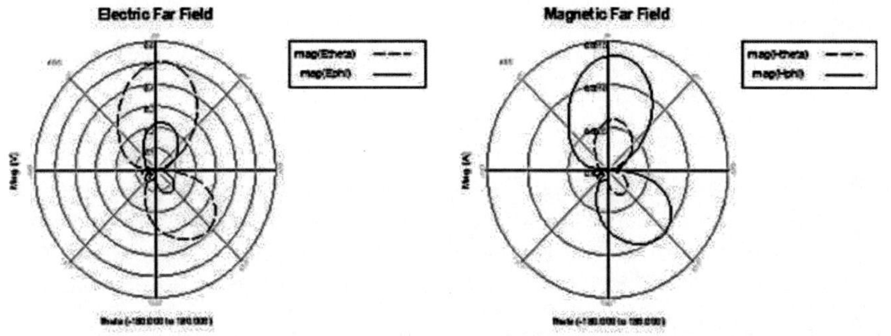

(a) f = 4.5 GHz

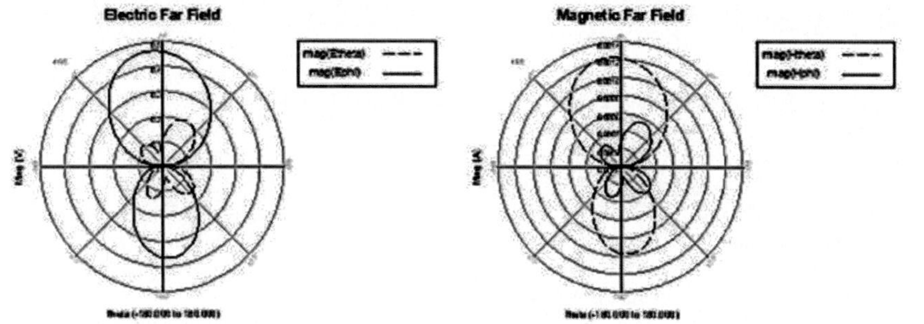

(b) f = 8.9 GHz

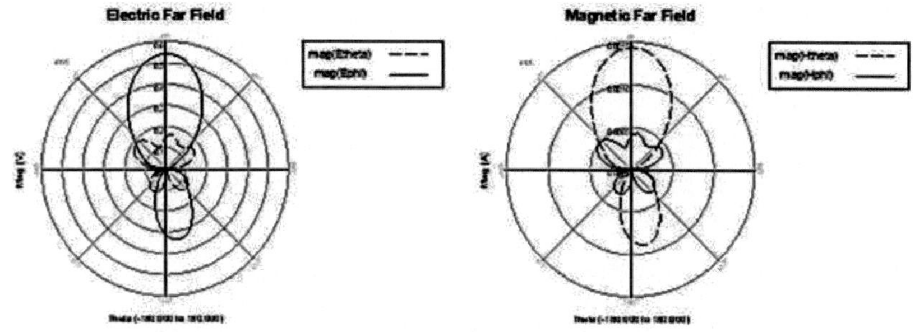

(c) f = 9.9 GHz

Fig. 9 a–c E-field and H-field for 4 × 4 MIMO antenna

5 Conclusion

The compact two-port and four-port ultra-wide band MIMO antennas have been presented in this dissertation. The proposed MIMO antenna has been designed with circular neutralizer-based decoupling structure. The obtained bandwidth of two-element antenna system is 95.11% in the frequency band 3.1–10.6 GHz and the bandwidth of four-element antenna system is 96.78% in frequency band 4–10.2 GHz. The observed gain and radiation efficiency are 3 dB and varying between 70.05 and 75.55% percent throughout the whole frequency spectrum. The corresponding isolation measurements are -26 and −29 dB for 1 × 2 and 2 × 2 antenna systems, respectively. As a result, the suggested UWB MIMO antenna successfully meets the required MIMO characteristics for wireless applications like WLAN, cellular, and satellite applications. The work can be expanded in the future to increase antenna efficiency and compactness.

References

1. Suriya I, Anbazhagan R (2019) Inverted-A based UWB MIMO antenna with triple-band notch and improved isolation for WBAN applications. AEU Int J Electron Commun 99:25–33
2. Ali WAE, Ibrahim AA (2017) A compact double-sided MIMO antenna with an improved isolation for UWB applications. AEU Int J Electron Commun 82:7–13
3. Prabhu P, Malarvizhi S (2019) Novel double-side EBG based mutual coupling reduction for compact quad port UWB MIMO antenna. AEU Int J Electron Commun 109:146–156
4. Zhang S, Pedersen GF (2016) Mutual coupling reduction for UWB MIMO antennas with a wideband neutralization line. IEEE Antennas Wirel Propag Lett 15:166–169
5. Pandit S, Mohan A, Ray P (2018) A compact four-element MIMO antenna for WLAN applications. Microw Opt Technol Lett 60:289–295
6. Khan, M. S. et al. A WLAN band-notched compact four element UWB MIMO antenna. Int. J. RF Microw. Comput. Eng. 30, e22282 (2020).
7. Palanisamy P, Subramani M (2021) Design of metallic via based Octa-Port UWB MIMO antenna for IoT applications. IETE J Res 1–11. https://doi.org/10.1080/03772063.2021.1892540
8. Yang Z-X, Yang H-C, Hong J-S, Li Y (2016) A miniaturized triple band-notched MIMO antenna for UWB application. Microw Opt Technol Lett 58:642–647
9. Hassan MM et al (2020) A novel UWB MIMO antenna array with band notch characteristics using parasitic decoupler. J Electromagn Waves Appl 34:1225–1238
10. Yu K, Li Y, Liu X (2018) Mutual coupling reduction of a MIMO antenna array using 3-D novel meta-material structures. Appl Comput Electromagn Soc J 33
11. Mathur R, Dwari S (2018) Compact CPW-Fed ultrawideband MIMO antenna using hexagonal ring monopole antenna elements. AEU Int J Electron Commun 93:1–6
12. Alibakhshikenari M et al (2019) Mutual-coupling isolation using embedded metamaterial EM bandgap decoupling slab for densely packed array antennas. IEEE Access 7:51827–51840
13. Kumar A, Ansari AQ, Kanaujia BK, Kishor J (2019) A novel ITI-shaped isolation structure placed between two-port CPW-fed dual-band MIMO antenna for high isolation. AEU Int J Electron Commun 104:35–43
14. Liu L, Cheung SW, Yuk TI (2013) Compact MIMO antenna for portable devices in UWB applications. IEEE Trans Antennas Propag 61:4257–4264
15. Tiwari RN, Singh P, Kanaujia BK, Srivastava K (2019) Neutralization technique based two and four port high isolation MIMO antennas for UWB communication. AEU Int J Electron Commun 110:152828

Energy Detector and Diversity Techniques for Cooperative Spectrum Sensing

M. Vamshi Krishna, M. V. D. Prasad, M. Ashok Kumar, and K. V. Kalyan Chakravarthy

Abstract Spectrum sensing is identified as a pivotal aspect for the cognitive radio network to efficiently utilize the unused radio frequency band. In cognitive radio networks, each cognitive radio user senses the spectrum hole, and in reply, forward their decisions, whether primary user is present or absent to the common receiver. In this research article, the cooperative spectrum sensing is calculated and analyzed using energy detector (ED) by selection combining (SC) and maximal ratio combining (MRC) diversity techniques at each cognitive radio user, and it is proved that by increasing number of branches (Antennas) at each cognitive radio user, maximal ratio combining (MRC) is giving better performance by optimal reduction of the total error in a cognitive radio network.

Keywords Maximal ratio combining (MRC) · Energy detector (ED) · Cognitive radio (CR) · Selection combining (SC)

1 Introduction

In the present trend, increasing wireless applications requires a radio spectrum, as the already existing radio spectrum allocated to specific applications, so there is a problem of spectrum scarcity. Based on the metrics by survey of Federal Communication Commission (FCC), the dedicated assigned spectrum band is not effectively utilized [1]. To meet the challenges of present wireless communication applications, FCC had introduced a novel technology cognitive radio (CR). The main function of cognitive radio (CR) technology is spectrum sensing. To derive better performance and results spectrum sensing, one of the most used and simplest detector is the energy detector (ED), which was proposed by the "urkowitz" [2]. To use effectively the unused radio spectrum by the secondary user (SU) without causing interference to the primary user (PU), spectrum sensing techniques have been employed. The common problems which will arise during spectrum sensing are multipath fading,

M. Vamshi Krishna (✉) · M. V. D. Prasad · M. Ashok Kumar · K. V. Kalyan Chakravarthy
Department of ECE, Dhanekula Institution of Engineering & Technology, Vijayawada, India

P. S. R. Chowdary et al. (eds.), *Evolution in Signal Processing and Telecommunication Networks*, Lecture Notes in Electrical Engineering 839,
https://doi.org/10.1007/978-981-16-8554-5_17

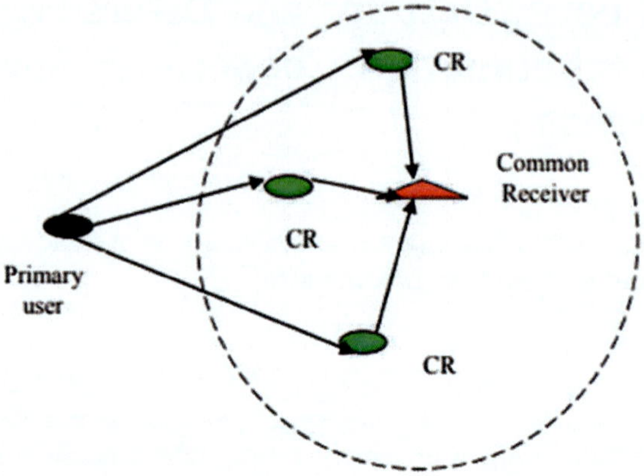

Fig. 1 Spectrum sensing in the cognitive radio network

hidden terminal problems; due to these problems, the cognitive radio (CR) may fail to identify a primary user (PU); the result of that causes interference to primary user (PU). To nullify these effects and improve the overall characteristics of the system, cooperative spectrum sensing is used. In cooperative spectrum sensing in preference to taking available selection by way of the every cognitive radio (CR) in a community, all of the cognitive radios (CRs) send their selections to the fusion center (Common receiver) [3] which is access point (AP) in a local area network (LAN) or base station (BS) in a mobile network [2] (Fig. 1).

In [4], the spectrum sensing with strength detector has been developed the usage of variety strategies. In [5], choicest range of cognitive radios is required to perform cooperative spectrum sensing for the enjoyable of given errors surely changed into proposed. In this paper using that optimality technique, cooperative spectrum sensing was performed by employing energy detector with integration of diversity techniques. It is shown that the total error in a cognitive radio network can be reduced by using diversity techniques.

2 Spectrum Sensing

To perform spectrum sensing, two binary hypothesis testing will be used. One is H_0, and other one is H_1. H_0 indicates only noise is present; H_1 indicates along with noise, primary user signal (licensed user) present. Let $Y[n]$ be the received signal which is defined as.

$Y[n]$ is defined as received signal

$$Y[n] \underline{\underline{\text{def}}} \begin{cases} \eta\eta[n] & H_0 \\ X[n] & H_1 \end{cases} \quad (1)$$

$n = 1, 2, 3, \ldots, N$ total number of received samples where $\eta[n]$ is noise of signal, and $X[n]$ is primary user signal.

Generally, two types of hypothesis testing errors will be considered in the spectrum sensing [6].

Type1 error: probability of false alarm

$$P_{\mathrm{F}} = \mathrm{Prob}\{H_1|H_0\} \tag{2}$$

Type2 error: probability of miss detection

$$P_{\mathrm{MD}} = \mathrm{Prob}\{H_0|H_1\} \tag{3}$$

P_{F} says that probability of primary user is present but actually primary user absent (noise signal present).

P_{MD} says that probability of primary user is absent but actually primary user present (licensed user present).

The effect of high P_{F} causes inefficient utilization of vacant spectrum, as well as the effect of high P_{MD} causes interference to licensed users.

3 Energy Detector

Energy detector is the maximum extensively used spectrum sensing technique, due to its decreased implementation complexity. The presence of a spectrum hollow is detected through comparing the measured electricity in the direction of a suitable threshold; it truly is fairly prone to the noise floor.

In the energy detector from the received signal samples, the probability of false alarm and probability of miss detection are mentioned below [4]

$$P_{\mathrm{F}} = \Gamma_u(N\lambda, N) \tag{4}$$

$$P_{\mathrm{MD}} = 1 - \Gamma_{\mathrm{u}}\left(\frac{N\lambda}{1+\mathrm{SNR}}, N\right) \tag{5}$$

where λ is a threshold; $\Gamma_{\mathrm{u}}(.,)$ is the upper incomplete gamma function mentioned as [7]

$$\Gamma_{\mathrm{u}}(a, n) \triangleq \frac{1}{r(n)} \int_a^\infty x^{n-1} \mathrm{e}^{-x} \mathrm{d}x \tag{6}$$

and probability of detection

$$P_{\mathrm{D}} = 1 - P_{\mathrm{MD}} \tag{7}$$

4 Diversity Techniques

(i) Maximal Ratio Combining

In MRC technique, the overall received SNR at the receiver side can be improved by placing multiple numbers of antennas [2]. The received SNR is expressed in terms of the signal energy per symbol E_s as [4]

$$\mathrm{SNR} = \frac{E_s}{N_0 \mathrm{BT_s}} \tag{8}$$

Assume ideal noise power spectral density $\frac{N_0}{2}$ on each branch and pulse shaping $\mathrm{BT_s} = 1$.

In no fading case [4], the probability of false alarm

$$P_{\mathrm{FMRC}} = P_{\mathrm{F}} \tag{9}$$

Probability of miss detection

$$P_{\mathrm{MDMRC}} = 1 - \Gamma_{\mathrm{u}}\left(\frac{N\lambda}{1 + \mathrm{MSNR}}, N\right) \tag{10}$$

where M is the number of receiving antennas.

Probability of detection

$$P_{\mathrm{DMRC}} = 1 - P_{\mathrm{MDMRC}} \tag{11}$$

The performance of energy detector using MRC technique under no fading case can be observed from Figure 2.

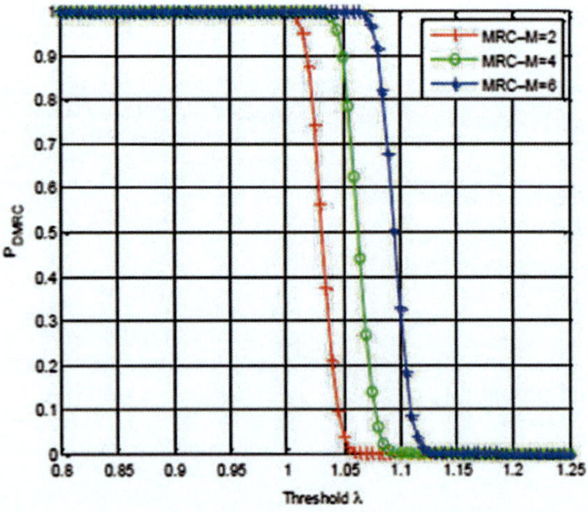

Fig. 2 P_{DMRC} versus detection threshold λ, number of samples $N = 10{,}000$, SNR $= -18$ dB

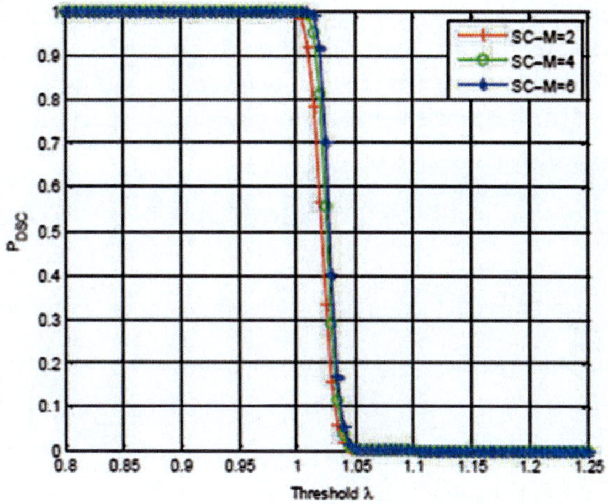

Fig. 3 P_{DSC} versus detection threshold λ, number of samples $N = 10{,}000$, SNR $= -18$ dB

(ii) Selection Combining
In selection combining technique, the received signal which has highest SNR can be selected and fed to the combiner [2]

The probability of false alarm

$$P_{\text{FSC}} = \prod_{i=1}^{M} \Gamma_{\text{u}}(N\lambda, N) \tag{12}$$

Probability of miss detection

$$P_{\text{MDSC}} = \prod_{i=1}^{M} \left[1 - \Gamma_{\text{u}}\left(\frac{N\lambda}{1 + \text{SNR}_i}, N \right) \right] \tag{13}$$

Probability of detection

$$P_{\text{DSC}} = 1 - \prod_{i=1}^{M} \left[1 - \Gamma_{\text{u}}\left(\frac{N\lambda}{1 + \text{SNR}_i}, N \right) \right] \tag{14}$$

The performance of energy detector using SC technique under no fading case can be observed from Figure 3.

5 Cooperative Spectrum Sensing

In cooperative spectrum sensing, every cognitive radio person feel the spectrum and is derived to a decision whether or no longer number one person is present or absent then ahead their selections to the common place receiver [2]. At the not unusual receiver, distinct types of vote casting guidelines are used: AND rule, OR rule, *p* out of *L* rule where *L* is the total number of cognitive radio users in a network.

The false alarm probability cooperative spectrum sensing is given by [2],

$$T_{\mathrm{f}} = \mathrm{Prob}\{H_1|H_0\} = \sum_{l=P}^{L} \binom{L}{l} (P_{\mathrm{F}})^l (1 - P_{\mathrm{F}})^l \tag{15}$$

and miss detection probability of cooperative spectrum sensing is given by

$$T_{\mathrm{m}} = \mathrm{Prob}\{H_0|H_1\} = 1 - \sum_{l=P}^{L} \binom{L}{l} (P_{\mathrm{F}})^l (1 - P_{\mathrm{F}})^l \tag{16}$$

Total error $= T_{\mathrm{f}} + T_{\mathrm{m}}$.

To minimize total error, the optimum value of *P* is given by [5]

$$P_{\mathrm{opt}} = \min\left(L, \left\lceil \frac{L}{1+\alpha} \right\rceil\right) \tag{17}$$

where $\alpha = \frac{\ln \frac{P_F}{1-P_M}}{\ln \frac{P_M}{1-P_F}}$ and $\lceil . \rceil$ denotes ceiling function.

6 Simulation Results and Discussion

Figure 2 shows that probability of detection versus detection threshold in MRC combining, when number of receiving branches (Antennas) $M = 2$, $M = 4$, and $M = 6$ at SNR $= -18$ dB, number of samples = 10,000. From Figure 2, it is derived that the probability of detection is improving when number of branches is increasing from $M = 2$, $M = 4$ to $M = 6$ that means large wide gap is existing among curves.

Figure 3 shows that probability of detection versus detection threshold in SC combining, when number of receiving branches (Antennas) increasing from $M = 2$, $M = 4$, and $M = 6$ at SNR $= -18$ dB, number of samples = 10,000. From Figure 3, we can read the values of probability of detection is improving when number of branches is increasing from $M = 2$, $M = 4$ to $M = 6$. But, compared to MRC technique in SC technique, the gap among curves is narrow that means the improvement of probability of detection in SC technique is somewhat less compared to MRC technique.

Figure 4 shows the total error (overall cooperative spectrum sensing Error) versus detection threshold in MRC-ED, SC-ED, ED techniques at an SNR $= -18$ dB when number of receiving branches $M = 2$, number of samples $N = 10{,}000$, and total number of cognitive users in a cognitive radio network $L = 40$ users are considered. From Figure 4, it is observed that the minimum total error (overall cooperative

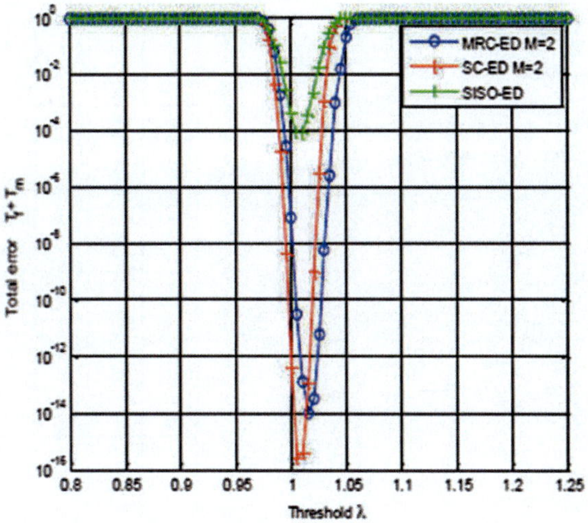

Fig. 4 Total error versus threshold λ, number of samples $N = 10{,}000$, SNR = −18 dB, number of cognitive radio users $L = 40$, number of receiving branches $M = 2$

spectrum sensing error) in a cognitive radio network is given by the SC technique which is $2.437e^{-16}$.

Figure 5 shows the total error (overall cooperative spectrum sensing Error) versus detection threshold in MRC-ED, SC-ED, ED techniques at an SNR = −18 dB when number of receiving branches $M = 4$, number of samples $N = 10{,}000$, and total number of cognitive users in a cognitive radio network $L = 40$ users are considered. From Figure 5, it is observed that the minimum total error (overall cooperative spectrum sensing error) in a cognitive radio network is given by the MRC technique which is $1.551e^{-50}$. Here, the error in MRC is improved due to increasing of receiving

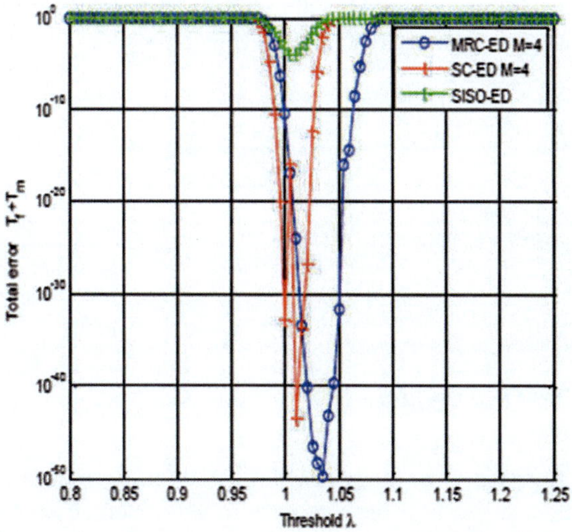

Fig. 5 Total error versus threshold λ number of samples $N = 10{,}000$, SNR = −18 dB, number of cognitive radio users $L = 40$, number of receiving branches $M = 4$

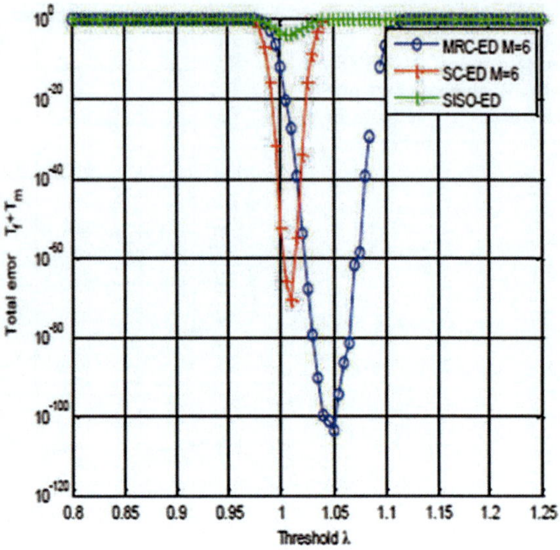

Fig. 6 Total error versus threshold λ number of samples $N = 10{,}000$, SNR = −18 dB, number of cognitive radio users $L = 40$, number of receiving branches $M = 6$

branches from $M = 2$ to $M = 4$ which makes increase overall SNR in MRC at receiver side.

Figure 6 shows the total error (overall cooperative spectrum sensing error) versus detection threshold in MRC-ED, SC-ED, ED techniques at an SNR = −18 dB when number of receiving branches $M = 6$, number of samples $N = 10{,}000$, and total number of cognitive users in a cognitive radio network $L = 40$ users are considered. From Figure 6, it is observed that the minimum total error (overall cooperative spectrum sensing error) in a cognitive radio network is given by the MRC technique which is $2.558e^{-104}$. Here also, the error in MRC is improved compared to Figure 5; this is due to increasing of receiving branches from $M = 4$ to $M = 6$, which makes increase overall SNR in MRC at receiver side.

Minimum – total error Min $(T_f + T_m)$	Number of receiving branches $M = 2$	Number of receiving branches $M = 4$	Number of receiving branches $M = 6$
ED_{min}	$8.427e^{-05}$	$8.427e^{-05}$	$8.427e^{-05}$
$(SC - ED)_{min}$	$2.437e^{-16}$	$3.735e^{-44}$	$2.473e^{-71}$
$(MRC - ED)_{min}$	$1.022e^{-14}$	$1.551e^{-50}$	$2.588e^{-104}$

7 Conclusion

To enhance the usage of radio spectrum, without disturbing to primary user, cooperative spectrum sensing may be used. Here, each cognitive radio user senses the spectrum hole using energy detector (ED) by employing MRC and SC techniques. In

MRC technique, the output SNR from all the receiving branches (Antennas) linearly combined. Thus, the total spectrum sensing error in the cognitive radio network was minimized by using MRC technique at each CR by providing more number of receiving branches.

References

1. Urkowitz H (1967) Energy detection of unknown deterministic signals. IEEE Proc 55(4):523–531
2. Letaief KB (2007) Cooperative spectrum sensing. Cogn Wirel Commun Netw
3. Sun H, Laurenson D, Wang C-X (2010) Computationally tractable model of energy detection performance over slow fading channels. IEEE Commun Letters 14(10):924–926
4. Chen Q, Gao F, Nallanathan A, Xin Y (2008) Improved cooperative spectrum sensing in cognitive radio. In: Proceedings of IEEE VTC 2008 Spring, 2008, pp 1418 –1422
5. Chen Q, Motani M, Wong W-C, Nallanathan A (2011) Cooperative spectrum sensing strategies for cognitive radio mesh networks. IEEE J-STSP 5(1):56–67
6. Wang C-X, Chen H-H, Hong X, Guizani M (2008) Cognitive radio network management: tuning in to real-time conditions. IEEE Veh Technol Mag 3(1):28–35
7. Vamshi SR (2020) Channel estimation techniques for OFDM and GFDM: A review. Test Eng Manage 83:17143–17149. Publication Issue: March–April 2020
8. Digham FF, Alouini M-S, Simon MK (2003) On the energy detection of unkown signals over fading channels. In: Proceedings of IEEE ICC, pp 3575–3579

Classification of Non-fluctuating Radar Target Using ReliefF Feature Selection Algorithm

Jagan Mohana Rao Pathina and P. Rajesh Kumar

Abstract Target classification is one of the essential functions in radar systems. The key characteristic used for the target classification is the radar cross section (RCS), obtained by processing the scattered signals from the target. The dedicated machine learning models with common geometrical structures like sphere, circular cylinder, frustum and circular disc are realized to identify the simple and complex targets. A large collection of feature set is obtained by using the MODWPT feature extraction. To remove the extraneous and unnecessary features from the feature set, ReliefF optimal feature selection is proposed. The feature subset thus obtained is given to different classifiers, namely SVM and KNN, and their performance is observed.

Keywords Target classification · ReliefF feature selection · MODWPT · SVM and KNN

1 Introduction

The target classification plays a prominent role in the modern radar systems, and it has a great significance in military and civil aspects [1]. The main objective of the target classification is to distinguish targets with the help of features extracted from the target and by comparing the similarities in the features of the test target with the trained or known targets.

It is well-known fact that data plays a very crucial role in artificial intelligence and machine learning problems. Obtaining real-time data for conducting the experimental study in this research domain is cumbersome. The dedicated machine learning models are developed with common geometrical structures to identify the simple and complex targets. Combination of simple geometric structures is used to model the complex targets. The target classification is done by extracting the features of the target using feature extraction method and classifying using the classification methods [2].

J. M. R. Pathina (✉) · P. R. Kumar
Andhra University College of Engineering, Visakhapatnam, India

P. S. R. Chowdary et al. (eds.), *Evolution in Signal Processing and Telecommunication Networks*, Lecture Notes in Electrical Engineering 839,
https://doi.org/10.1007/978-981-16-8554-5_18

The performance can be improved with the help of feature selection. To avoid the effects of false alarm of signal components due to noise, a multiscale analysis technique, namely maximal overlap discrete wavelet packet transform (MODWPT), is used for feature extraction. MODWPT collects abundant amount of features. It is very essential to have important features of the targets for their classification. Extraneous and unnecessary data will tend to decrease the performance of the classification. To select an optimal feature subset with key features of the targets from the feature set, the feature selection called ReliefF algorithm is proposed.

2 Data Set

Classification of target requires the measurement of aspect angle along with the target properties (like geometry, material type and azimuth and elevation angles) which includes RCS. RCS is the measure of ability of target to backscatter radar's electromagnetic signals in the direction of radar receiver.

$$\sigma = \frac{\text{Power radiated towards source per unit solid angle}}{\text{incident power density}/4\pi} \text{m}^2 \tag{1}$$

The RCS of simple targets depends on physical size and wavelength of signal, and for complex targets, it depends on aspect angle, frequency and polarization [3]. To test the effectiveness of the proposed method, four canonical targets of geometry with circular cylinder, sphere, frustum and circular disc are analysed (Table 1).

Where

r represents the sphere radius,

$k = \frac{2\pi}{\lambda}$, λ represents wavelength.

Y_n represents the spherical Bessel of the second kind of order n,

J_n represents the spherical Bessel of the first kind of order n and

$H_n^{(1)}$ represents the Hankel function of order n, $H_n^{(1)} = J_n(kr) + jY_n(kr)$.

RCS mathematical models for circular cylinder, frustum and disc are given in Table 2 and their respective shapes in Fig. 1.

Here, the cone angle is, $\tan\alpha = \frac{r_2 - r_1}{H}$.

Table 1 RCS mathematical models for sphere at different wavelength

Region	RCS of sphere
Rayleigh	$\sigma = 9\pi r^2 (kr)^4 r \ll \lambda$
Optical	$\sigma = \pi r^2 r \gg \lambda$
Mie	$\frac{\sigma}{\pi r^2} = \left(\frac{j}{kr}\right) \sum_{n=1}^{\infty} (-1)^n (2n+1) \left[\left(\frac{kr J_{n-1}(kr) - n J_n(kr)}{kr H_{n-1}^{(1)}(kr) - n H_n^{(1)}(kr)} \right) - \left(\frac{J_n(kr)}{H_n^{(1)}(kr)} \right) \right]$

Table 2 RCS for circular cylinder, disc and frustum

Shape	RCS for normal backscattered incidence	RCS for non-normal backscattered incidence
Circular cylinder	$\sigma\theta_n = \frac{2\pi H^2 r}{\lambda}$	$\sigma = \frac{\lambda r \sin\theta}{8\pi(\cos\theta)^2}$
Frustum	$\sigma\theta_n = \frac{8\pi\left(z_2^{\frac{3}{2}} - z_1^{\frac{3}{2}}\right)^2}{9\lambda} \frac{\sin\alpha}{(\cos\alpha)^4}$	$\sigma = \frac{\lambda z \tan\alpha}{8\pi \sin\theta}(\tan(\theta - \alpha))^2$
Disc	$\sigma\theta_n = \frac{4\pi^3 r^4}{\lambda^2}(\theta = 0\circ)$	$\sigma = \frac{\lambda r}{8\pi \sin\theta (\tan\theta)^2}$

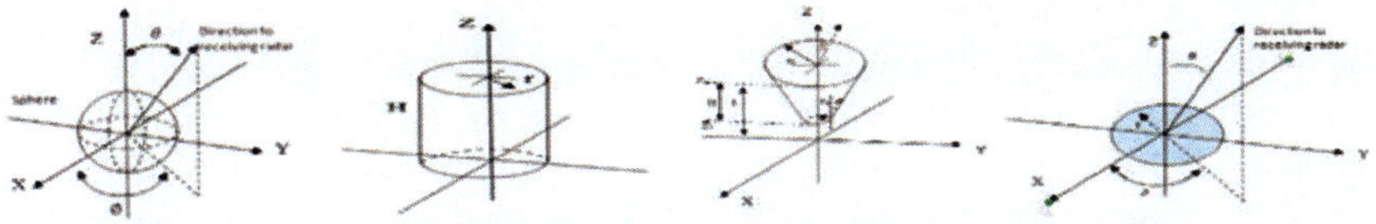

(a) (b) (c) (c)

Fig. 1 **a** Sphere, **b** Circular cylinder, **c** Frustum and **d** Circular flat plate (disc)

RCS of complex targets is computed by any combination of the simple targets coherently. Since targets like ships, aircrafts are complex in shape, the analytical methods like geometry optics and physical optics become difficult to apply it. A lot of electromagnetic approaches have been developed. These require a lot of computation and are a bit cumbersome for real-time targets [4]. So, we used numerical computation methods like method of moments (MoM) or finite element analysis (FEM).

The complex target models are developed using these simple scatters like sphere, cylinder, frustum and circular disc. These scatters are placed at the four vertices of a square in the xy-plane. Using these, the complex targets are developed.

The complex targets RCS pattern is the coherence which is the combination of simple targets and is represented by

$$\sigma = \left| \sum_p \sqrt{\sigma_p e^{i\phi_p}} \right|^2 \quad (2)$$

where ϕ_p represents the relative phase and σ_p represents RCS of pth scatter, respectively (Fig. 2).

3 Feature Extraction

Wavelet transforms are used to perform the multiscale analysis of a signal. However, it is difficult to achieve temporal resolution and frequency resolution through this at

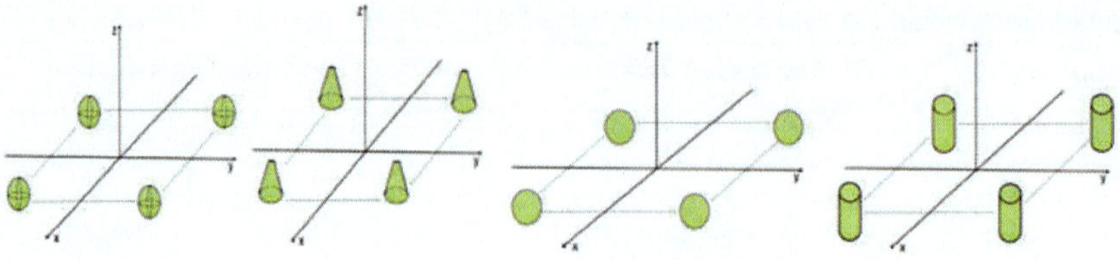

(a) (b) (c) (d)

Fig. 2 Formation of complex targets using four simple scatters. **a** Complex sphere, **b** Complex frustum, **c** Complex disc and **d** Complex cylinder

a same interval of time. MODWPT is a non-orthogonal transform and used in finite frequency partition of signal. Unlike DWT, it does not require down-sampling and no limitation on the signal length N [5]. MODWPT also decomposes the high-frequency part of the signal preserving the energy of the original signal. It uses quadrature mirror filters to partition the signal into equal sub-bands resulting in wavelet and scaling coefficients by the high pass and low pass filters, respectively. The signal is partitioned into 2^j sub-bands at the jth level, and frequency range of the sub-band in the jth level at the nth node is $\left[\frac{f_s n}{2^{j+1}}, \frac{f_s(n+1)}{2^{j+1}}\right]$. The properties of these high pass filter $h(n)$ and low pass filter $g(n)$ are given in a mathematical form as:

$$\sum_{n=-\infty}^{\infty} g(n) = \sqrt{2}, \sum_{n=-\infty}^{\infty} g^2(n) = 1, \sum_{n=-\infty}^{\infty} g(n)h(n) = 0 \tag{3}$$

$$\sum_{n=-\infty}^{\infty} h(n) = \sqrt{2}, \sum_{n=-\infty}^{\infty} h^2(n) = 1, \sum_{n=-\infty}^{\infty} h(n)g(n) = 0 \tag{4}$$

The decomposition of signal using MODWPT is given by

$$s_j^{2z}(k) = \frac{1}{\sqrt{2}} \sum_{n=-\infty}^{\infty} g(n) s_{j-1}^{z}(k-n) \tag{5}$$

$$s_j^{2z+1}(k) = \frac{1}{\sqrt{2}} \sum_{n=-\infty}^{\infty} h(n) s_{j-1}^{z}(k-n) \tag{6}$$

The reconstructed coefficients of MODWPT at any jth level are given by (Fig. 3)

$$a_j^{2z}(k) = \frac{1}{\sqrt{2}} \sum_{n=-\infty}^{\infty} \tilde{g}(n) s_{j-1}^{2z}(k-n) \tag{7}$$

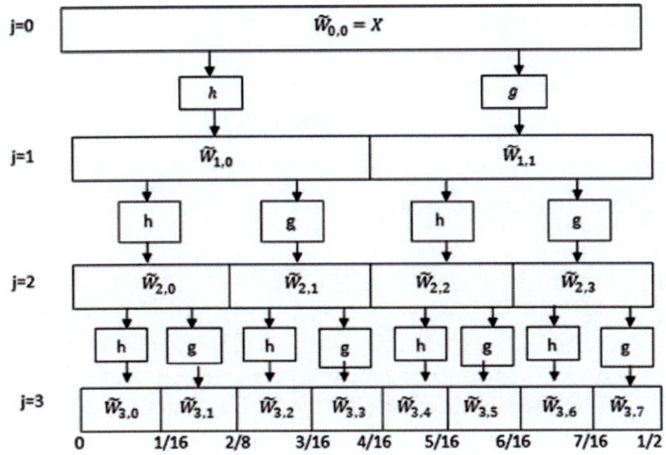

Fig. 3 Decomposition of signal into MODWPT coefficients $\widetilde{W}_{j,n}$ with levels $j = 1, 2, 3$ and frequency index n ranging from 0 to $2^j - 1$

$$a_j^{2z+1}(k) = \frac{1}{\sqrt{2}} \sum_{n=-\infty}^{\infty} \tilde{h}(n) s_{j-1}^{2z+1}(k-n) \tag{8}$$

4 Feature Selection

The main function of the feature selection is to select a feature subset with the main key features of the target, remove the irrelevant, redundant data and help in increasing the accuracy of the classification. ReliefF feature selection is proposed to select an optimal feature subset and improve the accuracy of the target classification. ReliefF algorithm is used to solve the multiclassification and can also deal with missing and noisy data [6]. ReliefF algorithm ranks the importance of predictors. It sets weight according to the recognition of the feature of the sample with the adjacent feature samples and rearranges the data in the order of weights.

ReliefF algorithm initially sets all the weights of the predictors to 0 and then a sample R is selected from the training sample, and k neighbour samples are found in the same and different classes. Randomly, multipoints are selected to get the feature weights and are compared within class distance and between class distances from neighbour samples [7]. This procedure is repeated for all feature dimensions to obtain the high accurate weight of each feature.

Weights of the predictors are given by.
If $\boldsymbol{x}_p, \boldsymbol{x}_q$ belong to same class:

$$\boldsymbol{W}_j^i = \boldsymbol{W}_j^{i-1} - \frac{\Delta_j(\boldsymbol{x}_p, \boldsymbol{x}_q)}{\boldsymbol{m}}.\boldsymbol{d}_{rq} \tag{9}$$

If $\boldsymbol{x}_p, \boldsymbol{x}_q$ belong to different classes:

$$\boldsymbol{W}_j^i = \boldsymbol{W}_j^{i-1} - \frac{\boldsymbol{P}_q}{1 - \boldsymbol{p}_p}\frac{\Delta_j(\boldsymbol{x}_p, \boldsymbol{x}_q)}{\boldsymbol{m}}.\boldsymbol{d}_{rq} \tag{10}$$

$\boldsymbol{W}_j^i$ is the weight of the predictor F_j at the ith iteration.
$\boldsymbol{P}_q$ is the probability of the class to which $\boldsymbol{x}_q$ belongs, and $\boldsymbol{p}_p$ probability of the class to which $\boldsymbol{x}_p$ belongs.
$\Delta_j(\boldsymbol{x}_p, \boldsymbol{x}_q)$ is the difference between the two samples $\boldsymbol{x}_p, \boldsymbol{x}_q$.

For discrete samples:

$$\Delta_j(\boldsymbol{x}_p, \boldsymbol{x}_q) = \begin{cases} 0 & \textbf{if } \boldsymbol{x}_{pj} = \boldsymbol{x}_{qj} \\ 1 & \textbf{if } \boldsymbol{x}_{pj} \neq \boldsymbol{x}_{qj} \end{cases} \tag{11}$$

$\boldsymbol{m}$ is the number of iterations given by the updates.

The greater the feature weight, the higher is the rank, and it contributes more to the classification of the target. The features at the bottom of the ranking are reduced, and a reduced feature subset is obtained with important features of the targets.

5 Classifiers

The reduced feature set is given to the classifiers for classification of targets. The classifiers are trained using the 70% of the data set, and rest of the 30% is used for testing. The classifiers performance is analysed.

5.1 Support Vector Machine

The main intention is to attain a decision boundary which can separate the training data. If the separation is not possible with a linear hyperplane, the classifier maps the data into high-dimensional feature space using some pre-defined functions (kernel) [8]. These points will define the separating line better by calculating margins.

For a training set of (x_i, c_i), with x features, c class labels and i = 1, 2, …, K, SVMs are constructed from the following mathematical optimization procedure [9]:

$$\text{minimize}\left[\frac{1}{2}\left(w^T w\right) + C\sum_{j=1}^{k} \epsilon_j\right] \text{subject to } C_j\left(w^T \phi(x_i) + b\right) \geq 1 - \epsilon_j, \epsilon_j \geq 0.$$

where

ω is the decision plane orientation vector.
b is bias.
ε_i is marginal slack variable.
ϕ is mapping function.
C is penalty parameter of error term.

The expected classification error for unseen test samples is minimized by finding an optimum linear hyperplane by SVM [10].

KNN.

Finds the k training data (nearest neighbours) that is closest to the test example and assigns most occurring class from these samples to the test example [11]. The nearest neighbours are found by using the Euclidean distance criterion:

The Euclidean distance between any pair $x_1 = \left(x_{1,1}, x_{1,2}, \ldots, x_{1,k}\right)$ and $x_2 = \left(x_{2,1}, x_{2,2}, \ldots, x_{2,k}\right)$ is given by

$$d(x_1, x_2) = \sqrt{\sum_{j=1}^{k}\left(x_{1,j} - x_{2,j}\right)^2} \tag{12}$$

6 Experimental Results

Procedure for experimental study is as follows:

1. Synthesized data generation
2. Feature extraction
3. Classification methods.

6.1 Data Generation

In machine leaning, the data plays a vital role. Obtaining the real-time data from the targets is cumbersome. So, machine learning models are developed with the help of synthesized *I/Q* samples and are used to recognize the simple targets. The combination of these simple geometric structures can be used for the complex target recognition.

Synthesized data is the data being created in the software domain using various varieties of mathematical models which replicate the conditions of practical circumstances. Firstly, the RADAR operating parameters like the set of elevation and azimuthal scan angles, operating frequency are chosen. Secondly, a desired simple geometrical shape is considered, and its geometrical parameters (radius, height, etc.,) are fixed. The theoretical mathematical equations corresponding to the RCS patterns of the chosen geometrical shape are to compute the theoretical RCS pattern by passing these parameters. The obtained RCS pattern is then applied to a backscatter RADAR target model to simulate a finite number of echo returns from different aspect angles, which changes from sample to sample.

By varying the dimensions, for each simple geometrical shape, ten different structures are considered, and their corresponding RCS patterns are synthesized. A total of 40 (4 × 10) RCS patterns are obtained for 4 shapes. For each geometrical target, 250 motion profiles are simulated and thereby for each geometrical shape, 2500 (250 × 10) motion profiles are simulated. In total, 20 scans are performed, and the corresponding echoes are collected. In each scan, 50 different aspect angles are considered. It results in 50 echoes per scan and 1000 (20 × 50) per target. Therefore, 1 motion profile of a target contains 1000 echo returns. A total of 10,000 motion profiles are generated for 40 targets of 4 geometrical structures. In a similar way, the motion profiles for complex target models are also obtained.

6.2 Feature Extraction

The MODWPT feature extraction will partition the signal into 2^j sub-bands at the jth level and with equal energy levels. To lessen the feature set and improve the performance of the classifier, a ReliefF optimization is used for feature selection. It finds the weight of the samples in the feature set and assigns a rank to them. The samples with more weight are selected and are used for classification. This helps to improve the accuracy of classification for multiclasses and also in presence of noise. The obtained feature set is partitioned into 70% training data and 30% as the testing data. The classifiers are trained with the help of this training data and are able to discriminate the unknown (test) targets using the known (trained) targets. The machine learning classification models, namely SVM and KNN, are developed. By tuning the various model parameters, the performance of the classifier is analysed.

Results are obtained in the scatter plot which shows the accuracy of the classification and errors for different feature values, and also, the estimation of the feature weights of the predictors is plotted.

From the obtained confusion matrix during analysis, it is observed that SVM with ReliefF gives an accuracy of 85.60% and KNN with ReliefF gives an accuracy of 93.6% (Figs. 4 and 5).

From the obtained confusion matrix during analysis, it is observed that SVM with ReliefF algorithm for complex geometry gives an accuracy of 68.9% and KNN gives an accuracy of 92.4%.

The overall performance metrics are given as follows.

The overall performance metrics for simple targets using KNN and ReliefF algorithm are shown in Table 3.

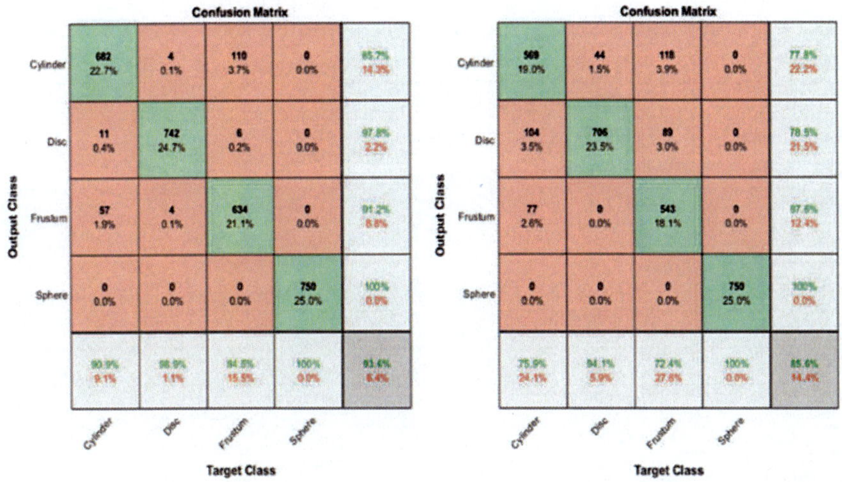

(a) (b)

Fig. 4 **a** Confusion matrix of simple geometry of KNN + ReliefF feature selection. **b** Confusion matrix of simple geometry of SVM + ReliefF feature selection

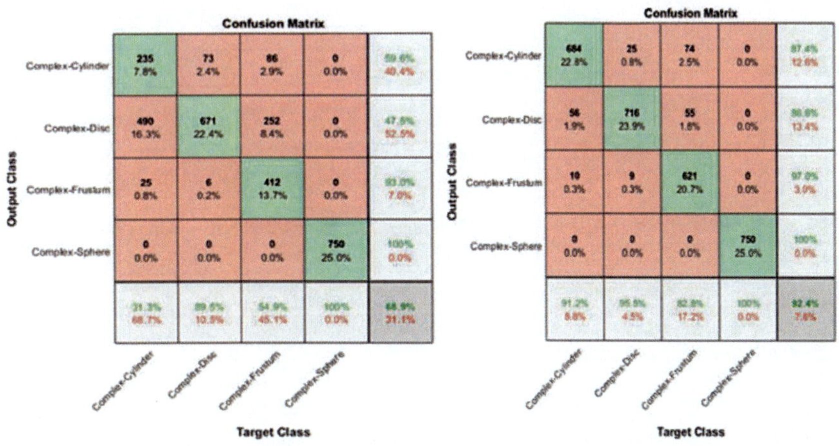

(a) (b)

Fig. 5 **a** Confusion matrix of complex geometry of KNN + ReliefF feature selection. **b** Confusion matrix of complex geometry of SVM + ReliefF feature selection

The overall performance metrics for simple targets using SVM and ReliefF algorithm are shown in Table 4. All the metrics are high for simple sphere target.

The overall performance metrics for complex targets using KNN and ReliefF algorithm are shown in Table 5. All the metrics are high for the complex sphere target.

Table 3 Overall performance of the classifiers KNN with ReliefF optimization for simple targets

Metrics	Cylinder	Disc	Frustum	Sphere
Precision	0.8568	0.9776	0.9122	1
Specificity	0.9491	0.9918	0.9727	1
Sensitivity	0.9093	0.9893	0.8453	1
$F1$-score	0.8823	0.9834	0.8775	1

Table 4 Overall performance of the classifiers SVM with ReliefF optimization for simple targets

Metrics	Cylinder	Disc	Frustum	Sphere
Precision	0.7784	0.7853	0.8758	1
Specificity	0.9250	0.9061	0.9634	1
Sensitivity	0.7587	0.9413	0.7240	1
$F1$-score	0.7684	0.8563	0.7924	1

Table 5 Overall performance of the classifiers KNN with ReliefF optimization for complex targets

Metrics	Complex cylinder	Complex disc	Complex frustum	Complex sphere
Precision	0.8736	0.8658	0.9703	1
Specificity	0.9547	0.9488	0.9921	1
Sensitivity	0.9200	0.9547	0.8280	1
$F1$-score	0.8924	0.9081	0.8935	1

Table 6 Overall performance of the classifiers SVM with ReliefF optimization for complex targets

Metrics	Complex cylinder	Complex disc	Complex frustum	Complex sphere
Precision	0.5964	0.4749	0.9300	1
Specificity	0.9202	0.6531	0.9816	1
Sensitivity	0.3133	0.8947	0.5493	1
$F1$-score	0.4108	0.6202	0.6907	1

The overall performance metrics for complex targets using SVM and ReliefF algorithm are shown in Table 6. All the metrics are high for complex sphere target.

7 Conclusions

The simple and complex targets models are developed using a dedicated machine learning technique. For the developed target model, returns are generated which depend on the RCS of the target. MODWPT feature extraction is applied to the obtained raw data set. To further reduce the computation time and increase the accuracy of the classification, Relieff feature selection is used to select a feature subset with greater weights. This reduced feature set is given to the different classifiers, namely SVM and KNN, and their performance is observed through the scatter plots. It is observed that KNN with ReliefF classifier resulted in the best accuracy of 93.6% for simple geometry and 92.4% for complex geometry compared to SVM.

References

1. Skolnik MI (2001) Introduction to radar systems, 3rd edn. McGraw-Hill
2. Khodjet-Kesba M (2014) Automatic target classification based on radar backscattered ultra wide-band signals. Université Blaise Pascal - Clermont-Ferrand II
3. Knott EF (1988) Radar cross section. In: Brooker E (ed) Aspects of modern radar. Artech House, Norwood. MA
4. Mahafza BR. Radar systems analysis and design using Matlab, 2nd edn.
5. Comparison between discrete wavelet transform and maximal overlap discrete wavelet transform as an analysis tool for H.264/AVC video. In: Ali HHSM, Sharif SM (eds) ICCCEEE, 2018
6. Kononenko I, Simec E, Robnik-Sikonja M (1997) Overcoming the myopia of inductive learning algorithms with RELIEFF. Retrieved from CiteSeerX
7. Wang Z, Zhang Y, Chen Z (2016) Application of ReliefF algorithm to selecting feature sets for classification of high resolution remote sensing image. In: IEEE international geoscience and remote sensing symposium (IGARSS)
8. Ying F, Xing W (2017) Radar signal recognition based on modified semi-supervised SVM algorithm. IEEE IAEAC, 2017
9. Yang W, Jiaguo L, Changyao Z (2007) Algorithm of target classification based on target decomposition and support vector machine. IEEE
10. Bhateja V, Gautam A, Tiwari A, Satapathy SC, Nhu NG, Le DN (2018) Haralick features-based classification of mammograms using SVM. In: Information systems design and intelligent applications. Springer, Singapore, pp 787–795
11. Comparison: KNN & SVM algorithm (2017). Int J Res Appl Sci Eng Technol (IJRASET) 5

A Deep Comprehensive 3D Modelling for the Prediction of Coronavirus on Medical Scans of COVID-19 Patients

Avutu Sai Meghana, K. Thirupathi Rao, and N. Sarada

Abstract The global pandemic of coronavirus disease 19 (COVID-19) left many lives in question with the virus, and many people are being suffered every day. In this hectic situation, each patient needs to get attended by the doctor for the scans and determine if they are positive with the virus to move on to the diagnosis stage. It is hard to attend each patient when many people are being tested and effected. It is time where machine learning and artificial intelligence are used. A lot of efforts were made by the researchers to diagnose the effected patients; using 2D images in present, we present an observation that helps detect COVID-19 with the use of 3D images. In this paper, we have worked with 3D image data and have found one of the best models among Inception v3, ResNet-50, VGG16, EfficientNet B0, MobileNet v2.

Keywords COVID-19 · EfficientNet B0 · MobileNet v2 · VGG16 · Inception v3 · ResNet-50

1 Introduction

In today's world, people are best recognised by health than their passion and lifestyle. Nowadays, everyone is doing different tasks looking for good health than temporary pleasures in life. Even though we try hard to get saved, many diseases do not knock our door and get in. Many dreadful diseases are taking human lives without any notice to the person like cancer. Today, our world is fighting against one such disease: an unwanted guest in our life, the corona. Ever since the first case and spread of the disease in China has been announced, it did not take long for the virus to spread over the European, African, Asian countries leading its way worldwide. The reports

A. S. Meghana (✉) · K. Thirupathi Rao · N. Sarada
Koneru Lakshmaiah Education Foundation, Vaddeswaram, Andhra Pradesh, India
e-mail: 195036006@kluniversity.in

K. Thirupathi Rao
e-mail: kthirupathirao@kluniversity.in

P. S. R. Chowdary et al. (eds.), *Evolution in Signal Processing and Telecommunication Networks*, Lecture Notes in Electrical Engineering 839,
https://doi.org/10.1007/978-981-16-8554-5_19

found that 4-5 days of time is required for the virus to show any symptoms and extend up to 15 days. The virus's common symptoms are fever, dry cough, tiredness and some least common and serious symptoms like breathing issues, chest pain and loss of speech. It has been a hard task for the doctors to test all the patients approaching them with mild and serious symptoms. To provide better understanding of the affected patients, doctors can use AI and machine learning algorithms to detect virus existence while comparing the images or information specific to a COVID-19 patient's health. Since it mainly affects the breathing system by infecting the lungs, the lung scans are beneficial for detection. So, a vital rage of research is being done using the scan images. We investigated different approaches using deep learning algorithms, machine learning, AI and many more by using the available COVID-19 data.

2 Survey

As we already knew that to apply a deep learning model, we need a large amount of data that may or might not be labelled. Since the disease is new, many researchers needed large data to identify the virus. Waheed et al. [1] use an auxiliary classifier GAN (AC-GAN) for the data augmentation and high-resolution image generation. Wang et al. [2] have implemented a COVID-19 pneumonia lesion segmentation network called COUPLE-Net to identify the damages done to any organs, and it is mostly to find the existence of the virus by the damage done in the chest scans. Patch-based convolution neural network approach is used to solve the systematic collection of the chest X-ray data [3] that uses a relatively smaller number of parameters for detection. Hu et al. [4] have proposed a deep learning model with usage of 3D volumetric chest CT of COVID-19 which is being used. To analyse CT images in multiple views [5], they have proposed a diagnosis pipeline to classify COVID-19 and community-acquired pneumonia (CAP).

Another AI model is implemented by Zheng et al. [6] for predicting COVID-19 which is based on a retrospective approach where LSTM network is used to estimate the infection rate and extract the features from news through NLP. Wang et al. [7] have proposed an experimental study on using Inception v3 transfer learning model by considering pulmonary images to extract the features, and different classifiers are used to classify the pulmonary images. Hartanto and Wibowo [8] have proposed an Android-based application to detect the skin cancer in its early stages while utilising a smartphone. Carrer et al. [9] proposed an unsupervised automatic method, which helps inspect and localise the pleural line present in the image data using HMM and Viterbi models. Wang et al. [10] research used 3D chest CT images to detect infected with COVID-19 or some other viruses, and they have proposed a model that has two 3D ResNets. Roy et al. [11] have proposed a deep network which has been derived from the spatial transformer network, which helps detect the disease level. Fan et al. [12] have proposed a lung infection segmentation deep network (Inf-Net), which can identify the CT's infected regions. Shi et al. [13] have proposed detecting the COVID-19 virus using artificial intelligence mechanism on the medical imaging. The whole research has a series of AI models which are using both X-ray and CT scan images. Machine learning is not restricted to medical field. It is widely used in diverse domains, like football game [14] and drowsiness detection [15].

3 Methodology

After knowing the researchers' different models, we identified various perspectives to detect the virus's existence. Having that knowledge as a base, we have made a comparison among different algorithms on both 2D and 3D data. This paper has restricted to comparison with a few most prominent models like EfficientNet, VGG16, Inception v3 and ResNet-50 on 2D images dataset, along with other algorithms like MobileNet v2, EfficientNet b0 and Inception v3 on 3D medical images dataset.

3.1 Datasets

3.1.1 2D Images

The COVID-19 CT scan 2D dataset contains the images of COVID-29 lung scans and healthy non-COVID-19 patient images. The dataset is split to training, testing and validation. The dataset contains about 3500 images of both COVID-19 and non-COVID-19 patients. The pre-processing in the 2D data includes the arrangement of the available images to spit into train, test and validation. The ratio was 7:2:1 which was arranged to train the models. The images in the dataset look, as mentioned in Figs. 1 and 2.

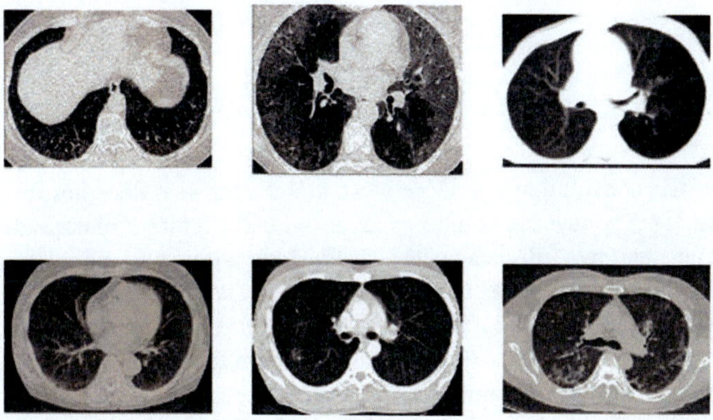

Fig. 1 2D dataset images

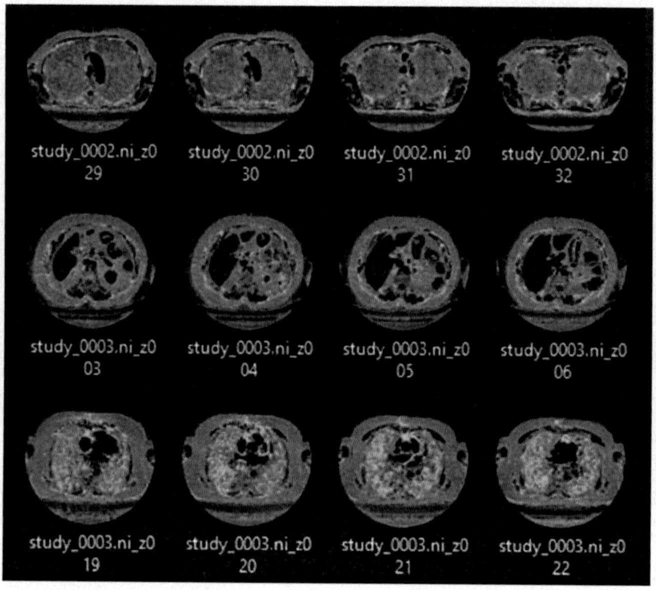

Fig. 2 3D nii images

3.1.2 3D Images

When it comes to the dataset containing 3-dimensional medical images dataset, it contains medical nii images segregated in the form of folders. Each folder contains about 100 medical images which sums the dataset with 500. The folders with images are segregated in such a way that each folder has a specific level of COVID-19 positive images in it. The 500 nii images are first categorised to train, test and validation after converting them to 2D which give us around 8193 images that can be trained to get much better outputs. The images in the dataset look, as mentioned below.

3.2 VGG16

It is one of the convolution neural network architectures. It does not focus on the parameters, but they have convolutional layers with 3 × 3 filters, max-pooling layers of 2 × 2 filters and two fully connected layers along with softmax for output. There are 16 layers for this model as the name specifies. There is some padding with all the convolution layers in the model. The activation functions are used to change the values. For example the usage of ReLU activation function restricts the negative values. This model takes an input of 224 × 224 size images. Each model uses the image generator that labels the data easily. It also had different features to mould the input to the desired model wanted. The features it has are rescaling, zoom, flip, etc.

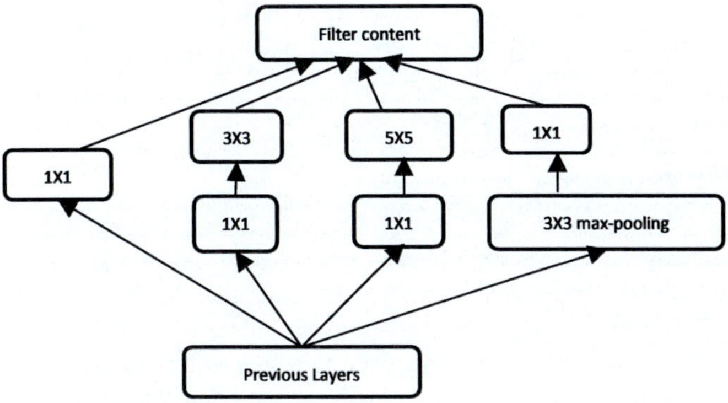

Fig. 3 Architecture of Inception v3

3.3 Inception v3

The images are not always clear and have the correct positions. Let us consider an example of different pictures of animals, humans or even plants. The images vary in sizes, and even though the images are of the same size, the object might not be in the same position. Some images objects are small and huge background and vice versa in other images. In such cases, the filters needed are different. It is hard for a neural network with a particular filter to find the right object using a single filter. So, because the right kernel choice is hard, the solution for this came up with multiple filters or kernels. The basic model has around three different kernels (Fig. 3).

Since the deep learning networks execution is much more expensive, the authors of this model have added a 1×1 filter at the start of the other 3×3 and 5×5 convolution layers. Later, they improved this model by cutting the 5×5 filters with the two 3×3 filters to increase the computational speed, making this the Inception v3 model. Another v3 model is that the filters are designed to be 1XN or NX1 form to factorise the convolutions, leading to a much more cost-reducing model. The model used RMSProp optimiser along with the batch norm in the auxiliary classifiers.

3.4 ResNet50

It is one of the classic neural network models used as a backbone for image data. Most of the deep learning models are stacked with layers, but this model has the most important feature to skip the layers. The skip's main reason is to ensure that the model's next layers would perform better than the other lower layers. The skipping concept can be explained, as shown in Fig. 4.

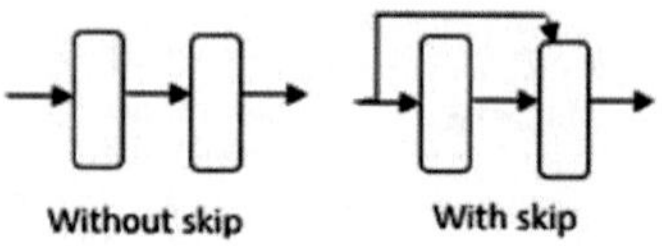

Fig. 4 ResNet50

We can analyse the difference between the general model and a ResNet model by the above figure. In the general deep learning model, the convolution layers are connected. When we consider the ResNet model, there is a skip option where one convolution layer's input can also be given to another layer using the skip option. This model has the batch norm along with the convolution. The ResNet-50 contains three stages, and each stage has a convolutional and another block called identity. Each convolution block has three convolutional layers, and the identity block also has three convolutional layers. In total, the ResNet has about 48 convolutional layers.

3.5 MobileNet v2

MobileNet v2 is released by the engineers of Google which is a deep neural network model that works in identification, detection, segmentation and classification problems. This model works best when the parameters are low. It is much effective when we consider object detection. It is built on the idea of depth-wise separable convolutions as the building blocks which help in finding the deep features in the data. The architecture of MobileNet contains the convolutional layers along with a linear bottle neck in between the layers of convolutional and the final layers. The convolutional layers have 32 filters followed by a residual bottleneck with 19 layers. The usage of linear layers is very crucial and helps to preserve the information from destroying as the nonlinear layers cause destruction in most of the information and hurt the performance (Fig. 5).

The MobileNet v1 has a depth-wise convolution layer along with a 1 × 1 convolutional second layer which was also called pointwise convolutional layer. While the MobileNet v2 has three different layers with 1 × 1 convolution with ReLU 6 as its first layer and a depth-wise convolution as its second and a 1 × 1 convolution as a third layer. The major difference between the v1 and v2 can be the fast computations in v2 comparing with the other and requires only fewer parameters to get accurate

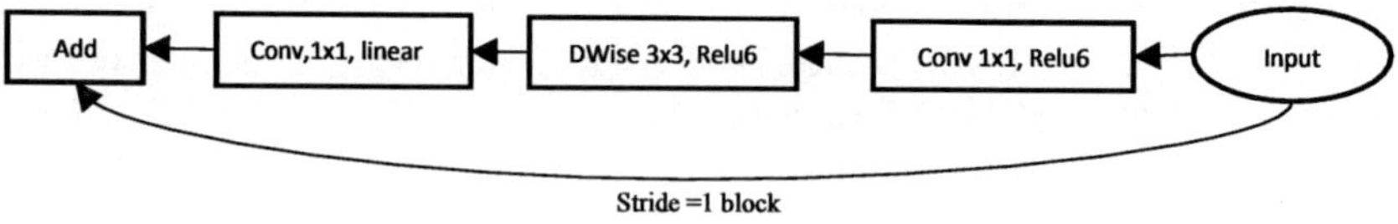

Fig. 5 MobileNet v2

results. The v2 or v1 both have a building block, and there are residual steps means few steps are having shortcuts moving one layer to other without passing through the intermediate layers.

3.6 *EfficientNet b0*

This model was from Google which was mentioned in one of the Google's publication papers in 2019. According to the paper, it has better accuracy as it reduces the parameters that help in considerable increase in the efficiency of the models. The compound scaling method acts like a copy of the baseline method in its functionality to amplify the predictive capacity. The MBConv with squeeze and excitation optimisation acts as a building block for this model. This is quite like the residual block of MobileNet v2. The first layer is a 1×1 convolutions followed by a 3×3 depth-wise and pointwise convolutions which reduce the number of feature maps. The depth associated with the network helps in finding more complex features which can be achieved by simply stacking the convolutional layers on top of another. This kind of deep networks has a problem of vanishing gradient, but the methods of skip connections help in solving the problem. Width scaling a network is much helpful for the model to learn finer features.

4 Results and Discussion

Firstly, let us consider the accuracy of the three different image classifier models VGG16, Inception v3 and ResNet-50. After training these models with the COVID-19 CT scan data, the scores are given in Table 1. These are the scores of the 3 different models for 50 epochs on 3265 images. As interpreted in the table, the model's accuracy at these epochs is 62% for VGG16, 66% for Inception v3 and 75% for ResNet-50. The scores of losses, validation loss and accuracies are also achieved as above.

As we can see from the table, the VGG16 model has an average accuracy of 62% with just 50 epochs, but the loss is equivalently higher than the accuracy with 64% (Figs. 6 and 7).

Table 1 Scores for 50 epochs on 2D image dataset

Models	Accuracy score	Loss	Val acc	Val loss
VGG16	0.62	0.64	0.54	0.87
Inception v3	0.66	0.63	0.54	1.88
ResNet-50	0.75	0.76	0.45	2.77

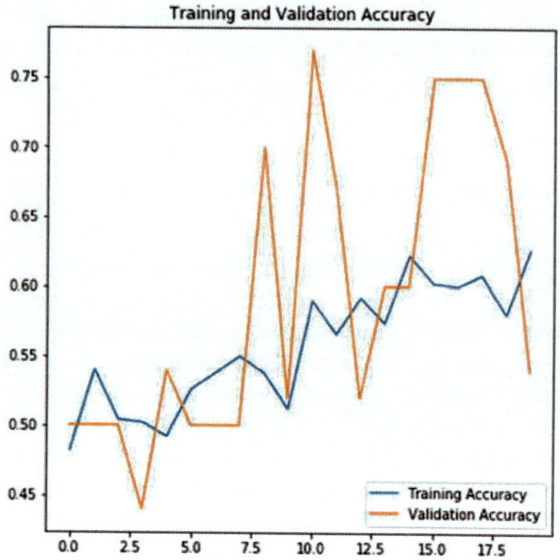

Fig. 6 VGG16 accuracy graph on 2D image dataset

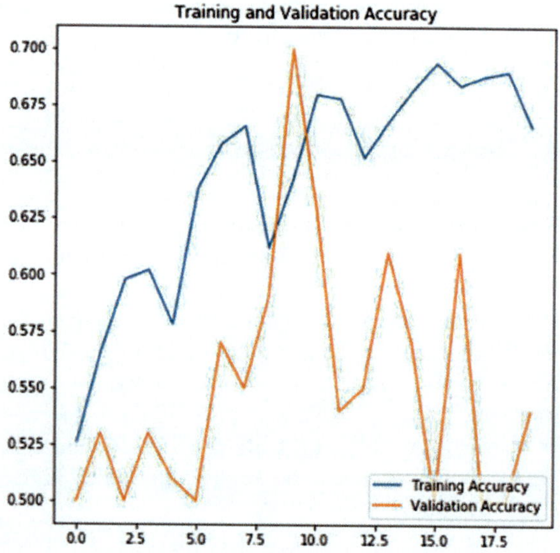

Fig. 7 Inception v3 accuracy graph on 2D image dataset

When we consider the other models like Inception v3, which has a decent accuracy of 66%, the loss is equivalent to the accuracy. Still, unlike VGG16, it is less than the accuracy that the model gained, while ResNet-50 got 75% which is the highest compared to all the other models. Since the loss and accuracy are good, the model is a good fit (Fig. 8).

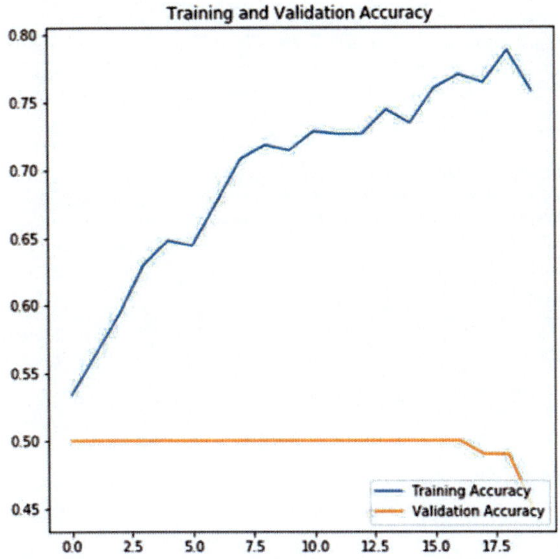

Fig. 8 ResNet-50 accuracy graph on 2D image dataset

Finally, ResNet-50 has scored the highest and has a very drastic change in the accuracy with the epochs. This seems like accuracy and epochs are directly proportional, and the model is quite a good fit for the CT image data. The accuracy-level deviation in the other two models, namely VGG16 and Inception v3, might be due to the model unable to select the right features from the given images. We can also observe the higher fluctuations in the accuracy graphs of the models which indicate the compatibility issue with the data. Later, we carried out the comparison of models with 3D images by converting them to 2D. The scores of the models are listed as below.

From Table 2, we can understand that the models were doing well when we consider 3D images. Of the three models, MobileNet v2 and EfficientNet have scored the same level of accuracy, while the Inception v3 is a bit lagging when we compare it with other models in terms of accuracy. The accuracy graph of the models is as in Fig. 9.

Table 2 Scores for 50 epochs on 3D image dataset

Models	Accuracy score	Loss	Val acc	Val loss
MobileNet v2	0.75	0.51	0.55	0.88
Inception v3	0.72	0.56	0.60	0.71
EfficientNet B0	0.75	0.50	0.62	0.74

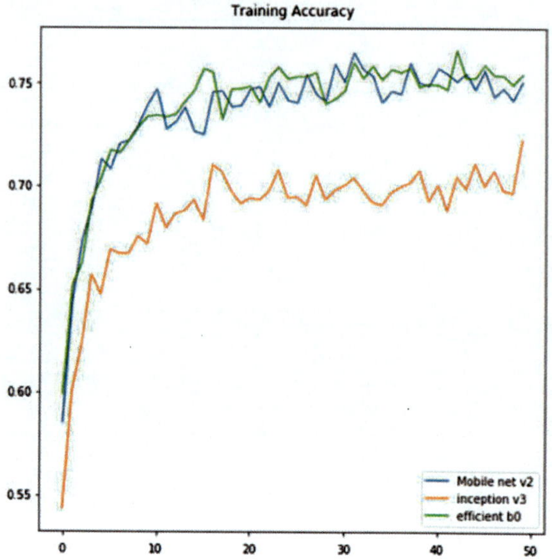

Fig. 9 Accuracy graph for the model on 3D image dataset

By examining the graph, it is evident that the three models have drastically increasing accuracy. When the epochs were 20, the models Inception v3 and EfficientNet B0 are having similar fluctuations, and similarly, the fluctuation in accuracy got matched with MobileNet v2 and Inception v3 at 10 epochs, but it is not exactly same as the MobileNet model reached a higher accuracy at the 10 epochs, but Inception v3 was not too high in accuracy at that stage, but it showed a good increase in accuracy when compared to its previous iterations. The prediction from models is also mentioned where we can find the performance of the models. The prediction was done for all the three models, and the classification was justified by specifying the correct labels as green and the wrong labelled prediction as red in colour (Fig. 10).

From the prediction of MobileNet v2, we can get to understand that from the 20 images for prediction, the model was able to classify 90% of the images with the correct labels, but only 6 of the images were classified wrong as we can see images with wrongly classified that are labelled in red colour. From Fig. 11, we can understand that from the predictions of Inception v3, it is evident that the model is not good with prediction as most of the images were classified to be incorrect. If we come across with the prediction of EfficientNet B0, it is also having a similar prediction ratio as the MobileNet v2 and is very much similar when we compare with the accuracy levels.

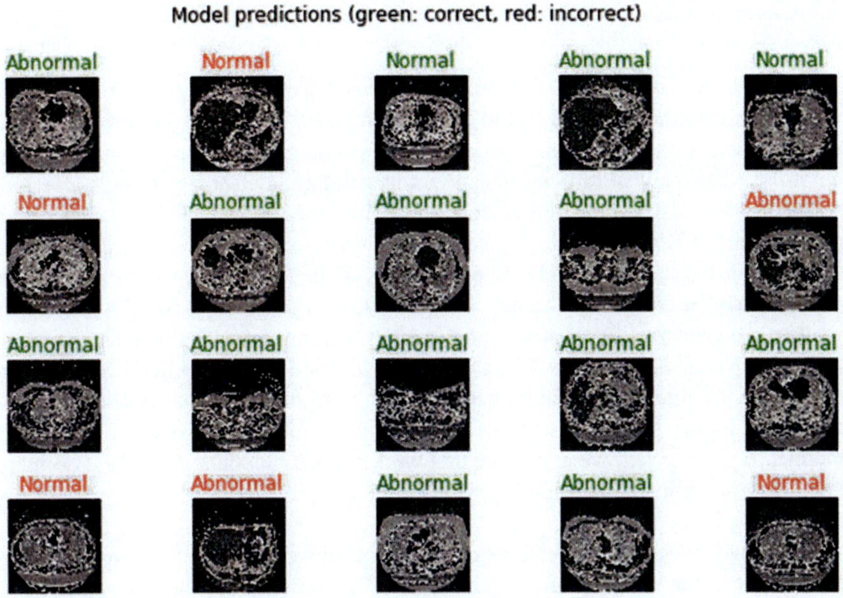

Fig. 10 MobileNet v2 prediction on 3D image dataset

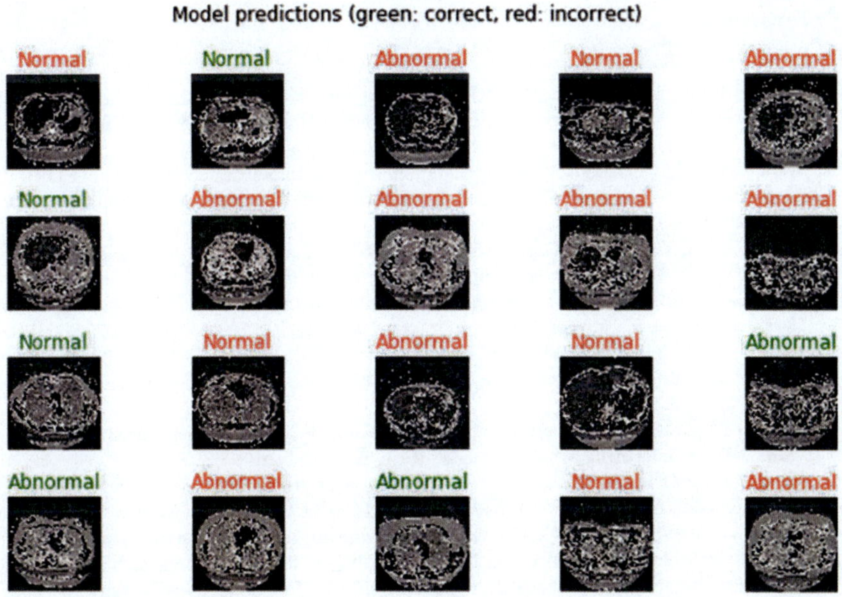

Fig. 11 Inception v3 prediction on 3D image dataset

5 Conclusion

Since the spread of a potentially unknown disease like coronavirus has struck the world with different challenges and lifestyle changes, it is highly important to find a way to detect it to make sure everyone is safe. Using machine learning, it is very helpful for the medical professionals to understand and detect if patient is affected with corona or not. When considering the performance of models in the paper, the performance of ResNet-50 has shown a phenomenon accuracy when compared with all the other models when working on 2D data, while MobileNet v2 has performed outrageously in terms of accuracy and prediction when compared with the other models. The accuracy level of Inception v3 has increased drastically at same epoch level from 2D and 3D images. The accuracy of the model is high when we applied it on 3D images. Since 3D images are having high resolution, the images will be much clear with highly important features which help the models to work in an efficient manner. By this paper, we could say that working with 3D data in medical domain is much more efficient, considering its resolution.

Declaration.

"We have taken permission from competent authorities to use the images/data as given in the paper. In case of any dispute in the future, we shall be wholly responsible".

References

1. Waheed A, Goyal M, Gupta D. CovidGAN: Data augmentation using auxiliary classifier GAN for improved covid-19 detection. IEEE Access 8. https://doi.org/10.1109/ACCESS.2020.2994762
2. Wang G, Liu X, Li C (2020) A noise-robust framework for automatic segmentation of COVID-19 pneumonia lesions from CT images. IEEE Trans Med Imaging. https://doi.org/10.1109/TMI.2020.3000314
3. Oh1 Y, Park S, Ye JC (2020) Deep learning COVID-19 features on CXR using limited training data sets. IEEE Trans Med Imaging. https://doi.org/10.1109/TMI.2020.2993291
4. Hu S, Gao Y, Niu Z, Weakly supervised deep learning for COVID-19 infection detection and classification from CT images. IEEE Access. https://doi.org/10.1109/ACCESS.2020.3005510
5. Kang H, Xia L, Yan F (2020) Diagnosis of coronavirus disease 2019 (COVID-19) with structured latent multi-view representation learning. IEEE Trans Med Imaging. https://doi.org/10.1109/TMI.2020.2992546
6. Zheng N, Fellow, IEEE, Du S, Member, IEEE, Wang J, Member, IEEE (2020) Predicting COVID-19 in china using hybrid AI model. IEEE Trans Cybern
7. Wang C, Chen D, Hao L (2019) Pulmonary image classification based on inception-v3 transfer learning model. IEEE Access 7. https://doi.org/10.1109/ACCESS.2019.2946000
8. Hartanto CA, Wibowo A (2020) Development of mobile skin cancer detection using faster R-CNN and MobileNet v2 model. In: IEEE 2020 7th international conference on information technology, computer, and electrical engineering (ICITACEE) **DOI:** https://doi.org/10.1109/ICITACEE50144.2020.9239197
9. Carrer L, Member, IEEE, Donini E, Student Member, IEEE, Marinelli D (2020) Automatic pleural line extraction and COVID-19 scoring from lung ultrasound data. IEEE Trans Ultrasonics Ferroelectr Freq Control. https://doi.org/10.1109/TUFFC.2020.3005512

10. Wang J, Bao Y, Wen Y (2020) Prior-attention residual learning for more discriminative COVID-19 screening in CT images. IEEE Trans Med Imaging. https://doi.org/10.1109/TMI.2020.2994908
11. Roy S, Menapace W, Oei S (2020) Deep learning for classification and localization of COVID-19 markers in point-of-care lung ultrasound. IEEE Trans Med Imaging. https://doi.org/10.1109/TMI.2020.2994459
12. Fan D-P, Zhou T, Ji G-P (2020) Inf-Net: automatic COVID-19 lung infection segmentation from CT images. IEEE Trans Med Imaging. https://doi.org/10.1109/TMI.2020.2996645
13. Shi F, Wang J, Shi J (2020) Review of artificial intelligence techniques in imaging data acquisition, segmentation and diagnosis for COVID-19. IEEE Rev Biomed Eng. https://doi.org/10.1109/RBME.2020.2987975
14. Rajesh P, Bharadwaj, Alam M, Tahernezhadi M (2020) A data science approach to football team player selection. In: 2020 IEEE international conference on electro information technology (EIT).https://doi.org/10.1109/EIT48999.2020.9208331
15. Babitha D, Ismail M, Chowdhury S, Govindaraj R, Prakash KB (2020) Automated road safety surveillance system using hybrid CNN-LSTM approach. Int J Adv Trends Comput Sci Eng 9(2). https://doi.org/10.30534/ijatcse/2020/13292202

Feature Extraction of ECG Signal Using Variational Mode Decomposition

Boni Shanmukh and R. Shanmughasundaram

Abstract Variational mode decomposition (VMD) method is used for feature extraction of ECG signal. MATLAB/Simulink is used for simulating the VMD. The performance of VMD is compared with EMD and then from simulation results, it is observed that VMD performs better than EMD with respect to the statistical properties of the decomposed intrinsic mode signal. The features obtained after decomposition of ECG signal can be used for effective diagnosis of arrhythmia, which occurs due to irregularities in heart beats.

Keywords Electrocardiogram (ECG) · Empirical mode decomposition (EMD) · Variational mode decomposition (VMD) · Signal-to-noise ratio (SNR)

1 Introduction

The electrocardiogram (ECG) has been the most popular diagnosis for its obvious reasons like non-invasive and capability to record the cardiac activities. It can be sensed from the surface of the skin. With the basic features of the ECG signal like P, QRS, and T waves, it is possible to evaluate the functioning of the heart. The morphological properties of the waves constitute duration, slope, and amplitude. With the help of these details of the waves, it is possible to analyze the cardiac activity, which has been a usual practice by the cardiologists. It is significant to mention about the common heart disease known as arrhythmia (ARTH) which occurs due to the faulty electrical conduction of probes that gradually leads to irregular heartbeats. The empirical data from the ECG has to be thoroughly analyzed to diagnose the disease related to the heart. In medical terms, it is a cumbersome task to be conclusive on arrhythmia using the details of the ECG signals. Hence, it is essential to use

B. Shanmukh (✉) · R. Shanmughasundaram
Department of Electrical and Electronics Engineering, Amrita School of Engineering, Amrita Vishwa Vidyapeetham, Coimbatore, India

R. Shanmughasundaram
e-mail: r_shanmughasundaram@cb.amrita.edu

P. S. R. Chowdary et al. (eds.), *Evolution in Signal Processing and Telecommunication Networks*, Lecture Notes in Electrical Engineering 839,
https://doi.org/10.1007/978-981-16-8554-5_20

algorithms and techniques to analyze an ECG signal for efficient diagnosis of the heart. This type of diagnosis involves the use of computing and intelligent decision support system to assist the doctors which helps in enhancing the diagnostic accuracy [1]. The empirical mode decomposition (EMD) is formulated by Huang et al. [2]. It detects and decomposes the signal into several principal modes, usually a signal that is completely described using the Fourier spectrum. The algorithm iteratively finds the local minima or maxima of a signal. These details are used to estimate the lower/upper peaks of the signal. These are interpolated to extract the extreme values following which the removal of average of the envelopes takes place similar to a low pass filter. This typically isolates the higher-frequency components referred as modes of a signal. This strategy is continued iteratively even on the extracted low pass centerline. A few experiments have showcased that EMD is sharing essential similarities with wavelets and adaptive filter banks [3]. It is possible to obtain the principal modes of the signal as result of this algorithmic approach. The resultant decomposition phenomenon is highly sensitive to the peak determining techniques and the interpolation process along with the effective termination criteria [4]. This drastically degrades the robustness of decomposition [4, 5]. The variational mode decomposition (VMD) performs better when the signal is sensitive to noise. Some applications of signal decomposition in which the EMD and VMD methods have been adopted are audio engineering [6], climate analysis [7], neuromuscular signals of medical and biology [8–11], and phonocardiogram signals [12].

2 Variational Mode Decomposition

VMD is used to decompose the input signal and provide several sub-signals known as modes. These type of modes typically exhibit sparsity properties. It implies that each mode k is compact and can be determined along with the decomposition. Hilbert transformation is used to realize the unilateral frequency spectrum and to calculate the minimum phase responses of respective signal. Transition to baseband should be done which involves mixing the exponent while tuning the corresponding estimated center frequency. The constrained variational problem is given as

$$\min_{u_k, w_k}\left\{\sum_k \left\| \partial_t\left[\left(\delta(t)+\frac{j}{\pi t}\right)*u_k(t)\right]\mathrm{e}^{-jw_k t}\right\|_2^2\right\} \text{s.t.} \sum_k u_k = f \tag{1}$$

The constraints are realized in several ways. Both the quadratic penalty terms along with the Lagrangian multiplier are used to overcome the unconstrained problem to obtain minima and maxima values. The Lagrangian function is given as follows [13, 14]:

$$L(u_k, w_k, \lambda) = \alpha \sum_k \left\| \partial_t\left[\left(\delta(t)+\frac{j}{\pi t}\right)*u_k(t)\right]\mathrm{e}^{-jw_k t}\right\|_2^2$$

$$+\left\|f-\sum u_k\right\|_2^2+\left\langle \lambda, f-\sum u_k\right\rangle \tag{2}$$

The method of sequential sub-optimization of central frequencies using Lagrangian is referred as alternate direction method of multipliers (ADMM). Hence, the VMD algorithm should be included in all the sub-optimizations yielding ADMM and the Fourier domain wherever necessary. The detailed algorithm and procedure of VMD are found in [16, 17]. The VMD algorithm is seen in Fig. 1.

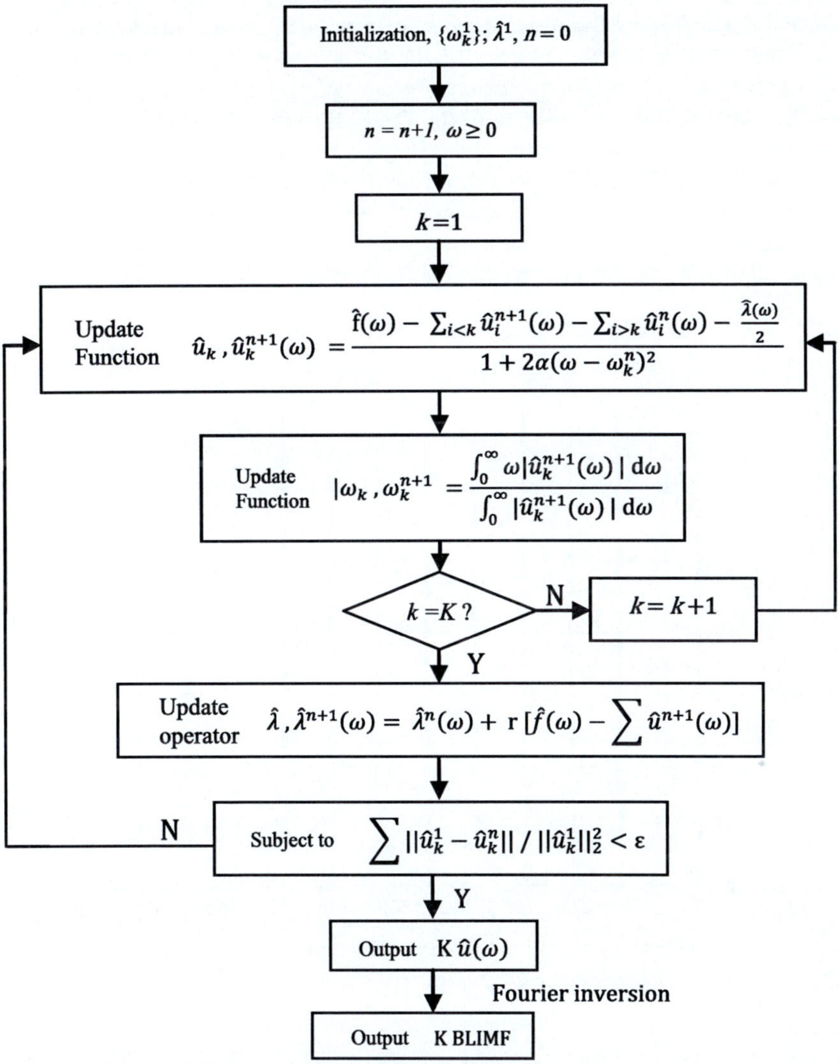

Fig. 1 Variational mode decomposition flowchart [15]

3 Feature Extraction Process

To study the feature extraction properties of VMD, an ECG signal with 1000 samples is considered as shown in Fig. 2. This dataset is taken from data world [18]. Gaussian noise is added to the signal to test the decomposition efficiencies of both EMD and VMD. The signal is processed using ensemble method to obtain all the required normalized data for processing with its centralized frequencies, bandwidth coefficients thus obtaining their mode properties.

The VMD and EMD algorithms are implemented in MATLAB/Simulink. Figures 3 and 4 show the instantaneous frequency (IF) modes obtained for both EMD and VMD. It is observed from the simulation results that the VMD-related IF modes show a good result in reduction of noise with less distortions in each modes while compared to the IF modes in EMD. The EMD and VMD are compared based on the statistical properties of decomposed signals like mean, standard deviation, and signal-to-noise ratio (SNR). The comparison of VMD and EMD in terms of statistical properties and SNR ratio is shown in Tables 1 and 2, respectively. The SNR value is calculated for all the four intrinsic mode signals of VMD and EMD. From the tabulation, it can be observed that the performance of VMD is good compared to EMD.

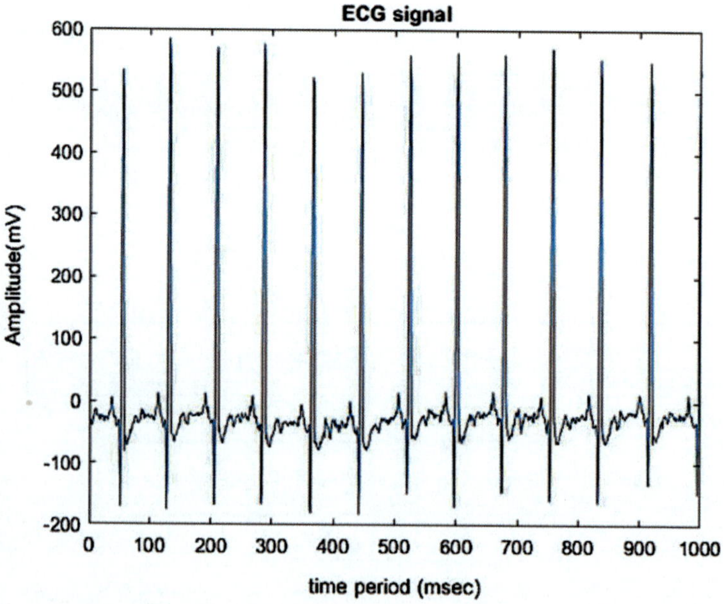

Fig. 2 Input ECG signal

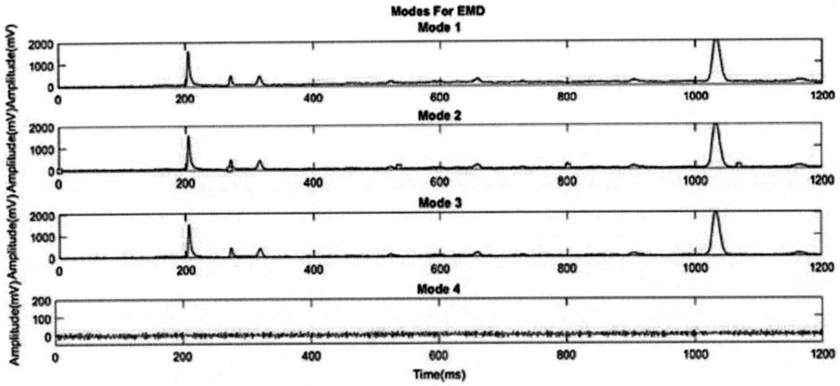

Fig. 3 Modes obtained for EMD

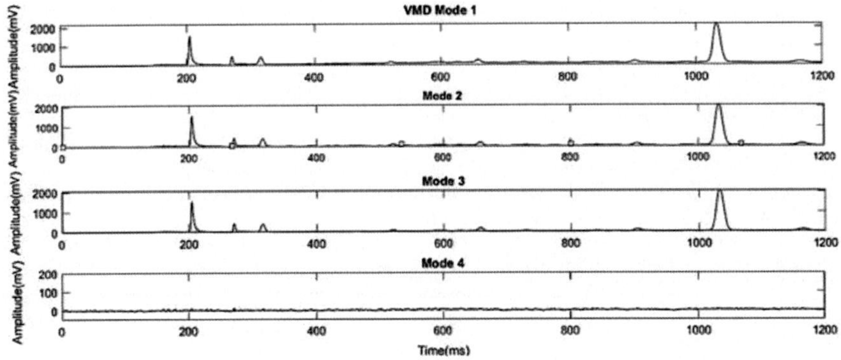

Fig. 4 Modes obtained for VMD

Table 1 Comparison of VMD and EMD in terms of statistical properties

Central tendencies	VMD	EMD
Mean	8.2086	7.3260
Standard deviation	113.5558	101.469

Table 2 Comparison of VMD and EMD in terms of SNR

Mode level	SNR	
	VMD	EMD
Mode 1	4.0642	2.7678
Mode 2	5.8816	3.8314
Mode 3	6.2769	6.1570
Mode 4	13.896	13.349

4 Conclusion

In this paper, VMD and EMD algorithms are implemented in MATLAB/Simulink for the feature extraction of an ECG signal. The algorithms are compared based on the statistical properties of the decomposed intrinsic mode signals. From the simulation results, it is observed that VMD performs better compared to EMD in terms of noise reduction. The features obtained after decomposition of ECG signal can be used for effective diagnosis of arrhythmia, which occurs due to irregularities in heart beats.

References

1. Thomas M, Das MK, Ari S (2015) Automatic ECG arrhythmia classification using dual tree complex wavelet based features AEU-Int. J Electron Commun 69:715–721
2. Huang NE, Shen Z, Long SR, Wu MC, Shih HH, Zheng Q, Yen N-C, Tung CC, Liu HH (1998) Theempirical mode decomposition and the Hilbert spectrum for nonlinear and non-stationary time series analysis. Proc Roy Soc A: Math Phys Eng Sci 454(1971):903–995
3. Flandrin P, Gonçalvès P, Rilling G (2005) EMD equivalent filter banks, from interpretation toapplications. In Hilbert-Huang transform and its applications, pp 57–74
4. Rilling G, Flandrin P, Gonçalvès P (2003) On empirical mode decomposition and its algorithms. In: IEEE-EURASIP workshop on nonlinear signal and image processing
5. Rilling G, Flandrin P (2008) One or two frequencies? The empirical mode decomposition answers. IEEE Trans Signal Process 56(1):85–95
6. Klügel N (2012) Practical empirical mode decomposition for audio synthesis. In: International conference on digital AudioEffects (DAFx-12), no 2, pp 15–18
7. Barnhart BL, Eichinger WE (2011) Empirical mode decomposition applied to solar irradiance, global temperature, sunspot number, and CO_2 concentration data. J Atmos Solar Terr Phys 73(13):1771–1779
8. Assous S, Humeau A, L'huillier J-P (2005) Empirical mode decomposition applied to laser Doppler flowmetry signals: diagnosis approach. In: IEEE Engineering in medicine and biology conference (EMBC), vol 2, pp 1232–1235
9. Andrade AO, Nasuto S, Kyberd P, Sweeney Reed CM, Van Kanijn FR (2006) EMG signal filtering based on empirical mode decomposition. Biomed Signal Process Control 1(1):44–55
10. Liu S, He Q, Gao RX, Freedson P (2008) Empirical mode decomposition applied to tissue artifact removal from respiratory signal. In: IEEE engineering in medicine and biology conference (EMBC), pp 3624–3627
11. Mostafanezhad I, Boric-Lubecke O, Lubecke V, Mandic DP (2009) Application of empirical modedecomposition in removing fidgeting interference in Doppler radar life signs monitoring devices. In: IEEE engineering in medicine and biology conference (EMBC), pp 340–343
12. Sujadevi VG, Soman KP, Kumar SS, Mohan N, Arunjith AS (2017) Denoising of phonocardiogram signals using variational mode decomposition. In: 2017 International conference on advances in computing, communications and informatics (ICACCI), 2017, pp 1443–1446. https://doi.org/10.1109/ICACCI.2017.8126043
13. Swathi ON, Ganesan M, Lavanya R (2017) R peak detection and feature extraction for the diagnosis of heart diseases. In: 2017 International conference on advances in computing, communications and informatics (ICACCI), 2017, pp 2388–2391. https://doi.org/10.1109/ICACCI.2017.8126204
14. Soman KP, Sachin Kumar S, Mohan N, Poornachandran P (2019) Modern methods for signal analysis and its applications. In: Kumar R, Wiil U (eds) Recent advances in computational intelligence. Studies in computational intelligence, vol 823. Springer, Cham. https://doi.org/10.1007/978-3-030-12500-4_17

15. Hu C, Wang Y, Yang J, Zhang S (2018) Fault detection in a multistage gearbox based on a hybrid demodulation method using modulation intensity distribution and variational mode decomposition. Appl Sci 8(5). https://doi.org/10.3390/app8050696
16. Dragomiretskiy K, Zosso D (2014) Variational mode decomposition. IEEE Trans Signal Process 62:531–544
17. Wang Y, Markert R, Xiang J, Zheng W (2015) Research on variational mode decomposition and its application in detecting rub impact fault of the rotor system. Mech Syst Signal Process 60:243–251
18. OrestiBanos, The MHEALTH (Mobile Health) dataset is devised to benchmark techniques dealing with human behavior analysis based on multimodal body sensing. Dataset. Data.world. https://data.world/uci/mhealtdataset/workspace/project-mmary?agentid=uci&datasetid=mhealth-dataset.

Plant Disease Identification Based on Leaf Images Using Deep Learning

Hardev Mukeshbhai Khandhar, Chintan Bhatt, Dac-Nhuong Le, Harshil Sharaf, and Wathiq Mansoor

Abstract Plant diseases need the maximum amount of attention currently as ever to ensure that food systems are in situ within the next season. There are serious implications for biological science and also the atmosphere more generally. Deep learning technologies produce opportunities for revamping operations, employment management, production, and forestry management. This paper focuses on the utilization of MobileNetV2 in the image process on mobile devices. This advanced version of MobileNet is specifically curated to make it useful and compatible with mobile applications, and it also permits efficient inference along with the utilization of operations existing in all neural frameworks. The given approach resulted in close to 100% accuracy on the given dataset with minimum number of iterations. It is also inferred that a good accuracy score can be achieved in real-time mobile devices since the model uses less computations to make accurate predictions. This approach hence focuses on "Next-Gen" revolutionary on-device computer vision network.

Keywords Crop disease · Machine learning · Deep learning · Convolutional neural networks · Smart agriculture

H. M. Khandhar · C. Bhatt · H. Sharaf
Charotar University of Science and Technology, Changa, Gujarat, India
e-mail: 18ce043@charusat.edu.in

C. Bhatt
e-mail: chintanbhatt.ce@charusat.ac.in

H. Sharaf
e-mail: 18ce121@charusat.edu.in

D.-N. Le (✉)
Haiphong University, Haiphong, Vietnam
e-mail: Nhuongld@dhhp.edu.vn

W. Mansoor
University of Dubai, Dubai, UAE
e-mail: wmansoor@ud.ac.ae

P. S. R. Chowdary et al. (eds.), *Evolution in Signal Processing and Telecommunication Networks*, Lecture Notes in Electrical Engineering 839,
https://doi.org/10.1007/978-981-16-8554-5_21

1 Introduction

The agricultural sector plays a big and strategic role in the economic development of a country like India. The commercial revolution and agricultural developments are complementary to each other and are reciprocally supporting each input and output. If agriculture fails to satisfy the rising demands of food merchandise, it is found to affect the expansion rate of the economy adversely. Modern farming, i.e., farming using advanced technologies is very important for tackling certain challenges within the agricultural sector like food security, property, productivity, and alternative environmental impacts [1]. With the exponential rise of the population within the country, a need, therefore, arises to keep up equivalent time accessibility yet as high biological process quality across the country keeping in mind to guard the natural ecosystems with property farming techniques as well. Reduction of agricultural land because of the rise in urbanization is projected to drive the expansion of the smart agriculture market.

India as an agriculturally predominant country loses close to 35% of the annual crop yield because of varied plant diseases. Early identification of those diseases continues to be a tough task due to the dearth of correct infrastructure and experience. The present methodology for disease detection is just eye observation by specialists through which identification and detection of plant diseases are done. Timely sickness detection in crops remains a difficult task for the farmers since they do not have a plethora of choices except consulting fellow farmer friends or the government helplines.

This paper is broadly classified into five key sections, namely the introduction that focuses on the current scenario and the need for deep learning technology in agriculture. Second is a literature review, where the authors have read and replicated some previous research work done to better understand the current challenges and to articulate a proper approach in the new era. The third section is the methodology that focuses on the proposed work and how it uses MobileNetV2 architecture as an aid to computer vision. The proposed methodology is currently in a dormant state and in progress to make a robust classifier. The fourth section, namely experimental results includes the analysis of the results obtained and what tests are done to confirm the accuracy of the model. The fifth and last section is the conclusion that includes a discussion about the challenges and essential future scope of the project work. The motivation of this work comes from a vision to help the farmers who form a strong base of the Indian economy. The authors have essentially contributed to checking the feasibility and ease of using the mobile devices which will help the farmers take better and informed decisions without any delays. It is also necessary that the proposed approach does not compromise the performance metrics of the model.

2 Literature Review

In the simplest terms, image processing can be described as each possible action performed on a picture. People, in general, have the inherent ability to spot and extract objects from a picture whereas a computer cannot. For a machine, the image is just sort of a matrix of pixels. Deep learning techniques are needed to acknowledge the objects in from this matrix. The early detection of plant sicknesses is essential for effective control and management encouraging the financial rate and consequently, the environmental effect of their treatment is reduced. If plant illnesses are not dealt with of their early ranges, the production price may be significantly magnified since the ailment will propagate to the overall crop product [2]. The regular way is to rent skilled agriculturists who reveal the flora, however, this could now not constantly be attainable for farms set in isolated rural regions. Furthermore, the fee may not be affordable for farmers with small-scale agricultural lands.

"Deep learning constitutes a recent, modern technique for image processing and data analysis, with promising results and large potential. As deep learning has been successfully applied in various domains, it has recently entered also the domain of agriculture [3]." "A plant disease diagnosis method that can be implemented with the resources of a mobile phone application, that does not have to be connected to a remote server, is presented and evaluated on citrus diseases. It can be used both by amateur gardeners and by professional agriculturists for early detection of diseases. The features used are extracted from photographs of plant parts like leaves or fruits and include the color, the relative area and the number of the lesion spots [4, 5]."

Image segmentation is the method of separating or grouping a picture into completely different elements. These elements commonly correspond to one thing that humans will simply separate and think about as individual objects [6]. Computers do not have any means of showing intelligence recognizing objects, and then many alternative strategies are developed to section pictures. The segmentation method relies on numerous options found within the image. This may be color data, boundaries, or phase of a picture.

Table 1 illustrates a comparison of several deep learning approaches in terms of various performance metrics. The memory-intensive and highly computation-intensive properties of deep learning approaches restrict their applications in portable devices. At the same time, the compression and acceleration of network models will reduce the classification accuracy [7–10]. Therefore, as compared to above-mentioned traditional neural network architectures, it is believed that MobileNetv2 is the "Next-Gen" revolutionary on-device computer vision network. MobileNets are small, low-latency, low-power models parameterized to meet the resource constraints of a variety of use cases. They can be built upon for classification, detection, embeddings, and segmentation.

Table 1 Comparison of DL architectures and their performance metrics

DL architecture/algorithms	Performance metrics	References
AlexNet, GoogleLeNet	Accuracy (99.35%)	[11]
MobileNetV2, YOLOv3	F1 score (0.93), accuracy (91.32%)	[12]
VGG16	Accuracy (90.40%)	[13]
DenseNet	Accuracy (99.75%)	[14]
AlexNet, VGG16, VGG 19,	F1 score (97.14), Accuracy (97.86%)	[15]
SqueezeNet, GoogleLeNet		
MobileNetv2	Accuracy > 90%	[16]

3 Methodology

The methodology applied here is simply an attempt to test the MobileNetV2 model on an Android device with a diverse set of images. The earlier work focused on tests being conducted on tomato leaves only. However, this is an attempt to generalize the model and test it on a wide variety of leaves.

3.1 Our Approach

The proposed methodology is a unique way to help farmers detect the diseases from handheld devices in real-time. This makes them independent in terms of help required to find abnormality in the health of their crops. In this experiment, deep learning is applied on the available leaf images dataset. An image processing and classification rule together with reasoning strategies are planned here to spot the diseases in natural conditions through a mobile device. Three different devices have been used to capture and take a look at the model practically. The general disease detection methodology represented during this research experiment is aimed at mobile phone devices. The planned methodology uses extraction of leaf features like the variety of pores, color changes in relative areas, etc. It becomes a simple methodology for the end user since there is no demand to be acquainted with the source code or the deep learning model design.

The technique shown here differs from preceding works in the following approaches: (i) Keras follows best practices for reducing psychological feature load. It additionally offers consistent and straightforward APIs (application programming interfaces), minimizing the number of user actions needed for common use cases. It provides clear and actionable feedback upon user error. (ii) As compared to different pre-trained models, MobileNetV2 is helpful because it is specifically curated for mobile devices. (iii) By using this version of pre-trained model, the load on any mobile device is less, eventually resulting in quicker throughputs. And (iv) higher accuracy and lesser training time as compared to different models (Fig. 1).

Data Extraction: The process of crawling and analyzing data from numerous knowledge sources in an exceedingly distinct pattern is outlined as data extraction. The data is further processed that involves adding another information and processes within the data workflow.

Data Validation: In elementary language, data validation is the method to ensure that the information has undergone cleansing to ensure that quality information is retrieved, that is, data is correct and useful. This method checks the accuracy and validity of information supply before using it.

Data Augmentation: It is a method that allows us to considerably increase the information diversity for model training, while not creating efforts to gather new knowledge. This is often used to train giant neural networks. This artificially expands the dimensions of the training dataset. The Keras deep learning neural network library will fit models using image data augmentation via ImageDataGenerator class.

Model Fitting: This means that we are creating our algorithmic rule to learn the relationship between predictors and outcomes so that we will estimate the future values of the end result. The best-fitted model can have a selected set of characteristics that defines this problem.

MobileNetV2: MobileNetV2 improves the innovative and overall performance of cellular devices based on more than one responsibility and benchmarks also as across a spectrum of numerous model sizes. It is a powerful function extractor for object detection and segmentation. The MobileNetV2 models are a lot quicker as compared to MobileNetV1. It uses two times fewer operations, has higher accuracy, and desires thirty percent fewer parameters.

Flutter: Flutter is a free and open-source mobile UI framework created by Google and released in May 2017. In a few words, it permits you to create a native mobile application with only one codebase. This implies that you simply will use one programming language and one codebase to form two completely different apps (for iOS and Android). Google Flutter is actually smart whether it is in terms of development or sustainability of your mobile application. It helps deliver authentic standards as seen fit for an android or iOS and helps scale back development time while achieving better UI than any previous cross-development platform.

4 Results and Discussion

4.1 Dataset

This dataset includes over 54,303 images curated by experts consisting of unhealthy as well as healthy leaves of crop plants through the existing online platform PlantVillage. It contains 38 directories of different species and categories containing colored images of healthy and infected crop leaves.

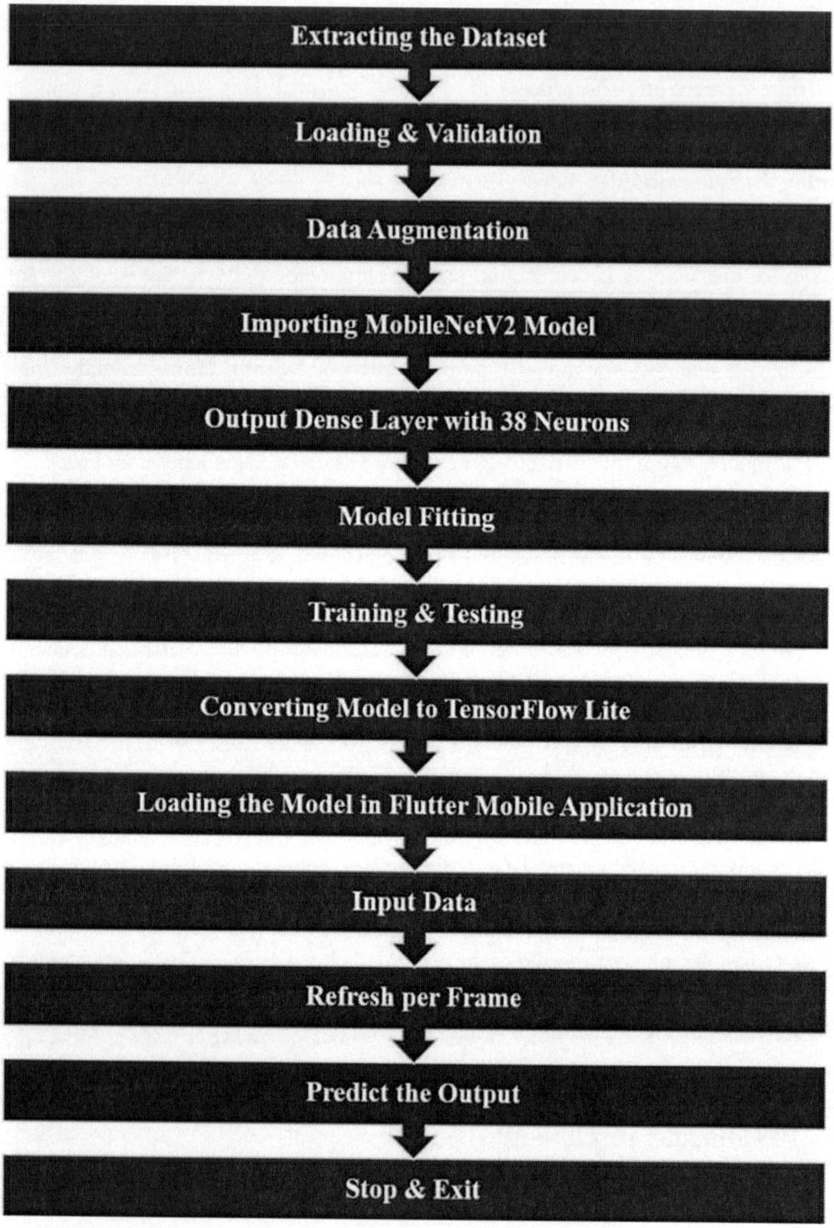

Fig. 1 Methodology of the experiment performed

Table 2 Performance metrics of the methodology used

Model name	Accuracy (in %)	Precision (in %)
MobileNetV2	99.96	99.71
InceptionV2 Resnet	99.27	98.84
InceptionV3	93.99	90.19
NASNet Mobile	92.51	93.78

4.2 *Experimental Results*

The complete experimental analysis was done on Google Colaboratory with enhanced and robust GPU. The results as shown in Fig. 4 were obtained by applying the methodology on a set of images. The best model was MobileNetv2 which could train efficiently in just 50 epochs with a batch size of 64 images and at a learning rate of 0.01 (Fig. 2).

Moreover, as attempts were made to try and improve the model, the loss function helps to measure how good the prediction model does in terms of being able to predict the expected outcome. Figure 3a, b represent the model accuracy and model loss for the best performing model (MobileNetv2), respectively.

Table 2 shows the various metric values for all the models under experiment. MobileNetV2 has given the best results as compared to alternative models. This model was able to classify all the test images accurately. InceptionV2 ResNet has additionally given correct results. However, MobileNetV2 has lesser number of hidden layers and it is specifically curated to run with efficiency on a mobile device atmosphere with a lower quantity of stress on the mobile operating system. The ultimate model was converted to TensorFlow lite to make it compatible with mobile devices and flutter environment. Hence, it is inferred that MobileNetv2 was the model with best performance. Hence, it can be concluded that the proposed approach was successful in testing the model on real-time data in mobile devices.

One of the challenges faced during the experimentation was existence of two extra classes in the classification report generated. The report depicted the presence of two label classes that were not a part of the test experiment. Some other notable challenges also include presence of complex background, uncontrolled capture conditions, and varied light conditions [7].

5 Conclusion and Future Work

With this experiment, the authors are convinced that in this modern technological era, it is completely possible for the farmers to be self-sufficient and their mobile devices can be equipped with path-breaking technology to detect and prevent plant diseases. The dataset was augmented as well as labeled. Finding out new datasets

Fig. 2 Results obtained from the methodology applied on a set of images

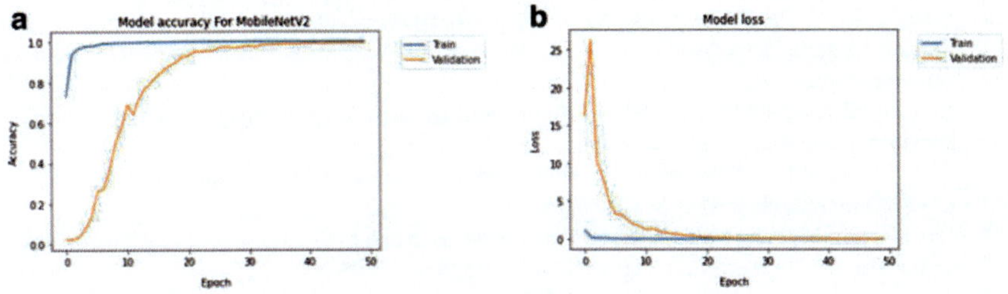

Fig. 3 Left to right: **a** model accuracy and **b** model loss

is a difficult task, especially in the Indian scenario. Moreover, there is a limited availability of the labeled and classified dataset. The challenge of collecting and preparing a customized dataset persists and hence results in limited experimental analysis. However, the authors are further planning to work on unsupervised learning algorithms so that it is possible to find an alternative for non-labeled dataset as well. The future scope of this paper includes development of a robust algorithm keeping in view of a larger number of experiments and more possible postures of leaf images as well as exposure to different light conditions, not just limited to generic methods.

References

1. Singh D, Jain N, Jain P, Kayal P, Kumawat S, Batra N (2020) PlantDoc: a dataset for visual plant disease detection. In: Proceedings of the 7th ACM IKDD CoDS and 25th COMAD, pp. 249–253
2. Saradhambal G, Dhivya R, Latha S, Rajesh R (2018) Plant disease detection and its solution using image classification. Int J Pure Appl Mathe 119(14):879–884
3. Kamilaris A, Prenafeta-Boldú FX (2018) Deep learning in agriculture: a survey. Comput Electron Agric 147:70–90
4. Petrellis N (2019) Plant disease diagnosis for smart phone applications with extensible set of diseases. Appl Sci 9(9):1952
5. Bhatia GS, Ahuja P, Chaudhari D, Paratkar S, Patil A (2019) Plant disease detection using deep learning. In: International conference on computer networks and inventive communication technologies. Springer, Cham, pp 408–415
6. Barbedo JGA (2016) A review on the main challenges in automatic plant disease identification based on visible range images. Biosys Eng 144:52–60
7. Bhateja V, Tripathi A, Sharma A, Le BN, Satapathy SC, Nguyen GN, Le DN (2016) Ant colony optimization based anisotropic diffusion approach for despeckling of SAR images. In: International symposium on integrated uncertainty in knowledge modelling and decision making. Springer, Cham, pp 389–396
8. Ashour AS, Dey N, Le DN (2017) Biological data mining: techniques and applications. Min Multimedia Doc 1(4):161–172
9. Le BN, Le DN, Nguyen GN (2016) Optimizing selection of PZMI features based on MMAS algorithm for face recognition of the online video contextual advertisement user-oriented system. In: International symposium on integrated uncertainty in knowledge modelling and decision making. Springer, Cham, pp 317–330

10. Wang W, Hu Y, Zou T, Liu H, Wang J, Wang X (2020) A new image classification approach via improved MobileNet models with local receptive field expansion in shallow layers. Comput Intell Neurosci
11. Mohanty SP, Hughes DP, Salathé M (2016) Using deep learning for image-based plant disease detection. Front Plant Sci 7:1419
12. Liu J, Wang X (2020) Early recognition of tomato gray leaf spot disease based on MobileNetv2-YOLOv3 model. Plant Methods 16:1–16
13. Wang G, Sun Y, Wang J (2017) Automatic image-based plant disease severity estimation using deep learning. Comput Intell Neurosci 2017:2917536
14. Too EC, Yujian L, Njuki S, Yingchun L (2019) A comparative study of fine-tuning deep learning models for plant disease identification. Comput Electron Agric 161:272–279
15. Türkoğlu M, Hanbay D (2019) Plant disease and pest detection using deep learning-based features. Turk J Electr Eng Comput Sci 27(3):1636–1651
16. Zaki SZM, Zulkifley MA, Stofa MM, Kamari NAM, Mohamed NA (2020) Classification of tomato leaf diseases using MobileNet v2. IAES Int J Artif Intell 9(2):290

An Improved Unsharp Masking (UM) Filter with GL Mask

Akshita Shukla, Vikrant Bhateja, Alka Singh, and Akansha Singh Rathore

Abstract Lung cancer is the fatal lung disease which involves uncontrolled growth of cells in tissue of lungs. Screening of the lung cancer can be carried out by CT-scan images. Enhancement of CT-scan images is necessary to deal with anomalous details and low-quality pixels. This paper therefore proposes an enhancement filter using unsharp masking (UM) approach with Grunwald–Letnikov (GL) fractional derivative mask for contrast and sharpness improvement of CT-scan images. This technique enhances the non-visible or partially visible regions. The quality of enhanced image has been evaluated using measure of enhancement (EME) as quality metric. The obtained results show significant enhancement than conventional UM without inciting noise.

Keywords UM · GL mask · CT-scan images · EME

1 Introduction

Lung cancer is one of the most common cancer-related life-taking diseases. Symptoms of lung cancer are not observable at initial stage which may leads to more dangerous situation. However, if diagnosed at early stage, the life of the patient can be saved[1]. CT-scan imaging is considered as one of the best imaging techniques for screening of lung cancer as they are more reliable, has better infection visibility, better diagnosis capability and non-invasive nature. Along with specific diagnostic features, CT-scan images also comprise low-quality pixels, undesirable distortions and other several anomalous details. This may lead to false detection of lesion region. So, there is a need for enhancement so as to minimize these challenges [2]. The work so far studied and reviewed focuses on various enhancement techniques. Gaussian and Gabor filter for smoothing and enhancement is used in the work of Kavitha et al.

A. Shukla · V. Bhateja (✉) · A. Singh · A. S. Rathore
Department of Electronics and Communication Engineering, Shri Ramswaroop Memorial College of Engineering and Management (SRMCEM), Lucknow, Uttar Pradesh 226028, India

Dr. A. P. J. Abdul Kalam Technical University (AKTU), Lucknow, Uttar Pradesh 226031, India

P. S. R. Chowdary et al. (eds.), *Evolution in Signal Processing and Telecommunication Networks*, Lecture Notes in Electrical Engineering 839,
https://doi.org/10.1007/978-981-16-8554-5_22

[2], respectively. These filters have limitations that these reduce details of the image and take more time to compute. Adaptive histogram equalization (AHE) is another common technique used for enhancement, but it has drawback of noise amplification [3]. Hwa et al. [4] put forward modified Canny edge detection method that is CEED-Canny. This method includes Canny edge detection with local morphological contrast enhancement. It blurs the image. Morphological filtering technique for enhancement with suitable structuring element is also discussed in Refs. [5–7]. For preprocessing operations, an erosion filter and cylindrical nodule enhancement filter (CNEF) are used in work of A. Teramoto et al. [9]. One of the most common approaches is UM, which makes use of scaled high pass filtered response added to input image to give enhanced image as output [8, 10, 11]. However, it undergoes overshooting of undesirable artifacts and amplification of noise. Combination of fractional derivative-based UM along with Laplacian mask is used for enhancing the visual quality of image. This combination improves the texture, edge and contrast of image [12]. This paper proposes an enhancement technique using improved UM with GL fractional derivative mask for contrast enhancement and edge improvement. Rest of the paper is organized in following sections: Sect. 2 focuses on general outline of UM and GL mask along with improved UM. Section 3 deals with image quality assessment, encapsulates the results, performance parameters and discussion of the obtained outcomes of the work. Section 4 includes conclusion which comprises summary of the proposed work.

2 Proposed Unsharp Masking Filter

2.1 Background of Conventional UM

UM is the technique used for contrast and edge enhancement of the input image. In this technique, the input image is first passed through high pass filter. The filtered image is then scaled and further added to input image to obtain edge enhanced sharp image [10, 11].

Major limitations of UM are as follows:

- This technique amplifies the area with high contrast which leads to overshooting of undesirable artifacts [11].
- The high pass filter used in conventional UM amplifies the existing noise in the image which results in visually distorted image [11].

2.2 Grunwald–Letnikov (GL) Mask

Grunwald–Letnikov (GL) mask is the fractional derivative mask which is the extension of Euler method [12]. The GL-based derivative of a function $z(t)$ is given in Eq. (1).

$$D^{\mu}z(t) = \lim_{h \to 0} \frac{1}{\Gamma(\mu)h^{\mu}} \sum_{k=0}^{\left(\frac{t-a}{h}\right)} \frac{\Gamma(\mu + k)}{\Gamma(k+1)} z(t - kh) \tag{1}$$

where h denotes sampling period such that $\left(\frac{t-a}{h}\right)$ must be an integer, and the range of k lies between 0 and $\left(\frac{t-a}{h}\right)$.

The mask of the Grunwald–Letnikov (GL) with $\mu = 0.15$ and intensity factor = 1 using Eq. (1) is generated for the proposed work.

2.3 Improved UM Filter

This paper proposes an efficient way of enhancement using improved UM with GL mask. This modified technique overcomes the drawback of conventional UM that are overshooting of artifacts and noise amplification problem [11–13]. In the proposed work, the high pass filter is replaced by GL fractional derivative mask so as to improve the performance.

The image acquired from database is firstly converted to grayscale image. GL mask is applied on the resultant image which is further scaled with scaling factor. Value of scaling factor must lie between 0 and 1. The obtained image is then added to the grayscale image so as to get the enhanced image as depicted in Fig. 1. The algo-

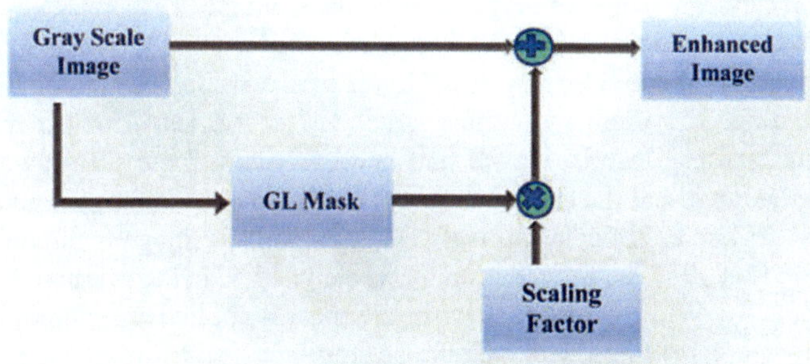

Fig. 1 Proposed design methodology for an improved UM filter [10]

rithm for the proposed enhancement using improved UM with GL mask is explained below:

Algorithm 1: Procedural Steps for Enhancement using Improved UM with GL Mask

BEGIN
Step 1: *Input* CT-scan image of lungs [*I*]
Step 2: *Process* Convert [*I*] from RGB to grayscale [*f*]
Step 3: *Apply* GL mask to [*f*]
Step 4: *Define scaling factor* as $0 < s < 1$
Step 5: *Multiply* [*f*] with scaling factor [*s*]
Step 6: *Add* [*I*] and [*s*]
END

3 Results and Discussions

3.1 Image Quality Assessment

Enhancement of image is done to ameliorate the interpretability of details present in image which makes a difference in giving superior input for advance utilize. The quality of the enhanced image can be estimated using certain quality assessment parameters. In this proposed work, quality measure of enhanced images is done through measure of enhancement (EME) [6]. EME is the parameter which makes measure of contrast enhancement. Higher values of EME indicate the betterment of contrast enhancement of original image [5, 6, 12].

3.2 Experimental Results

Acquisition of CT-scan images of lung cancer for the work is done using Radiopaedia database. It is wiki-based educational Web resource which contains reference articles, radiology images, and patient cases. It is largest freely available radiology-related resources. Radiopaedia share the clinical-grade CT-scan images for the research purpose. This dataset comprises axial and coronal view of the CT-scan images recorded with and without the contrast agent [15]. The acquired image is converted from RGB to gray scale. In conventional UM, the resultant image is filtered using high pass filter [16, 17]. The filtered output is scaled and added to original image to get final output [10]. While in proposed work, GL mask is applied on obtained image. The resultant image is scaled with suitable scaling factor and further added to the grayscale image to get ameliorated result. The results of conventional and modified UM along with original CT-scan images are shown in Figs. 2 and 3.

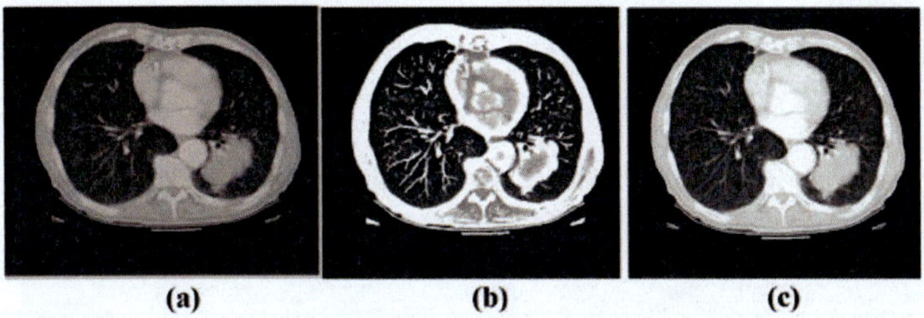

Fig. 2 **a** Original test image#1 [15], **b** conventional UM output of test image#1 and **c** improved UM with GL mask output of test image#1

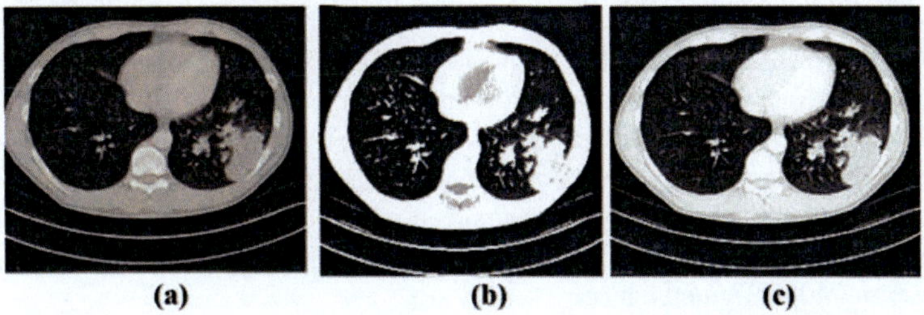

Fig. 3 **a** Original test image#2 [15], **b** conventional UM output of test image#2 and **c** improved UM with GL mask output of test image#2

Table 1 Values of EME

CT-scan images	Original	Conventional UM filter	Improved UM with GL filter
Test image#1	2.4096	4.3458	6.2207
Test image#2	8.6729	9.7867	11.7685
Test image#3	4.6277	5.7400	6.6372

The image quality assessment of enhanced image is carried out using EME as described in Sect. 3.1, and their respective values are listed in Table 1.

3.3 Discussion

Figures 2 and 3 show two CT-scan images, namely test image#1 and test image#2 and their respective enhanced output with conventional UM and improved UM with GL, respectively. Test image#1 shows the cancerous lesion on the right side of lungs. Test image#2 comprises lesion at the bottom part attached with lining of the lungs. The cancerous regions are not so clear, and also, both images are blurred. So, the details of the cancerous region are not efficiently interpretable. When conventional UM is applied to both the images, artifacts are strongly enhanced along with cancerous lesion. The image possesses over-brightness. Hence, both the images are visually unpleasant. However, when improved UM with GL mask filter is applied, sharpness and contrast of the images are increased. Also, the images are visually balanced. The increased values of EME also indicate that the enhanced result obtained by proposed work is better than original images and conventional UM output images.

4 Conclusion

In this paper to get ameliorated enhanced output, GL-based UM is implemented. In the proposed technique, there is a replacement of high pass filter with GL mask to improve the enhanced result of conventional UM. This method overcomes over-brightness and noise amplification problem of conventional UM. The increase in percent of EME indicates the betterment of obtained enhanced image by proposed filter than the image obtained using conventional UM in terms of noise level, contrast and sharpening. In future work, the enhanced image can be further segmented and undergo features extraction and classification process so as to categorize lung cancer into benign and malignant cancer.

References

1. Kavitha MS, Shanthini J, Sabitha R (2019) ECM-CSD: an efficient classification model for cancer stage diagnosis in CT lung images using FCM and SVM techniques. J Med Syst 43(3):73–81
2. Lakshmanaprabu SK, Mohanty SN, Shankar K, Ramirez G (2019) Optimal deep learning model for classification of lung cancer on CT images. Future Gener Comput Syst 92:374–382
3. Stark JA (2000) Adaptive image contrast enhancement using generalizations of histogram equalization. IEEE Trans Image Process 9(5):889–896
4. Hwa SKT, Bade A, Hijazi MA (2020) Enhanced Canny edge detection for COVID-19 and pneumonia x-ray images. In: Proceedings of 1st IOP conference series on materials science and engineering. Johor, Malaysia, pp 12–16
5. Alankrita AR, Shrivastava A, Bhateja V (2011) Contrast improvement of cerebral MRI features using combination of non-linear enhancement operator and morphological filter. In: Proceedings of 3rd IEEE International conference on network and computational intelligence (ICNCI). Zhengzhou, China, pp 182–187
6. Arya A, Bhateja V, Nigam M, Bhadauria AS (2019) Enhancement of brain MR-T1/T2 images using mathematical morphology. In: Information and communication technology for sustainable development, vol 933, pp 833–840. Springer, Singapore
7. Raj A, Alankrita, Srivastava A, Bhateja V (2011) Computer aided detection of brain tumor in magnetic resonance images. Int J Eng Technol 3(5):523–532
8. Tiwari DK, Bhateja V, Anand D, Srivastava A, Omar Z (2018) Combination of EEMD and morphological filtering for baseline wander correction in EMG signals. In: Proceedings of 2nd International conference on micro-electronics, electromagnetics and telecommunications. Singapore, pp 365–373

9. Teramoto A, Hiroshi F (2013) Fast lung nodule detection in chest CT images using cylindrical nodule-enhancement filter. Int J Comput Assist Radiol Surg 8(2):193–205
10. Deng G (2010) A generalized unsharp masking algorithm. IEEE Trans Image Process 20(5):1249–1261
11. Bhateja V, Misra M, Urooj S (2017) Human visual system based unsharp masking for enhancement of mamographic images. J Comput Sci 21:387–393
12. Kaur K, Jindal N, Singh K (2020) Fractional derivative based unsharp masking approach for enhancement of digital images. Multimedia Tools Appl 80(3):3645–3679
13. Bilcu RC, Vehvilainen M (2008) Constrained unsharp masking for image enhancement. In: Proceedings of International conference on image and signal processing. Berlin, Heidelberg, pp 10–19
14. Singh K, Saxena R, Kumar S (2013) Caputo-based fractional derivative in fractional Fourier transform domain. IEEE J Emerg Sel Top Circ Syst 3(3):330–337
15. CT-Scan Database is available at: http://radiopaedia.org. Last updated on 20 Sept 2020
16. Gupta R, Bhateja V (2012) A new unsharp masking algorithm for mammography using non-linear enhancement function. In: Proceedings of International conference on information systems design and intelligent applications. Visakhapatnam, India, pp 779–786
17. Bhateja V, Misra M, Urooj S (2018) Unsharp masking approaches for HVS based enhancement of mammographic masses: a comparative evaluation. Future Gener Comput Syst 82:176–189

Segmentation of Microscopy Images Using Guided Filter and Otsu Thresholding

Ankit Yadav, Vikrant Bhateja, Disha Singh, and Bhavesh Kumar Chauhan

Abstract Microscopy images are acquired by capturing the microscopic view of blood sample under a microscope using a camera. The image quality is not that reliable further making bacteria segmentation a challenging task. Guided image filter (GIF) is reliable for contrast enhancement and noise reduction. Otsu thresholding (OT) has been a suitable algorithm for image segmentation. This paper proposes a combinative approach of aforesaid methods for segmentation of bacterial cells in microscopy images. The image quality assessment (IQA) of the enhanced image is evaluated using parameters like standard deviation (SD) and enhancement measure estimation (EME). This combinative approach produces better segmentation results, and the same is proved by IQA.

Keywords Microscopy images · GIF · OT · EME

1 Introduction

Bacterial species identification is necessary because the biological information of the microorganisms is very essential in the field of medicine, veterinary science, farming, biochemistry, and food industry. Although many microorganisms are useful in our daily lives (such as streptococcus for fermentation of dairy and vegetable products), many diseases are caused by them (often they are infectious). Therefore, identifying them becomes obligatory so that the infection level can be diagnosed and cured properly [1]. Microscopes are used as microorganisms are not visible to the naked eye. Microscopy images are termed as the images of the microorganisms captured under a microscope. Microscopy images are acquired using samples (blood, oral cavity or urine) from the human body using which a slide is prepared and further placed under a microscope for capturing an image. Further, with the help of these images,

A. Yadav · V. Bhateja (✉) · D. Singh · B. K. Chauhan
Department of Electronics and Communication Engineering, Shri Ramswaroop Memorial College of Engineering and Management (SRMCEM), Lucknow, Uttar Pradesh 226028, India

Dr. A. P. J. Abdul Kalam Technical University (AKTU), Lucknow, Uttar Pradesh 226031, India

P. S. R. Chowdary et al. (eds.), *Evolution in Signal Processing and Telecommunication Networks*, Lecture Notes in Electrical Engineering 839,
https://doi.org/10.1007/978-981-16-8554-5_23

microorganisms are detected and counted by the pathologist [2]. Because of poor visibility, contrast, acquired noise, etc., the credibility of these images is questioned for segmentation purpose. For increasing the accuracy of segmentation, enhancement of the microscopy images is necessary. After enhancement, these images can be segmented to obtain only the meaningful regions of the image. To achieve better results, preprocessing techniques are used. In the works [3–5], multi-scale retinex (MSR) is used for the contrast enhancement of the image. Dynamic range compression and color constancy are obtained by this method. The GIF [6] is used for enhancement as well as noise filtering in the microscopy images. It mostly emphasizes on the various uses of GIF such as noise improvement, contrast enhancement, and feathering. Segmentation using OT followed by morphological operations and counting of RBCs and WBCs is carried out in Ref. [7]. In Ref. [8], OT is used for segmentation of written text from images, but the image preprocessing techniques can be improved to achieve better results.

In this paper, GIF and OT are used together, to produce better segmentation results as GIF provides a very reliable results in image enhancement. The organization of the rest of the paper is in the following sections: Sect. 2 describes the general overview of GIF and OT; In Sect. 3, discussion is on the IQA, summary of the results, performance metrics, and discussion of the outcomes of this work. In Sect. 4, conclusion of the work is discussed.

2 Proposed Method for Segmentation of Microscopy Images

2.1 *Guided Image Filter (GIF)*

GIF is a widely used bilateral filter because of its extraordinary edge-preserving property, and also, its efficiency is very high [9]. Considering the guidance image and a specified window, the pixel values are calculated by GIF using a linear model. GIF has three controlling parameters which are regularization parameter, window size, and multiplication factor. The values of these parameters are decided according to the image selected as the input image. Because of the suitable values of these parameters, gradient reversal artifacts are prevented in the GIF, thus making it more reliable than the conventional bilateral filters. GIF can be used in a large number of applications such as smoothening, enhancement, feathering, and noise removal. [10]. If the guidance image and all the controlling parameters are selected and used appropriately, GIF can be used for making the image smoother and more structured. The values of the controlling parameters differ for different types of images depending on the size of the objects present in the image. Therefore, to avoid any kind of distortions in the input image, the values of the controlling parameters should be selected with utmost care. In this work, GIF is used with window size of 16, regularization parameter as 0.01, and multiplication factor as 6. The algorithm for the GIF is given below:

Algorithm 1: Procedural steps for enhancement using GIF

BEGIN
Step 1: Input $p =$ test image
Step 2: Input $i =$ guidance image
Step 3: Input $e =$ regularization parameter
Step 4: Input $r =$ window size
Step 5: $\text{mean}_p = f_{\text{mean}}(p)$
Step 6: $\text{mean}_i = f_{\text{mean}}(i)$
Step 7: $\text{corr}_i = f_{\text{mean}}(i * i)$
Step 8: $\text{corr}_{ip} = f_{\text{mean}}(i * p)$
Step 9: $\text{var}_i = \text{corr}_i - \text{mean}_i * \text{mean}_i$
Step 10: $\text{cov}_{ip} = \text{corr}_{ip} - \text{mean}_i * \text{mean}_{ip}$
Step 11: $a = \text{cov}_{ip}/(\text{var}_i + e)$
Step 12: $b = \text{mean}_p - a * \text{mean}_i$
Step 13: $q = \text{mean}_a * i + \text{mean}_b$
Step 14: Output the result q
END

2.2 *Otsu Thresholding*

Segmentation on the preprocessed image is applied to retain the bacteria present in the image and remove the unwanted background. The process in which the image is partitioned into multiple segments to achieve a more meaningful image which is easier to analyze is called segmentation [11]. In this work, OT is used for segmenting the preprocessed image to extract the meaningful regions of the input image. The preprocessed images are converted to gray level followed by binarization. OT works according to the mathematical Eq. (1) [12]:

$$p(x, y) = \begin{cases} 255 \; q(x, y) \geq t \\ 0 \quad q(x, y) \leq t \end{cases} \tag{1}$$

where $p(x, y)$ is the value of pixel of the result at (x, y); $q(x, y)$ is the value of input image pixel, and t is the optimum threshold value.

OT is an automatic region-based segmentation algorithm which largely depends on the value of threshold selected. For achieving reliable segmentation, special attention has to be given while selecting an optimal value of threshold [13]. In our proposed method, OT segmentation is applied at a threshold value of 0.9. The threshold value was tested at a very large set of values, but at $t = 0.9$, the result of thresholding was giving the best results which were very clear in appearance without any distortions.

The above explained methods for preprocessing and segmentation are applied together for achieving desirable results. GIF produces very dependable results for image enhancement which is further used for segmentation using OT.

3 Results and Discussions

3.1 *Image Quality Assessment*

In the above discussed methods, contrast enhancement is carried out by GIF and followed by segmentation using OT. Thus, quality assessment of the output at each step for verification of the result is necessary. Hence, contrast enhancement of the response of GIF is evaluated by calculating EME [14] and SD [15] for both input and the output images as they are absolute values. Higher the value of EME and SD, better is the result. The quality assessment of the segmentation process is carried out by calculating SD of the segmented image which is again an absolute value. Higher value of SD claims the better results.

3.2 *Experimental Results*

The above discussed simulation is performed on the dataset of microscopy images of bacterial cells taken from DiBAS dataset [16] because it consists of bacterial microscopy images with bacterial cells of varying size. In the initial stage, some of the images from the dataset are selected as test images. The GIF is applied for contrast enhancement as discussed in Sect. 2.1. Following it OT is applied as discussed in Sect. 2.2. Response of the simulation is shown in Figs. 1 and 2. In Fig. 1, Test image#1 is shown along with the results at all the stages of this work, while Fig. 2 shows rest of the test images along with the final output used for segmentation.

The IQA parameters as discussed in Sect. 3.1 of all the test cases at all the stages are calculated. These corresponding values are compared in Table 1.

3.3 *Discussions*

Figure 1 depicts the test image 1 and the responses at all the proceeding steps, while Fig. 2 shows the test image 2 and test image 3 and the segmented image generated by OT in the final stage. The original test image 1 consists of the bacterial cells, but they are hazy in appearance, thus making the segmentation process less reliable and the result produced on this image loses its information. The GIF response is clear in appearance, and the bacterial cells are clearly distinguished from each other. The

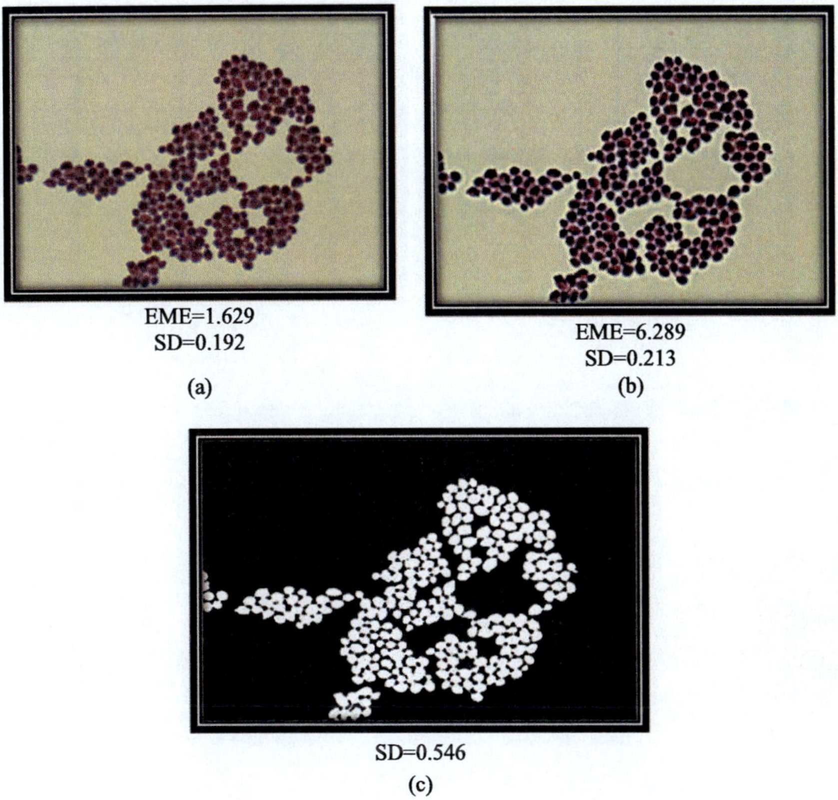

Fig. 1 **a** Original test image#1, **b** GIF response, **c** segmentation using OT

contrast of the image is also improved as proved by the IQA parameters depicted in Table 1. Segmentation is carried out in this image which retains only the required information in the picture while removing the rest of it. Similarly, test images 2 and 3 also undergo the same procedure to give the segmented image, respectively. Table 1 compares IQA parameters at all the stages of processing.

4 Conclusion

In this paper, an improved approach for segmentation of bacterial cells in an image is presented in which the image preprocessing is carried out by GIF which is used for both contrast enhancement as well as noise filtering. Further, segmentation is carried out by OT. The resultant image after GIF shows very good visual characteristics as depicted in Fig. 1 and is further proved by IQA parameters such as EME and SD. The assessment of segmentation is proved by comparing the SD of original and the OT

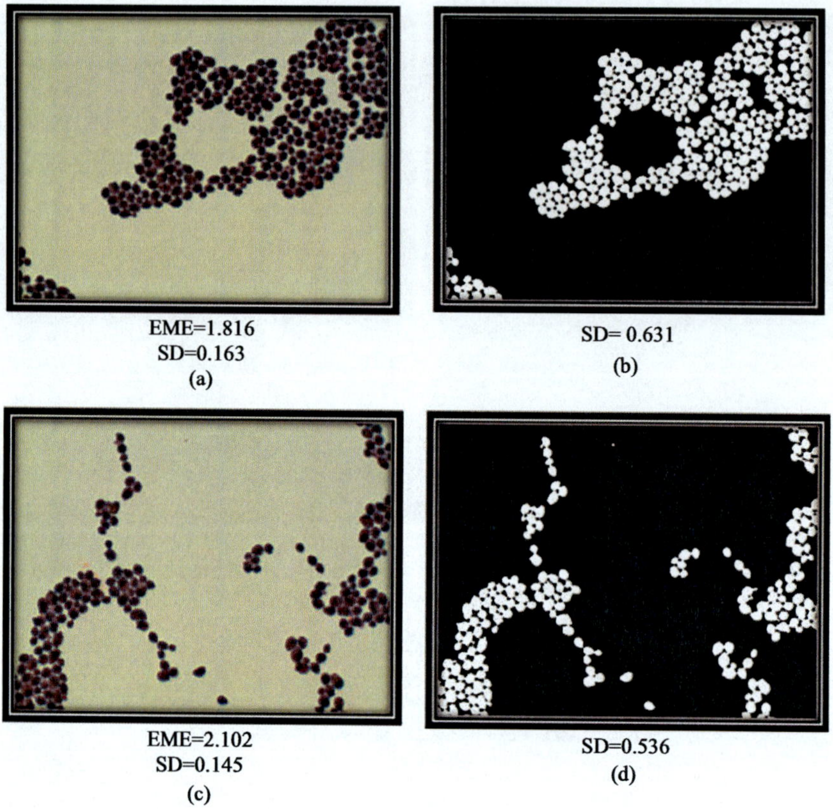

Fig. 2 **a** Original test image#2, **b** bacteria segmentation using OT, **c** original test image#3, and **d** bacteria segmentation using OT

Table 1 IQA at all the stages of processing

Test cases	Original image		GIF response		OT response
	EME	SD	EME	SD	SD
Test case#1	1.629	0.192	6.289	0.213	0.546
Test case#2	1.816	0.163	7.325	0.201	0.631
Test case#3	2.102	0.145	5.869	0.223	0.536

responses. Therefore, in this proposed segmentation method, the visual quality of the image is greatly improved which produces better results with segmentation. The images after segmentation can further be used for feature extraction and classification of bacterial species for the diagnosis of diseases. This segmented image can also be used for counting of the microorganisms which are present in the image for diagnosis of the disease. Hence, this work provides with a significant and dependable result.

References

1. Zielinski B, Oleksiak AS, Rymarczyk D, Piekarczyk A (2020) Deep learning approach to describing and classifying fungi microscopic images. PLoS ONE 1–21
2. Nizar A, Yigit A, Isik Z, Alpkocak A (2019) Identification of leukemia subtypes from microscopic images using convolutional neural network. Diagnostics 9(3):1–11
3. Jobson DJ, Rahman Z, Woodell GA (2001) Spatial aspect of color and scientific implications of retinex image processing. Visual Inf Process Int Soc Opt Photonics 4388:117–128
4. Rahman ZU, Jobson D, Woodell J, Woodell GA (2004) Retinex processing for automatic image enhancement. J Electron Imaging 13(1):100–110
5. Barnard K, Funt B (1997) Analysis and improvement of multi-scale retinex in color and imaging. Soc Imaging Sci Technol 1997(1):221–226
6. He K, Sun J, Tang X (2010) Guided image filtering. In: Proceedings of European conference on computer vision. Berlin, Heidelberg, pp 1–14
7. Win KY, Choomchuay S (2017) Automated segmentation of cell nuclei in cytology pleural fluid images using OTSU thresholding. In: Proceedings of International conference on digital arts, media and technology (ICDAMT). Chiang Mai, Thailand, pp 14–18
8. Oliver N, Morse B, Barret W (2011) A recursive Otsu thresholding method for scanned document binarization. In: Proceedings of IEEE workshop on applications of computer vision (WACV). Kona, HI, USA, pp 307–314
9. Awasthi N, Katare P, Gorthi SS, Yalavarthy PK (2020) Guided filter based image enhancement for focal error compensation in low cost automated histopathology microscopic system. J Biophotonics 13(11):1–23
10. Sharma A, Bhateja V, Sinha AK (2015) Synthesis of flash and no-flash image pairs using guided image filtering. In: Proceedings of 2nd International conference on signal processing and integrated networks (SPIN). Noida, India, pp 768–773
11. Gupta A, Tripathi A, Bhateja V (2013) Despeckling of SAR images in contourlet domain using a new adaptive thresholding. In: Proceedings of 3rd IEEE international advance computing conference (IACC). Ghaziabad, India, pp 1257–1261
12. Taquee A, Bhateja V, Shankar A, Srivastava A (2018) Combination of wavelets and hard thresholding for analysis of cough signals. In: Proceedings of second world conference on smart trends in systems, security and sustainability (WorldS4). London, UK, pp 266–270
13. Raj A, Srivastava A, Bhateja V (2011) Computer aided detection of brain tumor in magnetic resonance images. Int J Eng Technol 3(5):523–533
14. Bhateja V, Nigam M, Bhadauria AS, Arya A, Zhang EY (2019) Human visual system based optimized mathematical morphology approach for enhancement of brain MR images. J Ambient Intell Humanized Comput 1–9
15. Krishan A, Bhateja V, Sahu A (2014) Medical image fusion using combination of PCA and wavelet analysis. In: Proceedings of International conference on advances in computing, communications and informatics (ICACCI). Delhi, India, pp 986–991
16. The bacterial image dataset (DIBaS) is available online at: http://misztal.edu.pl/software/databases/dibas/. Last visited on 10 Dec 2020

Contrast Enhancement of CT-Scan Images of Lungs Using Morphological Filters

Alka Singh, Vikrant Bhateja, Akansha Singh Rathore, and Akshita Shukla

Abstract Lung cancer is one of the most fatal pulmonary diseases. Its screening can be done by multiple techniques such as X-ray imaging and computer tomography (CT) scan imaging. The CT-scan images have much better reliability and feature characteristics as compared to other screening techniques. Sometimes the traits and features of these images cannot be recognized due to the dire quality of the image attained. This paper proposes the use of combination of morphological operators of morphological filtering (MF), like open and bottom-hat function which can be helpful in reducing the noise of the image and also sharpens the image. The background details and visibility are also improved. The image quality assessment (IQA) of the enhanced image is done by the use of measure of enhancement (EME). The escalating values of the IQA convey that the noise from the image has been removed, the characteristics of the image appear to be sharper and retain the important features of the image.

Keywords CT-scan imaging · MF · EME

1 Introduction

Lungs are one of our body's most important organs. They are a set of soft and porous organs positioned on either side of the chest that are filled with air. The lungs are responsible for providing oxygen to the blood through respiration [1]. The lung cancer is hard to detect at an early stage due to the fact that lesion is imperceptible at lower stage which results in delay with the medical treatments required to cure the disease, which may further lead to death. This can be avoided by detecting the lung cancer at a very early stage [2]. Amongst all the screening techniques, the CT-scan imaging provides more details, better feature visibility and also improves the characteristics

A. Singh · V. Bhateja (✉) · A. S. Rathore · A. Shukla
Department of Electronics and Communication Engineering, Shri Ramswaroop Memorial College of Engineering and Management (SRMCEM), Lucknow, Uttar Pradesh 226028, India

Dr. A. P. J. Abdul Kalam Technical University (AKTU), Lucknow, Uttar Pradesh 226031, India

P. S. R. Chowdary et al. (eds.), *Evolution in Signal Processing and Telecommunication Networks*, Lecture Notes in Electrical Engineering 839,
https://doi.org/10.1007/978-981-16-8554-5_24

of the image than the other techniques [3]. These CT-scan images also have some drawbacks like overlapping, variation in the density of nodules, obstructions, etc.

The details of the lung cancer can be impacted by the undesirable distortion and low-quality pixels that are attained through the screening. To overcome this, the contrast enhancement of the acquired image is done. Work related to enhancement of CT-scan images can be carried out by different methods such as threshold filter—it is simple and has less parameters [3], Gaussian filter reduced the noise and also acts as a high pass filter [1], histogram equalization technique reduces the noise and also reduces the details [4], and erosion filter is also a part of MF which increases the quality of image [5]. Further contrast enhancement can be done by using HSO algorithm and Otsu-based system which is a complex methodology [3], and Gabor filter reduces the details of the image [1], CNEF [2] and MF with modified Canny edge detection [5]. The problems in pre-existing method are that complexity of the algorithm increases and the important details of the images can get reduced. To minimize these, MF is used as it is a simple and fast technique for contrast enhancement [6]. This paper uses the combination of morphological operators for the enhancement of the CT-scan images. The structuring of the rest of the paper is in the following sections: Sect. 2 describes the outline of MF and proposed methodology; Sect. 3 discusses image quality assessment, reviews the results and discusses outcome of work; Sect. 4 briefs the conclusion of the work.

2 Proposed Design Methodology

2.1 Overview of Morphological Filter (MF)

MF is a nonlinear filter with various underlying operations whose features are set by the structuring properties of the entities [7]. These features can be enhanced on the basis of the type of structuring element used. There is multiple structuring element like rectangle, diamond, line, disk, etc. Here, on the basis of the characteristics of lesion, the disk-shaped structuring element is used. A structuring element is a matrix that defines the neighborhood and brackets the pixel in the image being prosecuted [8]. The morphological operators are erosion as mentioned in Eq. (1), dilation as mentioned in Eq. (2), opening as mentioned in Eq. (3) and closing as mentioned in

Table 1 List of morphological operators [9]

Morphological operations	Formulas
Erosion	$(A \ominus B)$ (1)
Dilation	$(A \oplus B)$ (2)
Opening	$A \circ B = (A \ominus B) \oplus B$ (3)
Closing	$A \bullet B = (A \oplus B) \ominus B$ (4)

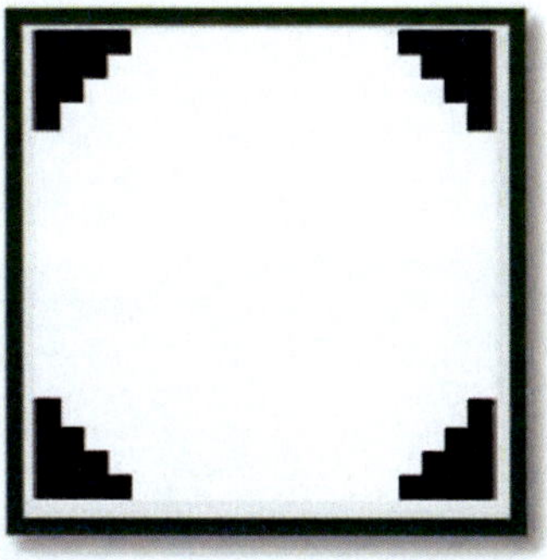

Fig. 1 Disk structuring element of radius 10

Eq. (4). These functions are tabulated in Table 1. The bottom-hat and top-hat filtering are the combination of closing and opening, respectively, with the input image [9].

2.2 Proposed MF Combination

This paper involves the refinement of the attained CT-scan images with the use of MF. To implement this technique, firstly the image is acquired from the data set [5, 6]. Then, the acquired image is converted into the grayscale image which also converts it into a two-dimensional (2-D) image which can be used for the application of further operations. Then, the combination of open function and bottom-hat function is used [10]. Here, two separate structuring elements are taken into consideration as shown in Fig. 1.

The open function is used to increase the background details, i.e., it increases the brightness of the objects and hence improves its visibility. The attained image is then subtracted from the grayscale image and then again added to the grayscale image to attain the important features of the object [7]. The bottom-hat acts as a low pass filter (LPF) which in turn reduces the brightness of the image and highlights the darker region present in there. Contrast adjustment is then applied to the image to make it more visually perceivable. This output and the output from the open function are then subtracted, and the final enhanced image is attained [11]. This process is shown below by a flow diagram in Fig. 2.

2.3 IQA

Modification in the values of pixel to increase the quality of the image and increasing its details is called enhancement. But in certain cases, the enhancement may also result in the increment of noise or it may be overexposed or underexposed. Therefore, evaluation of the enhanced image is obligatory [12]. There are many parameters

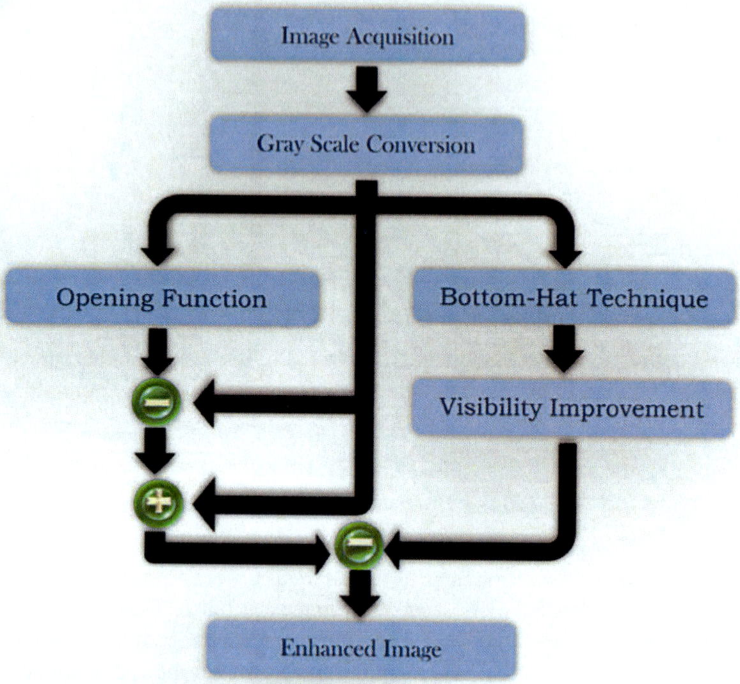

Fig. 2 Flow diagram of the proposed methodology

available on which the quality of the enhanced image can be assessed. EME is one of the important IQA functions that is used in this paper [13].

Algorithm 1: Algorithm to attain enhancement through MF

Begin
Step 1: Input CT-scan image acquisition [*I*]
Step 2: Process conversion of the input image [*I*] into grayscale [*G*]
Step 3: Define two structuring elements, one for open and the other for bottom-hat filtering
Step 4: Compute the open function of the grayscale image [*G*], and then, subtract it from the same [*O*]
Step 5: Compute adding image [*O*] and the grayscale image [*G*] to get [*O*1]
Step 6: Compute the bottom-hat transform of the grayscale image, and adjust its contrast [B]
Step 7: Output subtracting image attained from [*O*1] and [B]
End

---

3 Results and Discussion

3.1 Experimental Values

Following through the proposed methodology, firstly the image acquisition is done from the database which is taken from lung image database consortium (LIDC) [14]. This image is then converted into the grayscale image. Then, the structuring element is assigned for both open and bottom-hat functions, respectively. Then, open function and bottom-hat function are applied. The result of both the functions is then subtracted to attain the final image [12]. The final result is discussed in Fig. 3 for Test_Image#A and in Fig. 4 for Test_Image#B. The IQA values of the test images are discussed in Table 2. These values will show a huge increment in their values which in turn specify the quality of the enhancement [15].

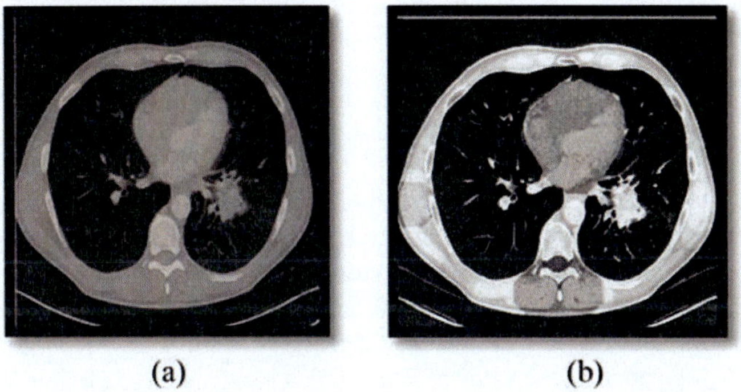

Fig. 3 Enhancement process of lung cancer image, **a** original image Test_Image#A, and **b** the enhanced image attained Test_Image#A [14]

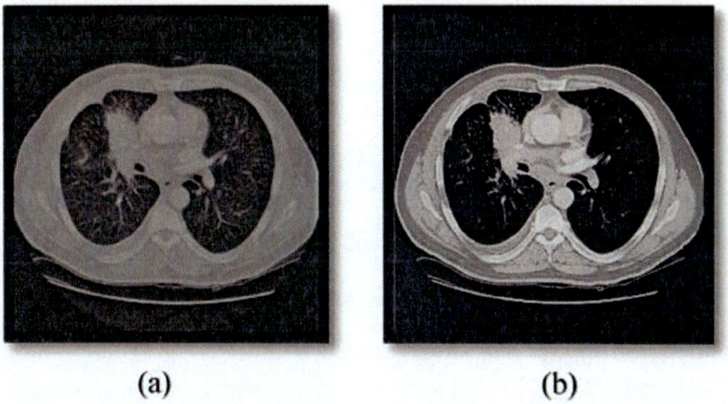

Fig. 4 Enhancement process of lung cancer image, **a** original image Test_Image#B, and **b** the enhanced image attained Test_Image#B [14]

Table 2 Values of IQA obtained for test image, using proposed methodology

Images	EME of original image	EME of enhanced image
Test_Image#A	5.5331	8.3215
Test_Image#B	5.0908	9.0854
Test_Image#C	4.2395	8.0854
Test_Image#D	4.3256	7.3727
Test_Image#E	5.2028	9.6214

The above comparison of the original image and the enhanced image shows that the details of the image have been increased. The visibility of the lesion has increased and hence will be easier to detect [10]. The increment in the EME shows that the image is refined and now has a very specific and detailed representation of the characteristics of the image. This is the result after curtailing the background features. The dark features of the background are brightened, and the overexposed area becomes stalled.

4 Conclusion

An improved approach toward contrast enhancement of the image has been discussed in this paper. The original CT-scan images consist of many distortions, low-quality pixels, etc., which in this paper are overcome by the use of morphological operators, i.e., the combination of bottom-hat function and open function. This combination helped in improving the attributes and features of the image [13]. This can be confirmed by considering the value of EME which is escalated to a higher value after enhancement. The higher values confirm that the quality of the image has increased drastically which in itself is remarkable. This enhanced image can be further used for the future segmentation purposes.

References

1. Kavitha MS, Shanthini J, Sabitha R (2019) ECM-CSD: an efficient classification model for cancer stage diagnosis in CT lung images using FCM and SVM techniques. J Med Syst 43(3):73–81
2. Teramoto A, Hiroshi F (2013) Fast lung nodule detection in chest CT images using cylindrical nodule-enhancement filter. Int J Comput Assist Radiol Surg 8(2):193–205
3. Rajinikanth V, Dey N, Raj ANJ, Hassanien AE, Santosh KC, Raja NSM (2020) Harmony-search and Otsu based system for coronavirus disease (COVID-19) detection using lung CT scan images. arXiv preprint arXiv:2004.03431
4. Nithila EE, Kumar SS (2017) Automatic detection of solitary pulmonary nodules using Swarm intelligence optimized neural networks on CT images. Eng Sci Technol Int J 20(3):1192–1202

5. Hwa SKT, Bade A, Hijazi MA (2020) Enhanced Canny edge detection for COVID-19 and pneumonia x-ray images. In: Proceedings of 1st IOP conference series on materials science and engineering. Johor, Malaysia, pp 12–16
6. Hassanpour H, Samadiani N, Salehi SM (2015) Using morphological transforms to enhance the contrast of medical images. Proc Egypt J Radiol Nucl Med 46(2):481–489
7. Verma R, Mehrotra R, Bhateja V (2013) A new morphological filtering algorithm for pre-processing of electrocardiographic signals. In: Proceedings of the fourth international conference on signal and image processing (ICSIP). India, pp 193–201
8. Bhateja V, Devi S (2011) A novel framework for edge detection of microcalcifications using a non-linear enhancement operator and morphological filter. In: Proceedings of IEEE 3rd international conference on electronics computer technology (ICECT), vol. 5. Kanyakumari, India, pp 419–424
9. Arya A, Bhateja V, Nigam M, Bhadauria AS 2020 Enhancement of brain MR-T1/T2 images using mathematical morphology. In: Information and communication technology for sustainable development. Advances in intelligent systems and computing, vol 933, pp 833–840
10. Chaddad A, Tanougast C (2016) Quantitative evaluation of robust skull stripping and tumor detection applied to axial MR images. Brain Inf 3(1):53–61
11. Verma R, Mehrotra R, Bhateja V (2013) An integration of improved median and morphological filtering techniques for electrocardiogram signal processing. In: Proceedings of IEEE 3rd international conference advance computing. Ghaziabad, India, pp 1223–1228
12. Tiwari DK, Bhateja V, Anand D, Srivastava A, Omar Z (2018) Combination of EEMD and morphological filtering for baseline wander correction in EMG signals. Proceedings of 2nd international conference on micro-electronics, electromagnetics and telecommunications. Springer, Singapore, pp 365–373
13. Bhateja V, Nigam M, Bhadauria AS, Arya A, Zhang EYD (2019) Human visual system based optimized mathematical morphology approach for enhancement of brain MR images. J Ambient Intell Humanized Comput 1–9
14. CT-Scan images of lungs with lung cancer are available at: https://wiki.cancerimagingarchive.net/display/Public/LIDC-IDRI. Last updated on 24 May 2021
15. Franchi G, Fehri A, Yao A (2020) Deep morphological networks. Pattern Recogn 102:107246

Review of Different Binarization Techniques Used in Different Areas of Image Analysis

Shreya Pandey and Jyoti Bharti

Abstract In the last few years, the image binarization area has progressed to a large degree. In this paper, a thorough survey regarding the techniques for image binarization used in different areas is done. For analysis and comparison of the methodologies of binarization used in document, medical and video text images, a variety of traditional methodologies along with the latest works are analyzed. Binarization is one of the most critical processes of nearly all the systems of image processing. Most of the image processing systems are designed in which, the binarization is used to determine the region of interest from a given image which is intended for a particular purpose. A summary of the classical methodologies contained in the literature is discussed in this section. Several scholars have recommended a variety of methods for binarizing document images, video text images, and medical images over the past many years.

Keywords Document enhancement and binarization · Convolutional neural networks · Deep learning · U-Net architecture · MRI · Thresholding

1 Introduction

The two main distinctions of digital images are grayscale images and binary images. A single pixel is assigned an intensity value of 0 and 255 in a grayscale image, where it can take only one of the two values in a binary image, either 0 or 1. Image binarization is defined as the technique for converting a grayscale picture to a binary image. In several research fields, it has numerous applications, particularly for analysis of document images, medical image processing, and processing of other forms of captured images. Binarization is usually the initial step in multi-phase image processing and in applications of image analysis. In the document analysis process, document enhancement and document binarization are the important preprocessing steps. Noise, spots, bleed through, contrast variance, irregular lighting, blurriness,

S. Pandey (✉) · J. Bharti
Department of Computer Science and Engineering, Maulana Azad National Institute of Technology, Bhopal, India

P. S. R. Chowdary et al. (eds.), *Evolution in Signal Processing and Telecommunication Networks*, Lecture Notes in Electrical Engineering 839,
https://doi.org/10.1007/978-981-16-8554-5_25

etc., often led to degradation of historical documents, which makes it impossible to correctly recognize the text and store it in digital form. Document binarization is the way to deal with the above issues and increase the everlasting quality of document images to restore their natural appearance. Another important research area is medical image processing. Almost every medical image analysis methods need to create an initial image in a binary form. The benefit of providing a binary image is, it decreases the data complexity and makes the segmentation process of images easier.

2 Literature Review

Some of the most popular applications of binarization and the binarization methods used for these applications are outlined in this paper.

2.1 Document Image Processing

2.1.1 Non-machine Learning-Based Methods

Many approaches for addressing the binarization of deteriorated records have been proposed in the past. They are primarily categorized into two categories—global thresholding and local thresholding—which are based on the type of approach used. Ostu's method [1] is the most commonly used method which uses global thresholding approach. This algorithm considers that the image on which thresholding is performed, consists of two classes of pixels (background and foreground pixels) and finds the optimal threshold value which partitions those two classes in such a way that their cumulative distribution (intra-class variance) is small. In this approach, the threshold value which reduces the intra-class variance which is defined as the weighted sum of variances of the two classes is searched exhaustively. Minimizing variance within the class is similar to maximizing variance between the classes. This method produces acceptable results when the large number of pixels is alike to each other in each of the classes.

Global thresholding methods give wrong results while distinguishing the objects where background is inconsistent because of bad lighting conditions. If the document illumination is not uniform, global binarization techniques tend to leave a trail of noise around the page's edges. In order to resolve these issues, local thresholding methods have been suggested for the document binarization as shown in Fig. 1. In local thresholding, for each of the pixel, a local threshold is determined using the information found in the local neighborhood of the pixels.

A threshold value is determined at each pixel in local adaptive thresholding algorithms, which makes use of certain local statistics including range, surface-fitting parameters, and variance of the pixel neighborhood. Niblack's technique [2] slides

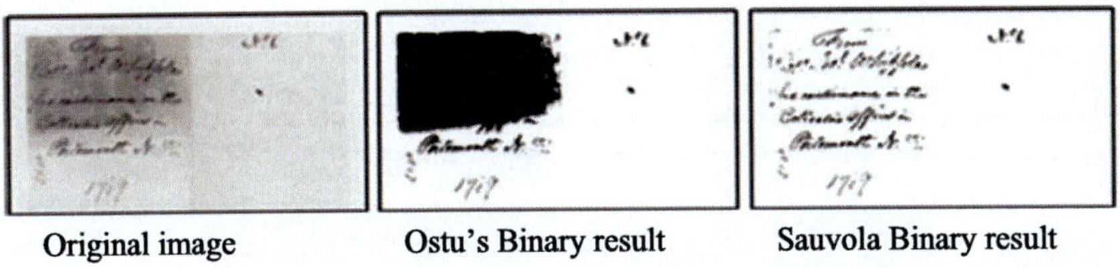

Fig. 1 A comparison of Ostu and Sauvola methods is shown in this example [2]

a rectangular window across the gray-level image and doing it, this calculates pixel-wise threshold. The method acclimates the threshold value in-line with the local mean as $m(i, j)$ and the standard deviation as $\sigma(i, j)$ and thus calculates a window of size $w \times w$. The threshold value T is represented as shown in Eq. (1):

$$T(i, j) = m(i, j) + k\sigma(i, j). \tag{1}$$

where k is constant value, which defines how much of the entire edge of printed object is preserved and possessed a value lying between 0 and 1. At $k = -0.2$ and $w = 15$, this approach produces good outcomes. The k value and the sliding window size determine the binarization quality.

The method given by Sauvola et al. [3] uses a technique based on local variance. It is a refinement of Niblack's method [2], particularly whenever the background contains big variations, light texture, stains, and the documents that are poorly and irregularly lighted. It complies with the significance of the standard deviation. For instance, threshold gets lowered when it comes to text on a stained or dirty paper. The threshold value is computed by using Eq. (2):

$$T(x, y) = m(x, y)\left[1 + k\left\{\frac{\partial(x, y)}{1 - \partial(x, y)} - 1\right\}\right] \tag{2}$$

The typical values of $R = 128$ and $k = 0.5$ give the satisfactory result. Here, m is mean value and σ is standard deviation of the whole window, and k is a bias which varies from Niblack's method whenever the contrast in nearby neighborhood pixels is low. In this case of low contrast, the threshold value $T(x, y)$ goes lower than the mean value, therefore essentially eliminating the analogously darker regions of background. The parameter such as k regulates the threshold value calculated in the nearby window so that higher the k value, lower is the threshold value compared to local mean. The R value was found to have a very limited impact on the output, whereas it was greatly affected by the k values and the window size. A good binary image will be generated by the best mix of the sliding window and the k value.

Bernsen [4] used the local gray range methodology. Here the range obtained between the minimum and maximum pixel values inside the local window is deployed for determining the threshold. Also in this technique, the local threshold $T(x, y)$ at pixel position (x, y) is computed inside a window size of $w \times w$ by using Eq. 3:

$$T(x, y) = 0.5\left(I_{max(ij)} + I_{min(i,j)}\right) \tag{3}$$

where $I_{max(i,j)}$ is maximum gray value and $I_{min(i,j)}$ is the minimum gray value inside the local window, with below condition on the contrast

$$C(i, j) = \left(I_{max(i,j)} + I_{min(i,j)}\right) >= 15 \tag{4}$$

The threshold value is fixed at midrange value in this procedure, which is the mean value of maximum and minimum gray values in a nearby window size of $w \times w$. The good outcomes are given by a value of $w = 31$. Nonetheless, if the contrast $C(i, j)$ is lower than a particular threshold, i.e., 15, then that neighborhood, on the basis of value of $T(x, y)$, is seen to constitute only single class, either background or foreground. Unlike Niblack and Sauvola, the threshold value is not influenced by any bias.

In doing binarization, all the above approaches were taking a large time in computation. A new updated algorithm [5] is given which reduces the computation time required for local threshold estimation, which provides an easy means of calculating local threshold by using some equations. Binarization can be greatly accelerated by making use of the integral sum image for determining the local sum due to which the computational time is independent on the dimensions of window.

An input image's (I) integral sum image (G) is an image where the intensity values at each pixel location are equivalent to the addition of all the pixels' intensity values which are above and to the left of that position in the original input image. After finding the integral sum image G, the local sum $s(x, y)$ of any window size can be calculated in one pass without relying on the size of the window as shown in Eq. 5:

$$\begin{aligned} s(x, y) = {} & [g(x + d - 1, y + d - 1) + g(x - d, y - d)] \\ & - [g(x - d, y + d - 1) + g(x + d - 1, y - d)] \end{aligned} \tag{5}$$

where $d = \text{round}(w/2)$, $m(x, y)$ is the local arithmetic mean at position (x, y) which is average of all the pixels inside the window of size $w \times w$ of the image I. This is computed as shown in Eq. 6:

$$m(x, y) = s(x, y)/w^2 \tag{6}$$

In this way by using the integral sum image, local mean can be computed accurately in one pass without having to depend on nearby window size. This technique needs to calculate the mean deviation $\partial(x, y)$ and local mean to compute the local threshold and local standard deviation is not required. The threshold is calculated from Eq. 7:

$$T(x, y) = m(x, y)\left[1 + k\left\{\frac{\partial(x, y)}{1 - \partial(x, y)} - 1\right\}\right] \tag{7}$$

Here, $\partial(x, y) = I(x, y) - m(x, y)$ is local mean deviation and k is a bias which is used to regulate the adaptation level by varying the threshold value. It ranges between 0 and 1. The local sum can be easily determined by firstly computing the integral sum image. After then mean can be computed easily with arithmetic operation by using the local sum, without having to depend on the window size as given in Eq. (2).

Now the local mean deviation is calculated simply by only subtracting the mean value from the pixel under consideration. Therefore, this method can binarize more quickly than others because the local mean and the local mean deviation calculation takes less time.

In all the above methods [2–6], performance of document binarization varies at different window sizes. Large window size is required for Niblack's and Bernsen's techniques. These two approaches are ineffective with smaller window size for certain images, for instance, with 5×5 window size, the modified and Sauvola's methods provide excellent results as compared to Niblack's and Bernsen's methods as shown in Fig. 2. Sauvola, Niblack's, and Bernsen's methods have computational time complexity. $O(w^2 \times n^2)$ for an image of size $n \times n$, whereas the modified method has $O(n^2)$ only (similar to global threshold).

2.1.2 Machine Learning-Based Methods

In the last few years, the diversity of binarization methodologies focusing on supervised learning approaches has expanded significantly. Deep neural networks (mostly CNNs) with different architectures are commonly used by them and best among them have beaten the traditional method and performed much better than classical methods like Ostu and Sauvola. Deep learning offers a large number of benefits relative to traditional approaches since in a broad training dataset, thousands of parameters of neural networks are made to learn. Therefore, unlike the conventional methods in which a binary value is predicted for each of the pixel in input image, nowadays the neural networks are trained to learn the difference between input and expected output (GT). Also, the output with each iteration is refined.

In Ref. [7], firstly image is enhanced and then binarization is performed on enhanced image. The author applied the CNN to input image directly, enabling the neural network to iteratively identify the degradations and iteratively remove those degradations. This technique uses two approaches to iteratively correct the image: (1) feed the input image into the same neural network, often known as "Recurrent Refinement (RR), for fine enhancement," and (2) train a new network having same or different architectures, known as "Stacked Refinement (SR)." After image refinement, Ostu's method is applied for binarization. For using Otsu's global threshold method, background of the enhanced images obtained from previous step should be uniform. The background was then made uniform, and the images were sampled having different scales on the images which are being tested. This method iteratively refined the output of the degraded input. The output of each iteration was averaged so that the risk of losing weak and thin ink traces could be minimized. After this,

(a) (b)

(c) (d)

Fig. 2 Comparison of different binarization approaches on the text document: (a) Updated method with k = 0.07 (b) Sauvola's method with k = 0.07 (c) Niblack's method with k = -3 and (d) Bernsen's method [5]

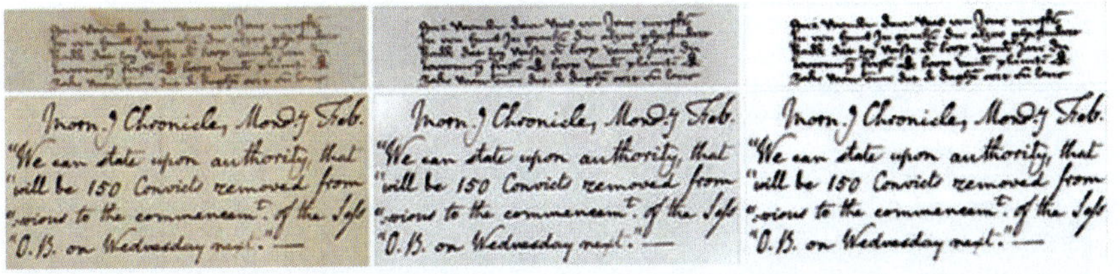

Fig. 3 Some of the illustrations of the enhanced and binarized form of sample document images taken from the H-DIBCO 2014 data set. The original images are shown in left column, the enhanced images by using the SR method are shown in middle column, the binarized images based on enhanced images are shown in the right column [7]

Ostu's method is used for binarization. The combination of all these above methods (fusion + scaling + uniform + Ostu) was named as deepOstu method (Fig. 3).

In Ref. [8], the author considered neural network-based solution. They used U-Net architecture. This architecture contains two paths. First path called contraction path (encoder) describes the context in the image. The encoder is only a conventional stack of convolutional and maxpooling layers. Second path is the symmetrical expanding path (decoder) which enables precise localization by using transposed convolutions.

Thus, it is a fully convolutionary end-to-end (FCN) network; i.e., it just consists of convolutional layers and does not contain any dense layer, which allows it to accept variable size images. This architecture can be applied for a huge range of domains in image segmentation and binarization area (Fig. 4).

In Ref. [], the author has done binarization of musical documents. Apart from traditional problems like erratic lightening or source degradation, musical documents suffer from high degree of heterogeneity. Therefore, the standard methods of document binarization, which were originally developed for text documents are not suitable for the musical scores. There are different specific explanations for this lack of generalizability, but they are primarily due to the high degree of heterogeneity of notation and style of musical documents. In the case of musical documents, binarization approach is based on training a convolutional neural network in which each image pixel is classified as either foreground or background. Hence, the approach basically consists of inspecting each of the image pixel to identify it as either background or foreground. For doing this, the representative data of each interested pixel is used, and a CNN is trained for differentiating between the foreground and background.

2.2 *Binarization of Medical Images*

Binarization is used in the medical image processing in order to classify the region of interest (ROI) from the given image. For medical image processing, binarization is essential. Most of the medical image processing methods require to yield a binary form of original input image. The benefits of producing a binary image are that it

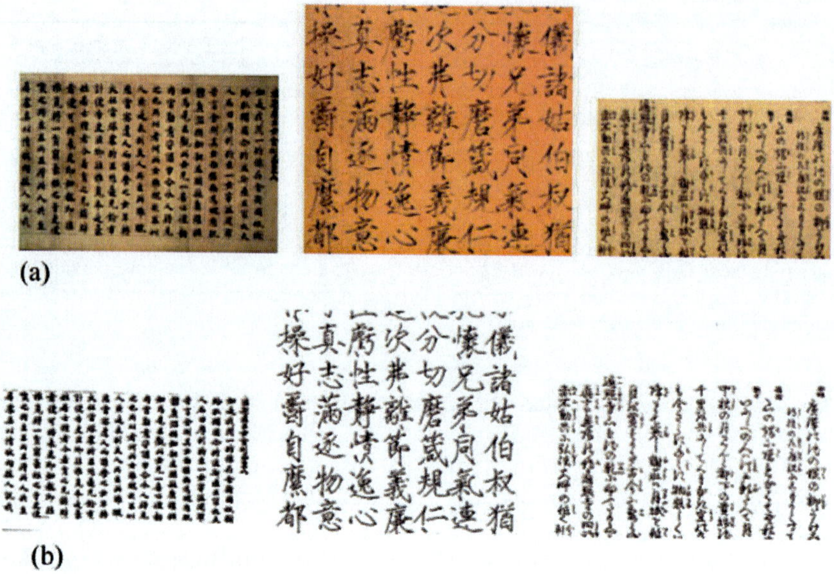

Fig. 4 Examples of documents with hieroglyphs with above binarization results [8]: **a** Source image, **b** resultant binarized image [8]

decreases the data complexity and simplifies the image segmentation process which is required for detecting the brain tumor using MRI.

Using MRI images brain tumor identification requires two processes: preprocessing and post-processing. For example in Ref. [10], the various steps for detecting brain tumor are shown in Fig. 5.

In Ref. [10], Ostu's method is used for thresholding purpose. The thresholding-based technique developed by Otsu is frequently used to segment a picture for subsequent processing, such as feature analysis and for binary image transformation. Here, Otsu's segmentation was applied just on the resulting blue and green channels images and the red channel was neglected since it was noticed during research that only blue and green channels are necessary for the detection of brain tumor present in the RGB grayscale image. After this, by adding the blue and green channels the resultant image was generated, and after this, the complement of resulting image was computed. On the complemented image, Otsu's method was performed, and the image is converted into binary form. In Ref. [11], thresholding was performed just before the image segmentation. On the basis of detection of the highest input image pixel, the required threshold for the portion of brain tumor was determined. The highest intensity image pixels were identified as tumor portions. The threshold was adjusted in such a way that its value was 30 pixels smaller as compared to maximum pixel value. The image pixels with value more than threshold value are labeled with value 1, and rest are assigned the value 0. The pixels having value "one" are identified as tumor portion, and others are identified as portions of normal image. After this step, level

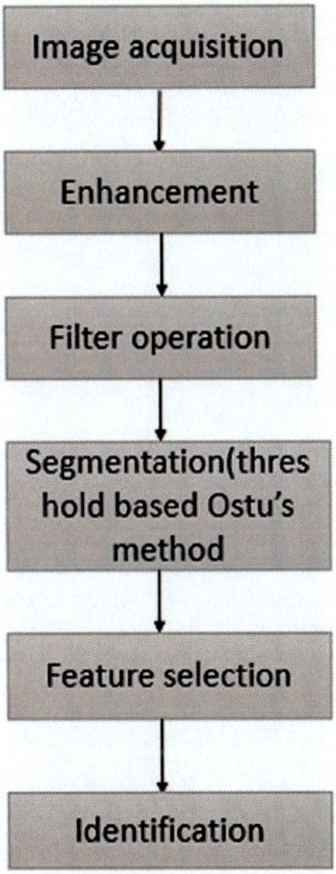

Fig. 5 Brain tumor identification system

set segmentation was done. The thresholding process was used to find out the initial seed required for the level set segmentation. The tumor segment acquired from the thresholding technique was used as the first step in the segmentation process.

In Ref. [12], the author has described a new method for medical image binarization which is based on square wave representation. A square wave is a kind of waveform, in which the input signal consists of two levels +1 (foreground) and −1 (background). In this method, using the current intensity value, a local threshold value was computed at each interval. Using this local threshold value, the image pixel was then labeled with a value +1 or −1. The steps used in this method are summed up as follows:

Step 1: Assume $f(x, y)$ as the intensity values of medical image taken as input.
Step 2: Calculate $f_t(x, y)$ by using Eq. 8 given below:

$$T(x, y) = m(x, y)\left[1 + k\left\{\frac{\partial(x, y)}{1 - \partial(x, y)} - 1\right\}\right] \tag{8}$$

Step 3: Recur steps (i) to (iii) for each of the f_t (x, y) value in f_t.

(i) Calculate the local threshold value Ti with the help of following Eqs. 9 and 10.

$$T_i = 2 \times f_t(x, y) \times \left(\frac{d}{100}\right) \tag{9}$$

$$d = \frac{s}{N(f)} \times 100 \tag{10}$$

(ii) Assume t^l as the normalized value of f_t(x, y) whose value lies in range of 0 and 2π in f_t.

(iii) Create a binary image g by comparing t^l and T_i and by using Eq. 11 given below:

$$g(x, y) = \begin{cases} 1 \text{ if } f_t^l(x, y) \le T_i \\ 0 \quad \text{otherwise} \end{cases} \tag{11}$$

The main benefit of this technique is that the performance of this method can be evaluated even without availability of ground truth, with the help of region non-uniformity (NU) measure. The NU metric assess the internal quality of the segmented value and is not dependent on ground-truth data. NU measure is defined as:

$$\text{NU} = \sigma_0^2/\sigma_T^2 \tag{12}$$

where σ_T^2 is variance of entire image and σ_0^2 is the variance of foreground. It is anticipated that in the best case (well-segmented image) NU measure will be near to 0, while in worst case it is 1 (Fig. 6).

The results obtained with the use of square wave method are equal to or even performing better than Otsu's method, but the key difference between the two is that less computational complexity is needed for square wave method. The drawback of square wave method is that for noisy images it does not give good results.

Niblack's and Sauvola's techniques also give better results in medical image binarization. In Niblack's technique, the neighborhood size should be lower enough

Fig. 6 Binarized images derived by thresholding. Column 1 represents original image, column 2 represents binarized image obtained by the square wave binarization method, and column 3 shows binarized image obtained by Otsu's method [12]

Image	Using Niblack Technique	Using Sauvola Technique
Image1	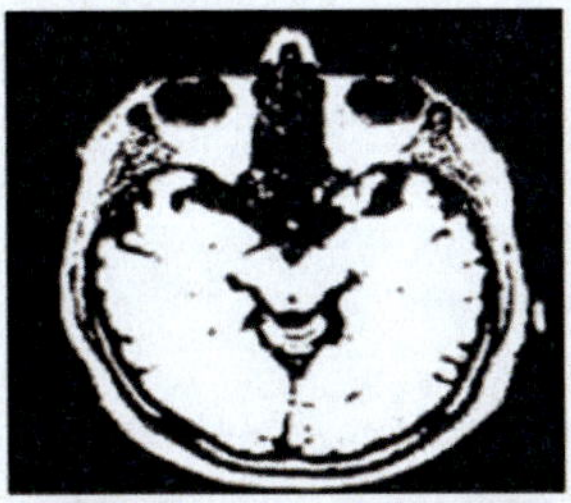	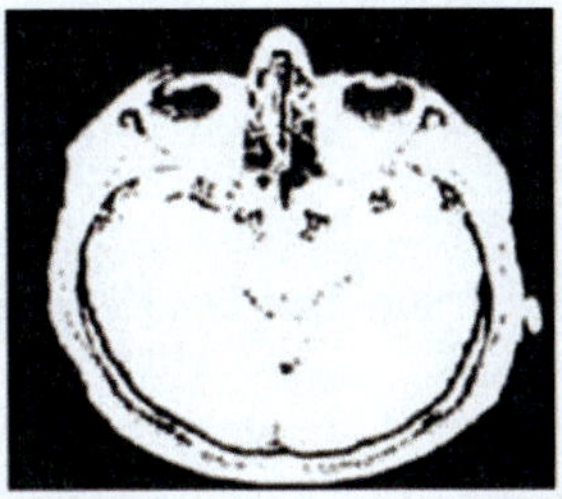
Image 2	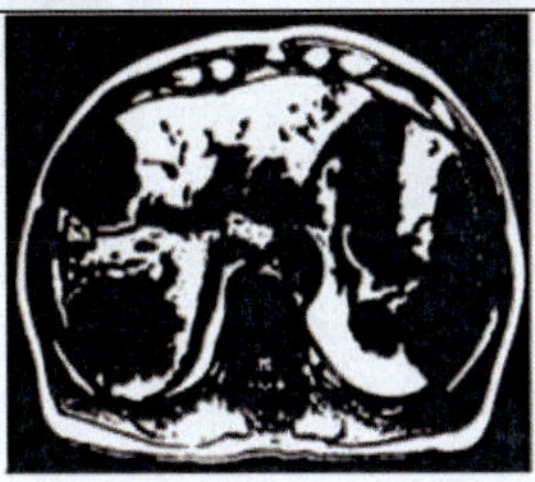	
Image 3	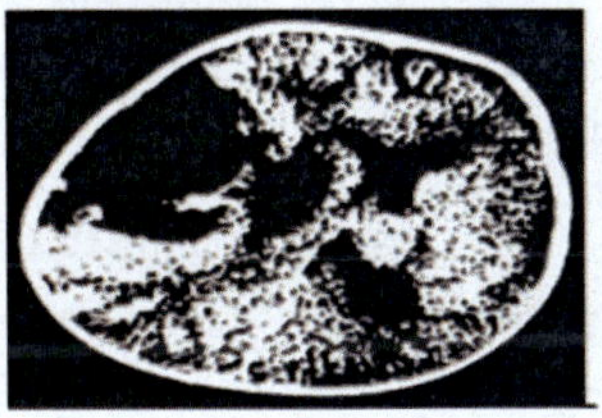	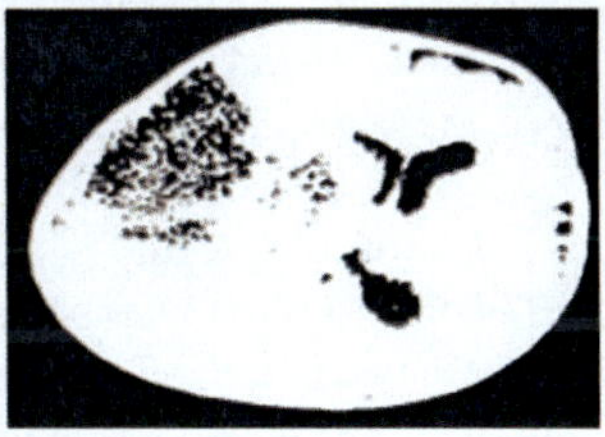

Fig. 7 Comparison of Niblack's and Sauvola's methods on medical images [13]

to retain the local information, but at same time it should be big enough to minimize the noise. In Ref. [13], Niblack's and Sauvola's thresholding algorithm is applied on medical images. In Fig. 7, Niblack's and Sauvola's thresholding techniques are compared. Niblack's method minimizes the background noise more effectively as compared to Sauvola method. The efficiency of these algorithms was calculated using the PSNR segmentation parameter, which reveals that Niblack's algorithm performs better than the Sauvola algorithm.

2.3 *Binarization of Scene Images and Video Text Images*

Text present in the scene images contains information which is vital and is then exploited in various content-based videos and in other image applications. Video text recognition is also an important matter for indexing of visual knowledge contained in video archives. Because of the presence of low resolution, complex background, variations in font and font sizes, and several text orientations, it has become a challenge to create an automated method for identifying video texts written on signboards, names of street, room numbers, names of building, and names of hotel.

2.3.1 Non-machine Learning Methods

In Ref. [14], the author presented a novel image binarization scheme based on variance for the automatic segmentation of text taken from lower resolution scene images. A colored image is basically made up of RGB planes, i.e., red (R), green (G), and blue (B) planes. Each plane is moved through the binarization phase separately. The various steps involved in this method are (Fig. 8):

1. Firstly, from grayscale image, the variance matrix is determined, which marks the change in the intensity values of the image pixels.
2. Using Ostu's binarization process, we binarize the variance matrix and divide the image in two groups, one with high variance values and another with low variance values.
3. After then, by making use of each of the group, two grayscale images are formed (one taken from the black region and other taken from the white region of the binarized image obtained in previous step). Binarizing the two above-obtained images (one from the black region and other from the white region) independently would produce much more even binarization since there are no fluctuating pixel intensities in them.
4. These two binarized images each from black and white region are then combined to form binarized image for a particular plane.
5. Finally, to get the final binarized image, binarized images from each plane are mixed. Figure 9(a) shows the binarized image of variance matrix and is computed by moving a window of size 5×5 over the image.

Images in Fig. 10a (corresponding to white region) and Fig. 10b (corresponding to black region) are combined to form binarized image for R plane as shown in Fig. 10e. In the same way, white and black pixels of G plane are combined to form binarized image for G plane. In the same way, binarized image for B plane is obtained. Finally,

Fig. 8 **a** Input colored image, **b** R plane, **c** G plane, and **d** B plane

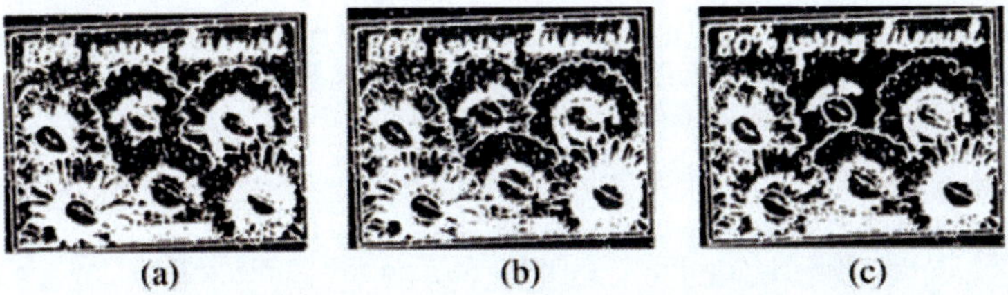

Fig. 9 Binarization of variance matrices over **a** R plane, **b** G plane, and **c** B plane

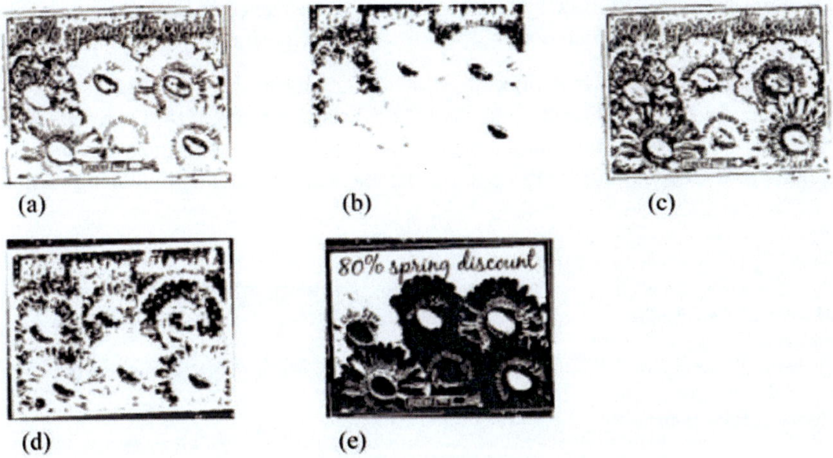

Fig. 10 Binarized image of grayscale corresponding to **a** White pixel in 9(a), **b** Black pixel in 9(a), **c** White pixel in 9(b), **d** Black pixel in 9(b), **e** Merged image of 10(a) and 10(b) (binarized image of R plane) [14]

binarized images from each of the R, G, B planes are merged to obtain final image (Fig. 11).

In Ref. [15], the author suggested an efficient approach for binarizing video texts, based on edge detection. With the help of this method, texts having complex or low

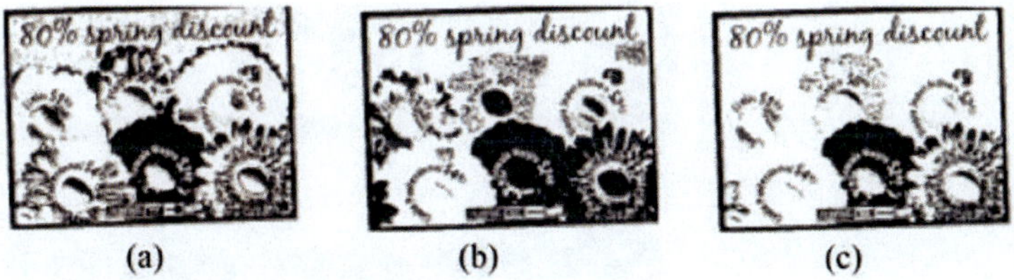

Fig. 11 **a** Binarized image of G plane, **b** binarized image of B plane, and **c** final binarized image (merged 10(e), 11(a), 11(b)) [14]

contrast background can be efficiently separated from one other and thus can be accurately identified by the OCR software.

The various steps of this method are as follows:

1. Using Canny edge detector, draw a clear boundary around text for video text.
2. Select a seed that is located on the text's inner side.
3. The flood filling algorithm is then performed to the block by using the seeds which are obtained in the previous step which now serve as the starting nodes. Flood fill algorithm is applied to fill up the contour formed in earlier step. This filling up of contour will make the text identifiable to OCR.
4. Remove the false edges. In the text block, there may be some background objects edges, which may preclude OCR from accurately performing the segmentation.

The outcomes of this technique are shown in the following figures (Fig. 12).

This technique significantly improves the precision of video text OCR detection having a low contrast rate and a complex background. This approach can quickly eliminate false edges, unlike other approaches and hence poses no problem for further OCR processing.

Fig. 12 **a** Detection of edge, **b** Selection of seeds. The selected seeds are indicated by green color. **c** Flood filling. **d** Removal of false edges and color reverse [15]

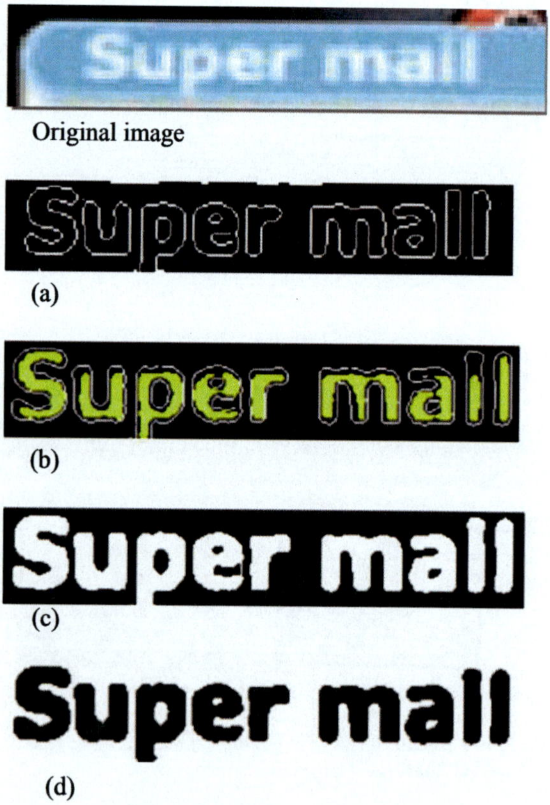

2.3.2 Machine Learning-Based Methods

In Ref. [16], for color text images, the author has presented a novel automated binarization scheme, which was based on supervised learning. In this method, the author proposed a neural architecture based on a convolutionary neural network architecture, referred to as convolutional text binarizer (CTB). This proposed scheme automatically learns to do binarization from a training collection of synthesized text images and their matching intended binary images which are based on a particular architecture of convolutionary neural networks, without making any predictions or using tunable parameters. CTB comprises of five different heterogeneous layers as shown in Fig. 13. Each layer consists of feature maps that are the outcomes of convolution, subsampling, up-sampling, or inverse convolution operations. Performing these automatically trained operations and integrating them ensures that robust features are extracted, which leads to the automated creation of binary image.

In Ref. [18], the author has developed a new system for identifying the video texts by performing binarization with the help of a Bayesian classifier. This method mainly involves three steps as shown in Fig. 14:

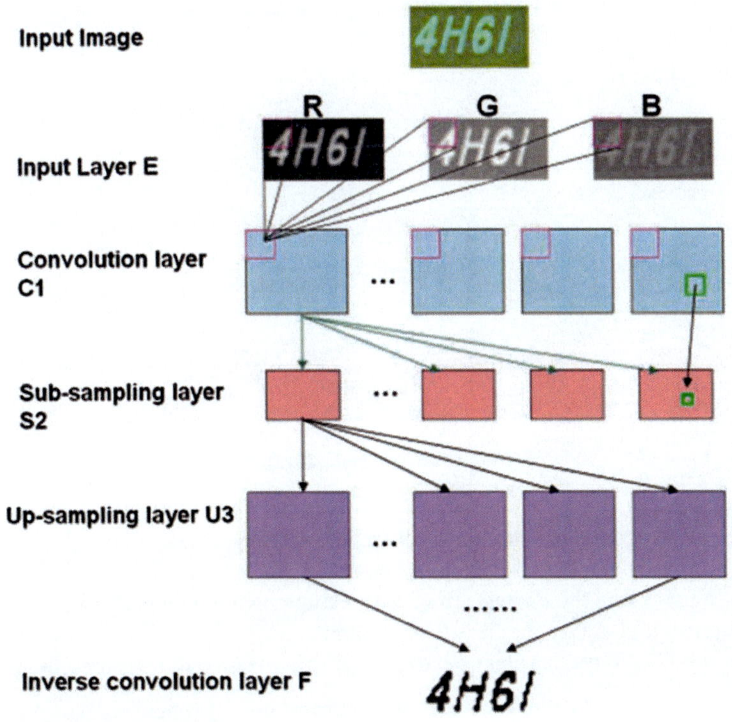

Fig. 13 CTB architecture [16]

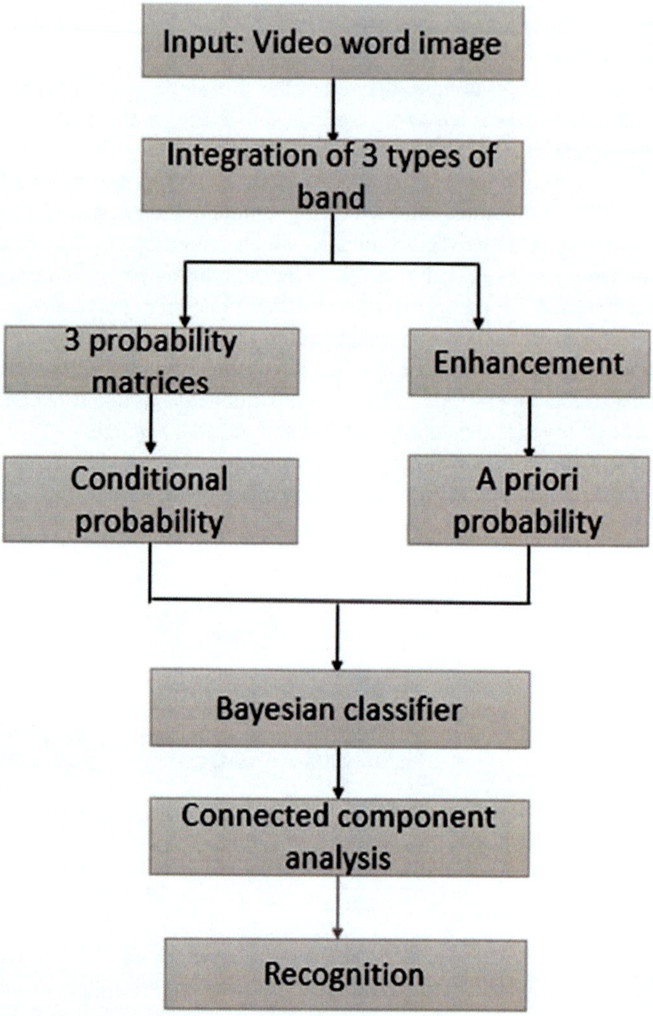

Fig. 14 Flowchart of Bayesian classifier-based method [18]

1. Combining gradient sub-bands, color, and wavelet in a peculiar order so as that the image content gets enhanced,
2. Without using the knowledge of data, estimating conditional probabilities and a priori probability for classification, and
3. Applying a Bayesian classifier to perform binarization of texts present in video.

The Bayesian classifier may produce binarization outcomes with some disconnections due to the presence of complex background and low resolution of videos. Connected component analysis is done to overcome such disconnections, and the gaps are filled by using Canny edge operator (Table 1).

Table 1 Comparative analysis of various algorithms

Method (Reference No.)	Strengths and weaknesses
[2] (document images)	Unlike Ostu's method, it gives good results even in the presence of poor quality of source document. Does not work well in the presence of uneven illumination
[3]	It shows good adaption in various defect types such as noise, illumination, and resolution change. It does not give satisfactory results in situations where background contains low resolution
[6]	Computational time independent of window size. Hence, computational complexity is low as compared to [2] and [3]. It performs well only for particular type of degradation
[7]	Performs much better on deteriorated document images as compared to [2, 3, 6]. It works well only for small patches of input, but large smears cannot be eliminated completely. It considers thin or weak strokes as deteriorations and removes them but actually they belong to original text
[8]	They took most of the degradations into account as compared to [8] as well as the issues involving page folding, document edges, or layout elements. It is really a time taking procedure, and therefore, it is required to simplify the final network
[12] (medical images)	There is no requirement of image histogram or any complex function for computing threshold value. It requires lower computational complexity. For noisy images, it does not yield good results. This can be removed by applying eliminating noise before binarization
[13]	Niblack's algorithm reduces background noise more effectively as compared to Sauvola algorithm. Thus, unlike the case of document binarization, results of Niblack also are better than Sauvola in case of medical image binarization
[14] (video text images)	This method is very efficient for image having low resolution
[15]	Texts with complex background or low contrast can be well-separated from each other and hence can be accurately recognized by OCR software. This method removes false edges easily. There is need to improve efficiency
[16]	Works for identifying colored text areas in video images. This method gives good results even in case of complex background, low resolution or video coding artifacts does not give effective results for scene images
[18]	Gives good results for both video texts and scene texts. Works well for multi-oriented texts. Gives good recognition rate at character and pixel levels. Sometimes output of this method does not well preserve the shapes of characters

3 Some Other Areas Where Image Binarization is Used

3.1 *Face Detection*

Binarization is a useful method for determining the position of eyes and lips. Binarization is used in almost every approach as being one of the processing phase for producing a binary mask that identifies the facial region within an image.

3.2 *Hand Gesture Recognition*

Binarization is sometimes employed in segmenting the region of hands for recognition of hand gesture. Gesture recognition is a crucial area of research toward building better HCI systems.

3.3 *Fingerprint Recognition*

In fingerprint pictures, binarization aids in detecting the position of ridge bifurcation, the existence of arch, and loops. In the preprocessing steps, image binarization is employed to find out the areas of ridges and valleys within a finger impression.

3.4 *Iris Recognition*

Iris recognition for human recognition is a well-known biometric approach. The job of locating the iris is crucial. A scanned picture can be binarized with an appropriate threshold to determine the pupil's position. The positions of pupil boundary and the eyelids help in localizing the iris.

3.5 *Gait Recognition*

Silhouettes of human beings are examined for detecting cycle and for extracting shape feature for human recognition. Background removing and binarization are used to create binary silhouettes from a sequence of video frames in silhouette-based gait analysis.

4 Discussion and Conclusion

Binarization is defined as the process of converting a grayscale image into a black or white, called as binary image. Thresholding is the best method for performing binarization, i.e., labeling pixels to white (or 1) if the gray value is equal to or greater than the threshold or labeling to black (0) if the gray value is smaller (0). In this survey, we explored different areas where image binarization is used. We mainly focused in three areas—in analysis of historical documents, in medical image processing, and in processing the video text images for recognizing the characters and scene images. In analyzing historical documents and video text images, binarization is one of the major steps used. It is an indispensable part in the analysis and processing of documents and video text images. For effective document binarization, it is very essential that image should be clean, noise-free, and the degradation of image should be as minimum as possible. Therefore, before finally performing binarization, some of the researchers preprocess the document images [] for improving the quality of image. For document binarization and video text identification, we saw both the methods like traditional methods (thresholding) and machine learning methods. Due to the complicated background and low resolution of the video, it is difficult to maintain the shape and structure of each character. As a result, the methods available for document analysis may not be appropriate for recognizing video texts. The conventional methods of binarization used for documents analysis perform well for high-contrast graphic text with clear background, but not for scene text or video text because it suffers from inconsistent effects of lighting variations. In comparison with document analysis, the recognition rate of the video analysis methods is low. In medical image processing, binarization is used as a minor step, and therefore, mostly conventional methods are adopted for binarization. But if binarization step is improved with some robust techniques, then in further steps of medical image processing, the desired accuracy can be obtained.

The future approach should focus on combining the image improvement techniques and the image binarization methods for document analysis. There is need for some more effective methods for video text analysis which provides more accurate recognition of characters. In addition to this, there is also a need for machine learning-based techniques for image binarization that takes care of all types of degradations arising in document analysis and takes care of different variations arising in video images.

References

1. Sahoo P et al (1988) A survey of thresholding techniques. Comput Vision Graph Image Process 41:233–260
2. Sulaiman A, Omar K, Nasrudin MF (2019) Degraded historical document binarization: a review on issues, challenges, techniques, and future directions. J Imaging 5(4):48. https://doi.org/10.3390/jimaging5040048

3. Sauvola J, Pietikäinen M (2000) Adaptive document image binarization. Pattern Recogn 33(2):225–236
4. Bernsen J (1986) Dynamic thresholding of gray-level images. In: Proceedings of 8th international conference on pattern recognition. Paris, pp 1251–1255
5. Singh TR et al (2012) A new local adaptive thresholding technique in binarization. arXiv preprint arXiv:1201.5227
6. Saddami K, Afrah P, Mutiawani V, Arnia F (2018) A new adaptive thresholding technique for binarizing ancient document. In: 2018 Indonesian association for pattern recognition international conference (INAPR). Jakarta, Indonesia, pp 57–61. https://doi.org/10.1109/INAPR.2018.8627036
7. He S, Schomaker L (2019) Deep Otsu: document enhancement and binarization using iterative deep learning. Pattern Recogn 91:379–390
8. .Bezmaternykh PV, Ilin DA, Nikolaev DP (2019) U-Net-bin: hacking the document image binarization contest. Comput Opt 43(5):825–832. https://doi.org/10.18287/2412-6179-2019-43-5-825-832
9. Tensmeyer C, Martinez T (2017) Document image binarization with fully convolutional neural networks. In: Document analysis and recognition (ICDAR), 2017 14th IAPR International Conference on, 1. IEEE, pp 99–104
10. Shahriar Sazzad TM, Tanzibul Ahmmed KM, Hoque MU, Rahman M (2019) Development of automated brain tumor identification using MRI images. In: 2019 International conference on electrical, computer and communication engineering (ECCE). Cox'sBazar, Bangladesh, pp 1–4. https://doi.org/10.1109/ECACE.2019.8679240
11. Reddy D, Dheeraj, Kiran, Bhavana V, Krishnappa HK (2018) Brain tumor detection using image segmentation techniques. In: 2018 International conference on communication and signal processing (ICCSP). Chennai, 2018, pp 0018–0022. https://doi.org/10.1109/ICCSP.2018.8524235
12. Somasundaram K, Kalavathi P (2011) Medical image binarization using square wave representation. In: Balasubramaniam P (ed) Control, computation and information systems. ICLICC 2011. Communications in computer and information science, vol 140. Springer, Berlin, Heidelberg. https://doi.org/10.1007/978-3-642-19263-0_19
13. Senthilkumaran N, Vaithegi S (2016) Image segmentation by using thresholding techniques for medical images. Comput Sci Eng Int J 6:1–13. https://doi.org/10.5121/cseij.2016.6101
14. Ghoshal R, Saha A, Das S (2017) A variance based image binarization scheme and its application in text segmentation. In: Shankar B, Ghosh K, Mandal D, Ray S, Zhang D, Pal S (eds) Pattern recognition and machine intelligence. PReMI 2017
15. Zhou Z, Li L, Tan CL (2010) Edge based binarization for video text images. In: 2010 20th International conference on pattern recognition. Istanbul, Turkey, pp 133–136. https://doi.org/10.1109/ICPR.2010.41
16. Saidane Z, Garcia C (2007) Robust binarization for video text recognition. In: Ninth international conference on document analysis and recognition (ICDAR 2007). Parana, pp 874–879. https://doi.org/10.1109/ICDAR.2007.4377040. Li H, Doermann DS, Kia OE (2000) Automatic text detection and tracking in digital video. IEEE Trans Image Process 9(1):147–156
17. Lienhart R, Wernicke A (2002) Localizing and segmenting text in images and videos. IEEE Trans Circuits Syst Video Technol 12(4):256–268
18. Roy S, Shivakumara P, Roy PP, Pal U, Tan CL, Lu T (2015) Bayesian classifier for multi-oriented video text recognition system. Expert Syst Appl 42(13):5554–5566. ISSN 0957-4174

Deciphering Flash Flood-Induced LULC Dynamics Using Remote Sensing Technology and Statistical Indices

Rakesh Saur and Virendra Singh Rathore

Abstract Land use and land cover (LULC) dynamics in flash flood-affected areas is different than other flood-affected areas. Therefore, quantitative understanding of LULC dynamics in such areas is important. Statistical indices were applied to analyze LULC dynamics for the periods 1973–1990, 1990–2000, and 2000–2010. LULC classes were derived from satellite data. In order to get precise LULC change spots, analysis of LULC transformation was performed at grid levels (grid size 24 km^2). Out of nine LULC classes studied, three classes, that is, settlement, grass/grazing land, and inland wetland, showed significant change. Spatial dynamic degree and tendency status index changed randomly over all temporal study periods. Significant change was observed in grids located along river course and in settlement areas. Settlement class was increased by 78.20 km^2 (202.76%) area. Grass/grazing land and inland wetland classes were decreased by 100.62 km^2 (82.02%) and 15.81 km^2 (69.19%) area, respectively. The LULC change in this area can be attributed to shifting river courses, frequent flooding, sand deposition, population growth, and increased agriculture activities.

Keywords Flash flood · LULC change · Grid analysis · Spatial dynamic degree index · Tendency index

1 Introduction

LULC change in all over the world is very rapid which may be due to the terrain attributes coupled with changing climatic conditions. Rapid LULC change is linked to population growth, urbanization, intensive agriculture, mining, massive construction, deforestation, etc. LULC change detection analysis is important to understand

R. Saur (✉) · V. S. Rathore
Department of Remote Sensing, Birla Institute of Technology, Mesra, Ranchi, Jharkhand 835215, India

V. S. Rathore
e-mail: vsrathore@bitmesra.ac.in

P. S. R. Chowdary et al. (eds.), *Evolution in Signal Processing and Telecommunication Networks*, Lecture Notes in Electrical Engineering 839,
https://doi.org/10.1007/978-981-16-8554-5_26

the future change in LULC classes, consequential causes, and active processes behind the changes, and understanding of this helps to restore or prevent serious impact on terrestrial and environmental degradation. More importantly, due to growing population, the demand of land resources is increasing. Thus, restoring or understating the decadal change LULC [1, 2] and its nature, pattern, rate, or magnitude are important for better planning [3]. IGBP (2006) has observed that the LULC change rate is very rapid in last few decades.

Several studies have applied techniques like algebra, transformation, advanced models, GIS, and visual analysis to study LULC in the all over the world and reported massive LULC change in small watershed areas [4, 5] to continental scale [6–8]. Though, none of the classification techniques is considered to be 100% accurate. However, to improve LULC classification accuracy image fusion techniques, multi-temporal optical and microwave satellite data and indices such as NDVI, NDWI, and NDBI are used to extract particular LULC class [9–12]. Landscape pattern analysis gives insight understanding of spatial pattern and their ecological relationship between the LULC classes [13]. Various model indices, Markov chain, cellular automata, and sleuth, are also used to simulate and predict the future LULC [14–16].

Tianran and Wang [17] have attempted to analyze various LULC spatial change rate, transformation quantity, and direction (negative and positive) into different LULC classes using regional spatial dynamic degree index and tendency status index of single and all (comprehensive) classes. Result shows that land use change gradually transformed from equilibrium state of bilateral transformation to extremely unbalanced state of unilateral transformation LULC classes in the study area. Analysis of spatial dynamic degree of different LULC classes at sub-catchments level result shows land change dramatically cause of economic development, population growth, and national micro-policy to understand the influence of hydrological regime change in the basin [18, 19].

Understand the specific LULC class dynamics and change pattern in the area is very important to analyses. So, grid level analysis gives in the in-depth view to locate the change in the study area [20], and by categorizing the level of severity, policy maker can take the action according to server LULC changes in the study area.

Also, to study the anthropogenic and natural causes of LULC change, the spatial processes involved, e.g., function, direction, and factors leading to change in LULC dynamics and quantitative analysis of it, are important to study.

Therefore, present study is taken up to analyze and understands spatial dynamic and tendency of LULC change for different temporal intervals: 1973, 1990, 2000, and 2010 in the study area which is known for extreme natural hazard (flash flood). The study envisages the objectives: (a) assessment of spatial quantity change of LULC category, (b) observe spatial dynamic index change of LULC classes, and (c) observe tendency index of LULC classes of the study area.

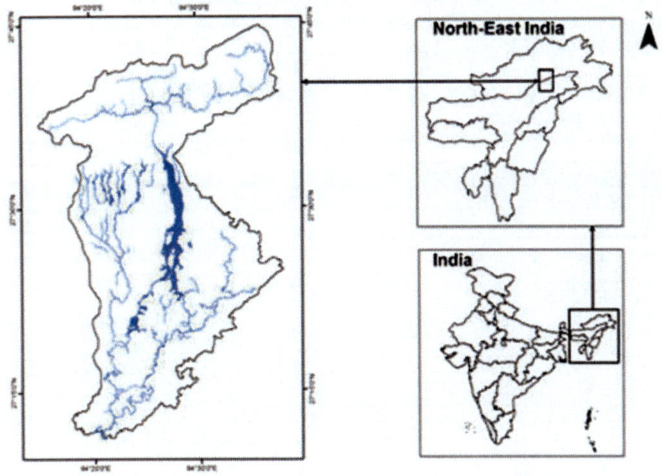

Fig. 1 Location map of the study area

2 Study Area

For this study, the Jia-dhal River basin was selected as study area. The basin is situated in the north-east region of India. The Jia-dhal River is a tributary of Brahmaputra River. The basin is severely affected by frequent flood [21, 22]. This was the main reason of selecting this basin area for study. The total basin area is about 1173.65 km^2 (Fig. 1), and it can be divided into two broad physiographic areas: hilly and plain. The river originates from the lower Himalayan ranges in West Siang district of Arunachal Pradesh and flows southward through the flood plains of Dhemaji district in Assam. The hills constitute about 401.84 km^2 (34.24%) area, whereas plain constitutes nearly 771.75 km^2 (65.76%) area in the basin. The area experiences sub-tropical monsoon climate in general. However, the upper catchment remains moist in all seasons with harsh winter and short summer (http://cgwb.gov.in/).

3 Methods

Study focused on the LULC change dynamics through satellite imagery and statistical indices was used to in-depth understanding of LULC changes of the study area. To assess the research objective, comprehensive methodology has been adapted (Fig. 2).

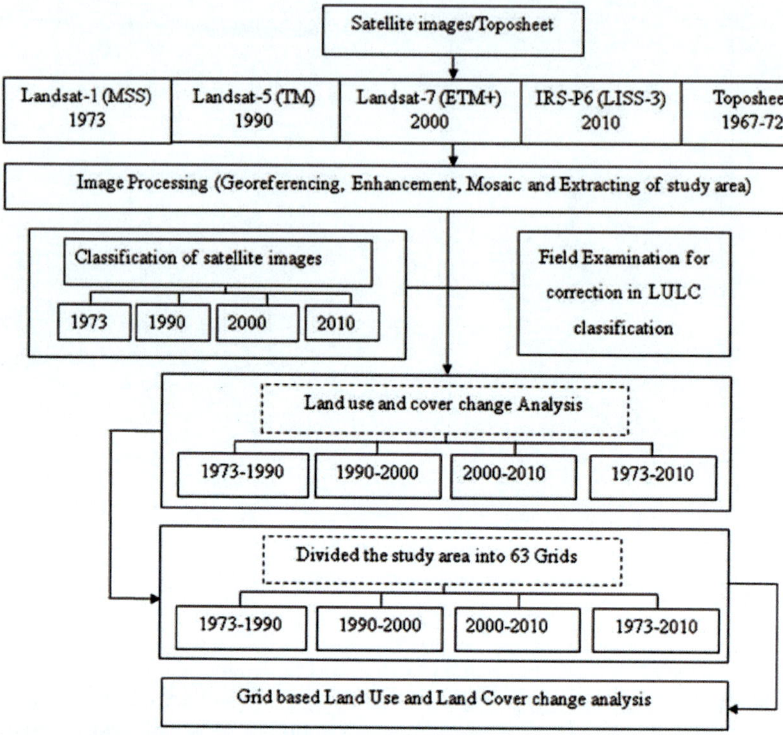

Fig. 2 Flowchart for deriving LULC change analysis

3.1 Land Use and Land Cover Classification

Satellite data of the years 1973, 1990, 2000, and 2010 and topographic maps (scale 1:50,000) (Fig. 2) were used in this study. Firstly, the topographic maps were geo-referenced, followed by satellite images (1973, 1990, 2000, and 2010) using image to image geo-referencing technique for accurate spatial adjustment between the satellite data. Then, spectral enhancement technique was applied on these images in Erdas imagine software for better visibility (Fig. 2), which is considered to be suitable for land use and cover feature class enhancement. For LULC spatial change rate and tendency analysis, the study area has been classified into nine LULC classes following the classification system (level-2) given by NRSA [23]. The classes considered are: sandy area, river, inland wetland, vegetation, settlement, grass/grazing land, agriculture land, scrub, and plantation. In order to achieve high accuracy in extracting LULC information from satellite images, instead of using classification algorithm, on-screen visual interpretation of image approach was adopted in this study. Different LULC classes were identified using the image interpretation keys (tone, texture, pattern, shape, size, and association) and indices—PCA, NDWI, and NDVI. While on-screen manual digitization of LULC classes, topographic maps surveyed in 1972

Table 1 Spatial dynamic degree and tendency index equation for LULC change

S. No.	Name and symbols of index	Expression
a	Spatial dynamic degree of single LULC change (Rss)	$\text{Rss} = \frac{\text{Ub}-\text{Ua}}{\text{Ua}} \times \frac{1}{T} \times 100\%$
	where Rss represents annual quantitative change of single LULC; Ub and Ua represent area of a category at the beginning and end of the study period, respectively; and T is duration of the study	
b	Spatial dynamic degree of comprehensive LULC change (Rts)	$\text{Rts} = \frac{\sum_{i=1}^{n} \lvert \Delta U_{\text{in}-i} - \Delta U_{\text{out}-i} \rvert}{2\sum_{i=1}^{n} U_{\text{ai}}} \times \frac{1}{T} \times 100\%$
	where Rts represents comprehensive LULC change, $\Delta U_{\text{in}-i}$ represents sum area of other LULC categories transformed to LULC category *i*, $\Delta U_{\text{out}-i}$ represents sum area of LULC category *i* transformed to other LULC categories, and $\sum_{i=1}^{n} U_{\text{ai}}$ represents sum area of various LULC categories at the beginning of the study period	
c d	Tendency index of single land use and land cover change (Ps) Tendency index of comprehensive land use and land cover change (Pt)	$\text{Ps} = \frac{\text{Rs}}{\text{Rss}} = \frac{\Delta U_{\text{in}} - \Delta U_{\text{out}}}{\Delta U_{\text{in}} + \Delta U_{\text{out}}}, \left\lvert \frac{\text{Rs}}{\text{Rss}} \right\rvert \leq 1$, or $-1 \leq P_s \leq 0$ $P_t = \frac{\sum_{i=1}^{n} \lvert \Delta U_{\text{in}} - \Delta U_{\text{out}} \rvert}{\sum_{i=1}^{n} \lvert \Delta U_{\text{in}} + \Delta U_{\text{out}} \rvert}, 0 \leq \frac{\text{Rt}}{\text{Rts}} \leq 1$, or $-1 \leq P_t \leq 0$
	where ΔU_{in} represent sum area of other LULC categories transformed to a particular LULC class and ΔU_{out} represent sum area of a particular LULC category transformed to other LULC category during the study periods. *Ps* represents tendency index of single LULC change, and *Pt* represents tendency index of comprehensive LULC change	

were also used for improving interpretation accuracy particularly in case of poor spatial resolution satellite image of 1973. The satellite data of different period (1973, 1990, 2000, and 2010) were then classified into the LULC classes (sandy area, river, inland wetland, vegetation, settlement, grass/grazing, agriculture land, scrub, and plantation).

3.2 Division of Study Area into Grids

Instead of analyzing LULC change in terms of area, for example, central, north, south, west, etc., grid-wise change analysis approach can be more effective, because in this study area the LULC change is very random and mainly due to frequent flooding and river course shifting. Thus, for better in-depth understanding of LULC change behavior and pattern during the years 1973–1990, 1990–2000, 2000–2010, and 1973–2010 in the perspective of spatial dynamic degree and tendency state of single and comprehensive LULC change, the entire study area was divided into grids. Each grid is comprised of 24 km^2 area (Fig. 3), as this is optimal size for this study

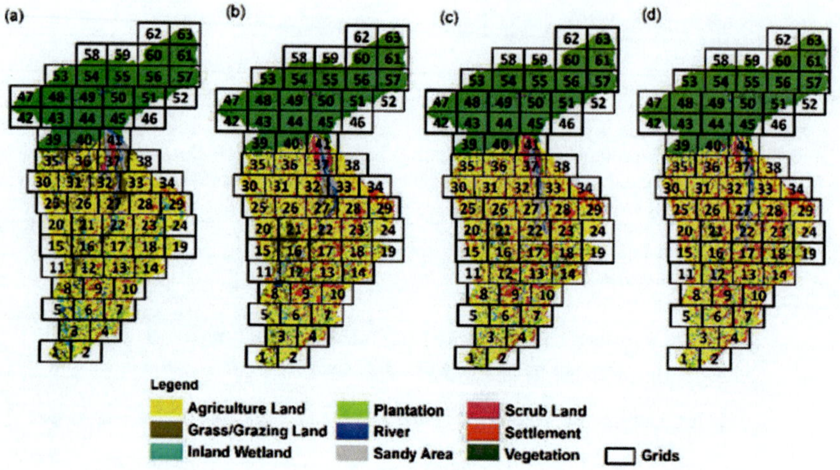

Fig. 3 LULC with grids of Jia-dhal River basin in different periods: **a** 1973, **b** 1990, **c** 2000 and **d** 2010

area to observe various classes of LULC change. The entire study area was divided into total 63 grids.

3.3 *Spatial Dynamic Degree and Tendency Index*

Spatial dynamic degree and tendency index describe quantity, rate, and direction (negative and positive) of LULC change. In this study, the spatial dynamic degree and tendency index for the periods 1973–1990, 1990–2000, 2000–2010, and 1973–2010 were calculated grid wise for the entire study area using equations (Table 1) given by Hu and Wang [17].

Tendency index of single LULC change (Ps) shows results into two ranges: $0 \leq P_s \leq 1$ and $-1 \leq P_S \leq 0$.

In the category $0 \leq P_s \leq 1$ if P_s approaches '0',

1. it means LULC growth of area is in increased direction,
2. area bilateral transformation is more frequent,
3. reflects the equilibrium state and
4. area transformation from other classes to this class is slightly higher than this class to other class.

If P_s approaches '1',

1. it means LULC transformation direction is mainly from other class to this class and
2. transformation is in un-equilibrium state.

In the category $-1 \leq P_S \leq 0$ if P_S approaches '0',

1. scale growth area is in decreased direction,
2. bilateral transformation is more frequent,
3. represents equilibrium state, and
4. transformation of the area from other classes to this class is slightly lower than this class to other classes.

If P_s approaches '−1',

1. it means land use transformation is mainly from this class to other class.
2. area transformation is extremely in un-equilibrium state and
3. land of this category gradually shrinks in scale.

Tendency index of comprehensive land use and land cover change (Pt), if approaches '0',

1. bilateral transformation between all the LULC class is very frequent and
2. reflecting tendency of balance transformation.

If it approaches '1',

1. the transformation direction of each LULC class is unilateral and
2. extremely imbalance (LULC class transformation mainly is from one class to another class or vice-versa).

Comprehensive spatial dynamic degree (Rts) and comprehensive tendency index (Pt) for all the 63 grids periods of 1973–1990, 1990–2000, 2000–2010, and 1973–2010 were calculated using to understand LULC change pattern. For each temporal period, comprehensive spatial dynamic degree values have been categorized into six equal interval categories (Table 2) and comprehensive tendency index into four equal interval categories (Table 3).

Table 2 Equal interval classes of comprehensive spatial dynamic degree index

S. No.	Comprehensive dynamic degree index classes	1973–1990	1990–2000	2000–2010	1973–2010
1	Very low spatial change rate (VLs)	0–0.49	0–0.55	0–0.27	0–0.32
2	Low spatial change rate (Ls)	0.50–0.99	0.56–1.10	0.28–0.55	0.33–0.65
3	Moderate spatial change rate (Ms)	1–1.48	1.11–1.65	0.56–0.82	0.66–0.97
4	High spatial change rate (Hs)	1.49–1.97	1.66–2.20	0.83–1.10	0.98–1.30
5	Very high spatial change rate (VHs)	1.98–2.46	2.21–2.75	1.11–1.37	1.31–1.62
6	Extremely very high spatial change rate (EVHs)	2.47–2.96	2.76–3.29	1.38–1.65	1.63–1.95

Table 3 Equal intervals classes of comprehensive tendency index

S. No.	Tendency comprehensive index classes	Range
1	Equilibrium (Eq)	0.00–0.25
2	Quasi-equilibrium (QEq)	0.25–0.50
3	Unbalanced state (Us)	0.50–0.75
4	Extremely unbalanced state (EUs)	0.75–1

4 Results and Discussion

4.1 Spatial Dynamic Degree Index Analysis of LULC Classes

During the period 1973–1990, scrub and plantation classes showed highest dynamic degree, with annual change rate of 8.95% and 8.12%, respectively, among all the analyzed LULC classes. Next highest dynamic degree was observed in the class settlement with annual change rate of 5.77%. Inland wetland class showed highest negative change rate of −3.68%. During the period 1990–2000, inland wetlands class showed highest positive change rate of 5.19%. LULC classes such as scrub, plantation, and settlement showed 2.16%, 2.96%, and 2.39% of annual change rate, respectively. During the period 2000–2010, LULC classes like inland wetland and scrub class contrary to the previous decade (1990–2000) showed a significant negative annual change rate of −4.65%and −4.00%, respectively. However, classes like plantation and settlement continued the positive change with annual change rate of 0.58% and 2.34%, respectively.

Interestingly, agriculture land showed a positive dynamic degree in all the temporal periods of study (Table 4), which is an indicative of continuous increasing demand of agriculture land for growing population in this area. However, grass/grazing land class showed a negative annual change rate in all the temporal periods of study indicating a continuous transformation of this to other LULC classes, mostly into agriculture land. Sandy area and river classes showed a very high degree of dynamics throughout the study period with annual change rate of −0.19% and 0.07%, 1.94% and −2.12%, and −0.55% and 1.28% during 1973–1990, 1990–2000, and 2000–2010, respectively, indicating frequent transformation of these classes into different LULC classes. During 1973–2010 (entire study period), the settlement class showed the highest annual change rate of 5.48% among all the LULC classes of study, which is due to continuous population growth.

4.2 Tendency Index Analysis of LULC Classes

For better understanding of LULC change trend in the area, tendency index (P_S) for LULC classes was computed (Table 4) and categorized into four types (Table

Table 4 Spatial dynamic degree and tendency index of LULC classes for the periods 1973–1900, 1990–2000, 2000–2010, and 1973–2010

LULC classes	Rss				Ps			
	(1973–1990)	(1990–2000)	(2000–2010)	(1973–2010)	(1973–1990)	(1990–2000)	(2000–2010)	(1973–2010)
Sandy area	−0.19	1.94	−0.55	0.25	−0.02	0.13	−0.05	0.05
River	0.07	−2.12	1.28	−0.26	0.02	−0.42	0.21	−0.12
Inland wetland	−3.68	5.19	−4.65	−1.87	−0.66	0.4	−0.56	−0.65
Vegetation	−0.01	0.09	−0.02	0.01	−0.04	0.37	−0.07	0.06
Settlement	5.77	2.39	2.34	5.48	0.96	0.83	0.91	0.96
Grass/Grazing land	−2.46	−5.13	−3.69	−2.22	−0.39	−0.55	−0.19	−0.78
Agriculture land	0.26	0.18	0.02	0.17	0.11	0.09	0.01	0.14
Scrub	8.95	2.16	−4	2.26	0.48	0.23	−0.69	0.31
Plantation	8.12	2.96	0.58	3.39	1	1	0.82	1
	Rts				Pt			
Comprehensive in the region	0.34	0.37	0.23	0.28	0.28	0.31	0.23	0.4

Rss—Spatial dynamic degree of single LULC class; Rts—Comprehensive spatial dynamic degree of LULC classes; Ps—Tendency index of single LULC class and Pt—Comprehensive tendency index of LULC classes

2). Agriculture land class showing tendency index (P_S) value is 0.11 during 1973–1990, 0.09 during 1990–2000, 0.01 during 2000–2010, and 0.14 during 1973–2010 (Table 4). Lulc class (Ps) approaches 0, Indicating an increased growth scale, bilateral transformation relationship between the classes are at high frequency; LULC class is in equilibrium state and transformation of other LULC classes in this LULC class is slightly higher than this LULC class to other LULC class. Settlement and plantation P_S value is 0.96 and 1 during 1973–1990, 0.83 and 1 during 1990–2000, 0.91 and 0.82 during 2000–2010, and 0.96 and 1 during 1973–2010, respectively. Which approaches 1, meaning LULC transformation direction is mainly from other LULC classes to this class of LULC class, transformation area is in un-equilibrium state and due to that there is sudden increase in the LULC class area found. P_S value is negative and close to - 0, as sandy area and vegetation classes value is 0.02, −0.04 in 1973–1990, −0.05, −0.07 in 2000–2010, respectively, representing a negative growth scale, bilateral transformation is more frequent and transformation of this LULC class is to mainly in other LULC classes and P_S values approaches −1 of inland wetland P_S value −0.56 during 1973–1990, and −0.66 during 2000–2010. Grass/grazing land class also showed −0.55 during 1990–2000 and −0.78 during 1973–2010 (Table 4), which shows that land use transformation is mainly from this LULC class of land to other LULC classes of land. The land transformation is extremely in un-equilibrium state.

4.3 Comprehensive Spatial Dynamic Degree Grid Wise

The entire study area was divided into 63 grids, each of 24 km^2 area. LULC change comprehensive spatial dynamic degree was calculated for all the grids (Table 5). The comprehensive spatial dynamic degree index values were categorized into six equal intervals (Table 2) for better understanding. During the period 1973–1990, grid: 37 experienced extremely very high spatial change rate (EVHs), grids: 35 and 36 experienced very high spatial changes rate (VHs), grids: 11, 31, 32, 33, and 41 experienced high spatial change rate (Hs), and grids: 5, 16, 25, 34, 38, and 40 experienced moderate spatial changes rate (Ms) (Figs. 4a, 6).

During the period 1990–2000, grids: 11 and 16 experienced extremely very high spatial change rate (EVHs), grid: 22 experienced very high spatial change rate (VHs), grids: 15, 27, and 41 experienced high spatial change rate (Hs), and grids: 4, 5, 12, and 32 experienced moderate spatial change rate (Ms) (Fig. 4b).

During the period 2000–2010, grids: 1, 22, 27, and 41 experienced extremely very high spatial change rate (EVHs), grids: 2, 3, 5, 6, 8, 14, 18, 28, 32, 33, 37, and 38 experienced moderate spatial change rates (Ms) of this category in this period number of grids were more as compare to previous two temporal study periods (1973–1990 and 1990–2000). Grids: 19 and 34 experienced very high spatial change rate (VHs), and grids: 12, 16, 21, and 36 experienced high spatial change rate (Hs) (Fig. 4c). During the entire study period (1973–2010), grid: 29 experienced extremely very high spatial change rate (EVHs). Grid: 41 experienced very high spatial change

Table 5 Comprehensive spatial dynamic degree index values in the grids

Grid No.	1973–1990	1990–2000	2000–2010	2000–2010	Grid No.	1973–1990	1990–2000	2000–2010	2000–2010
1	0.59	0.60	1.64	0.78	33	1.58	0.88	0.63	0.63
2	0.42	0.52	0.81	0.31	34	1.16	0.74	1.18	1.04
3	0.79	0.94	0.55	0.73	35	1.97	0.43	0.54	0.91
4	0.79	1.33	0.34	0.23	36	2.08	0.52	0.89	0.90
5	1.30	1.64	0.59	0.98	37	2.96	0.80	0.81	1.12
6	0.57	0.16	0.58	0.34	38	1.33	0.44	0.73	0.72
7	0.25	0.31	0.23	0.24	39	0.45	0.06	0.13	0.22
8	0.28	1.05	0.70	0.63	40	1.09	0.22	0.26	0.45
9	0.55	0.67	0.44	0.49	41	1.95	2.12	1.62	1.36
10	0.49	0.49	0.34	0.38	42	0.00	0.00	0.00	0.00
11	1.52	3.29	0.50	0.54	43	0.00	0.00	0.00	0.00
12	0.87	1.50	1.00	0.46	44	0.00	0.00	0.00	0.00
13	0.50	0.48	0.43	0.25	45	0.04	0.07	0.04	0.02
14	0.90	0.67	0.61	0.60	46	0.00	0.00	0.00	0.00
15	0.78	2.14	0.46	0.45	47	0.00	0.00	0.00	0.00
16	1.30	3.02	0.94	0.58	48	0.00	0.00	0.23	0.00
17	0.93	0.75	0.53	0.65	49	0.06	0.01	0.13	0.02
18	0.70	0.96	0.69	0.47	50	0.10	0.02	0.10	0.04
19	0.74	0.37	1.32	0.74	51	0.00	0.00	0.00	0.00
20	0.77	1.09	0.45	0.37	52	0.00	0.00	0.00	0.00
21	0.93	1.05	0.95	0.22	53	0.00	0.00	0.00	0.00

(continued)

Table 5 (continued)

Grid No.	1973–1990	1990–2000	2000–2010	2000–2010	Grid No.	1973–1990	1990–2000	2000–2010	2000–2010
22	0.55	2.22	1.54	0.59	54	0.00	0.00	0.01	0.00
23	0.43	0.41	0.30	0.33	55	0.00	0.00	0.02	0.00
24	0.58	0.41	0.41	0.35	56	0.00	0.00	0.00	0.00
25	1.03	0.65	0.34	0.57	57	0.00	0.00	0.00	0.00
26	0.86	0.59	0.18	0.66	58	0.00	0.00	0.00	0.00
27	0.87	2.01	1.65	0.88	59	0.00	0.00	0.00	0.00
28	0.53	0.84	0.66	0.46	60	0.00	0.00	0.00	0.00
29	0.63	0.39	0.45	1.79	61	0.00	0.00	0.00	0.00
30	0.78	0.36	0.38	0.37	62	0.00	0.00	0.00	0.00
31	1.75	0.55	0.41	0.90	63	0.00	0.00	0.00	0.00
32	1.91	1.37	0.60	1.29					

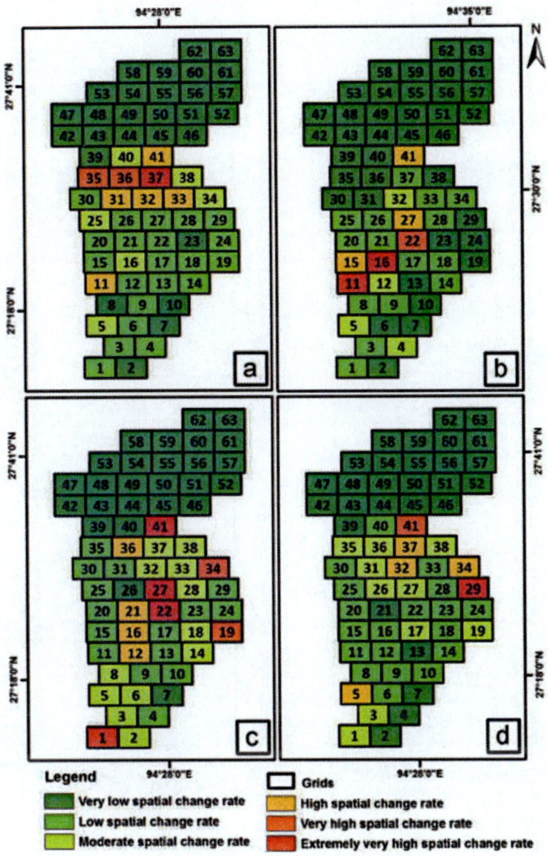

Fig. 4 Grid-wise LULC change based on comprehensive spatial dynamic degree index of Jia-dhal River basin for different periods, **a** 1973–1990, **b** 1990–2000, **c** 2000–2010, and **d** 1973–2010

rate (VHs), and grids—5, 32, 34, and 37 experienced high spatial change rate (Hs) (Fig. 4d).

4.4 *Comprehensive Tendency Index Grid Wise*

LULC classes' behavior can be better understood if comprehensive tendency index value is analyzed grid wise as explained above that the entire study area was divided into 63 grids (Fig. 5), each grid is of 24 km^2. The comprehensive tendency index values were classified into four equal intervals (Table 3). During the period 1973–1990, grids: 14, 20, 31, 34, 36, 38, 39, 40, 49, 50, and 54 showed extremely unbalanced state (EUs) (Fig. 7), grids: 2, 3, 4, 5, 6, 7, 9, 10, 11, 13, 15, 17, 18, 19, 21, 25, 26, 28, 30, 32, 33, 35, 37, 41, and 45 (Fig. 5a) showed unbalanced state (Us), grids: 1, 12, 16, 22, 23, 24, 27, and 29 showed quasi-equilibrium state (QEq), and grids: 8,

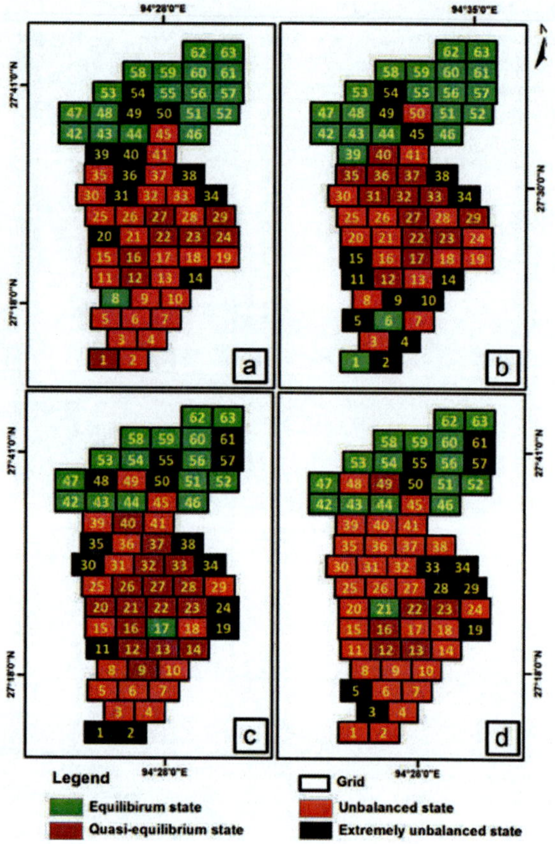

Fig. 5 Grid-wise LULC change based on comprehensive tendency index of Jia-dhal River basin for different periods, **a** 1973–1990, **b** 1990–2000, **c** 2000–2010, **d** 1973–2010

42, 43, 44, 46, 47 48, 51, 52, 53, and 55 showed equilibrium state (Eq). During the period of 1990–2000 (Fig. 5b), grid: 2, 4, 5, 9, 10, 11, 14, 15, 34, 38, 45, 49, and 54 showed extremely unbalanced state (EUs), grids: 3, 7, 8, 13, 16, 18, 19, 20, 21, 24, 25, 26, 28, 30, 41, and 50 showed unbalanced state (Us), grids: 12, 17, 22, 23, 27, 29, 31, 32, 33, 35, 36, 37, and 40 showed quasi-equilibrium state (QEq), and grids: 1, 6, 39, 42, 43, 44, 46, 47, 48, 51, 52, 53, 55, 56, 57, 58, 59, 60, 61, 62, and 63 showed equilibrium state (Eq). During the period 2000–2010 (Fig. 5c), grids: 1, 2, 11, 19, 24, 30, 34, 35, 38, 48, 50, 55, 57, and 61showed extremely unbalanced state (EUs), grids: 3, 4, 5, 6, 7, 8, 10, 15, 18, 25, 29, 31, 36, 29, 41, 45, and 49 showed unbalanced state (Us), grids: 9, 12, 13, 14, 16, 20, 21, 22, 23, 26, 27, 28, 32,33, 37, and 40 showed quasi-equilibrium state (QEq), grids: 17, 42, 43, 44, 46, 47, 51, 52, 53, 54, 56, 58, 59, 60, 62, and 63 showed equilibrium state (Eq). We also analyzed the value for entire the study periods 1973–2010, the grids: 3, 5, 19, 28, 29, 33, 34, 50, 55, 57, and 61 showed extremely unbalanced state (EUs), grids: 1, 2, 4, 6, 7, 8, 9, 10, 11, 14, 15, 17, 18, 20, 24, 25, 26, 27, 30, 31, 32, 35, 36, 37, 38, 39, 40, 41, 45, and 48 showed unbalanced state (Us), grids: 12, 13, 16, 22, 23, and 49 showed

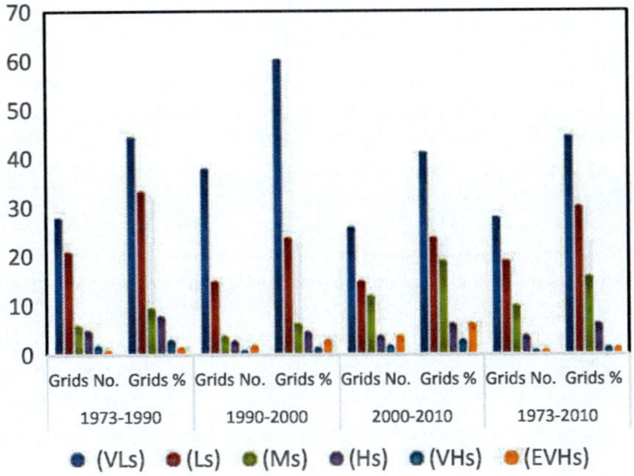

Fig. 6 Comprehensive spatial dynamic degree index LULC change of Jia-dhal River basin for different periods, **a** 1973–1990, **b** 1990–2000, **c** 2000–2010, and **d** 1973–2010. VLs-Very low spatial change rate, Ls-Low spatial change rate, Ms-Moderate spatial change rate, Hs-High spatial change rate, VHs-Very high spatial change rate, EVHs-Extremely very high spatial change rate

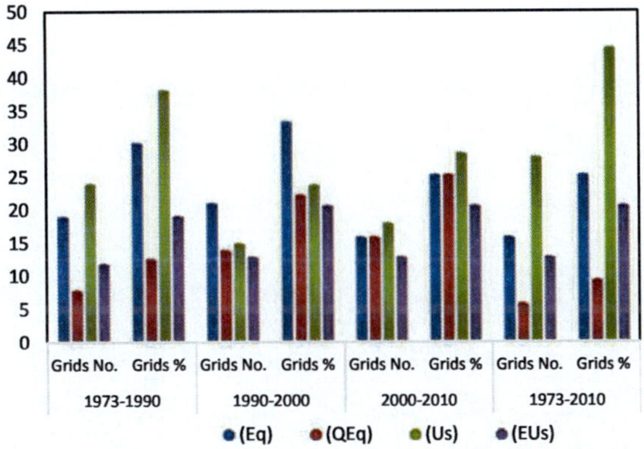

Fig. 7 Comprehensive tendency index LULC change of Jia-dhal River basin for different periods, **a** 1973–1990, **b** 1990–2000, **c** 2000–2010, and **d** 1973–2010. Eq-Equilibrium, QEq-Quasi equilibrium, Us-Unbalanced state, and EUs-Extremely unbalanced state

quasi-equilibrium state (QEq), and grid: 21, 42, 43, 44, 46, 47, 51, 52, 53, 54, 56, 58, 59, 60, 62, and 63 showed equilibrium state (Eq) (Fig. 5d) (Table 6).

4.5 *Driving Factor to Change the LULC Within the Grids*

Study area is divided into 63 grids, and LULC change indices were calculated for all the grids. Analysis shows that due to major impact of the flood disaster events, population pressure in the form of settlement area growth and agriculture activity for food security is the significant driving factors to changes LULC in the study area. Grid-wise analysis gives the in-depth view of high LULC changing location due to anthropogenic and natural events in different time periods (1973–1990, 1990–2000, 2000–2010, and 1972–2010). Analysis indicates during 1973–1990, grid no-36 and 37 are in VHs and EVHs category driving factor river flood, agriculture, and

Table 6 Comprehensive tendency index values in the grids

Grid No.	1973–1990	1990–2000	2000–2010	1973–2010	Grid No.	1973–1990	1990–2000	2000–2010	1973–2010
1	0.29	0.23	0.80	0.75	33	0.61	0.34	0.33	0.98
2	0.75	0.80	0.99	0.56	34	0.90	0.96	1.00	0.94
3	0.56	0.66	0.58	0.80	35	0.72	0.29	0.50	0.62
4	0.56	0.81	0.70	0.70	36	0.76	0.29	0.64	0.60
5	0.56	0.80	0.57	0.83	37	0.61	0.30	0.34	0.52
6	0.67	0.24	0.56	0.59	38	0.75	1.00	1.00	0.62
7	0.66	0.66	0.53	0.60	39	0.78	0.13	0.52	0.73
8	0.17	0.64	0.64	0.66	40	0.79	0.33	0.35	0.61
9	0.62	0.93	0.48	0.65	41	0.54	0.62	0.54	0.65
10	0.54	0.76	0.54	0.60	42	0.00	0.00	0.00	0.00
11	0.59	0.80	0.82	0.56	43	0.00	0.00	0.00	0.00
12	0.38	0.46	0.36	0.39	44	0.00	0.00	0.00	0.00
13	0.60	0.48	0.49	0.49	45	0.99	0.76	0.58	0.51
14	0.76	0.80	0.50	0.73	46	0.00	0.00	0.00	0.00
15	0.59	0.81	0.63	0.64	47	0.00	0.00	0.00	0.00
16	0.44	0.68	0.40	0.45	48	0.00	0.00	0.99	0.68
17	0.62	0.49	0.22	0.59	49	0.90	1.00	0.58	0.38
18	0.58	0.74	0.54	0.60	50	1.00	0.67	0.81	0.82
19	0.67	0.54	0.88	0.85	51	0.00	0.00	0.00	0.00
20	0.76	0.71	0.41	0.57	52	0.00	0.00	0.00	0.00
21	0.62	0.63	0.37	0.25	53	0.00	0.00	0.00	0.00

(continued)

Table 6 (continued)

Grid No.	1973–1990	1990–2000	2000–2010	1973–2010	Grid No.	1973–1990	1990–2000	2000–2010	1973–2010
22	0.31	0.46	0.30	0.49	54	1.00	1.00	0.24	0.15
23	0.36	0.43	0.40	0.42	55	0.00	0.00	0.79	0.78
24	0.46	0.57	0.76	0.58	56	0.00	0.00	0.00	0.00
25	0.75	0.50	0.65	0.64	57	0.00	0.00	1.00	1.00
26	0.58	0.52	0.44	0.92	58	0.00	0.00	0.00	0.00
27	0.29	0.44	0.49	0.57	59	0.00	0.00	0.00	0.00
28	0.55	0.65	0.43	0.80	60	0.00	0.00	0.00	0.00
29	0.46	0.38	0.63	0.82	61	0.00	0.00	1.00	1.00
30	0.66	0.70	0.92	0.52	62	0.00	0.00	0.00	0.00
31	0.79	0.44	0.51	0.74	63	0.00	0.00	0.00	0.00
32	0.57	0.45	0.44						

settlement activity. Whereas grid no-35 is in VHs found agriculture and settlement activity most influencing factor to change the LULC in these grids. During 1990–2000, study shows that grid nos. 11 and 12 significant causes due to agriculture and settlement area growth and grid no. 22 highly impacted from high flood events. At the same period, grid no. 4 found emerging new grass and grazing land and high growth of vegetation positive activity in the study area. During 2000–2010, grid nos. 41, 27, 22, and 1 is in EVHs category which is due to agriculture activity and settlement growth, and grid nos. 34 and 19 are in VHs category due to very high settlement growth also at the same time grid no. 12 is in Hs category due to high flood occurs and impacted the LULC class. During 1973–2010, grid no. 29 is in EVHs category and found that extreme high settlement growth and grid nos. 5, 32, 37, and 41 is in Hs and VHs categories due to very high agriculture and settlement growth in these grids of the study area.

4.6 Grid-Wise LULC Change Pattern Analysis

Erratic extensive LULC change is observed in different time periods of study in the grids. Grids 1 to 41 are situated in the plain area and from 42 to 63 are in hilly area.

Comprehensive spatial dynamic degree index of grids: 7, 42, 43, 44, 45, 46, 47, 48, 49, 50, 51, 52, 53, 54, 55, 56, 57, 58, 59, 60, 61, 62, and 63 (Fig. 4) is in the class of very low spatial change rate (VLs) during all the temporal study periods (1973–1990, 1990–2000, 2000–2010, and 1973–2010), which means LULC change was insignificant. Noticeably, all these grids except grid: 7 belong to hilly area. It is because there is lesser impact of forces (natural and anthropogenic) causing LULC change in the hilly area, in this part (Fig. 4a–d).

In grids: 1, 2, 3, 10, 14, 18, 21, 23, and 28, there was no variation in their comprehensive spatial dynamic degree index class during 1973–1990 and 1990–2000. However, later in the temporal period 2000–2010, LULC classes in these grids showed higher comprehensive spatial dynamic degree index, which indicates an abrupt LULC change in this period in these grids, because these grids are associated with river course and are located in the plain. Also, this area experienced flooding and shifting of river course, which caused change in LULC classes in these grids (Fig. 4a–c). Grids: 11, 31, 32, 33, 41, 35, 36, and 37 showed high LULC change (Fig. 4a) rate, as obvious from comprehensive spatial dynamic degree index (Hs, VHs, and EVHs) during 1973–1990, because these grids are located in the plain areas—near foot hills, where river enters from hills with high velocity and floods. Moreover, the area has high settlement density compared to other areas. Grids: 15, 27, 41 (Hs), 22 (VHs), 11, and 16 (EVHs) during 1990–2000 also showed high LULC change rate. These grids are located in the southern part of the basin (Fig. 4a, b). The high LULC change rate was due to embankment construction in the upper part of the basin as well as deposits of river load in this area. Grid: 12, 16, 21, and 36 (Hs), 19 and 34 (VHs), 1, 22, 27, and 41 (EVHs) during 2000–2010 showed high LULC change rate (Fig. 4c). These grids are not concentrated at one point, scattered in the plain area. This is because of random settlement growth, increased agricultural activities and embankment failure in the upper part of the study area.

5 Conclusions

During the period 1973–2010 (37 years), LULC change in the Jia-dhal river basin is significant. There was major inter class transformation between the LULC classes: grass/grazing land and inland wetland to agriculture land, agriculture land to settlement and sandy area. Significant change is found from 1973–2010 around, 78.20 km^2 increase with 5.48% area per year in the class settlement and agriculture land increased 31.59 km^2 with 0.21% area per year change and decrease was found in grass/grazing land (−100.62 km^2) with −2.22% area per year change. Spatial dynamic degree index and tendency index of classes were very high and inconsistent during the different temporal periods of study (1973–1990, 1990–2000, and 2000–2010). It means the LULC dynamics in this area is relatively faster, as it is driven by extreme pressure exerted by forces like river course shifting, rapid population growth leading to increased settlement density, and extensive agriculture.

The LULC change in the area was not very much directional, rather it was localized in different directions. Also, it was un-predictive. Therefore, grid-wise LULC change analysis was found to be more effective in terms of identifying localized changes and causes of it for sustainable management land resources.

The LULC change mainly in the area is attributed to the river course shifting due to embankment construction along the river bank in the foot hill region in the year 1974 and subsequently in the downstream area to limit the floods. Second important factor of causing LULC change is continuous increase in population.

References

1. Misra A, Murali M, Vethamony P (2013) Assessment of the land use/land cover (LU/LC) and mangrove changes along the Mandovi–Zuari estuarine complex of Goa, India. Arab J Geosci. https://doi.org/10.1007/s12517-013-1220-y
2. Zhao J, Guo W, Huang W, Huang L, Zhang D, Yang H, Yuan L (2012) Characterizing spatiotemporal dynamics of land cover with multi-temporal remotely sensed imagery in Beijing during 1978–2010. Arab J Geosci. https://doi.org/10.1007/s12517-013-1072-5
3. Khanday MY, Khan I, Javed A (2016) Watershed management on land use/land cover dynamics in semi-arid watershed, Central India. Arab J Geosci 9:458. https://doi.org/10.1007/s12517-016-2478-7
4. Berakhia RO, Oyanaa TJ, Adu-Prahb S (2014) Land use and land cover change and its implication in Kagera River Basin, East Africa. Afr Geogr Rev 209–231
5. Tsarouchi GM, Mijic A, Moulds S et al (2014) Historical and future land-cover changes in the Upper Ganges Basin of India. Int J Remote Sens 35(L9):3150–3176
6. Prestele R, Alexander P, Rounsevell MD et al (2016) Hotspots of uncertainty in land-use and land-cover change projections: a global-scale model comparison. Global Change Biol 3967–3983

7. Linderman M, Rowhani P, Benz D et al (2005) Land-cover change and vegetation dynamics across Africa. J Geophys Res 110:12104
8. Lambin EF, Geist HJ, Lepers E (2003) Dynamics of land-use and land-cover change in tropical regions. Ann Rev Environ Resources 28:205–241
9. Huang H, Chen Y, Clinton N et al (2017) Mapping major land cover dynamics in Beijing using all landsat images in Google Earth engine. Remote Sens Environ 202:166–176
10. Rujoiu-Mare M, Mihai B (2016) Mapping land cover using remote sensing data and GIS techniques: a case study of Prahova Subcarpathians. In: International conference environment at a crossroads: SMART approaches for a sustainable future. Procedia environmental sciences, vol 32, pp 244–255
11. Parihar N, Das A, Rathore VS et al (2014) Analysis of L-band SAR backscatter and coherence for delineation of land-use/land-cover. Int J Remote Sens 35(18):6781–6798
12. Parihar N, Rathore VS, Mohan S (2017) Combining ALOS PALSAR and AVNIR-2 data for effective land use/land cover classification in Jharia Coalfields region. Int J Image Data Fusion 8:130–147
13. Griffith JA (2004) The role of landscape pattern analysis in understanding concepts of land covers change. J Geogr Sci 14(1):3–17
14. Han H, Yang C, Song J (2015) Scenario simulation and the prediction of land use and land cover change in Beijing, China. Sustainability 7:4260–4279
15. Vázquez-Quintero G, Solís-Moreno R, Pompa-García M et al (2016) Detection and projection of forest changes by using the Markov chain model and cellular automata. Sustainability 8:236. https://doi.org/10.3390/su8030236
16. Chaudhuri G, Clarke KC (2013) The SLEUTH land use change model: a review. Int J Environ Resources Res 1:1
17. Hu T, Wang S (2015) Study on the dynamic of the land using based on the RS and GIS techniques. Int J Smart Home 9(10):15–26
18. Zhang T, Zhang X, Xia D, Liu Y (2014) An analysis of land use change dynamics and its impacts on hydrological processes in the Jialing River Basin. Water 6:3758–3782. https://doi.org/10.3390/w6123758
19. Hazarika N, Das AK, Borah SB (2015) Assessing land-use changes driven by river dynamics in chronically flood affected upper Brahmaputra plains, India, using RS-GIS techniques. Egypt J Remote Sens Space Sci 18:107–118
20. Saranya KRL, Reddy CS (2016) Long term changes in forest cover and land use of Similipal biosphere reserve of India using satellite remote sensing data. J Earth Syst Sci 125(3):559–569
21. Das PJ (2011) Building communities' capacity for flash flood risk management in the Jia-Dhal river basin Dhemaji district, Assam, India. Thesis Report
22. Das PJ, Bhuyan HK (2013) Policy and institutions in adaptation to climate change, case study on flood mitigation infrastructure in India and Nepal. ICIMOD Working Paper 2013/4
23. NRSC NRC-LULC-50K (2006) Manual national land use land cover mapping using multi-temporal satellite data. Remote Sensing and GIS Applications Area National Remote Sensing Centre, Indian Space Research Organisation (ISRO), Department of Space, Government of India, Hyderabad

Construing Crop Health Dynamics Using UAV-RGB based SpaceTech Analytics and Image Processing

Alok Bhushan Mukherjee, Nitesh Awasthi, and Govind Sharma

Abstract Phenomenal growth in global population is primarily responsible for disturbing land ecosystem services. Consequently, it affected food production due to surge in climate change induced hazards and subsequently created a huge gap between food demand and supply. Hence, a robust crop analytics framework is an absolute necessity for precision farming and climate-resilient agriculture. Deciphering crop health dynamics is an integral component of crop analytics. Therefore, this study aims to assess and quantify crop health dynamics using UAV/satellite remote sensing analytics. Various spectral indices such as Normalized Difference Vegetation Index (NDVI), Normalized Difference Moisture Index (NDMI), Bare Soil Index (BSI), and Red Edge Index were used in this study along with Greenness Index (GI) which was computed from UAV data. Derived indices from satellite imageries and UAV were studied in sync to assess crop health characteristics and possible crop yield scenario in hyperlocal scenario, that is, plot wise. Moreover, wheat ear distribution was assessed and counted in sample plots using image filtering techniques such as high pass (Laplacian filter) and low pass (Median filter) filters along with local maxima operation. Combining the results of crop health dynamics and wheat ear distribution/count enable users to have an indicative assessment of yield. Results of this study seems encouraging in the context of crop health and yield analysis.

Keywords Crop health dynamics · UAV-RGB analytics · Satellite remote sensing · NDVI · BSI · NDMI · Red Edge Index · Greenness Index · Laplacian filter · Median filter

---

A. B. Mukherjee (✉) · N. Awasthi · G. Sharma
Leads Research Laboratory (LeadsConnect Services Pvt. Ltd.), Noida, Uttar Pradesh, India
e-mail: alok.mukherjee@leadsconnect.in

N. Awasthi
e-mail: nitesh.awasthi@leadsconnect.in

G. Sharma
e-mail: govind.sharma@leadsconnect.in

P. S. R. Chowdary et al. (eds.), *Evolution in Signal Processing and Telecommunication Networks*, Lecture Notes in Electrical Engineering 839,
https://doi.org/10.1007/978-981-16-8554-5_27

1 Introduction

Construing crop phenology dynamics is central aspect of crop analytics framework. It helps in deciphering the complexities of crop physiology. Moreover, it may be instrumental in developing understanding about the optimal crop management practices in the context of climate conditions [1]. Reference [2] highlighted that increase in soil temperature may have adverse effects on crop phenology. In addition, its impact may further be on soil moisture, organic matter, and microbial activities. Reference [3] mentioned that there is challenge in applying remote sensing technology for studying crop phenology and crop growth dynamics in smaller landscapes. Therefore, Ref. [4] described the significance of UAV in collecting remote sensing data at higher resolutions. In addition, it is more flexible than satellite-based sensors. Reference [5] highlighted that application of UAV can be very effective for precision farming. Furthermore, use of UAV was demonstrated for assessing and quantifying different types of vegetation indices. The study performed by Ref. [6] was also in the direction of crop yield assessment and modeling. Reference [7] performed study related to biomass estimation. Plant height information and vegetation indices were combined for performing this study. The height of plant information was gathered from UAV-based multi-temporal crop surface models (CSMs) [7].

Reference [8] used multi-temporal UAV orthomosaic imageries for studying wheat growth status. Principle of image interpretation was also employed to perform this study. It was observed that canopy greenness of the wheat field reached peak in early July followed by the process of yellowing. Reference [9] highlighted that UAV is an efficient and effective way of obtaining significant field information. This further can be used for precision farming using remote sensing technology. It was also observed that UAV multispectral devices are suitable for quantitative wheat monitoring with better precision. Reference [10] highlighted that UAVs along with IoT (Internet of Things) were employed extensively for studying different aspects in recent years. Primarily, the aforementioned technology was used to identify pests and weed detection.

This study aims to demonstrate the utility of UAV-RGB backed satellite remote sensing analytics in assessing the dynamics of crop health. Objectives of the present study can be segregated into the following categories:

i. Assessment of the crop health dynamics of test area using satellite remote sensing analytics.
ii. Construing crop health dynamics in hyperlocal scenario (plot wise) using UAV-RGB analytics.
iii. To demonstrate the utility of combined framework comprises of satellite and UAV remote sensing technology in studying crop phenology dynamics.
iv. To assess the indicative crop yield scenario using hybrid framework comprised of UAV-RGB.

2 Study Area

The study area is bounded by 28.5426° north latitude and 79.4704° east longitude. Major crops in the study area are paddy, wheat, sugarcane, pearl millet, black gram, mustard toria, and lentil, respectively. Crop production area is high in Bareilly. It is also known for bamboo artwork, handicrafts, and furniture. Aforementioned aspects are also important for district economy. Therefore, farmers used expensive non-organic fertilizers which caused severe environmental health hazards in the study area. Hence, it is now required to understand that use of organic fertilizers should be encouraged to control the possibilities of environmental hazards due to the use of non-organic fertilizers. Moreover, the district is prone to drought, flood, hail-storm, heat wave, and cold wave occasionally.

3 Research Methodology

3.1 Collection and Pre-processing of Data

This section has been categorized into two segments; first, computation of indices for the entire test region using satellite imageries, and second, computation of indices for different sample plots using UAV data. Research methodology is shown in the Fig. 1. Spectral indices such as Normalized Difference Vegetation Index (NDVI), Normalized Difference Moisture Index (NDMI), Bare Soil Index (BSI), and Red Edge Index were prepared using Google Earth engine. These indices were prepared using Sentinel-2 satellite data series in a temporal window of 10–25 days. Next, Greenness Index was computed for sample plots using UAV-RGB data. The Greenness Index was also computed for the same temporal window (Table 1).

3.2 Classification of Data and Association Analysis

Identified factor layers were reclassified into four classes using equal interval method. Reclassification was done to identify segments in the test area corresponding to their possible characteristics. In addition, this may help in identifying regions which reflects similar characteristics when investigated in the context of whole test area and hyperlocal scenario. Moreover, descriptive statistics such as range, maximum, and minimum were quantified for each class of the classified spectral and greenness indices of the considered time interval. It was assessed to decipher the intrinsic characteristics of these indices. Next, characteristics and behavior of each index with respect to each other was assessed to understand the relationship among the indices.

Table 1 List of spectral and Greenness indices used in this study

S. No.	Index	Source	Computational formula	Significance
1	Normalized Difference Vegetation Index (NDVI)	Satellite imagery (Sentinel-2)	$\frac{(\mathrm{NIR}-R)}{(\mathrm{NIR}+R)}$	This helps in understanding the dynamics of vegetation health
2	Normalized Difference Moisture Index (NDMI)	Satellite imagery (Sentinel-2)	$\frac{(\mathrm{NIR}-\mathrm{SWIR})}{(\mathrm{NIR}+\mathrm{SWIR})}$	This index is effective in assessing the leaf water content
3	Bare Soil Index (BSI)	Satellite imagery (Sentinel-2)	$\frac{((R+\mathrm{SWIR})-(\mathrm{NIR}+B))}{((R+\mathrm{SWIR})+(\mathrm{NIR}+B))}$	It is used to identify the exposed soil
4	Normalized Difference Red Edge Index (NDRI)	Satellite imagery (Sentinel-2)	$\frac{(\mathrm{NIR}-\mathrm{RedEdge})}{(\mathrm{NIR}+\mathrm{RedEdge})}$	This index is effective in assessing the dynamics of dense vegetation
5	Greenness Index (GI)	UAV data	$\frac{(G)}{(R+G+B)}$	This index is effective in quantifying the greenness of vegetation which is an indicator of the status of crop physiology

3.3 Assessment of Crop Yield Scenario Using Hybrid Framework Comprised of Digital Image Processing Methods and Spectral Indices

Sample plot data was collected using UAV. The UAV data was processed using stage-wise image filtering. High pass filtering method was applied on UAV data followed by low pass filtering and local maxima analysis. Laplacian and median filtering methods were used in the high pass and low pass filtering groups, respectively. Next, local maxima analysis was performed on the result of median filtering to identify wheat ears pattern. Finally, resultant image of this process was classified using equal interval to identify dense clusters of wheat ears and the subsequent pattern. The wheat ears were counted in both ways to nullify any possibilities of inconsistencies, that is, firstly though manual counting and then, the counting process was automated. Based on the wheat ear count and the observed pattern, the possible scenario in terms of crop yield was assessed.

$$\text{Crop yield scenario} = f\begin{pmatrix}\text{computed spectral indices, computed greenness index,}\\ \text{wheatear count, wheatear cluster pattern}\end{pmatrix} \tag{1}$$

4 Results and Discussion

This section is categorized into two modules, that is, findings related to the process of assessing and quantifying crop phenology dynamics using spectral indices and Greenness Index based on satellite imageries and UAV data and application of digital image processing algorithms such as high pass filters, low pass filters, and local maxima analysis for counting wheat ears and assessing wheat ear pattern in the sample plots. In addition, this section explains the association analysis between the results which were obtained from the module pertaining to deciphering crop phenology at the whole test area level/plot level and the module related to estimating the wheat ear count using digital image processing methods.

4.1 Crop Phenology Dynamics Using Spectral Indices and Greenness Index

Findings of this segment/module, that is, representation of crop health using different indices for the whole test region, representation of crop health using different indices for sample plots, crop health representation of sample plots and frequency analysis of crop phenology are presented in Figs. 2, 3, 4, 5, and 6; Table 2, respectively. This is evident from figures that there is considerable difference in indices value for different time periods. This suggests that there is change in the intensity of crop maturity. In addition, it is clear from Table 2 that spectral indices such as Normalized Difference Vegetation Index (NDVI) and Normalized Difference Red Edge Index (NDRI) reflect similar characteristics.

4.2 Estimation of Wheat Ear Count pattern using Digital Image Processing Methods

Figures 7, 8, 9, 10, and 11 represent the results of wheat ear count estimation. However, Figs. 7, 8, 9, and 10 corresponds to the manual segregation of wheat ear clusters and the wheat ear count. Automated representation of wheat ear cluster segregation and the wheat ear count is presented in Fig. 11. Both the ways, that is,

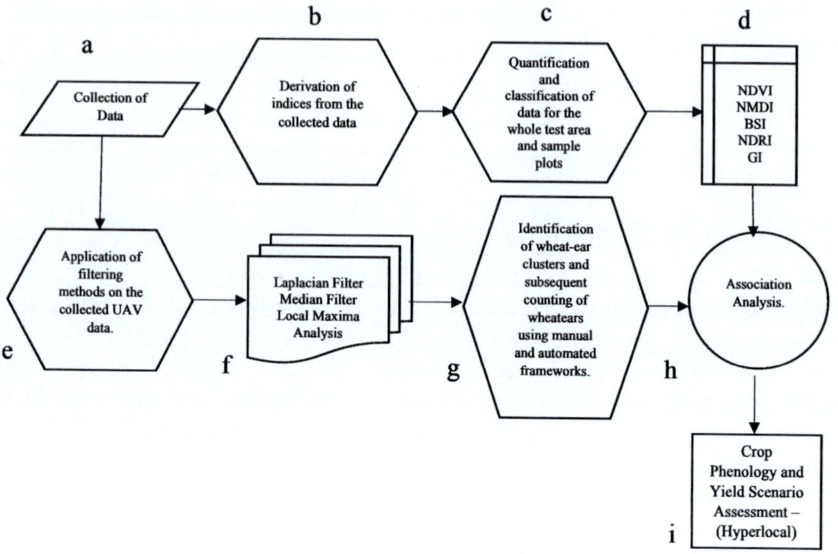

Fig. 1 Block diagram for crop phenology and yield scenario assessment at hyperlocal level using hybrid framework

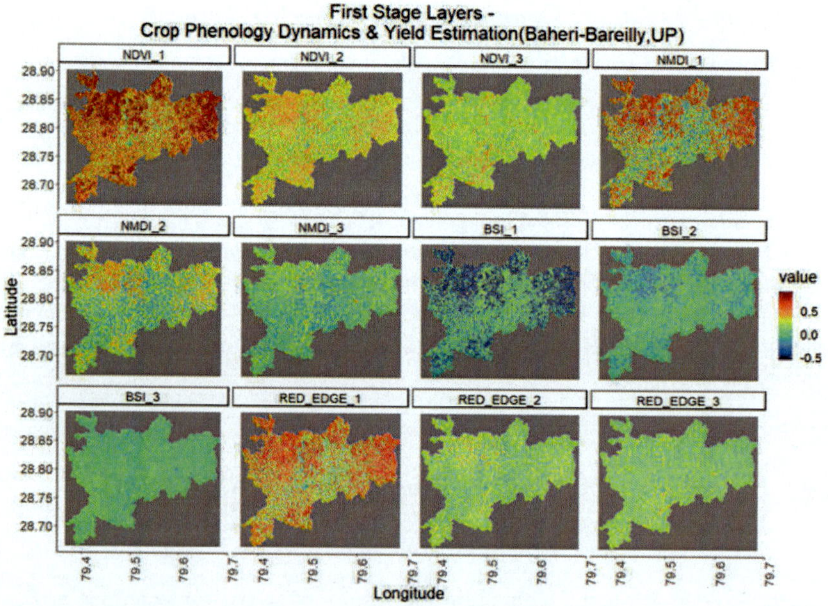

Fig. 2 Spectral indices in different time windows—whole test area

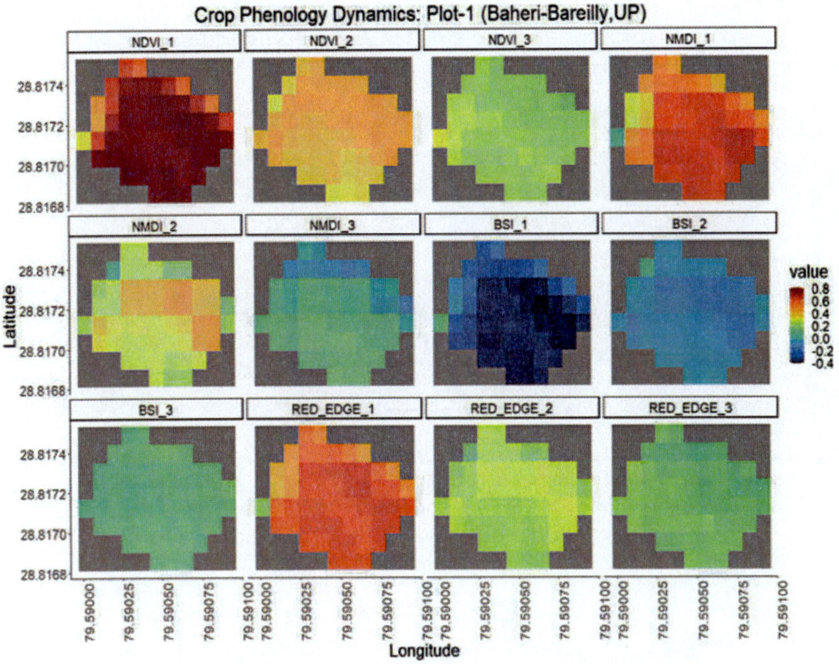

Fig. 3 Spectral indices in different time windows—sample plot 1

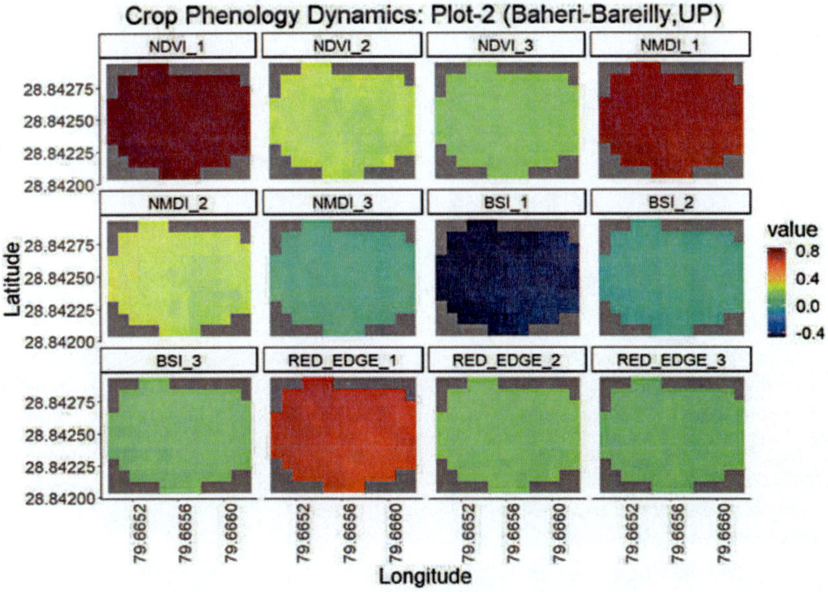

Fig. 4 Spectral indices in different time windows—sample plot 2

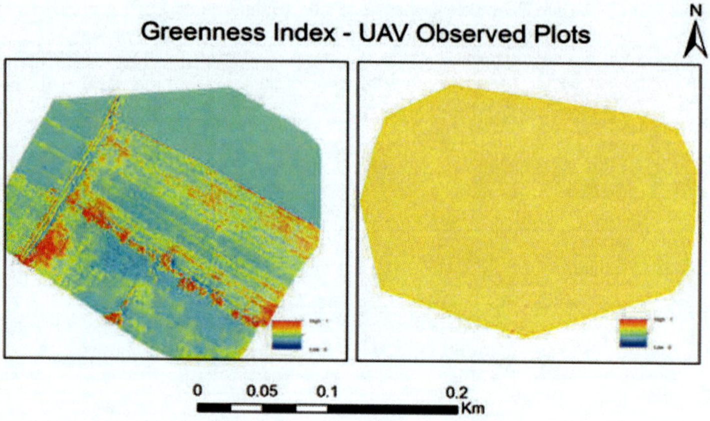

Fig. 5 Greenness Index—sample plots (Plot-1, Plot-2)

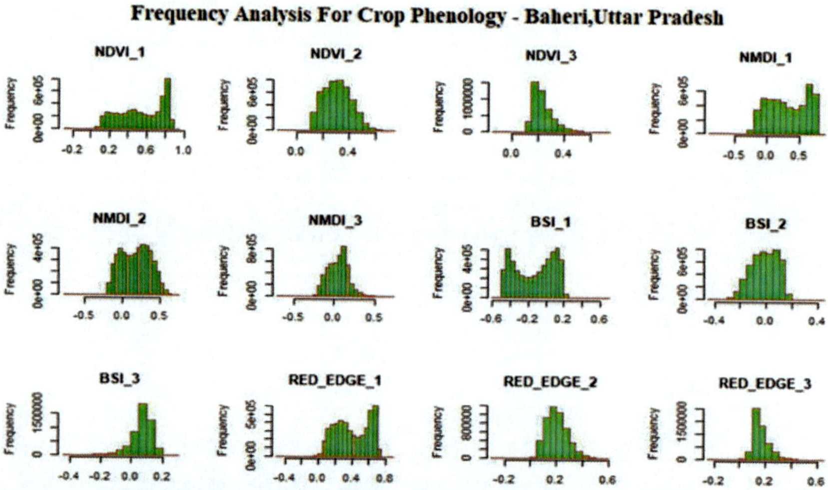

Fig. 6 Frequency analysis of spectral indices

manual and automated processing of wheat ear cluster segregation and the respective count was done to avoid the possibilities of any inconsistencies. Figures 7 and 9 correspond to cluster segregation, and Figs. 8 and 10 represent the manual count of wheat ear segregation.

In addition, careful observation of Figs. 8, 10, and 11 clearly indicates that plot 1 has better distribution of wheat ears than plot 2. Presence of wheat ears seems to be confined to a particular zone in the plot 2. In addition, when the pattern of wheat ears is studied in the context of spectral indices and greenness indices; this is clear that the crop health seems better in plot 1 than plot 2. As aforementioned, distribution of

Table 2 Value of spectral indices and greenness index in different time periods

Indices	Baheri block					
	Min			Max		
	T1	T2	T3	T1	T2	T3
NDVI	−0.23	−0.1	−0.11	0.91	0.7	0.73
BSI	−0.54	−0.42	−0.42	0.57	0.39	0.27
NDWI	−0.35	−0.14	−0.18	0.83	0.62	0.65
NMDI	−0.84	−0.42	−0.54	0.82	0.39	0.71
NDRI	−0.49	−0.26	−0.21	0.77	0.57	0.59
GI						
Indices	Plot 1					
	Min			Max		
	T1	T2	T3	T1	T2	T3
NDVI	0.33	0.32	0.14	0.81	0.44	0.29
BSI	−0.4	−0.15	0.03	−0.01	0.03	0.14
NDWI	0.39	0.29	0.22	0.72	0.39	0.31
NMDI	0.24	0.03	−0.14	0.67	0.42	0.15
NDRI	0.26	0.15	0.09	0.64	0.31	0.21
GI			0			1
Indices	Plot 2					
	Min			Max		
	T1	T2	T3	T1	T2	T3
NDVI	0.78	0.23	0.17	0.83	0.29	0.19
BSI	−0.47	−0.05	0.1	−0.41	0.01	0.15
NDWI	0.69	0.28	0.3	0.74	0.32	0.33
NMDI	0.69	0.22	−0.01	0.75	0.3	0.1
NDRI	0.62	0.14	0.1	0.68	0.18	0.14
GI			0			0.49

Fig. 7 Wheat ear cluster pattern—manual (sample plot 1)

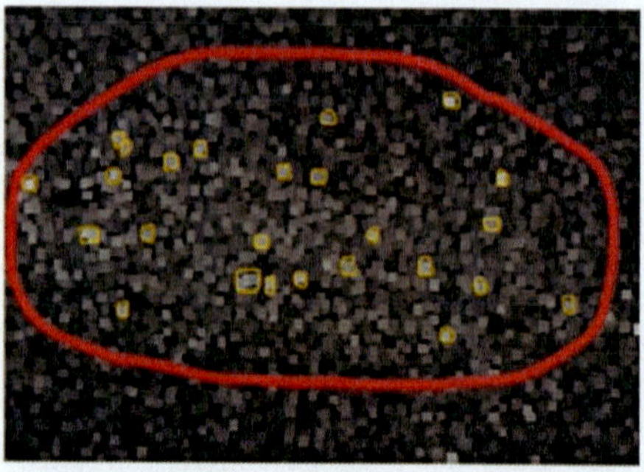

Fig. 8 Wheat ear count—manual (highlighted, sample plot 1)

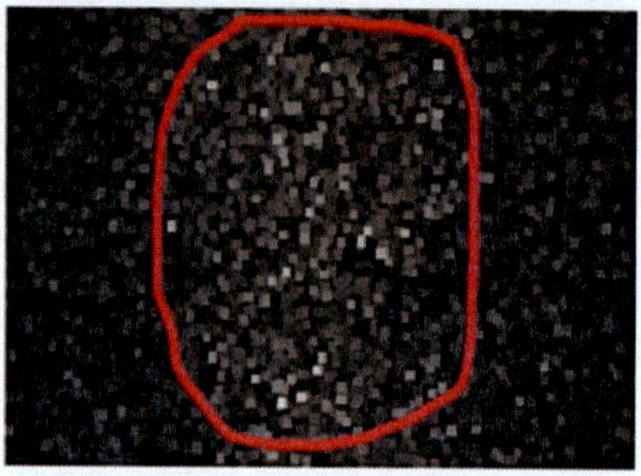

Fig. 9 Wheat ear cluster pattern—manual (sample plot 2)

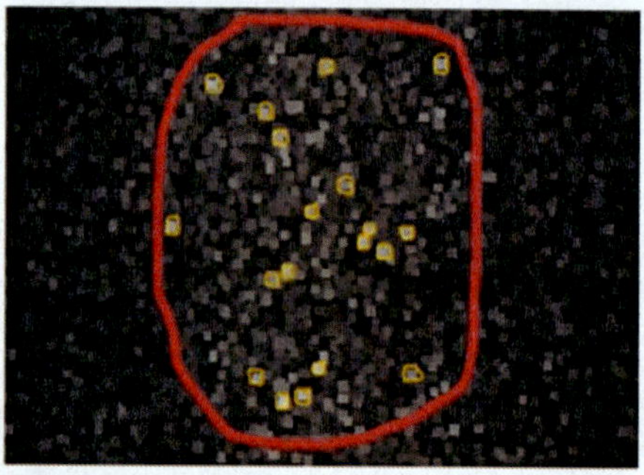

Fig. 10 Wheat ear cluster pattern—manual (highlighted, sample plot 2)

wheat ears cluster is better in plot 1, and the wheat ear count is also greater in plot 1. This suggests a strong possibility that the crop yield in plot may be higher than plot 2. However, this can be confirmed through the grain count in the respective plots which is not scope of this study.

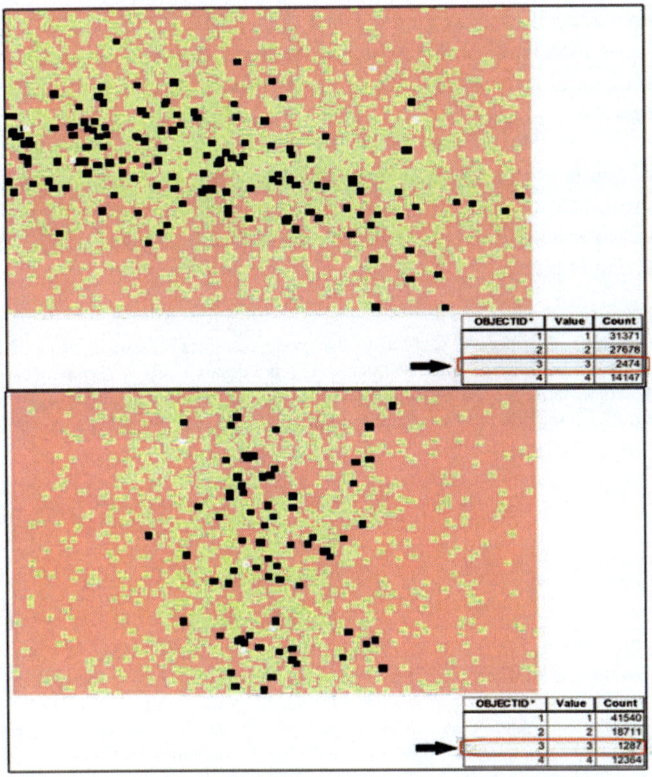

Fig. 11 Wheat ear cluster pattern and count estimation—automated (sample plot 1 and 2)

5 Conclusions

This study aims to study the dynamics of crop phenology and assess crop yield scenario using a hybrid approach. Hybrid approach corresponds to the application of both crop health analysis and wheat ear count estimation for generating a possible crop yield scenario at hyperlocal level. This is observed that sample plot 1 may have better yield than sample plot 2 due to its seemingly good crop health and better wheat ear distribution and count. Hence, this study can be concluded with the following points:

i. Application of hybrid approach can be significant in reducing the possibilities of inconsistencies.
ii. Using Normalized Difference Red Edge Index (NDRI) along with Normalized Difference Vegetation Index (NDVI) may be instrumental in tackling the limitation of NDVI.

iii. Framework which is comprised of Laplacian filters along with the use of low pass filter and local maxima analysis can identify the wheat ear distribution successfully. Wheat ear count estimation is also possible through this framework.
iv. Association analysis between crop phenology dynamics and wheat ear distribution/count assessment may be significant in assessing the possible crop yield scenario.
v. There is a strong likelihood that the sample plot 1 may have higher crop yield than sample plot 2.

Acknowledgements Authors are grateful to the group chairman of Leads Group, Mr. Navneet Ravikar for allowing and providing us a platform of Leads Research Lab, R&D unit of LeadsConnect Services Pvt. Ltd, to perform this investigation and for his inspiring and encouraging leadership. We are grateful to the anonymous reviewers for their suggestions and comments which will definitely help us to improve this work.

References

1. Liua L, Wallachb D, Lia J, Liua B, Zhanga L, Tanga L, Zhanga Y, Qiua X, Caoa W, Zhua Y (2017) Uncertainty in wheat phenology simulation induced by cultivar parameterization under climate warming. Eur J Agron 94:46–53. https://doi.org/10.1016/j.eja.2017.12.001
2. Ahmad S, Abbas G, Ahmed M, Fatima Z, Anjum MA, Rasul G, Khan MA, Hoogenboom G (2018) Climate warming and management impact on the change of phenology of the rice-wheat cropping system in Punjab, Pakistan. Field Crops Res 230:46–61. https://doi.org/10.1016/j.fcr.2018.10.008
3. Duncan JMA, Dash J, Atkinson PM (2015) The potential of satellite-observed crop phenology to enhance yield gap assessments in smallholder landscapes. Front Environ Sci 3. https://doi.org/10.3389/fenvs.2015.00056
4. Aasen H, Burkart A, Bolten A, Bareth G (2015) Generating 3D hyperspectral information with lightweight UAV snapshot cameras for vegetation monitoring: from camera calibration to quality assurance. ISPRS J Photogramm Remote Sens 108:245–259. https://doi.org/10.1016/j.isprsjprs.2015.08.002
5. Candiago S, Remondino F, De Giglio M, Dubbini M, Gattelli M (2015) Evaluating multispectral images and vegetation indices for precision farming applications from UAV images. Remote Sens 7:4026–4047. https://doi.org/10.3390/rs70404026
6. Bellundagi A, Singh GP, Prabhu KV, Arora A, Jain N, Ramya P, Ahlawat A (2013) Early ground cover and other physiological traits as efficient selection criteria for grain yield under moisture deficit stress conditions in wheat (Triticum aestivum L.). Indian J Plant Physiol 18(3):277–281. https://doi.org/10.1007/s40502-013-0047-6
7. Bendig J, Yu K, Aasen H, Bolten A, Bennertz S, Broscheit J, Bareth G (2015) Combining UAV-based plant height from crop surface models, visible, and near infrared vegetation indices for biomass monitoring in barley. Int J Appl Earth Obs Geoinf 39:79–87. https://doi.org/10.1016/j.jag.2015.02.012
8. Du M, Noguchi N (2017) Monitoring of wheat growth status and mapping of wheat yield's within-field spatial variations using color images acquired from UAV-camera system. Remote Sens 9(3):289. https://doi.org/10.3390/rs9030289

9. Lelong C, Burger P, Jubelin G, Roux B, Labbé S, Baret F (2008) Assessment of unmanned aerial vehicles imagery for quantitative monitoring of wheat crop in small plots. Sensors 8(5):3557–3585. https://doi.org/10.3390/s8053557
10. Nakshmi JVN, Hemanth KS, Bharath J (2020) Optimizing quality and outputs by improving variable rate prescriptions in agriculture using UAVs. Procedia Comput Sci 167:1981–1990. https://doi.org/10.1016/j.procs.2020.03.229

360° Video Summarization: Research Scope and Trends

Sanket S. Kulkarni and Ansuman Mahapatra

Abstract Video summarization provides short and precise representation of the original video clips preserving important information. Video summarization assists us in navigating through a long video sequence and retrieving the most important segments. Output summary can be static containing some key frames or dynamic comprising of short video clips. The aim of video summarization is representing the original video for easy comprehension of the visual information. There are various applications of video summarization which includes video surveillance, egocentric videos, news, movies, and sports videos. Watching a lengthy 360-degree video is an exhausting task for a human annotator and to identify the key events, especially when there are unlimited field of views (FoVs). 360-degree video summarization is a less-explored area. In this article, existing methods for 360-degree video summarization are reviewed and analyzed based on algorithms, key features, merits, and demerits.

Keywords 360° videos · Static video summarization · Dynamic video summarization · User-generated video summarization · Pano2Vid · Key frame extraction

1 Introduction

The 360-degree cameras are the omnidirectional camera systems where a view in each direction is captured with overlapping FoVs simultaneously. These videos are usually shot using either a single consumer camera having two wide angle lenses or a rig that has multiple camera lenses built into the system. Through the technique of video stitching, separate footage is merged into spherical video [1]. Users can communicate with the video during playback by pressing a mouse or inclining a mobile device in any direction in the video recording, then moving the video in

S. S. Kulkarni · A. Mahapatra (✉)
Department of Computer Science and Engineering, National Institute of Technology Puducherry, Karaikal, India
e-mail: ansuman.mahapatra@nitpy.ac.in

P. S. R. Chowdary et al. (eds.), *Evolution in Signal Processing and Telecommunication Networks*, Lecture Notes in Electrical Engineering 839,
https://doi.org/10.1007/978-981-16-8554-5_28

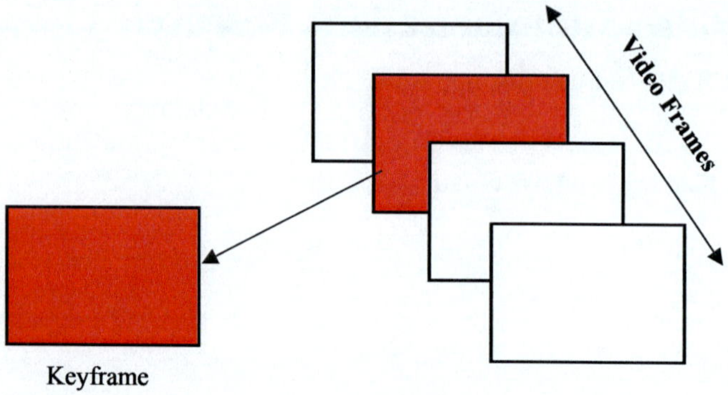

Fig. 1 Demonstrates the important key frames extracted from a video after summarization

parallel with the current view plane. Hardly any special display device is required to view the 360-degree video on a monitor, smartphone, or tablet [2]. Figure 1 illustrates that after summarization, it displays the most important key frames from a video that should be included in the final summary. The video summarization focuses on keyframes rather than all of the frames generated by the video, and it is dependent on the video's length. The classical video summarization techniques are classified as static and dynamic video summarization as depicted in Fig. 2. The static video summarization generates key frames from the input video, however the dynamic video summarization generates subshots.

The video frames found in 360-degree video are represented in equirectangular representation as shown in Fig. 3. The vehicle, as shown in Fig. 4, could be the most interesting part of that 360-degree video frame. Unlike classical video summarization we must summarize 360-degree videos both spatially and temporally by extracting the best viewpoint and finding the important frames in the initial video. The various steps in 360° video summarizations are as follows:

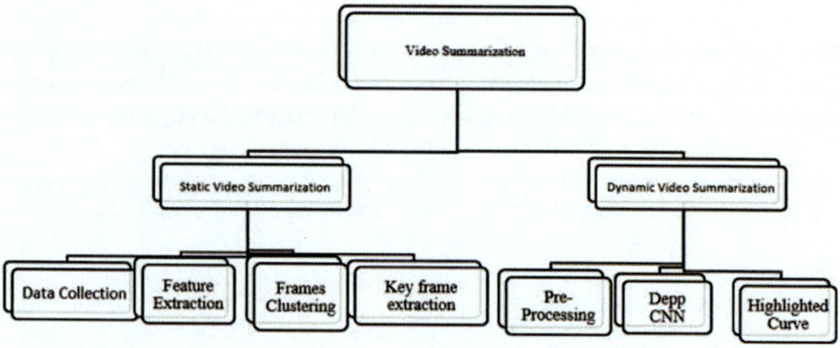

Fig. 2 Classification of classical video summarization techniques

Fig. 3 Image showing 360-degree view

Fig. 4 Illustrating field of view in 360 image

- For spatial summarization, use the normal field of view (NFOV) selection.
- Generating metashot of the entire video and selecting the subshots.
- Correlating to the forthcoming frame, the previously selected metashots are obtained for the upcoming summary.
- Then, the next stage is identifying the future metashots which are previously not selected (Fig. 5).

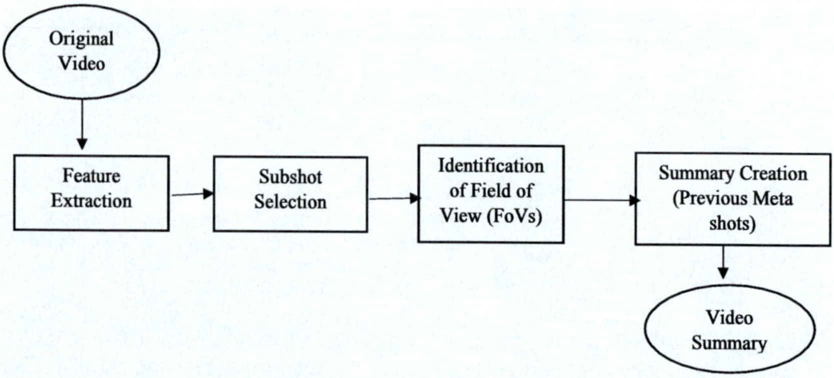

Fig. 5 Steps in video summarization

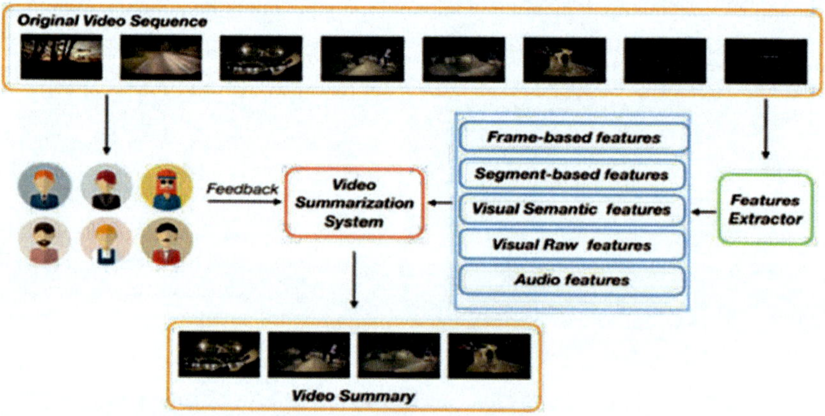

Fig. 6 Video summarization technique involving features [3]

1.1 Video Summarization Techniques

There are various techniques in video summarization which can be broadly categorized based on various aspects. This section mainly discusses on video summarization techniques and their applications.

- Feature-based video summarization: Color, motion, and voice are just a few of the features found in videos. This technique is particularly useful if the major focus of annotator is oriented toward the features. For instance, the user (annotator) wants to recognize colors, then it is preferable to use video summarization techniques that are color-based. Motion, color, evolving contents, expression, speech transcripts, objects, and other aspects are all considered in video summarization techniques based on the characteristics. Figure 6 demonstrates the various techniques of video summarization involving features.

- Video summarization using clustering: When we come across similar characteristics or activities within a frame, the most common technique is clustering. It also aids in the elimination of frames with erratic trends. Other video summarization methods allow for more efficient video browsing, but they also make it a tedious task to analyze the existing video summaries. Further this technique uses an approach called clustering that can be categorized in accordance with equivalent events, K-means, partitioning-based clustering approach, and spectral-based clustering approach.
- Key-shot identification approach: Initially, we perform the task of key-shot identification. Further present technique of video summarization is popularly used in case of YouTube videos information is mainly concerned about factors such as redundancy or identical sequences. This type of identifying sequences can be accomplished using co-occurrence detection that also includes determining similar kinds of data using keyframe pairs derived from different online videos. The clustering-based method can be used to speed up the key-shot identification process because equivalent keyframes usually have a small difference.
- Bag of Importance model: This technique of video summarization technique can be seen as a perspective with a collection of weighted features rather than ones which are equally significant. The mechanism utilizes the characteristics of intra-frame and inter-frame sequences in video by weighing the important individual features that make up the video as a whole. A video sequence must be projected to a low-dimensional sparse space to remove redundant visual features.

2 Challenges of Video Summarization

The video saves both spatial and temporal features. It is made up of frames, shots, and scenes and has a hierarchical structure. In user-generated videos, the video is recorded for a full length, but the task becomes tedious since the user wants to recognize only specific portions of video and not the entire video. In 360° videos, the vital role of video summarization is to use summarization to view only one particular region and not entire set. Since the annotators identify the portions of video which needs to be included, it is a very tedious task. The various other areas of video summarization include news videos where the specific person/user wants to know only a specific news regions like sports or politics, but the video has to be played to the end, then the content of news can be identified, and the user is satisfied which is enormous amount of time-consuming task. Video surveillance is one of the major applications of video summarization, wherein the CCTV footages are captured from civic spaces, but it becomes very difficult to identify fraudulent activities if there are multiple cameras which are placed in particular to spot or identify which section mainly focused on which event occurred. In case of sports videos (e.g., soccer and cricket), the viewer is mainly interested in key moments within the match such as stats or facts, but the viewer must focus for long period of time which makes it a tremendously time-consuming task. For the field of medicine endoscopic reports and X-ray reports, these are very huge data in videos, so using summarization makes the tasks easier. Though there are numerous challenges, researches have been carried out in order to solve the problems of video summarization the next section which mainly focuses on solutions to above-mentioned problems.

360° videos has a lot of storage requirements, has a limited field of view (FoV) due to the human visual system and display devices, has interactivity, smooth user navigation, and requires a lot of coding. Instead of traditional broadband televisions, video consumers are more interested in the immersive experience. The video's visible area (also known as the user's viewport) is displayed using a head-mounted display (HMD) with a high frame rate and resolution. Providing the end-user with all 360-degree frames in ultra-high resolution puts a lot of pressure on the service providers.

3 Literature Survey

There are many existing works on 2D video summarization in the article by Liu et al. [4] proposed an approach to use dissimilarity-based sparse modeling representative selection (DSMRS), the key idea is to extract the key frames instead of using the existing old models. They proposed the concept of accelerate values in order to exclude the outliers. Pablos et al. [5] suggested video summary that highlights the full video by using the actions of players as a reference point. There is a popular procedure within a framework known as deep neural network-based technique in order to extract the two types of characteristics which are defined based on gestures and to categorize video segments into most significant or non-significant segments is used. S. Bano and A. Cavallaro [6], this work was proposed to identify and synchronize structure automatically using the multicamera UGV. They suggested a chroma function which takes each query as an example in the video search event. The procedure used in this work mainly focusses on grouping the user generated videos which are belonging to the similar event and also to evaluate equivalent time change, and in contrast to the present recognition methods, this work focuses on notably determined classification threshold with the help of association and clustering for incoming videos.

Sahu and Chowdhury [7] mainly proposed graph-based video summarization this graph-based approach proposed a new measure for excluded arcs known as minimum spanning tree (MST) based on video resemblance, then further a graph is representing different frames in the video. Basavarajaih and Sharma [8] mainly focused on generic video summarization, wherein they use learned features from deep learning framework video frames that are clustered consecutive key frame selection that takes place through which it is used to summarize. Ping Li et al [9] recommended that by using every two frames, there exists a pairwise relations, it could determine the temporal dependency of a video summarization via global diverse attention which is a model that has been discovered in supervised, unsupervised, and semi-supervised scenarios.

Huang et al. [10], a CNN-based interactive attention network is put forward, in order to better capture the implicit interactive information between the query and a video. There are few works related to unsupervised video summarization which includes the works such as Yaliniz and Ikizler-Cinbis [11] work which focused to handle video summarization tasks, and an unsupervised method based on reinforcement learning was developed. The model is built on reinforcement learning and may not require any labels or user interaction. Li and Lei [12] used Kernel temporal segmentation algorithm for video extent of semantic similarity. In order to obtain video semantic representations, a model called video classification sub-network (VCSN) has been established. Huang, and Worring [13], this work focused on creation of end-to-end deep learning model for a text-visual embedding space for query-controllable video summarization. The dataset consisted of 190 videos with their corresponding score annotations with it.

Sridevi and Kharde [14], they took input videos and used shot boundary detection to split them into segments. In the case of a spatial stream, each video is divided into segments, which are then divided into frames. Spatial deep convolutional neural networks are used to extract features for each of these frames (DCNN). Sun et al. [15] included a feature extraction and key frame extraction technique called optimal threshold and maximum weight clustering. The frame density, inter-distance, and intra-distance are computed for generating the weights along with Maxi Weigh method which was put forward in order to cluster representative frame. Tejal Chavan et al. [16], they first extracted the key frames, then identified and checked for the distance between the objects to eliminate the video's idle scenes. Only useful frames and representative frames should be chosen; these frames will be used to regenerate the summarized video. Shruti Jadon and Mahmood Jasim [17] came up with an attempt to solve video summarization through unsupervised learning by employing traditional vision-based algorithmic methodologies for accurate feature extraction from video frames. Shuwen Xiao et al. [18] introduced a network called conventional hierarchical network based on a convolutional network and a global–local attention mechanism, the hierarchical attention network can generate video query-related summary in parallel.

The classical video summarization algorithms are mainly categorized in two groups (as depicted in Fig. 2.), i.e., static video summarization and dynamic video summarization. As shown in Fig. 1, static video summaries are mainly composed of a pair of key frames. The keyframes are the frames from the actual footage that contains most important information. Small video shots are used to accommodate important events in the original video in dynamic video summaries. 360-degree video summarization is a less-explored field with a limited number of papers (Table 1).

According to Su et al. [2], the goal is to guide an imaginary camera to obtain normal field of view (NFoV) video for a specified 360° video through choosing "where to look" at each time step in the panorama. The purpose of Pano2vid is to reduce the work of the videographer and the end viewer from quest of deciding on what to observe. Next, they proposed AUTOCAM approach. Hu et al. [19], a technique known as policy gradient technique is used for training pipelines by minimizing it, and the various steps involved are as follows: (1) With the help of viewing angle

Table 1 Existing works on 360-degree video summarization

Author	Algorithm/model	Key features	Drawbacks	Dataset
Hu et al. [19]	Deep RCNN	The view point has been repositioned from the current view angle to the next view angle	When there are salient or non-salient objects in the domain, it suffers	360° videos from 342 sports (basketball, parkour, BMX, skateboarding, and dance)
Su and Grauman [20]	AutoCam algorithm	A repetitive approach has been used to get a unique group of output trajectories	The derived videos appear to be user videos, which helps to provide useful information to a static audience	Pano2Vid dataset consisting of 86 360° videos crawled from 76 360 videos viewed by 58 subjects
Qiao et al. [21]	Multi-task DNN	Proposed a multitask DNN for video saliency detection (VS) in 360° videos	Only if each user's trajectory history is known can video saliency be used	Over 200 360° videos have been viewed across 30 subjects by Pano2Vid
Kang and Cho. [22]	Saliency estimation	They proposed a 360° video playback which is user friendly and easy to move	With the help of the camera path, changes in the user's position would be automatically updated	Pano2Vid, Sports 360° Video
Yu et al. [23]	AutoCam, RankNet	Detect optical flow and calculate saliency scores for video input	It is difficult to adjust the viewing plane of 360° videos since angle is constantly changing	Pano2Vid dataset, YouTube/Vimeo dataset
Lee et al. [24]	Past future memory network	The problem of sequences is the main focus with summaries in form of stories	They majorly provided a latent collective summary	Pano2Vid dataset and VIST dataset

and ground truth values which in turn computes regression loss, (2) It is possible to compute the loss function along with the transition in viewpoint, and (3) The major focus is on the foreground image object which can be improved with the increase in reward. Su and Grauman [20] proposed an algorithm called AUTOCAM which enables a common camera, prevents repetitive productions, and significantly more efficiently extracts output videos. During the course of 7 h an actual "in the wild" video, experimental results show that our generalized camera solved the problem of Pano2Vid dynamically, introducing the proper approach known as optimization makes it adaptable, and they proposed a proper technique to obtain "good" NFOV videos. With a set of metrics—that includes human editors—the proposed algorithm tries to get better results compared to the existing it also tries to bring step closer virtual photography. Qiao et al. [21], fixations are affected in 360° videos with the help of location of the viewport and objects. They proposed an MT-DNN approach for video saliency prediction. The major drawback is the methods of prediction of saliency on 360° videos which only concentrate on generating an entire panoramic frame saliency map and not the video saliency viewport map.

Kang et al. [22] focused primarily on a framework which includes both online and offline preprocessing stages, and 360° video navigation phase allows online processing for further work on 360° video summarization. The optical flow and saliency scores for an input video are computed during the preprocessing step. On the basis of this, the online video navigation step calculates and plays an online NFoV video of the optimal camera path that reflects user interaction. For enhanced user experience, for path planning, they used optical flow-based camera, salience-aware path updating, and adaptive control of the temporal window size. The issue of the past future memory network was addressed according to Lee et al. [24].

Despite the volatile increase in the 360° videos, its summarization has been still a less-explored area. There are few works such as AutoCam framework, deep360 pilot, and a model of the Composition View Score (CVS). In particular, the AutoCam framework solves the issue of Pano2Vid, which captures 360° videos and generates normal field of view (NFoV). The next stage is to provide the trajectories of the NFOV camera similar to those captured by a videographer that are generated and taken from a normal NFOV camera. In order to select NFOV shots from each frame in the 360° videos, the CVS framework first uses a deep ranking model for spatial summary, which is then applied to the temporal domain to produce a spatiotemporal highlight video. It is difficult to say whether the CVS framework, which was created with the problems of temporal summarization in mind, would be successful.

4 Conclusion and Future Work

360° video summarization plays a vital role in various fields such as video surveillance, entertainment industries, sports, and news. Identifying any salient feature by a video annotator is a difficult task as it is a cumbersome job to watch the entire video. 360-degree videos allow the user to capture the entire scene. In 360-degree videos, we have to identify an interesting viewpoint as well as identify key events in the video. Therefore, the 360-degree video summarization algorithm is more complex than normal FoV videos.

In this paper, we have discussed various existing 360-degree video summarization methods. From the literature, it is found that there are very few propositions on video summarization systems. Therefore, there is a scope of research in this field as the 360-degree video content is growing every day due to the availability of cheap 360-degree cameras for consumers [25].

In the development of new deep learning algorithms and architectures, video summarization techniques can be efficiently automated. In general, a deep generative model and variational auto-encoder are considered to be reasonably good methods for static and dynamic summarization approaches to 360-degree video summary techniques, which include a deep generative model and variational auto-encoder.

References

1. S. Priyadharshini and A. Mahapatra.:360° User-Generated Videos: Current Research and Future Trends, High Performance Vision Intelligence, PP. 117–135 (2020). DOI: https://doi.org/10.1007/978-981-15-6844-2_9
2. S. Y. Chuan, D. Jayaraman, and K. Grauman.: Pano2Vid: Automatic Cinematography for Watching 360° Videos. Asian Conference on Computer Vision, pp. 154–171(2016). Springer, arXiv:1612.02335v1
3. Jiang, Yudong, Kaixu Cui, Bo Peng, and Changliang Xu.: Comprehensive Video Understanding: Video summarization with content-based video recommender design. In Proceedings of the IEEE/CVF International Conference on Computer Vision Workshops, pp. 0–0. (2019). DOI: https://doi.org/10.1109/ICCVW.2019.00195
4. Y. Liu, H. Liu, Y. Liu, and F. Sun.: User-generated-video summarization using Sparse Modelling, International Joint Conference on Neural Network (IJCNN), pp.3909–3915, (2014). DOI: https://doi.org/10.1109/IJCNN.2014.6889581
5. Pablos TD, Nakashima Y, Sato T, Yokoya N, Linna M, Rahtu E (2017) Summarization of user-generated sports video by using deep action recognition features. Transactions on Multimedia 20(8):2000–2011. https://doi.org/10.1109/TMM.2018.2794265
6. Sophia B, Cavallaro A (2015) Discovery and organization of multi-camera user-generated videos of the same event. Inf Sci 302:108–121. https://doi.org/10.1016/j.ins.2014.08.026
7. Sahu A, Chowdhury AS (2021) First person video summarization using different graph representations. Pattern Recogn Lett 146:185–192. https://doi.org/10.1016/j.patrec.2021.03.013
8. Basavarajaiah M, Sharma P (2021) GVSUM: generic video summarization using deep visual features. Multimedia Tools and Applications 80(9):14459–14476. https://doi.org/10.1016/j.patrec.2021.03.013
9. Li P, Ye Q, Zhang L, Yuan L, Xu X, Shao L (2021) Exploring global diverse attention via pairwise temporal relation for video summarization. Pattern Recogn 111:107677. https://doi.org/10.1016/j.patcog.2020.107677
10. Huang, Jia-Hong, Luka Murn, Marta Mrak, and Marcel Worring. GPT2MVS: Generative Pre-trained Transformer-2 for Multi-modal Video Summarization. arXiv preprint arXiv:2104.12465 (2021). arXiv:2104.12465v1
11. Yaliniz, Gokhan, and Nazli Ikizler-Cinbis.: Using independently recurrent networks for reinforcement learning based unsupervised video summarization. Multimedia Tools and Applications 80, no. 12 (2021): 17827–17847. DOI: https://doi.org/10.1007/s11042-020-10293-x
12. Li, Zutong, and Lei Yang.: Weakly Supervised Deep Reinforcement Learning for Video Summarization with Semantically Meaningful Reward. In Proceedings of the IEEE/CVF Winter Conference on Applications of Computer Vision, pp. 3239–3247. (2021).
13. Huang, Jia-Hong, and Marcel Worring. Query-controllable video summarization. In Proceedings of the 2020 International Conference on Multimedia Retrieval, pp. 242–250 (2020). DOI: https://doi.org/10.1145/3372278.3390695
14. Sridevi, M., and Mayuri Kharde. Video Summarization Using Highlight Detection and Pairwise Deep Ranking Model. Procedia Computer Science 167 1839–1848 (2020). DOI: https://doi.org/10.1016/j.procs.2020.03.203

15. Sun, Yunyun, Peng Li, Yutong Liu, and Zhaohui Jiang.: Feature Extraction and Clustering for Static Video Summarization. In Research Square article(2021).
16. Chavan, Tejal, Vruchika Patil, Priyanka Rokade, and Surekha Dholay.: Superintendence Video Summarization. In 2020 International Conference on Emerging Trends in Information Technology and Engineering (ic-ETITE), pp. 1–7. IEEE, 2020.DOI: https://doi.org/10.1109/ic-ETITE47903.2020.150
17. Jadon, Shruti, and Mahmood Jasim. Unsupervised video summarization framework using keyframe extraction and video skimming. In 2020 IEEE 5th International Conference on Computing Communication and Automation (ICCCA), pp. 140–145. IEEE, 2020. DOI: https://doi.org/10.1109/ICCCA49541.2020.9250764
18. Xiao S, Zhao Z, Zhang Z, Yan X, Yang M (2020) Convolutional hierarchical attention network for query-focused video summarization. In Proceedings of the AAAI Conference on Artificial Intelligence 34(07):12426–12433. https://doi.org/10.1609/aaai.v34i07.6929
19. H.H Ning, Y.C Lin, M.Y Liu, H.T. Cheng, Y.J Chang, and M. Sun.: Deep 360 pilot: Learning a deep agent for piloting through 360 sports videos, IEEE Conference on Computer Vision and Pattern Recognition (CVPR), pp. 1396–1405, Honolulu, USA (2017). DOI: https://doi.org/10.1109/CVPR.2017.153 .
20. Su, Yu-Chuan, and Kristen Grauman.: Making 360 video watchable in 2D: Learning videography for click free viewing. Conference on Computer Vision and Pattern Recognition (CVPR), Honolulu, USA, pp. 1368–1376, (2017). DOI: https://doi.org/10.1109/CVPR.2017.150.
21. Qiao M, Xu M, Wang Z, Borji A (2020) Viewport-dependent Saliency Prediction in 360° Video. Transactions on Multimedia 23(9210):1–12. https://doi.org/10.1109/TMM.2020.2987682
22. K. Kang and S. Cho.: Interactive and automatic navigation for 360° video playback," ACM Transactions on Graphics (TOG), vol. 38, no. 4, pp. 1–11(2019). DOI: https://doi.org/10.1145/3306346.3323046.
23. Y. Yu, S. Lee, J. Na, J. Kang, and G. Kim.: A deep ranking model for spatio-temporal highlight detection from a 360° video, AAAI Conference on Artificial Intelligence vol. 32, no. 1, pp.7525–7533(2018).
24. S. Lee, J. Sung, Y. Yu, and G. Kim.: A memory network approach for story-based temporal summarization of 360° Videos, Conference on Computer Vision and Pattern recognition, pp. 1410–1419 (2018). DOI: https://doi.org/10.1109/CVPR.2018.00153
25. https://www.marketsandmarkets.com/Market-Reports/360-degree-camera-market-18216889.html

Performance Analysis of Underwater Acoustic Communication System with Massive MIMO-OFDM

Shaik Azeez and Bikramaditya Das

Abstract Underwater acoustic communication (UWC) is a game plan of sending and receiving information under the water. There are various procedures to experience such affiliations, yet each model has most generally using hydrophones. The UWC is intricate to convey in light of multipath propagation, time-varying nature, less to be had transmission capacity, and signal attenuation. In-wrinkling interest for transmission capacity, productivity, spatial variety, and execution of underwater acoustic communication has opened entryways for the utilization of multi-input and multi-output (MIMO). Multi-carrier modulation Along with OFDM offers high-rate transmission over long dispersive channels, while multiple-input multiple-output (MIMO) strategies increment the system capacity. A combination of MIMO and orthogonal frequency division multiplexing (OFDM) has end up being a promising answer for some situations in UWC. However, frameworks need to change the classification of modulation schemes dependent on underwater ecological conditions like seawater: temperature, saltiness, profundity, flows, waves, disturbance, and turbidity. In this paper, Web of underwater things (IoUT) gives the answer for separation dependent on the intricacy and generally execution of enhancements be confirmed dependent on various modulation schemes for feature investigation. The whole UWC channel recipients are created and checked by utilizing MATLAB with bellhop simulator software, noticing the likelihood of error, signal-to-noise ratio, and detection probability.

Keywords Underwater acoustic communication (UWC) · Multiple input multiple output (MIMO) · Orthogonal frequency division multiplexing (OFDM) · Channel estimation

S. Azeez · B. Das (✉)
Department of Electronics & Telecommunication Engineering, VSS University of Technology, Burla, Odisha 768018, India

P. S. R. Chowdary et al. (eds.), *Evolution in Signal Processing and Telecommunication Networks*, Lecture Notes in Electrical Engineering 839,
https://doi.org/10.1007/978-981-16-8554-5_29

1 Introduction

The underwater acoustic communication is truly expanding interest for some applications. The data transmission, spatial variety, and effectiveness of UWC are improving by the choice of procedures and improvement models. The electromagnetic, radiofrequency, and optical waves do not spread in seawater, alongside this sign has lessened firmly. The angle domain and frequency selective channel estimation algorithm with the MIMO model had been planned [1]. This calculation is isolated into two stages, from the start step nonzero tap positions are distinguished, and the following stage point space coefficients are perceived. This model adequately assesses the frequency selective channel for lessening the intricacy of UWC. The cutting-edge works identified with different UWC appraisal strategies like pilot-based, least square (LS), and small minimum mean square error (MMSE) models are investigated and close those future methods for channel estimation. The summed-up probability proportion indicators working is confirming the mistakes because of time varieties in the channel [2]. Specialists from the previous twenty years upgrade bit error rates and information rates. The OFDM is an elective procedure for single carrier modulation; due to its heartiness against multi-way nature, it rapidly tackles the inter-symbol interference. The smooth acknowledgment of the MIMO model improves the complex equalizer function.

The multipath communication debases the channel proficiency contrasted with single carrier transmission, and to relieve these issues, MIMO-OFDM underwater communication system approaches are conveyed over all channels. The underwater acoustic communication and its information rates have improved by applying the frequency trusted communication procedures. The models named as numerous phase-shift keying systems are executed; this procedure approves the better signal-to-noise ratio and better execution. The bit error rate in this investigation was determined as MFSK-FH 100 bps, and this framework improves the data transfer capacity effectiveness of an underwater mode [3].The counter multi-way, multi-carrier's communication method has been planned dependent on orthogonal chirp multiplexing innovation. This strategy additionally improves the hindrances, in homogeneity, and blurring parameters. The reproduction results beat the bit error rate and multipath superiority. The OFDM model gives the high robustness of multipath communication for UWC. Gatekeeper spans can address the issue of inter-carrier interference at the hour of transmission. This work is generally reasonable for underwater channel estimation to killing the breaks of UWC [4]. This design lessens the start to finish delay effectively with no energy utilization. The low transmission, signal-to-noise proportion, and defer issues are additionally confirmed for UWC [5]. The submerged acoustic networks are worked relying on the restriction model. Radio waves and optical waves were intensely weakened in water, and consequently acoustic signs are turned into a standard sign for UWC. The ecological factors immediately influenced the speed of sound waves; and some unacceptable acoustic signs will truly be creating the miss-situating results. Thusly, at any rate square blunder and balance remedy methods are carried out for adjusting the vulnerability of acoustic signs. The applied evenness revision and confinement calculation chronicles the incredible exhibition, and it is anything but at all delicate to the speed of an acoustic sign [6, 7].

2 Problem Statement

The bellhop simulator model is accustomed to assessing the channel reactions for chose for examination. The ocean profundity has changed from one meter to 10 m and the determination of transducers 1 to *n* contingent on the application. The distance between the hydrophone and n-transducer went from 100 to 1000 m. We can cross-check the communication between *m*th transducers to the nth hydrophone. The simulator itself contains the choice like the multipath determination; here 2–10 multipaths are chosen for operation rating. The essential channel delay is expected as 15 ms, and the conventional extra channel is picked for present and future examinations.

2.1 *Design Parameters of UWC*

Designing an efficient underwater communication system is only possible by massive MIMO-OFDM methods. These models are making efficient by using various optimization mechanisms [8–11]; in our approach for the UWC channel estimation, Doppler scaling factor has been calculated for perfect results. D_{p} = Doppler scaling factor θ_{p} = *z*th of the incident angle at selected acoustic signal [12].

$$D_{\mathrm{p}} = \left(\frac{v_{\mathrm{r}}}{c}\right)\cos\theta_{\mathrm{p}} \tag{1}$$

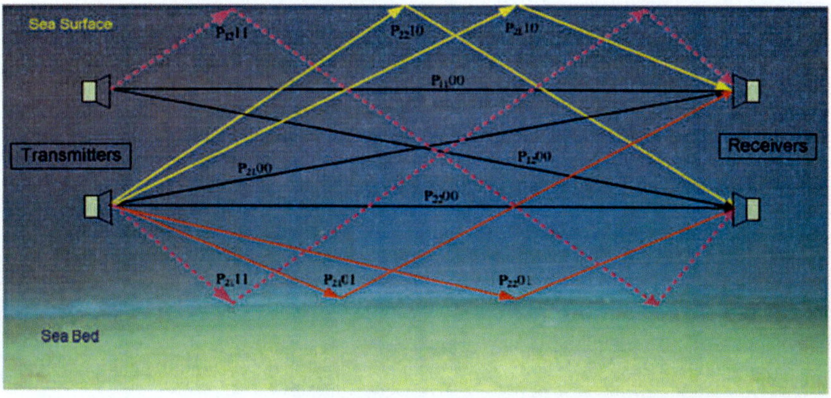

Fig. 1 Multipath channel effect in underwater communication

This mathematical equation calculates the Doppler shift velocity of transmitter or receiver, this Eq. 1 denotes the TX/RX speed, and '*c*' represents the speed of the acoustic wave in water [13, 14], and [15]. Generally, in UWC, the '*c*' is very high V_r, and this condition makes the D_p almost negligible, when the speed of the acoustic waves is less than more. Based on this type of theoretical and experimental calculations, differentiate sub-carries, resulting in that multipath communication is established. The impact of D_p for multi-carriers creates an inter-carrier interference [12, 16–19].

The nonlinear UWC spectrum degrades the performance of acoustic signals. The OFDM techniques were available for the reduction of PAPR depending on pre-coding selective mapping and false intervals.

3 Proposed Method

The advanced massive MIMO-OFDM model consists of Mt transducers and M receivers. The time duration has represented as $T'' = \mathrm{T} + Tg$ where T is the OFDM frame duration, T'' = overall time duration OFDM, and Tg = Guard interval.

$$F_k = F_c + k/T \quad K = -K/2 \ldots K/2 - 1 \tag{2}$$

where F_c denotes carrier frequency, F_k = subcarrier frequency, and $1/T$ = spacing between subcarriers. K = number of subcarriers and bandwidth B.W = K/T. $Eu(k)$ is the encrypted data symbol, $P(T)$ = pulse shaping filter. Using these parameters, calculate the Nth transducer transmitted signal.

$$\tilde{x}_N(t) = 2\mathrm{Re}\left\{\left[\sum_{k \in S_A} Sv, u[K] e^{j2\pi \frac{k}{T} t} p(t)\right] e^{j2\pi fct}\right\}, \quad t \in [0, T'] \tag{3}$$

The transmitter and receiver pair for multipath channel consists of Sv,u paths, utilizing these elements generating the impulse response can be calculated as

$$h_{v,u}(\tau, t) = \sum_{p=1}^{pv,w} A_{v,u} \delta\left(\tau - \left(\tau_{v,u} - D_p t\right)\right) \tag{4}$$

$A_{v,u}$ = Amplitude, $\tau_{v,u}$ = Time delay, and D_p = Doppler shift factor.

This mathematical computation clearly explains the impulse response of the present dynamic nature of underwater communication. The output signal at *v*th receiver and noise acceptance is given by

$$\tilde{y}_{v(t)} = \sum_{u=1}^{Nt} \sum_{f=1}^{fv,w} A_{v,u,f}\tilde{x}_u\big((1 + a_{v,u,f})t - \tau_{v,u}\big) + \tilde{n}_v(t) \tag{5}$$

In this equation, n_v is noise acceptance in output signal receiver model.

The BPSK modulation scheme is within a bit duration T_b, and the two different phase states of the carrier signal are represented as

$$s_1(t) \approx A_C \cos(2\Pi f_c t) \quad 0 \le t \le T_b \tag{6}$$

$$s_0(t) \approx A_C \cos(2\Pi f_c t + \Pi) \quad 0 \le t \le T_b \tag{7}$$

The QPSK signal within a symbol duration Tb is defined as

$$S(t) = \cos(2\Pi f_c t + \theta_n) \quad 0 \le t \le T_b \quad n = 1, 2, 3, 4 \tag{8}$$

where the signal phase is given by

$$\theta_n = (2n - 1)\Pi/4 \tag{9}$$

where in the 8 phase-shift keying (8PSK) is a form of phase modulation with eight phase states located at 0, π, $\pm$ $(\pi/4)$, $\pm$ $(\pi/2)$, and $\pm$ $(3\pi/4)$ radians in the IQ plane.

4 UWC Massive MIMO Hardware Design Model

In this segment, examining different difficulties and advantages of massive MIMO method for UWC transmission plans. The channel estimation is mostly relying on massive MIMO engineering [19–24]. The frequency, point, and stage-based MIMO-OFDM work have accomplished some improvement, but the vast majority of the cases essential advancements are required for UWC, and the new innovation predominantly focuses on profound learning procedures which are incredible potential in communication innovation. The profound learning-based underwater communication can settle the powerful idea of channel and time shifting boundaries needs to change [25–29] (Figs. 2 and 3).

5 Results

The overall exploration work can address the impediments of general MIMO underwater communication, for example, the information rates, dynamic nature of the channel, and probability of error. The accompanying clarification evil presence states

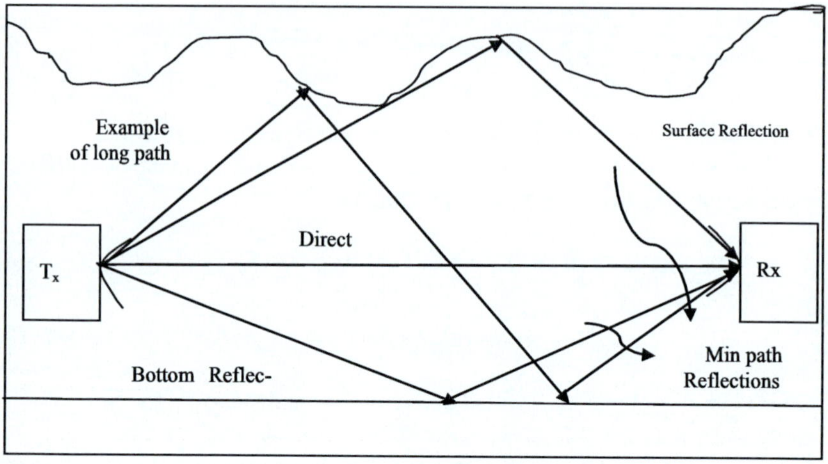

Fig. 2 System block diagram UWA communication

Fig. 3 FPGA device used at for the UWC project

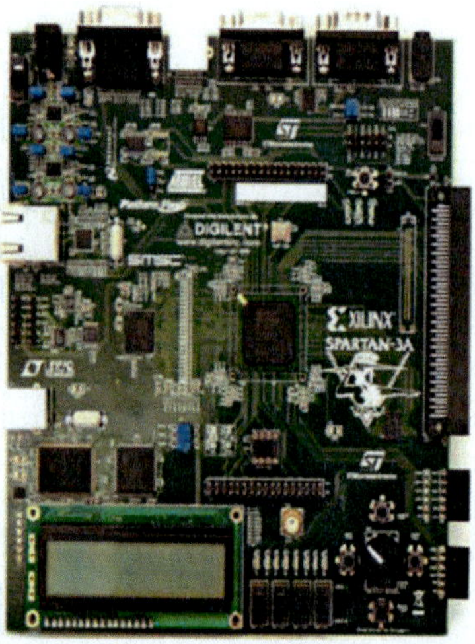

that benefits of UWC. The underwater acoustic communication channels smother the weighted Fourier partial frequency. The single carrier communication does not productive activities. The fractional Fourier change with crossover carrier communication builds up the multi-carrier transmission and reception. This powerful design mitigates the inter-carrier interference of UWC after the halfway Fourier transform demodulation strategies. Additionally, produced results are predominant to the UWC at high SNR and information rates. The underneath graph shows that oceanic square outline model and hard supported (Fig. 4).

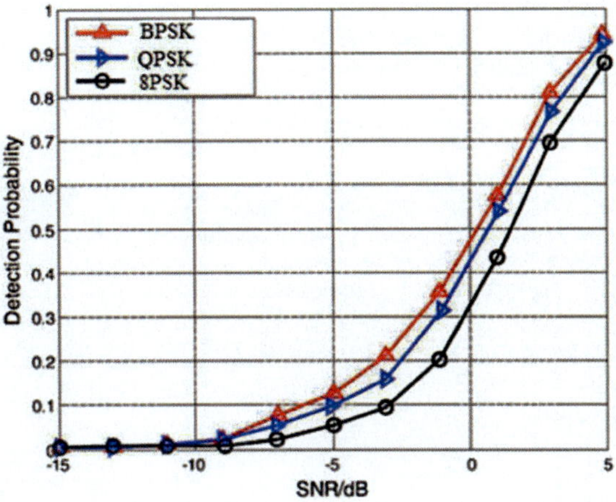

Fig. 4 Curve relationship of SNR versus detection probability

The epic methodologies of range signal discovery by the underwater communication method gives the relations of signal-to-noise proportion and detection probability. In this work, Fig. 5 exhibits that 8PSK accomplishes more identification likelihood contrast with QPSK, BPSK.

The number of observations and detection probability has to be increasing 8PSK by comparing with other modulation schemes (Fig. 6).

So, BER is needed to decrease for UWC channel such that proposed model is better at time variation and less bandwidth cases. Every receiving antenna has communicated with various channels in the UWC with 8PSK modulation scheme.

Figure 7 explains the number of user equipment per SE, and in this MMSE, un-correlation error is more compared to DAMR correlation. Using this analysis, proper MMSE and MR signals have to be selected for UWC communication channel.

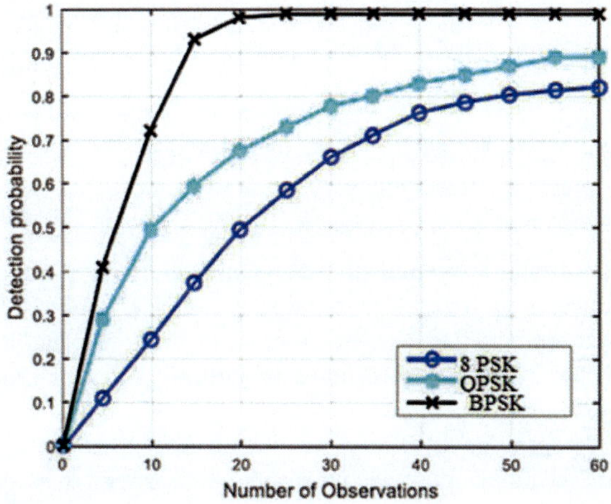

Fig. 5 Number of observation to detection probability

Fig. 6 BER to SNR with different channel model with 8PSK modulation scheme

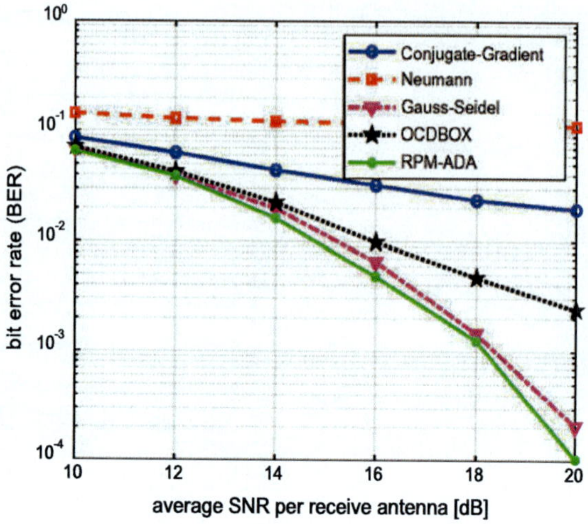

Fig. 7 MMSE-UE and SE analysis with 8PSK modulation scheme

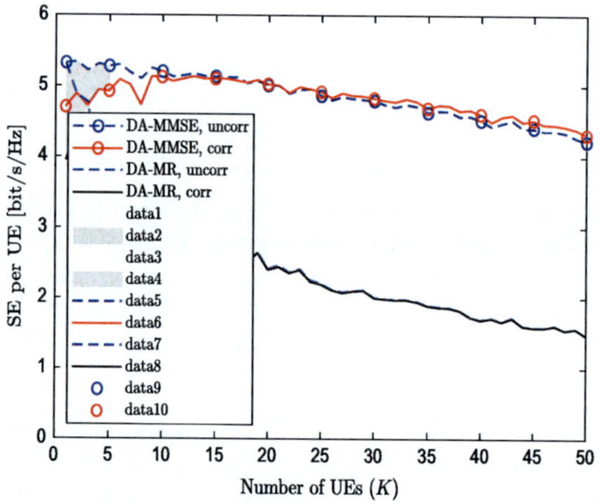

6 Conclusion and Future Work

In this investigation, utilizing diverse underwater modulation schemes, we broke down and reenacted different situations with various depths, distance, and BER esteems to make a reasonable correlation between the modulation schemes. As our reproduction results showed BPSK, QPSK, and 8 PSK methods accomplished better BER, signal-to-noise proportion, and detection probability. Consequently, picking these methods can assist with drawing out the underwater design lifetime by improving reliability and effectiveness. The massive MIMO configuration settles the channel multipath issues and subcarrier issues in UWC. In future work, we are focusing on massive MIMO-OFDM underwater acoustic communication-based highest Mary PSK to improve result dependent on SNR and bit error rate. The simulation results are noticed utilizing MATLAB programming.

Declaration We have taken permission from competent authorities to use the images/data as given in the paper. In case of any dispute in the future, we shall be wholly responsible.

References

1. Das B, Das S (2009) RAKE-MMSE time domain equalizer for high data rate UWB communication system. In: 2009 Annual IEEE India conference, Ahmedabad, India, 2009, pp 1–4. https://doi.org/10.1109/INDCON.2009.5409346
2. Khan MR, Das B (2021) Multiuser detection for MIMO-OFDM system in underwater communication using a hybrid binary spotted hyena optimizer. J Bionic Eng 18:1–11
3. Khan MR, Mahapatra S, Das B (2020) UWB Saleh-Valenzuela model for underwater acoustic sensor network. Int J Inf Technol 12(4):1073–1083
4. Das B, Tiwari S, Das S (2011) Performance study of discrete wavelet packet based MB-OFDM system for short range indoor wireless environment. In: 2011 International conference on devices and communications (ICDeCom),Mesra, India, 2011, pp 1–5. https://doi.org/10.1109/ICDECOM.2011.573856
5. Bocus MJ, Doufexi A, Agrafiotis D (2016) Performance evaluation of filter bank multicarrier systems in an underwater acoustic channel. In: IEEE 27th Annual international symposium on personal, indoor, and mobile radio communications (PIMRC), Valencia, pp 1–6,Spain, 2016
6. Li-Da D, Shi-Lian W, Wei Z (2018) Modulation classification of underwater acoustic communication signals based on deep learning. In: OCEANS—MTS/IEEE Kobe Techno-Oceans (OTO), pp 1–4, Kobe, Japan, 2018
7. Li C, Song K, Yang L (2017) Low computational complexity design over sparse channel estimator in underwater acoustic OFDM communication system. IET Commun 11(7):1143–1151
8. Pelekanakis K, Cazzanti L (2018) On adaptive modulation for low SNR underwater acoustic communications. In: OCEANS 2018 MTS/IEEE Charleston, Charleston, SC, USA, pp 1–6
9. Lawal B, Ali SSA, Awang AB (2010) Massive MIMO systems for underwater acoustic communication. In: 2020 IEEE international conference on underwater system technology: theory and applications (USYS), Penang, pp 159–164
10. Panda M, Das B (2020) Multi-agent system of autonomous underwater vehicles in octagon formation. In: Intelligent systems: proceedings of ICMIB 2020, 2020, vol 185, pp 125–138
11. Das B, Subudhi B, Pati BB (2014) Adaptive sliding mode formation control of multiple underwater robots. Arch control Sci 24(4):515–543
12. Das B, Subudhi B, Pati BB (2015) Employing nonlinear observer for formation control of AUVs under communication constraints. Int J Intell Unmanned Syst 3(2/3):122–155
13. Panda M, Das B, Subudhi B, Pati BB (2020) Adaptive fuzzy sliding mode formation controller for autonomous underwater vehicles with variable payload. Int J Intell Unmanned Syst
14. Panda M, Das B, Pati BB (2020) A hybrid approach for path planning of multiple AUVs. In: Innovation in electrical power engineering, communication, and computing technology, Singapore, 2020, pp 327–338. https://doi.org/10.1007/978-981-15-2305-2_26
15. Panda M, Das B, Subudhi B, Pati BB (2020) A comprehensive review of path planning algorithms for autonomous underwater vehicles. Int J Autom Comput, pp 1–32
16. Das B, Subudhi B, Pati BB (2016) Co-operative control coordination of a team of underwater vehicles with communication constraints. Trans Inst Meas Control 38(4):463–481

17. Das B, Subudhi B, Pati BB (2016) Cooperative formation control of autonomous underwater vehicles: an overview. Int J Autom Comput 13(3):199–225
18. Arikan T, Riedl T, Singer A, Younce J (2015) Comparison of OFDM and single-carrier schemes for Doppler tolerant acoustic communications. OCEANS—Genova, Genoa, Italy, pp 1–7
19. Altabbaa MT, Panayirci E (2017) Channel estimation and equalization algorithm for OFDM-based underwater acoustic communications systems. In: International conference on wireless communications, networking and mobile computing, pp 1–4
20. Kari D (2017) Robust adaptive algorithms for underwater acoustic channel estimation and their performance analysis. Digital Signal Processing. In: International conference on wireless communications, networking and mobile computing, pp 57–68
21. Qian C, Wang Z, Lu X, Wu Y, Qu F (2017) Sparse channel estimation for filtered multi tone in underwater communications. Proc of MTS/IEEE Oceans, pp 1–7
22. Kilfoyle DB, Preisig JC, Baggeroer AB (2015) Spatial modulation experiments in the underwater acoustic channel. IEEE J Oceanic Eng 30(2):406–415
23. Roy S, Duman TM, McDonald V, Proakis JG (2007) High-rate communication for underwater acoustic channels using multiple transmitters and space—time coding: receiver structures and experimental results. IEEE J Oceanic Eng 32(3):663–688
24. Das B, Das S (2010) Efficacy of multiband OFDM approach in high data rate ultra wideband WPAN physical layer standard using realistic channel models. IJCA 2(2):81–87. https://doi.org/10.5120/621-889
25. Khan MR, Das B, Pati BB (2020) Channel estimation strategies for underwater acoustic (UWA) communication: an overview. J Franklin Inst 357(11):7229–7265
26. Das B, Das S (2010) Interference cancellation schemes in UWB systems used in wireless personal area network based on wavelet based pulse spectral shaping and transmitted reference UWB using AWGN channel model. IJCA 2(2):88–92. https://doi.org/10.5120/620-890
27. Khan MR, Das B (2021) Multiuser detection for MIMO-OFDM system in underwater communication using a hybrid bionic binary spotted hyena optimizer. SpringerLink. https://doi.org/10.1007/s42235-021-0018-y. Accessed July 15, 2021
28. Das B, Subudhi B, Pati BB (2016) Co-operative control of a team of autonomous underwater vehicles in an obstacle-rich environment. J Mar Eng Technol 15(3):135–151
29. Panda M, Das B, Pati BB (2020) Global path planning for multiple AUVs using GWO. Arch Control Sci 30(1):77–100

Comparison of Feature Reduction Techniques for Change Detection in Remote Sensing

Gudivada Aparna, Kantamani Rachana, Koganti Rikhita, and Boggavarapu L. N. Phaneendra Kumar

Abstract The accuracy in prediction of change of land cover is important in growing urbanization. The changes on the surface of the earth occur due to disasters, deforestation, change in course of the river, urbanization, etc. If the change detection of the earth's surface is completed timely and accurately, then the connection and interaction between natural phenomena and humans are often better analyzed and understood as a result of which better management and use of resources can be done. Detection of change is most important for correct understanding of land cover changes with multi-temporal data. Main idea is to detect the regions that have changed in various images of an equivalent place taken at different timestamps. It has many growing applications in real world such as smart cities, disease treatment, and forestry. In this paper, a comparative study is made for the detection of change in remote sensing images using different machine learning approaches. Initially, the difference image is computed from the time stamped multispectral images. Then, feature reduction is employed with three approaches which use latent space, principal component, and a variation of it. The extracted components are then given to k-means clustering for detection of change, and change map is obtained. The results clearly show that the factor analysis combined with k-means detected the change effectively and also with less time when compared with other algorithms.

Keywords Principal component analysis · K-means · Multi-temporal images · Factor analysis · Sparse PCA

1 Introduction

Multi-temporal remote sensing images play a vital role in detecting and analyzing the major and minor changes in the areas of landmasses. The multi-temporal images are the satellite images which consist of 3–10 bands. It has wide variety of applications

G. Aparna · K. Rachana · K. Rikhita · B. L. N. Phaneendra Kumar (✉)
Department of Information Technology, Velgapudi Ramakrishna Siddhartha Engineering College, Vijayawada 520007, India
e-mail: phaneendra.b@vrsiddhartha.ac.in

P. S. R. Chowdary et al. (eds.), *Evolution in Signal Processing and Telecommunication Networks*, Lecture Notes in Electrical Engineering 839,
https://doi.org/10.1007/978-981-16-8554-5_30

which includes agricultural crop monitoring and urban planning management to evaluate natural disasters. The ratio of land cover change is very much important in new industrial areas. The changes on the surface of the earth occur due to deforestation, urbanization, etc. If the changes on surface of the earth are done frequently and accurately, then the relationship and interaction between natural phenomena and humans can be better analyzed, so that the better usage of resources can be achieved. The change detection helps the government and other organization to study and nature of the land cover. This paper proposed different combinations of machine learning algorithms to predict and estimate the change in images.

2 Literature Review

In [1], a methodology to address the binary classification problem of the temporal change detection. In [2], the authors have used the pseudo-invariant features for radiometric rectification along with feature reduction which is mainly used in the same area. In [3], k-means algorithm followed by the majority voting method which is used for change detection. A framework that combines the CNN and RNN is given in [4]. In [5], different deep learning models are used for change detection. In this [6], the adaptive parameters along with hierarchal clustering are used in four sets of SAR images. The adaptive parameters are useful for detecting changed areas and for automatic generation of training samples clustering is used. In this [7], different types of techniques are used for detecting changes which includes univariate image differencing, image regression, and image rationing. An approach based on Markov random fields (MRFs) used for image differencing [8] and also proposed an iterative method. In [9, 10], the object-based change detection model provides unique methods and approaches to capture detailed change information for high spatial resolution imagery and performed on the SAR images. In [11], proposed a combination of algorithms to improve accuracy. In [12], monitored the changes in land cover changes from three different years. A supervised based approach is used in [13] for efficient land cover identification. A combination of spectral features extraction is given in [14–19]. A variety of algorithms for images is proposed in [20–22].

3 Description of Dataset

For this work, Indian Remote Sensing Satellite Resourcesat-1, 2 with LISS-III sensor over the area Vijayawada bus stand, Andhra Pradesh is taken. Dataset comprises of four multispectral images. The images are stacked to get a single image. Figures 1 and 2 represent the images collected and stacked during 2009 and 2018, respectively.

Fig. 1 **a** Images from 2009, **b** images from 2018

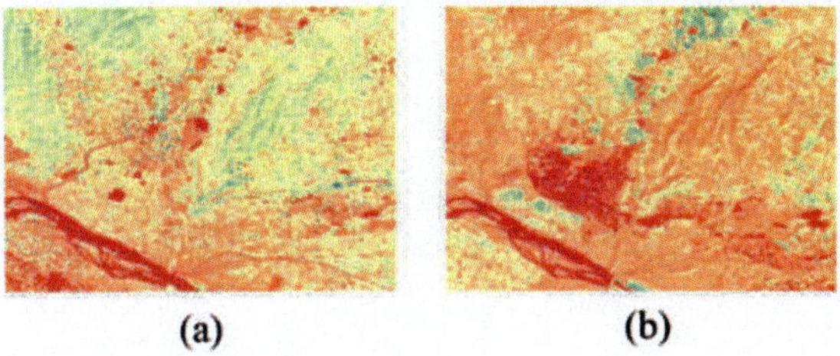

Fig. 2 Stacked image for the years **a** 2009, **b** 2018

4 Proposed Framework

This architecture diagram discusses about the methodology used in the present work which can be seen in Fig. 3.

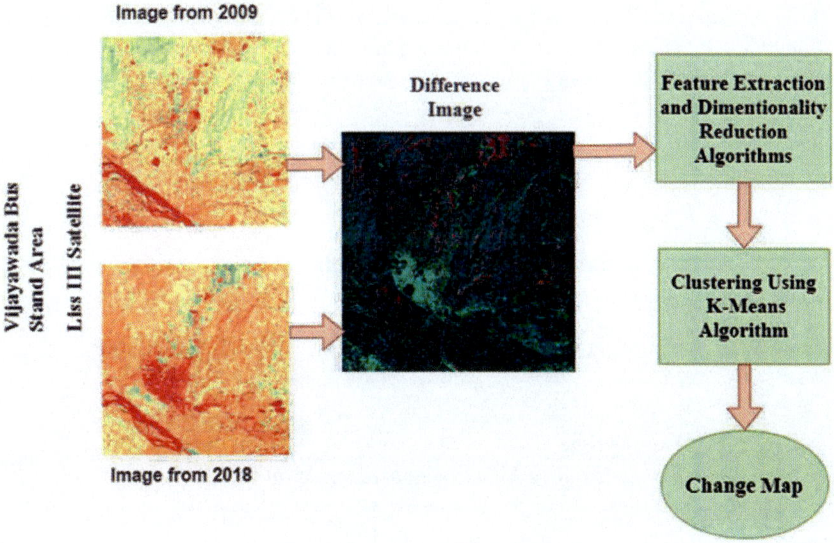

Fig. 3 Proposed architecture of change detection in remote sensing images

4.1 Input Dataset

The three bands, green, near-infrared, and red, are considered at two time scales. While processing images, the two images must be resized to have the same size.

4.2 Calculation of Difference Image

In this section, the difference for the two images is calculated by using the formula:

$$\text{Difference Image} = \text{abs}(\text{img1} - \text{img2}) \tag{1}$$

4.3 Feature Extraction and Dimensionality Reduction

Here, on multi-temporal images, three feature extraction methods are implemented.

4.3.1 Principal Component Analysis

In this section, dimensionality reduction technique, Principal Component Analysis (PCA) is proposed. Principal Component Analysis is used to examine the principal components by calculating mean vector, eigenvalues, and eigenvectors.

For any matrix, the mean feature vector can be calculated as

$$\mu = \frac{1}{p}\sum_{k=1}^{p} x_k \tag{2}$$

where x = feature matrix and p = number of patterns.

The co-variance array is computed as:

$$\text{Cov} = \frac{1}{p}\{x_k - \mu\}\{x_k - \mu\}^T \tag{3}$$

Here, T is the transpose.

To calculate the eigenvalues λ^T and eigenvectors v_i of the Cov.

$$\text{Cov}v_i = \lambda_i v_i \ i = 1, 2, \ldots p, \tag{4}$$

where p is feature number.

The eigenvectors are computed as:

(i) Sort the eigenvalues in decreasing order.
(ii) Opt for a threshold-θ
(iii) The number of values chosen to satisfy the relationship.

$$\left(\sum_{i=1}^{s} \lambda_i\right)\left(\sum_{i=1}^{q} \lambda_i\right)^{-1} \geq \theta \tag{5}$$

where s the value at which the value is chosen.
(iv) Final selection of eigenvectors

Finally, derive the feature matrix

$$P = V^T x \tag{6}$$

where V and x are components and feature vectors, respectively.

4.3.2 Factor Analysis

In this method, firstly, the correlation is calculated, then extraction of factors and rotation comes to a decision on the total number of underlying factors.

(i) A correlation matrix should be generated on all the input variables.
(ii) Identifying the irrelevant variables.
(iii) Correlation between variables is directly proportional, to the likeliness of sharing the common factors.
(iv) For the model to be appropriate, the variables must relate to each other.
(v) This step involves the factor extraction by using any kind of feature extraction method; here, the PCA is used again for feature extraction, where the objective of this step is to find out the factors, and these factors are estimated initially.
(vi) As explained, here the factors are extracted based on the relation between the variables.
(vii) The screen plot and eigenvalues are used to determine the factor number.
(viii) Consider those factors which have eigenvalues >1.
(ix) The screen plot shows the total variance associated with each factor.
(x) This step involves the factor rotation, rotation is done in order to make sure they are more meaningful, accurate, and easy to interpret. So, here we use varimax rotations.
(xi) The factors that are interpreted by identifying the largest values linking the factor to the measured variables in the rotated factor matrix.
(xii) Finally, select the factors.

4.3.3 Sparse PCA

In this section, dimensionality reduction technique, sparse PCA is proposed

(i) A unique technique is used in the statistical analysis and, particularly, while analyzing the multivariate data sets. It extends the method of principal component analysis (PCA) for the dimensionality reduction of data by introducing sparse structures to input variables.
(ii) Here, the sparsity is introduced and helps to find the combinations with only few variables as it cannot be done for the general algorithm.

4.4 K-Means Clustering

After getting the principal components or the other feature reduction techniques, k-means algorithm is used to divide the data into two clusters from which we can identify the changed and unchanged area in the image. For k-means, this paper uses the Euclidian distance.

4.5 Calculation of the Percentages Clustering

The percentage of change can be calculated from the final change map obtained. It can be calculated by using the formula,

$$\text{Change Percentage} = \text{np.count.nonzero(Changemap)} * \frac{100}{\text{Changemap}} \tag{7}$$

where change_map is the change map obtained after applying k-means algorithm.

5 Results and Observations

The obtained results and observations from different feature reduction methods are presented. Initially, the difference image is computed and is shown in Fig. 4. This figure is obtained by calculating the absolute difference between the two input images. This figure is then given to different feature extraction methods.

As a part of results, the difference image obtained from the two input images that are taken is shown in Fig. 4.

The change maps obtained with different feature reduction techniques as discussed in Sect. 4 followed by the clustering algorithm are shown in Fig. 5. Figure 5a shows the change map obtained through factor analysis with k-means, 5(b) and 5(c) represents

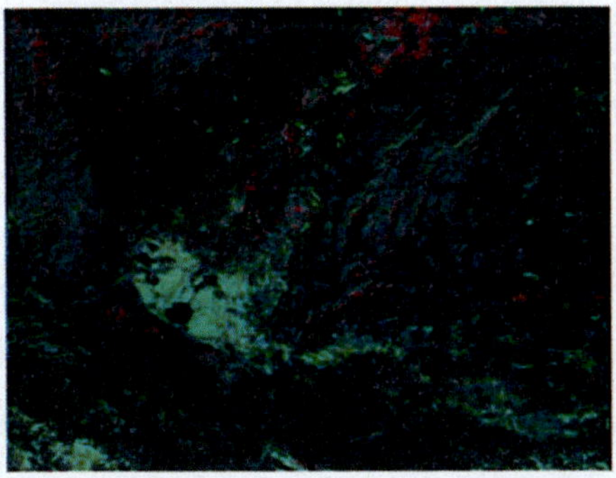

Fig. 4 Difference image

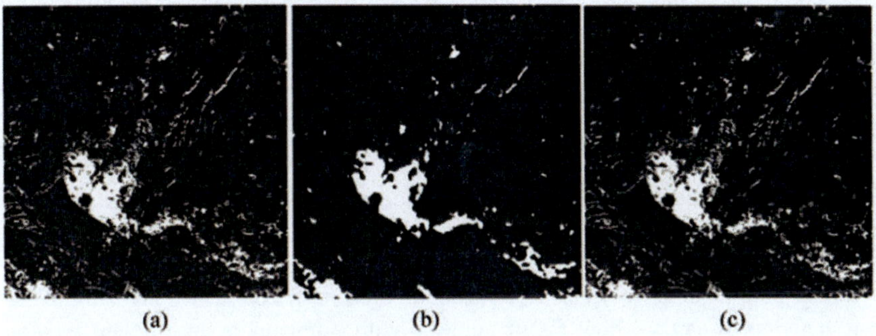

Fig. 5 **a** Factor analysis with k-means, **b** PCA with k-means clustering, and **c** sparse PCA with k-means

the change maps using PCA with k-means and sparse PCA with k-means algorithms, respectively.

From Fig. 5a, we can see the change map obtained when factor analysis is used for feature extraction and dimensionality reduction. Similarly, Fig. 5b, c shows the change map when applied PCA and sparse PCA, respectively, for feature extraction and dimensionality reduction.

From the Fig. 6a, it can be inferred that time taken for performing the algorithm with PCA + k-means, Sparse PCA + k-means and Factor analysis + k-means are 2.10, 1.75, and 1.74, respectively. Hence, the factor analysis combined with the k-means provided good change map in less time. Figure 6b clearly depicts the percentage of change obtained in the two time series images with different feature reduction techniques. It can be seen that with factor analysis followed by k-means detected the highest change of 10.5% in the two images comparative to PCA + k-means and SparsePCA + k-means which results in 7.32% and 10.31%, respectively.

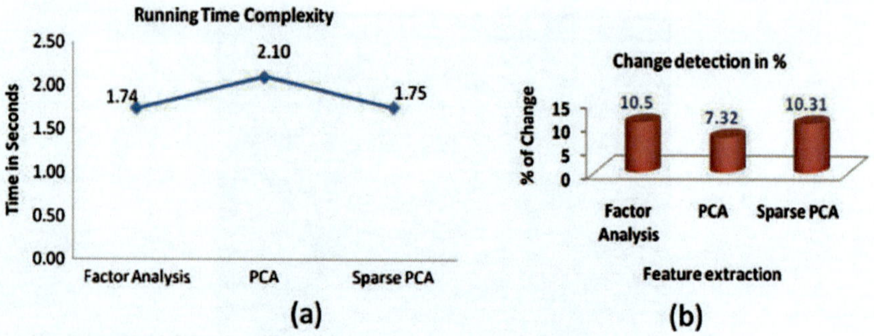

Fig. 6 **a** Running time analysis, **b** % of change

6 Conclusion

In this work, proposed different methodologies for finding the change map of same picture from same location in different time durations, and as a result, we found the percentage of change occurred considering the time duration. The results clearly show that the time taken for performing the algorithm is highest for sparse PCA and least for PCA, and the time taken for building the feature vector space is highest for PCA and least for sparse PCA. The time taken for clustering is higher in PCA compared to the other two and least in factor analysis. It can be seen that change percentage is higher and accurate in factor analysis compared to the other two algorithms. As a conclusion, it can say that factor analysis with K-means is an efficient method compared to other two methods. As an extension, we can also check with different change detection techniques (using deep learning) to monitor the changes.

References

1. Touati R, Mignotte M, Dahmane M (2020) Anomaly feature learning for unsupervised change detection in heterogeneous images. IEEE J Sel Top Appl Earth Obs Remote Sens, 13
2. Du Y, Teillet PM, Cihlar J (2002) Radiometric Normalization of multi-temporal high resolution satellite images with quality control for land cover change detection
3. Lv Z, Liu T, Shi C, Benediktsson JA (2017) Novel land cover change detection method based on k-means clustering and adaptive majority voting using BiTemporal remote sensing images. IEEE vol 7
4. Liu R, Cheng Z, Zhang L, Li J (2019) Remote sensing image change detection based on information transmission and attention mechanism, vol 7. IEEE
5. Khelifi L, Mignotte M (2020) Deep learning for change detection in remote sensing images: comprehensive review and meta-analysis. In: IEEE Natural Sciences and Engineering Research Council of Canada (NSERC)
6. Chen Y, Ming Z, Menenti M (2020) Change detection algorithm for multi-temporal remote sensing images based on adaptive parameter estimation, vol 8. IEEE
7. Singh A (1989) Review article digital change detection techniques using remotely-sensed data. Int J Remote Sens 10(6):989–1003

8. Bruzzone L, Prieto DF (2000) Automatic analysis of the difference image for unsupervised change detection. IEEE Trans Geosci Remote Sens 38(3):1171–1182
9. Chen G, Hay GJ, Carvalho LMT, Wulder MA (2012) Object-based change detection. Int J Remote Sens 33(14):4434–4457
10. Yousif O, Ban Y (2017) A novel approach for object-based change image generation using multitemporal high-resolution SAR images. Int J Remote Sens, 38(7)
11. Wang C, Wang X (2021) Building change detection from multi-source remote sensing images based on multi-feature fusion and extreme learning machine. Int J Remote Sens, 42(6), 30 Dec 2020
12. Tripathi S, Naik A, Patil S (2015) Analysis of change detection techniques using remotely sensed data
13. Bazi Y, Melgani F, Bruzzone L, Vernazza G (2009) A genetic expectation-maximization method for unsupervised change detection in multitemporal SAR imagery. Int J Remote Sens 30(24)
14. Vaddi R, Prabukumar M (2020) Hyperspectral image classification using CNN with spectral and spatial features integration. Infrared Phys Technol 107:103296
15. Phaneendra Kumar BLN, Prabukumar M (2020) Hyperspectral image classification using fuzzy-embedded hyperbolic sigmoid nonlinear principal component and weighted least squares approach. J Appl Remote Sens 14:1
16. Kumar B, Prabukumar M (2021) Whale optimization-based band selection technique for hyperspectral image classification. Int J Remote Sens 42:5109–5147. https://doi.org/10.1080/01431161.2021.1906979
17. Vaddi R, Prabukumar M (2020) Hyperspectral remote sensing image classification using combinatorial optimisation based un-supervised band selection and CNN. IET Image Process 14. https://doi.org/10.1049/iet-ipr.2020.0728
18. Vaddi R, Manoharan P (2020) CNN based hyperspectral image classification using unsupervised band selection and structure-preserving spatial features. Infrared Phys Technol 110:103457, ISSN 1350-4495. https://doi.org/10.1016/j.infrared.2020.103457
19. Phaneendra Kumar BLN, Prabukumar, M (2020) A new framework for hyperspectral image classification using Gabor embedded patch based convolution neural network. Infrared Phys Technol 110:103455. https://doi.org/10.1016/j.infrared.2020.103455
20. Bhateja V, Nigam M, Bhadauria A (2021) Region labeling based brain tumor segmentation from MR images. https://doi.org/10.1007/978-981-16-0878-0_81
21. Bishnu A, Bhateja V, Rai A (2021) Enhancement of synthetic-aperture radar (SAR) images based on dynamic unsharp masking. https://doi.org/10.1007/978-981-15-5400-1_81
22. Rai A, Bhateja V, Bishnu A (2020) Speckle suppression and enhancement approaches for processing of SAR images: a technical review. https://doi.org/10.1007/978-981-32-9690-9_74

Human Action Recognition in Videos Using Deep Neural Network

A. Hari Pavan, P. Anvitha, A. Prem Sai, I. Sunil, Y. Maruthi, and Vaddi Radhesyam

Abstract Identifying the various actions in video is important in computer vision research. The present paper aims to develop a model that would take the video input from different sources like security cameras and recorded videos and detect actions that are being performed in them by the people also commonly referred to as HAR or human activity detection. Based on series of observations in a video from frame to frame, the action can be identified. In the present work, deep convolutional neural networks (CNNs) is used to implement the model for HAR. The model can be applicable in several applications like human computer interaction, surveillance, elder people assistance, and patient monitoring system.

Keywords Human action recognition · Surveillance systems · Patient monitoring systems

1 Introduction

Human action recognition [9] is a quality computer [2] vision problem and has been well studied. The elementary goal is to examine a video to identify the actions taking place in the video. Originally, a video has a spatial aspect to it, i.e., the individual frames, and a temporal aspect, i.e., the ordering of the frames. Some actions (e.g., standing, running, etc.) can probably be identified by using just a single frame, but for more complex actions (e.g.., walking vs running, bending vs falling) might require more than one frame's information to identify it correctly. Detecting the human action in numerous videos is the main concern at this point. Generally, detecting the human action visually is very easy, but the computer detecting [8] the action is a task of risk. Before the machine detects the action, it should go under many processes to get the accurate results. Human action recognition is such where it detects the action in the video after going through all the steps of processing. It uses different algorithm giving different accuracy in the results.

A. Hari Pavan · P. Anvitha · A. Prem Sai · I. Sunil · Y. Maruthi · V. Radhesyam (✉)
Department of Information Technology, VR Siddhartha Engineering College, Vijayawada, India

P. S. R. Chowdary et al. (eds.), *Evolution in Signal Processing and Telecommunication Networks*, Lecture Notes in Electrical Engineering 839,
https://doi.org/10.1007/978-981-16-8554-5_31

2 Literature Review

In the last few decades, different researchers have addressed the problem of human action recognition using variety of the methods. In [1], human action recognition is executed using depth motion maps. From a given video input, the human activity is correct identified and conformed using robust classifier design.

An hierarchical model is designed by [2] to correctly identify human activity. Also focus on detailed study of relevant literature. Different levels of abstraction and combination of a variety of features are included in the work proposed by [3] in order to correctly identify the abnormal actions.

The other novel works include [4] which includes activity recognition from cloud of images and confirms the activity from the video frames, and in [5], HAR task is accomplished based on hardware sensors and reference as human skeleton.

The taxonomy of HAR using deep learning is shown in Fig. 1. The main two methods are based on depth maps and video streams as input. These methods use mainly CNN, RNN, and combined of both CNN and RNN.

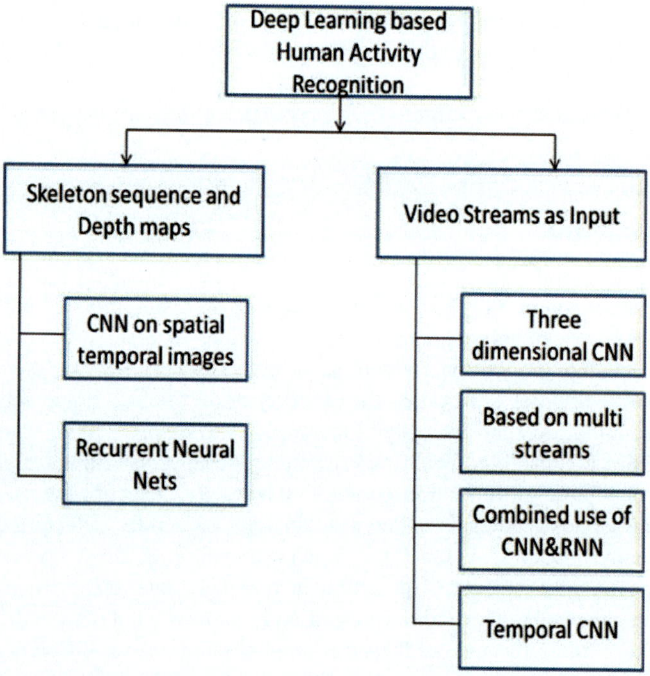

Fig. 1 Taxonomy of HAR using deep learning

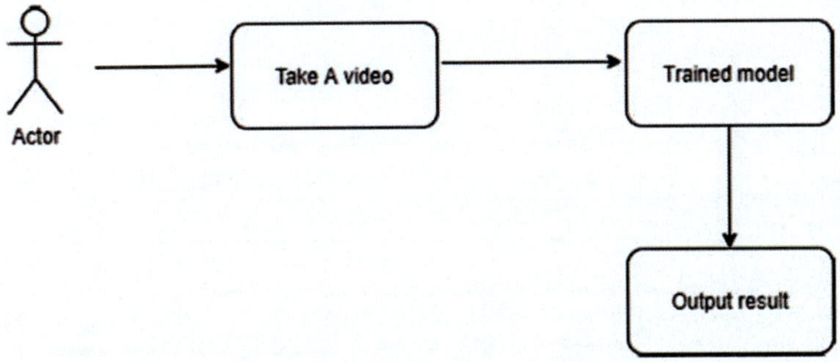

Fig. 2 Design methodology

3 Proposed Work

3.1 Project Design Methodology

See Fig. 2.

Actor: Here, actor is the user who watches the video and scrutinizes the output which the action is detected.

The user takes the real-world video as input and undergoes the preprocessing operation.

The trained model takes the processed video as input and detects the actions performed in the video.

The most frequently occurred action with highest probability is shown as the output result [6].

3.2 Architecture Diagram

See Fig. 3.

- The real-world video is taken as the input for this process to start.
- In preprocessing, the taken video input is divided into the frames with given dimensions.
- The frames are then trained to the model where there is different action predefined. The pretrained model is 'action_resnet' [7].
- The maximum probability action is detected from the model.
- The corresponding action's [8] class label is displayed as the output.

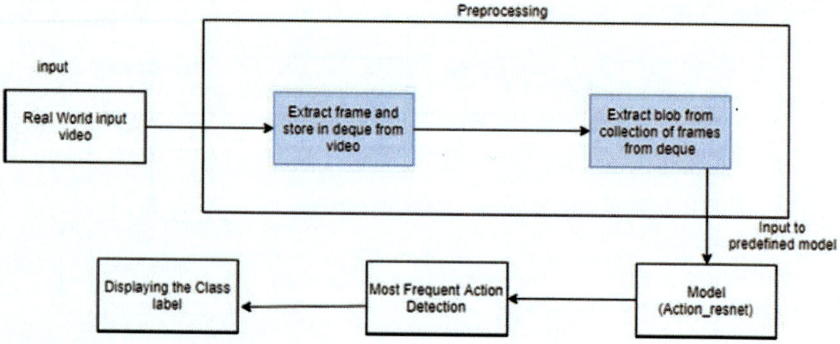

Fig. 3 Architecture diagram

3.3 *Overview of Algorithm Used*

- In the present work, the predefined ResNet50 is used for HAR. It is a pretrained model.
- The series of layers are included in the model like convolution layer, pooling layer, and fully connected layers.
- The model is shown in Fig. 4 with series of layers in between input and output layers.

4 Dataset and Tools

Datasets: We have taken real-world video as input; hence, we have no datasets to be described.

Tools: We have used Anaconda to run the code and to get the output for our project.

ANACONDA:

- Anaconda is an IDE used for a variety of applications in machine learning and data science.

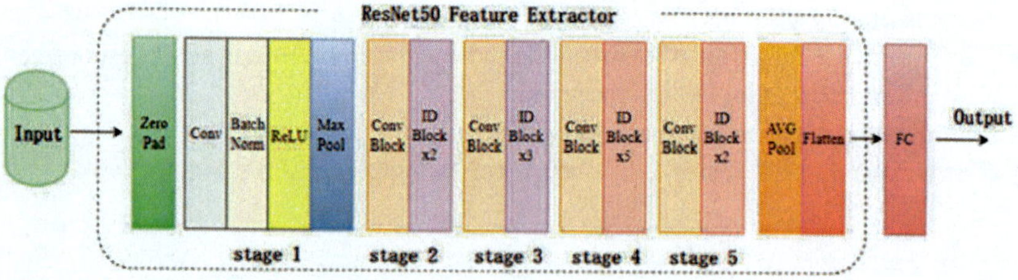

Fig. 4 Deep neural network architecture

- It is an open-source distribution used for Python and R programming.
- In spite of Python core language, Anaconda consists of more than 200 packages.
- JupyterLab, Jupyter Notebook, QtConsole, and Spyder are the available navigators in Anaconda [9].

Open CV:

- OpenCV is used for computer vision and machine learning applications.
- It is open source software which has more applications related to images and videos.
- Many real-time applications in human–computer interaction can be implemented using OpenCV.

5 Results and Observations

5.1 Description of Results and Observations

See Fig. 5.

5.2 Failed Recognitions

See Fig. 6.

Above cases are the failure recognitions occurred during the project execution phase. This could be because of the multiple object detection, view point variation, deformation, and occlusion.

5.3 Fieldwork Details

The model currently captures about 400 activities with about 78–95% accuracy. It shows many actions like—walking, jumping, skipping, etc.

Many such actions can be detected for the human action recognition in videos through the predefined model (resnet_model). Such models can further be trained on specific human actions and may be useful for detecting theft in real-time [10].

6 Societal Benefits

There are some societal benefits for the project. The elder people population is increasing. These people need assistance on regular intervals of time. Actions and

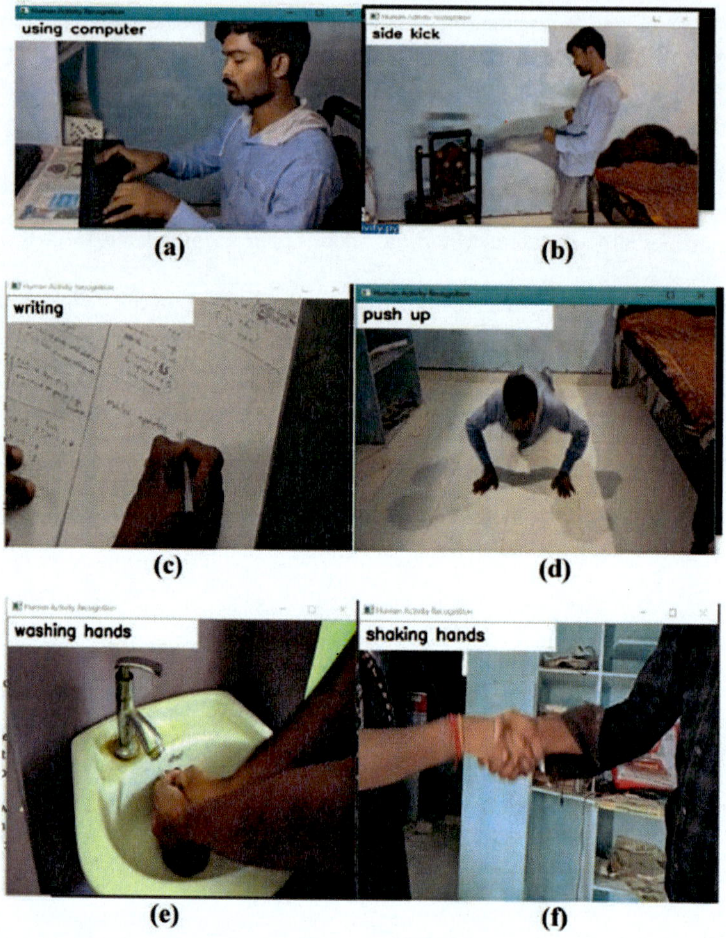

Fig. 5 Actions that are correctly recognized

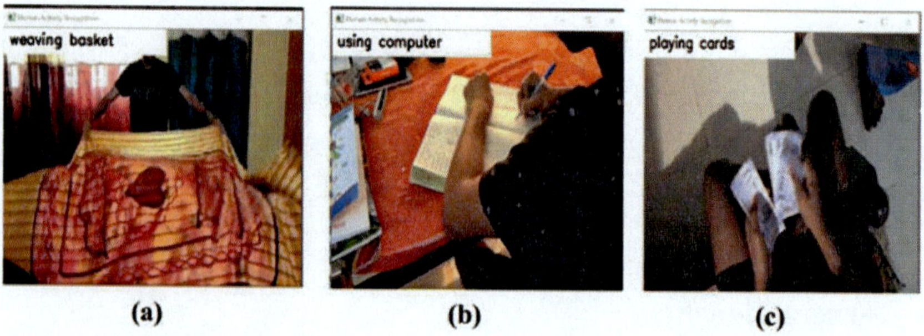

Fig. 6 Actions that are wrongly determined

gestures identification and continuous monitoring will help in providing good assistance systems. The abnormal behavior and action identification from video frames based on deep learning methods will be more useful to societal needs [11].

7 Conclusion and Future Work

In the current project, we learnt a lot about this developing field of study while researching and developing this project. During the development of this project to detect human activities from video feeds, we came across multiple approaches to help detect a subset of activities. As we all know that the accuracy of the models can never be 100%, so the results that are generated will always have a possibility of providing improper output as is the case with all machine learning classifications, so if we can increase the data to improve the overall efficiency of detection of human activity, we can do so in future models where a huge data set is used to train the developed models.

References

1. Zhenguo Shi, J. Andrew Zhang, Richard Xu, and Gengfa Fang "Human Activity Recognition Using Deep Learning Networks with Enhanced Channel State information" IEEE 2018.
2. J.K. Aggarwal and M.S. Ryoo. 2011. Human activity analysis: A review. ACM Comput. Surv. 43, 3, Article 16 (April 2011), 43 pages. DOI:https://doi.org/10.1145/1922649.1922653
3. Guo G, Lai A (2014) A survey on still image based human action recognition. Pattern Recogn 47:3343–3361. https://doi.org/10.1016/j.patcog.2014.04.018
4. Hasan, M. and Roy-Chowdhury, A. K. (2014). Continuous learning of human activity models using deep nets. In Computer Vision–ECCV 2014.
5. Karpathy, A., Toderici, G., Shetty, S., Leung, T., Sukthankar, R., and Fei-Fei, L. (2014). Large-scale video classification with convolutional neural networks. In Computer Vision and Pattern Recognition (CVPR), 2014 IEEE Conference on.
6. J. B. Yang, M. N. Nguyen, P. P. San, X. L. Li, and S. Krishnaswamy,"Deep Convolutional Neural Networks on Multichannel Time Seriesfor Human Activity Recognition," in Proceedings of the 24thInternational Conference on Artificial Intelligence, ser. IJCAI'15.AAAI Press, 2015, pp. 3995–4001.
7. Y. Chen and Y. Xue, "A Deep Learning Approach to Human ActivityRecognition Based on Single Accelerometer," in Systems, Man, andCybernetics (SMC), 2015 IEEE International Conference on. IEEE,Oct. 2015, pp. 1488–1492.
8. Hammerla N Y, Halloran S, Ploetz T. Deep, convolutional, and recurrentmodels for human activity recognition using wearables. arXiv preprintarXiv:1604.08880, 2016.
9. Jian Bo Yang, Minh Nhut Nguyen, Phyo Phyo San, Xiao Li Li, and Shonali Krishnaswamy. Deep convolutional neural networks on multichannel time series for human activity recognition. In IJCAI, 2015.
10. Xu C, He J, Zhang X, et al. Geometrical Kinematic Modeling on Human Motion using Method of Multi-Sensor Fusion[J]. Information Fusion, 2017, 41.
11. Charissa Ann Ronaoo and SungBae Cho. Evaluation of deep convolutional neural network architectures for human activity recognition with smartphone sensors. In Proc. of the KIISE Korea Computer Congress, pages 858–860, 2015.

Circular Ring Patch Antenna Array at 20.2 GHz with Circular Polarization in SATCOM Applications Using CST

S. V. Devika, S. Arvind, and S. Sudha

Abstract High directivity antennas which fetches for satellite communication applications are preferred to be array antennas. Providing greater directivity array of 'n' elements is designed in which circular ring's intrinsic geometry is chosen that leads to circularly polarized waves with high gain and increased radiation efficiency (Devika et al. in Indian J Sci Technol 9:38, 2016). This paper presents the introduction of phased array antenna, design of single micro strip element in the array, and methodology to achieve circular polarization which helps in satellite communication. This paper presents the circular patch structural design in CST studio that helps in improved performance characteristics of the antenna used for practical applications in satellite communications (Devika et al. in Indian J Sci Technol 9:38, 2016). This antenna provides outstanding performance regarding side lobe signal level, half power beam width (HPBW), gain bandwidth, cross polarization, and axial ratio bandwidth at the center frequency of 20.2 GHz.

Keywords Phased array · Satellite · Micro strip · Circular patch

1 Introduction

Phased array antennas are used in satellite communication applications which provides high directivity and beam width [1]. More number of elements leads to more directivity in the system. Irrespective of various structures in array, system like hexagonal, honey comb, and rectangular-phased array antennas provides high efficiency.

S. V. Devika (✉) · S. Arvind · S. Sudha
Hyderabad Institute of Technology and Management, CMR Technical Campus, Hyderabad, India

P. S. R. Chowdary et al. (eds.), *Evolution in Signal Processing and Telecommunication Networks*, Lecture Notes in Electrical Engineering 839,
https://doi.org/10.1007/978-981-16-8554-5_32

Table 1 Comparison table between grid type and area occupied by the respective grids considering a fixed number of 100 elements

For a fixed area of 0.0065 m^2	
Grid type	Number of elements contained
Honeycomb	100
Hexagonal	72
Rectangular	90

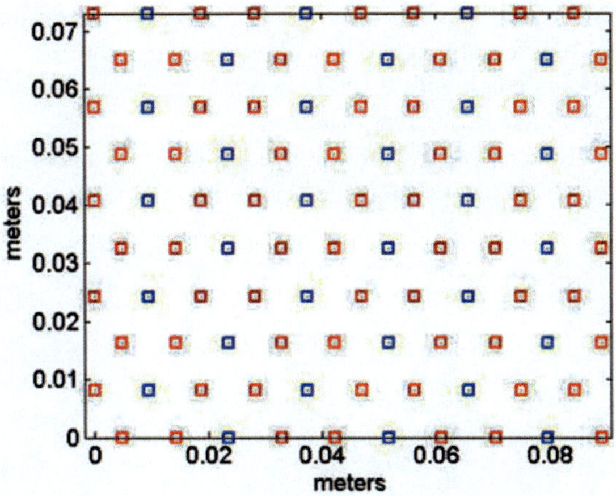

Fig. 1 100 element phased array in honeycomb grid with area of 0.0065 m^2

1.1 Honey Comb Grid Using MATLAB for 100 Array Elements in Phased Array

A total of 100 micro strip elements are placed in an antenna in honey comb model with half of the operating wavelength spacing. This structural design provides less height and more strength with area of 0.0065 m^2 with different grid types shown in Table 1.

Compared to other array structures, Honey comb design occupies minimum area with increased number of elements which is designed in MATLAB shown in the Fig. 1.

1.2 Design of Circular Ring Patch in CST

Circular patch ring is designed in CST studio. CST is the software simulation tool for antenna design. This tool is user friendly in designing micro strip antennas. [2] The basic design of circular ring patch contains two substrates placed one upon the other by allowing air gap between the substrates. Rogers RT5880 is chosen for substrate design. The relative permittivity of the substrate is 2.2. The designed antenna operates at a chosen frequency of 20.2 GHz because of availability of the receiver with the selected frequency placed at the campus. In order to achieve circular polarization with this patch, the bottom diameter of the patch is made bigger than the top patch. Figure 2 shows the basic patch design with top and bottom view with dimensions submitted in CST. The feed pin is connected to excite the bottom layer to achieve the desired outcome.

The diameter of the patch and substrates dimensions place a major role in generation of circularly polarized waves. The dimensions of bottom substrate depend on the ground plane dimensions [3]. Based on the bottom substrate, air gap is created with different height of 0,474 mm. Top circular patch is layered with bit variation to bottom patch with 9.12073 mm and diameter of 3.4670 mm. Figure 3 explains the multilayered patch as discussed.

To achieve circular polarization for satellite communication applications and to resonate the patch at center frequency 20.2 GHz, a small cut is done in the patch and

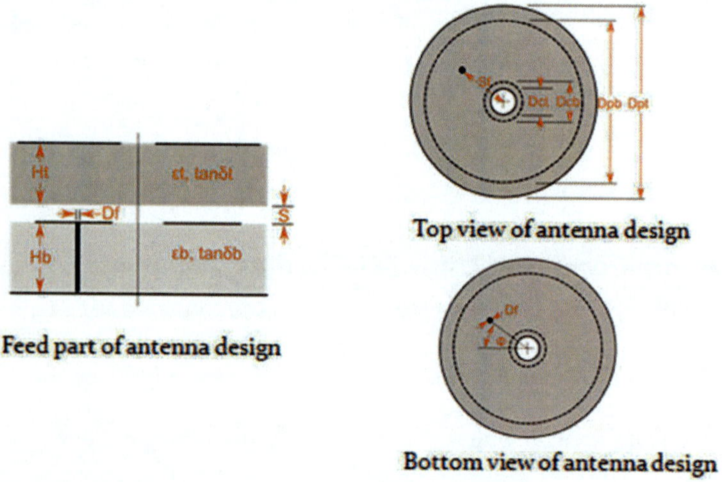

Fig. 2 Patch antenna design

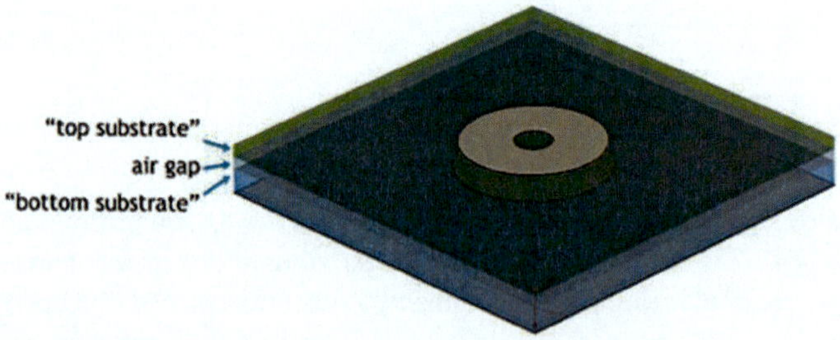

Fig. 3 Antenna substrate design

also small notch is introduced. This gave lot of variation in gain and efficiency while constructing an array with each circular patch element.

These structural modifications also reflected in various other parameters like return loss and the gain of the antenna.

Figure 4 presents the top view of the proposed model with difference in diameters of bottom and top substrates (Fig. 5).

The design of single antenna array is extended to array of four elements and verified with the results. The same can be extended to array of 'n' elements depending upon the application and desired directivity of an antenna. The cut in the patch is clearly visible on the patches which impacted on the increased gain of an antenna.

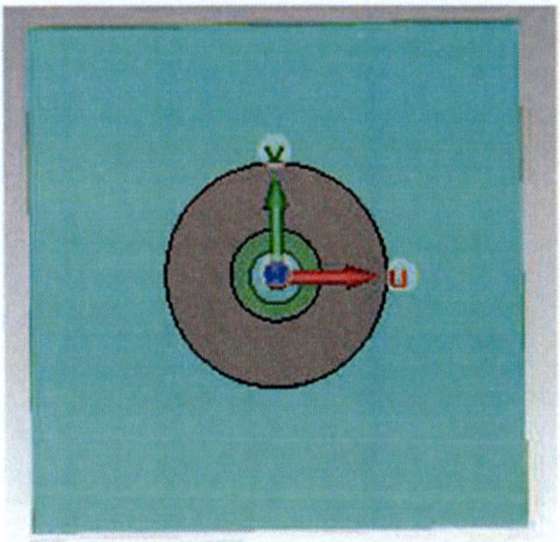

Fig. 4 Top view of proposed antenna

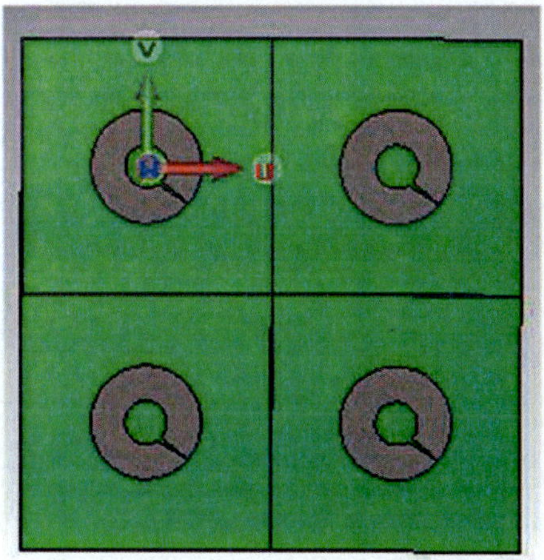

Fig. 5 Array of proposed antenna

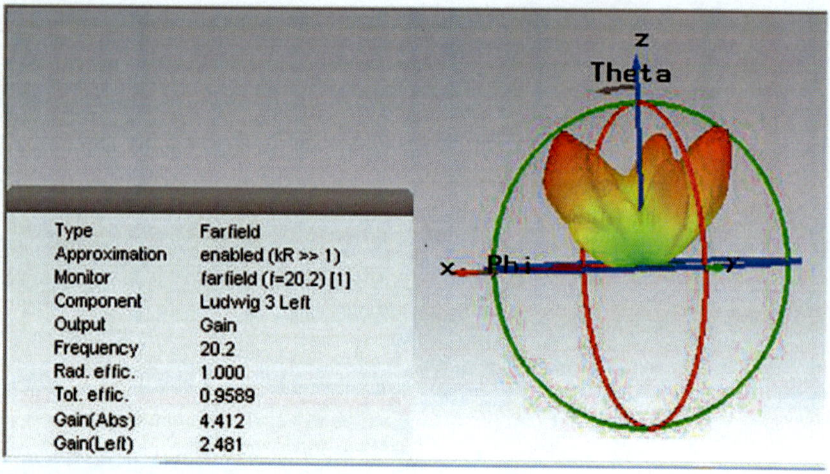

Fig. 6 LHCP of the proposed antenna

1.3 Polarization

Polarization is orientation of an antenna toward desired direction. Among linear and circular polarizations in communication, circular polarization has its own place in satellite communication applications.

The outcome of the proposed design is irrespective of the input given to the circular patch, and the output will be circular polarized wave [3]. If an incident wave is sent through the ground plane, the bottom level polarizes the vertical or capacitive part of the incident wave. Similarly, the second layer or air gap as per design polarizes the horizontal or inductive part of the signal and third layer which is top substrate polarizes the capacitive part again making a signal move in the circular polarization [3].

Figure 6 represents left-hand circular polarization of the designed antenna with gain (left) obtained to be 2.481 as far-field radiation. Similarly, Fig. 7 represents the right-hand circular polarization of the proposed antenna gain (right) of 2.554 dB. The obtained parameters of the designed antenna are presented in Table 2.

1.4 Gain(G) Versus Frequency

Gain is important terminology in an antenna which gives the information about the exact performance of an antenna and helps us in comparison with other antennas for various applications.

Gain is a vital parameter when it comes to design any antenna. Due to the relationship between gain and frequency of operation, the two different antennas with equal

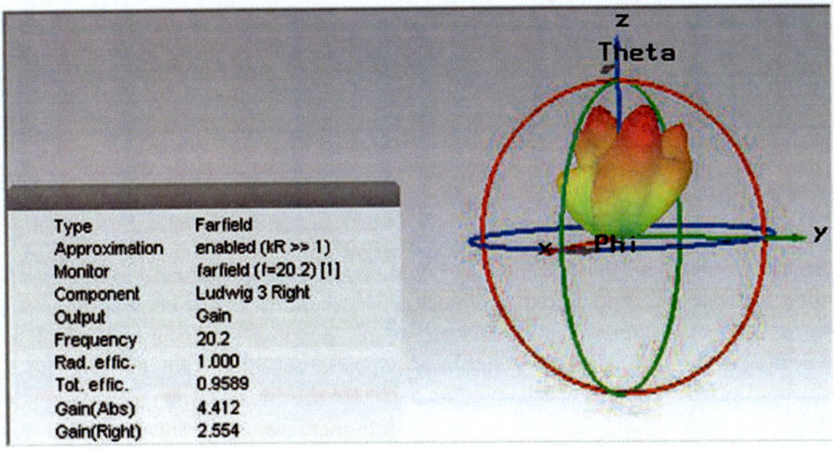

Fig. 7 RHCP of the proposed antenna

Table 2 Obtained antenna parameters

Name	Symbol	Value
Return loss	S_{11}	−14.67 dB
Voltage standing wave ratio	VSWR	1.45
Directivity	D	4.41
Gain	G	9.99

gain may vary in their size depending on the frequency of operation [3]. Figure 8 shows the gain achieved as 9.9995 at the center frequency of 20.2 GHz.

The above table represents the outcome of antenna array with four micro strip elements (circular patch). The return loss for all S-parameters was calculated, and average of −14.6 dB is achieved for an array [4]. The overall efficiency of designed antennas is observed as 100%.

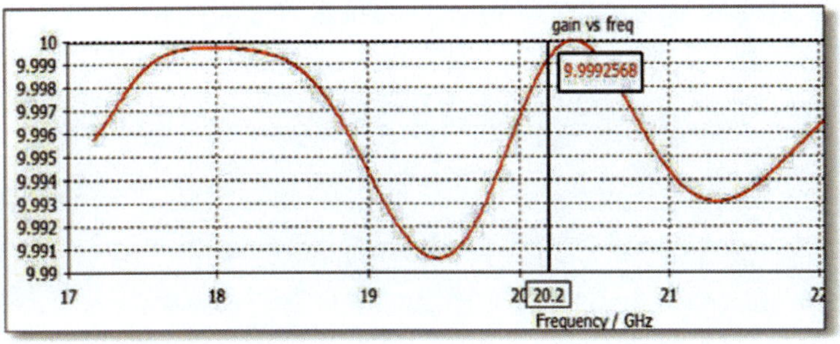

Fig. 8 Gain versus frequency of the designed antenna

Voltage standing wave ratio (VSWR) is observed to be 1.45 which is less than 2 for achieving better antenna performance. Respective directivity and gain of an array is observed to be 4.41 and 9.99 accordingly.

2 Conclusion

This paper concludes that in phased array antennas are used in satellite communication applications. Among the available array structures, honey comb structure is preferred as it occupies minimum space. The micro strip elements used in the array as designed to be circular ring patch to provide circular polarization waves which gives more directivity and efficiency. The design of the circular ring patch is done in CST studio with small truncation in the patch to increase the directivity. The corresponding results are presented at resonant frequency of 20.2 GHz to meet various applications of SATCOM.

Acknowledgements Authors want to thank Hyderabad Institute of Technology and Management, Hyderabad, and CMR Technical campus, Hyderabad, for supporting with necessary research infrastructure to carry out research work.

References

1. C. A. Balanis, Antenna theory: analysis and design, 3rd ed. Hoboken, NJ: Wiley Interscience, 2005.
2. Devika, S., Karki, K., Kotamraju, S. K., Kavya, K., & Rahman, M. Z. (2017). A New Computation Method for Pointing Accuracy Of Cassegrain Antenna In Satellite Communication. *Journal of Theoretical & Applied Information Technology*, *95*(13)
3. Devika SV, Kotamraju SK, Kavya KCS, Kumar VS, Suhas K, Vinu K, Anudeep B (2016) A circularly polarized Ka-band antenna for continuous link reception from GSAT-14. Indian J Sci Technol 9:38
4. Jung YB, Shishlov AV, Park SO (2009) Cassegrain antenna with hybrid beam steering scheme for mobile satellite communications. IEEE Trans Antennas Propag 57(5):1367–1372

Drowsiness Detection System for Drivers Using 68 Coordinate System

Cherie Vartika Stephen, Shreya Banerjee, and Rajat Kumar Behera

Abstract The study proposes a driver drowsiness detection using 68 coordinate system to avert accidents due to driver's fatigue and sleepiness. Driver's drowsiness and fatigue are one of the many significant reasons that can cause road accidents. Every year, there is an increase in the number of deaths and injuries globally because of it. A variety of studies have concluded that around 20% of road accidents are fatigue-related. People who drive trucks and heavy load vehicle especially at night feel sleepy and tend to drowse which result in different types of car accidents. The proposed drowsiness detection system helps to detect if a driver of a vehicle is drowsy and is a very useful system as it can help to prevent many such accidents. The proposed system calculates eye aspect ratio and mouth aspect ratio, grounded on the 68 coordinate system, and then, audio alert, and display screen warning on the screen if the driver is discovered to be drowsy.

Keywords Image processing · Face detection · Fatigue · Drowsiness · 68 coordinate system

1 Introduction

Human beings have forever created machines and come up with methods to facilitate and safeguard lives. With the evolution of technology, modes of transportation have developed, and the dependency has exponentially increased. Nowadays, everyone in the world uses some variety of transportation, and it is important to remain alert and active while driving.

Driver exhaustion may be a vital cause because of an increase in the number of cases of vehicle accidents. A report indicates that there are 1200 deaths and 76,000 injuries in a year that are often the resultant of accidents due to fatigue and tiredness of the driver, and approximately 20% of automobile accidents are because of the fatigue and drowsiness of the driver. Less attention heads the driver being distracted,

C. V. Stephen (✉) · S. Banerjee · R. K. Behera
School of Computer Engineering, Kalinga Institute of Industrial Technology (KIIT) Deemed To Be University, Bhubaneswar, India

P. S. R. Chowdary et al. (eds.), *Evolution in Signal Processing and Telecommunication Networks*, Lecture Notes in Electrical Engineering 839,
https://doi.org/10.1007/978-981-16-8554-5_33

and therefore, the chance of street accident goes high. Generally, fatigue is extremely difficult to evaluate or measure in contrast to alcohol or drugs that have simple prime indicators and tests easily accessible.

Possibly, the simplest answer to this problem is to spread awareness concerning drowsiness-related accidents and encouraging drivers to admit fatigue once required. However, due to need for employment and some extra money, drivers drive all night even with fatigue.

The improvement or innovations for recognizing or preventing drowsiness and fatigue of the driver is a vital check within the sector of accident preventing systems. Loss of the awareness due to the fatigue causes a number of changes within the human's body and activities like eye status and head position. These side effects and parameters enable to effectively measure the level of drowsiness.

The literatures relate to driver drowsiness detection have been explored. Ahonen et al. [1] used the technique of local binary pattern (LBP) to detect the face of the driver. LBP concludes local structures of an image effectively by comparing each pixel with its neighboring pixel. LBP divides the image extracted from the video into blocks and then generates LBP histogram from each block and forms feature histogram. But the main disadvantage of this technique is that it produces long histograms, which reduces the speed of the recognition process, and the binary data generated is noise sensitive. Redmon et al. [2] have implemented YOLO (You Only Look Once) to detect the driver's face. The face was extracted after which the Dlib toolkit is used to extract the facial features. The major drawback of the model was with incorrect localization.

To overcome the gaps, this study has undertaken for drowsiness detection of the driver by calculating eye aspect ratio and mouth aspect ratio, grounded on the 68 coordinate system, and then, audio alert and screen warning is displayed on the screen if the driver is discovered to be distracted.

2 Related Works

One of the main issues in developing an economically viable drowsiness detector is the method of acquiring correct drowsiness information. The drowsiness in a real environment cannot be experimented because of some safety reasons that cannot be ignored. Hence, a drowsiness detection system must be worked upon that can be tested under a secured laboratory setting. However, in a laboratory setting, the most reliable and informative data regarding the driver's drowsiness rely only on the approach in which the driver primarily gets into the drowsy state. The driver's drowsiness quotient is majorly dependent on the quality as well as the amount of their recent sleep cycle, at what the time of day that is possible, and therefore, the rise or increase during the duration of the driving task. During certain analysis experiments, the subjects had been completely sleep deprived, and in other cases, they were only partially sleep deprived. It was found that certain studies had appointed night shift employees as their subject, who were completely sleep deprived as the results were obtained next morning.

Kokonozi et al. [3] conducted an experiment in which the participants were under examination for 24 h before starting the experiment. They had to ensure that the participants were being deprived of sleep. Partially, in some of the experiments, they made the participants a little sleep deprived, by allowing them to sleep for not more than 12 h.

Peters et al. [4] had made an observation on a similar subject right through four consecutive days. Their level of sleepiness was tested based on no sleep deprivation at all, partial sleep deprivation, and by sleep depriving the subject completely. It was concluded that even in situations where they were partially deprived of sleep, the participants seemed to be drowsy after a while. So we can argue that it is the quality of the last sleep that could be a very important and crucial factor that will influence drowsiness.

Otamani et al. [5] made a confirmation that lack of sleep alone does not directly impact on the brain signals responsible for managing drowsiness, it is the time duration of the task, which has a major influence on it.

Ueno et al. [6] established a systematic process that takes advantage of the technology that processes images to study the photograph imagery of the face of the driver that would be taken with the help of a video camera. The diminishing level of the driver's alertness is detected on the basis of the degree to which the eyes of the driver are open or closed.

Forsman et al. [7] made an indication that the variability of the steering wheel gives a foundation for developing an economical and easy-to-install alternative technology for in-vehicle detection of drowsiness of the driver at levels of fatigue that were moderate.

Considering the above literatures, it can be concluded that none of the studies detect the drowsiness of the driver considering the eye status and head position, grounded on the 68 coordinate system.

3 Methodology

Fatigue and drowsiness of drivers or distraction like a phone call, talking to other passengers, is one of the few significant reasons causing road accidents, and they increase the number of deaths and injuries globally, every year. This calls for a need of an effective drowsiness detection system. To prevent these accidents, we have proposed a system that will caution the driver if the driver is distracted or is feeling drowsy. Drowsiness is detected by the status of the eye and head position of the driver. Facial landmarks detection is used for detection of distraction or drowsiness.

In this study, we have used the 68 coordinate system for detecting the parts of the face, extract the coordinates of eye and mouth and the concepts of eye aspect ratio (EAR) and mouth aspect ratio (MAR) to build the drowsiness detector.

To overcome the problems of drowsiness, we have come up with a simple solution devised in the structure of image processing. To conduct image processing, the study used OpenCV and DLib open source libraries and Python to execute the concept.

An infrared camera is utilized to constantly track the driver's facial landmark and motion of the eyes and mouth. Images are taken by the camera at a fixed frame rate of 20 frames per second (fps). These images are transferred to an image processing component which executes the face landmark detection to recognize driver's distractions and drowsiness. If the driver is discovered to be distracted, then an audio alert is given and a warning is shown on the screen.

3.1 Drowsiness Detection Conditions

The following are the proposed drowsiness detection conditions for this research.

1. If the eyes of drivers are closed for a definite amount of time, then it is considered that driver is drowsy and corresponding audio alert and screen warning is given to make the driver alert.
2. If the mouth of driver is observed to be open for the specific amount of time, then it can be concluded that the driver is yawning and an appropriate audio alert and screen warning is given to the driver.
3. If the driver's eyes are not on the road and the driver is distracted, then an appropriate audio alert and screen warning is given.

3.2 The 68 Coordinate System

The pretrained facial landmark detector present within the Dlib library is employed to approximate the placement of all the 68 (x, y) coordinates that map to the structures of the face. The indexes of the 68 coordinates are visualized as follows (refer Fig. 1).

These interpretations are a part of the 68 point iBUG 300-W dataset that the facial landmark predictor of the Dlib library has been trained on. In the above diagram, it can be observed that points are from 1 to 68. For the eye aspect ratio, we consider points 37–46. While for mouth aspect ratio, we consider points 61–68.

There are mainly two steps in facial landmark detection:

1. Face detection: This is to locate the human face and give a value in the form of coordinates of a rectangle (x, y, z, v).
2. Face landmark: When the face location has been obtained in an image, the points inside the rectangle are obtained.

The 68-face landmark model in the Dlib library shows the way to obtain the facial features like eyes, mouth, nose, etc. Sometimes, we might not want to detect all of the facial features which can be done by customized training of the Dlib's 68-landmark model.

Fig. 1 Indexes of the 68 coordinate

3.3 System Flowchart

Figure 2 depicts the flowchart of the proposed system. The flowchart starts with acquiring the video frame from the camera followed by detecting the face of the video frame. From the face, eyes and mouth are detected and tracked using 68 coordinate system. The EAR and MAR are calculated, and if the computed value results in drowsiness, the driver is alerted with audio and warned in the screen.

3.4 Approach

We used Dlib, a pretrained program trained on the HELEN dataset to detect human faces using the predefined 68 landmarks. After passing the video feed to the Dlib frame by frame, we are able to detect left eye and right eye features of the face. Now, we drew contours around it using OpenCV. The OpenCV provides the inbuilt library of image extraction from video. Using imutils, we choose the frame size as 900 (width = 900). Using Scipy's Euclidean function, we calculated sum of both eyes' aspect ratio which is the sum of two distinct vertical distances between the eyelids divided by its horizontal distance (refer Fig. 3). Each eye is depicted using six coordinates, beginning from the left-corner of the eye, and then going clockwise.

$$\text{EAR} = \frac{||p2 - p6|| + ||p3 - p5||}{2||p1 - p4||} \ldots \tag{1}$$

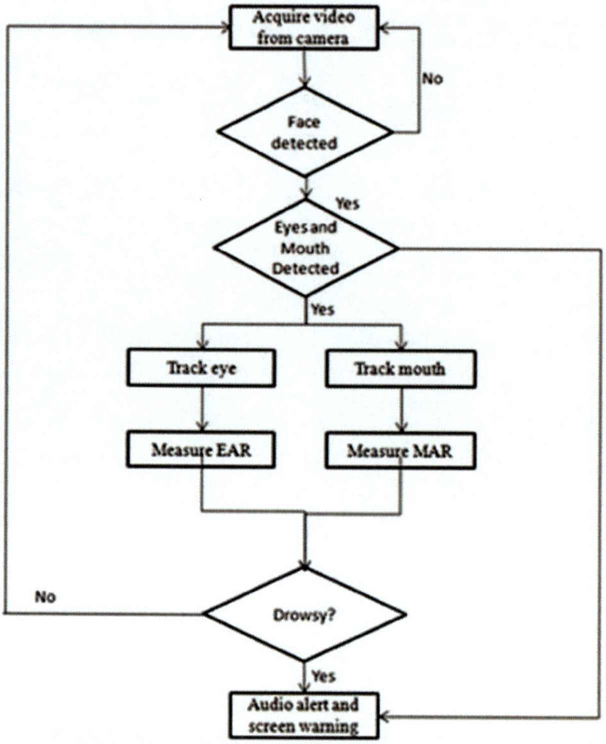

Fig. 2 Proposed system flowchart

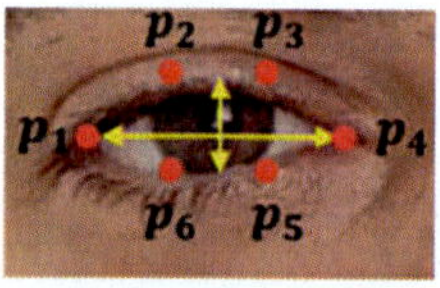

Fig. 3 EAR calculation (extracted from http://datahacker.rs)

In Eq. (1), the numerator calculates the distance between the vertical eye coordinates and the denominator calculates the distance between horizontal eyes coordinates, weighting the denominator correctly because there is one set of horizontal coordinates but two sets of vertical coordinates. The EAR is an important concept because it helps in predicting whether the eye is closed, open, or there is a blink. The EAR is nearly constant when the eyes are open, but quickly falls to zero when a person blinks and becomes nearly zero when the eye is closed.

Then, it is checked if the aspect ratio value is less than 0.25 (0.25 was chosen as a base case after valid tests). If it is less than threshold, an alarm is sounded and the user is warned.

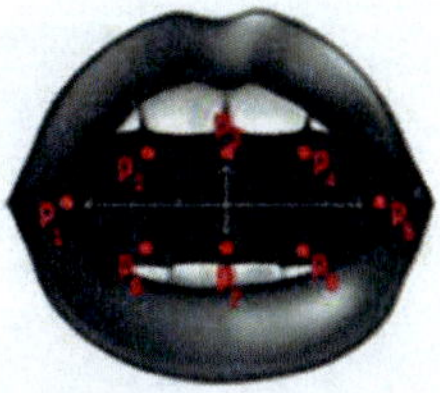

Fig. 4 MAR calculation (extracted from http://www.towardsdatascience.com)

In a similar way to decide the yawning factor, the mouth aspect ratio (refer Fig. 4) is computed as follows:

$$\text{MAR} = \frac{|p2 - p8| + |p3 - p7| + |p4 - p6|}{3 * |p1 - p5|} \ldots \quad (2)$$

Then, we check if the aspect ratio value is more than 0.35 (0.35 was chosen as a base case after some tests). If it is more than a threshold, an alarm is sounded, and the user is warned.

In this model, the video continuously captures the images, and therefore, we have considered constant time threshold values. If the eyes are closed for more than 2 s, then the alarm rings. If the mouth is open for more than 2 s, then only the alarm rings. If the face is distracted for more than 3 s, then the alarm rings for distraction.

3.5 Test Cases

Table 1 represents the test cases considered for this study.

Table 1 Test case conditions

Test ID	Test case conditions	Expected behavior	Expected result
T01	Straight face, good light, open eyes	Non-drowsy	Non-drowsy
T02	Straight face, good light, droopy eyes	Drowsy	Drowsy
T03	Straight face, good light, closed mouth	Not yawning, hence not drowsy	Not drowsy
T04	Straight face, good light, open mouth	Yawning, hence drowsy	Drowsy
T05	Not straight (turning in different directions)	Distracted	Distracted

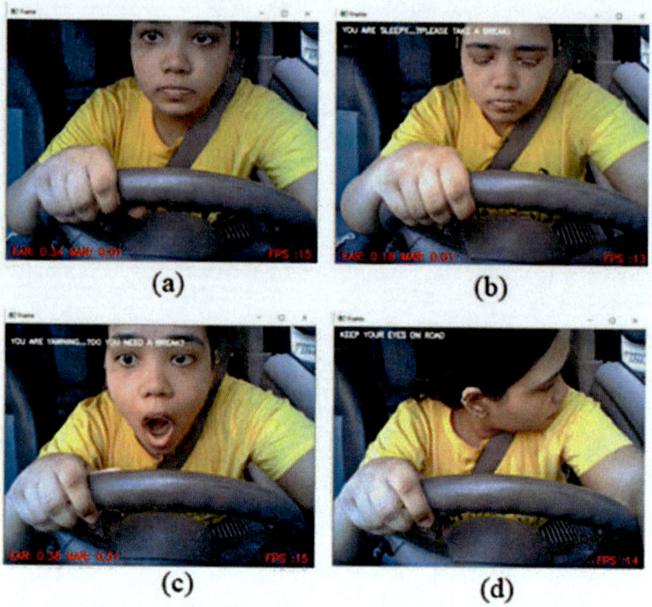

Fig. 5 Real time snapshots of drowsiness detection system

3.6 Result and Analysis

From the below figures, the result and analysis indicate that:

- In reference to Fig. 5a, the eye aspect ratio is 0.34 (No alarm since EAR > 0.25), hence not drowsy.
- In reference to Fig. 5b, the eye aspect ratio is 0.18 (alarm rings since EAR < 0.25), hence drowsy.
- In reference to Fig. 5c, the mouth aspect ratio is 0.51 (alarm rings since MAR > 0.35). Yawning, hence drowsy.
- In reference to Fig. 5d, no detection hence distracted.

4 Discussion

Severe incidents are often caused by driver fatigue and sleep, and therefore, a drowsiness detection system was proposed to detect a driver's drowsiness as accurately as possible. The combination of eye aspect ratio and mouth aspect ratio improves the accuracy of the system, resulting in a significant reduction in false alert and warning on the screen. Finally, the project can be used as an important method for predicting the emergence of drowsiness in drivers, as well as avoiding potential road accidents. The proposed system would assist in increasing driver safety, with a particular emphasis on long-haul truck drivers, night time drivers, people traveling long distances alone, and people who are sleep deprived.

The limitations of the system are: (i) Dependence on ambient light: The poor lightning condition makes detection of the eyes difficult even when the face is correctly detected. The error from this result can be avoided by the use of infrared backlights which helps to solve the low lightning problem, (ii) Optimum range: The webcam must be located at an optimum distance in order to avoid scenarios where either the face is too close to the webcam (less than 30 cm) or where the face is more than 70 cm away. Optimum range is therefore considered 40–70 cm in order to detect drowsiness, (iii) Hardware requirements: The system configuration consisted of 1.6 GHz and 1 GB RAM Pentium dual core processor. The ideal situation for running this would be using a higher configuration, where the level of detection of drowsiness will be faster as compared to an inferior set up. Therefore, a system if dedicated hardware in real-time applications need to be used to avoid frame buffering or slow detection level, (iv) Orientation of face: The tilting of the face beyond a certain level causes problems in detection of the face and in turn affecting the detection of the eyes. This scenario can be avoided by using tracking functions to detect any sort of movement or rotation of the object and also with the help of a trained classifier for tilted face profile and for tilted eyes, (v) Poor detection with spectacles: A major drawback of almost all eye detection systems designed so far is that they are unable to correctly detect the eyes, in situations where the driver wears spectacles, (vi) Problem with multiple faces: When there is more than one face detected by the webcam, it leads to error in the results. However, the problem is not of much concern as the study aim to detect the drowsiness of a single driver.

5 Conclusion

This research aims to devise a solution to detect the drowsiness of a driver and warn a driver before any accident occurs, decreasing the number of deaths and injuries to a great extent. We have designed and implemented a prototype of a drowsiness detection system using OpenCV that detects drowsiness and alerts the driver in real time. This system has been successfully tested, its limitations have been identified, and a future plan of action established.

6 Future Scope

Like any other research, this study also has some limitations. First, the model can be improved incrementally by using other parameters like blink rate, yawning, state of the car, etc. If all these parameters are used, it can improve the accuracy by a lot.

Second, we plan to extend this study by adding a sensor to track the heart rate in order to prevent accidents caused due to sudden heart attacks to drivers.

Third, the same model and techniques planned to be used for various other uses like Netflix, and other streaming services can detect when the user is asleep and stop the video accordingly. It can also be used in application that prevents user from sleeping.

References

1. Ahonen T, Hadid A, Pietikainen M (2006) Face description with local binary patterns: Application to face recognition. IEEE Trans Pattern Anal Mach Intell 28(12):2037–2041
2. Redmon, J., Divvala, S., Girshick, R., & Farhadi, A. (2016). You only look once: Unified, real-time object detection. In Proceedings of the IEEE conference on computer vision and pattern recognition (pp. 779–788).
3. Kokonozi, A. K., Michail, E. M., Chouvarda, I. C., & Maglaveras, N. M. (2008, September). A study of heart rate and brain system complexity and their interaction in sleep-deprived subjects. In 2008 Computers in Cardiology (pp. 969–971). IEEE.
4. Suryaprasad, J., Sandesh, D., Saraswathi, V., Swathi, D., & Manjunath, S. (2013, November). Real time drowsy driver detection using haarcascade samples. In CS & IT Conference Proceedings (Vol. 3, No. 8). CS & IT Conference Proceedings.
5. Miyaji M (2014) Method of drowsy state detection for driver monitoring function. International Journal of Information and Electronics Engineering 4(4):264
6. Romdhani, S., Torr, P., Scholkopf, B., & Blake, A. (2001, July). Computationally efficient face detection. In Proceedings Eighth IEEE International Conference on Computer Vision. ICCV 2001 (Vol. 2, pp. 695–700). IEEE.
7. Tabrizi, P. R., & Zoroofi, R. A. (2008, November). Open/closed eye analysis for drowsiness detection. In 2008 first workshops on image processing theory, tools and applications (pp. 1–7). IEEE.

Improve PV System with MPPT Technique by Using Chaotic Whale Optimization Algorithm

Krishna Kumar Pandey, Chandan Banerjee, and Vineet Kumar Tiwari

Abstract Solar energy has grown in popularity because of nonconventional energy resource. Therefore, the layout of PV cells has aroused fascinate of investigators round the globe. There are two major issues in this sector: the lack of a suitable idea for characterizing solar cells and the scarcity of information on photovoltaic panels. This scenario has an effect on the effectiveness of photovoltaic modules (panels). The current vs. voltage features are utilized to model the properties of solar cells. Taking these values into account, the modeling process entails solving complex nonlinear as well as multimodal objective factors. To recognize the variables of photovoltaic cells as well as frames, various systems have been suggested. The majority of each other frequently fail to find the best solutions. The CWOA is proposed in this paper for estimating the variables of solar cells. The suggested technique has the key benefit of utilizing chaotic maps to calculate and quickly apply the input variables of the optimization technique. This scenario is advantageous in complicated situations because the suggested methodology enhances their opportunity to discover for the right approach during the automated manner. The suggested technique can maximize detailed as well as multimodal optimization problem.

Keywords Power generation · MPPT technique · Chaotic WOA · PV system

1 Introduction

Solar radiated energy is one out of the most dependable as well as effective renewable energy sources. It is smooth, emission-free, and dependable, allowing utilized power via PV panels a popular research study [1]. Furthermore, the majority of investigation has focused on maximizing the outcome of the photovoltaic system. As a result, the PV array's output power is affected by climatic changes and exhibits solitons I–V features [2]. A MPP emerges from the PV curve's knee, where energy is greatest [3]. MPPT methods are used to obtain the most electrical output from the PV origin.

K. K. Pandey (✉) · C. Banerjee · V. K. Tiwari
Department of Electrical Engineering, United College of Engineering & Research, Prayagraj, UP, India

P. S. R. Chowdary et al. (eds.), *Evolution in Signal Processing and Telecommunication Networks*, Lecture Notes in Electrical Engineering 839,
https://doi.org/10.1007/978-981-16-8554-5_34

Because MPP differs with altering atmospheric conditions, MPPT algorithms detect the operating voltage of PV equivalent to MPP [4].

This paper is organized as follows: Representation of PV Module and Basic WOA is introduced in Sects. 2 and 3. In Sect. 4, literature survey of chaotic Whale optimization algorithm using MPPT technique. Sections 5 and 6 show the methodology and results, and Sect. 7 show conclusion of the article.

2 Mathematical Model of a Solar Cell

A photovoltaic (PV) cell is a nonlinear direct current (DC) control signal. Its electrical output power varies with thermal temperature as well as irradiation. The circuit model of an excellent PV panel is depicted in Fig. 1. This optimistic method is effective enough to comprehend PV features and the PV cell's sensitivity to modifying weather systems. The numerical formulation of the PV cell is described in [5] as follows:

$$I = I\text{pv} - Io\left[\exp\left(\frac{V - I.Rs}{a.Vt}\right) - 1\right] - \frac{V + I.Rs}{Rp} \tag{1}$$

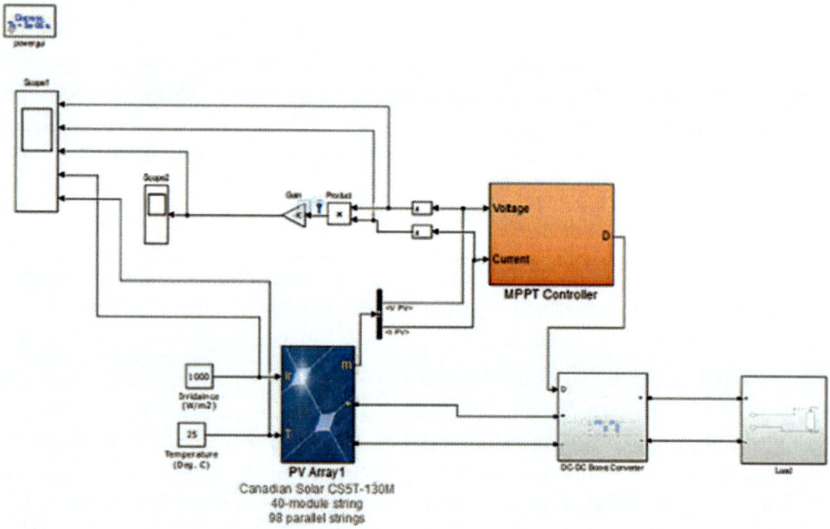

Fig. 1 Simulink model of proposed work

3 The Basic WOA

WOA is a smarter optimization technique that simulates humpback whale for aging behavior [6]. The method categorizes humpback whale actions into three groups: encircling quarry (prey), bubble-net attacking, and searching for hunt.

3.1 *Encircling Prey*

In WOA, it is thought that the existing objective function is the focus prey or is near to the prey in numerous solving issues where the optimization algorithm is not widely understood. Entities in humpback giant whale groups will tries to modify their areas in the path of the hunt once the prey position has been determined.

$$\vec{X}(t+1) = \vec{X}\text{prey}(t) - \vec{A}.\vec{D} \tag{2}$$

$$\vec{A} = 2\vec{r}.\vec{a} - \vec{a} \tag{3}$$

$$\vec{D} = \left|\vec{C}.\vec{X}\text{prey}(t) - \vec{X}(t)\right| \tag{4}$$

$$\vec{C} = 2\vec{r} \tag{5}$$

3.2 *Bubble-Net Attacking*

A bubble-net attack occurs when humpback giant whales swim concurrently around their hunt across a helical path in a narrow circle. For both compression bracketing method and the spiral design are used in WOA to replicate their actions. The selection probability of participants is in both ways during the optimization problem. Furthermore, the precise scientific design is given:

$$\vec{X}(t+1) = \begin{cases} \vec{X}\text{rand}(t) - \vec{A}.\vec{D}, & p < 0.5 \\ \vec{D}.e^{bl}.\cos(2\pi l) - \vec{X}\text{prey}(t) & p \geq 0.5 \end{cases} \tag{6}$$

$$\vec{D} = \left|\vec{X}\text{prey}(t) - \vec{X}(t)\right| \tag{7}$$

3.3 Searching for Prey

WOA releases the population's position during the hunting phase using the following formula [7]:

$$\vec{X}(t+1) = \vec{X}\text{rand}(t) - \vec{A}.\vec{D} \tag{8}$$

$$\vec{D} = \left|\vec{C}.\vec{X}\text{rand}(t) - \vec{X}(t)\right| \tag{9}$$

4 Literature Survey

Yashar et al. [8] enhanced whale optimization algorithms (EWOA) are suggested to cope with variable recognition issue of a WDPS device. The suggested EWOA efficiently solves the early concurrence problem of WOA by dividing each group in different subpopulations also revising the location for every whale as per the situation of the best representative in its present subpopulation, the role of the highest award of the other subpopulation and the location of the finest neighbor representative. In addition, fractional chaotic maps are inserted in the EWOA investigation procedure to enhance its classification accuracy. For verification purpose, the proposed approaches are used to detect specific WDPS parameters, where different statistics analyzes and comparative are made with other latest state-of-the-art methodologies. Simulation outcomes confirm that the approaches have less variation in parameter calculation, more convergence speed, and higher precision in contrasted with other approaches [8].

Dallel et al. [9] uses an enhanced variety of the SALP. Swarm algorithm named chaotic SALP swarm algorithm (CSSA) for the prediction of PV cell parameters in both diode designs (single and double. The CSSA outlook advantages from the properties of chaotic maps and has the responsibility of improving a good balance among exploration and exploitation processes. The implementation of the developed CSSA is contrasted to 14 well-known algorithms. Experimental demonstrated that the suggested algorithm has the ability to find ideal path with a precise assessment of the current vs. voltage characteristics such as high real solar cells [9].

Maniraj and Peer Fathima [10] A new WOA method has been developed that has the capability to achieve peak power described in solar PV panels under various weather zones. In addition, the suggested WOA algorithm was evaluated in the MATLAB/Simulink model and a comparative analysis with distinct MPPT algorithms, namely perturb and observe (PO), gray wolf optimization (GWO). The outcomes clearly indicated the suggested WOA approach providing more than 99.6% efficiency with high tracking speed and minimum payback period under PSC [10].

Singh and Shimi [11] Solar PV panels have set of a popular renewable energy resource because they generate clean and green energy and require little upkeep. Even so, as environmental conditions change, less connection is supplied. The output current changes from the maximum power point (MPP) to a different one. The MPP tracking (MPPT) system is planned to keep record of changing irradiance but keep the panels running at MPP. The MPPT controller produces a complex duty ratio for the DC–DC converter using an equation. This paper shows a MATLAB simulation of the MPPT PV method using perturb and observe (P&O) method. Irradiance variations are considered in the simulation design when the temperature stays constant. A contrast is made among non-MPPT and MPPT performance. The results show that the MPPT algorithm improves the system's efficiency in different climatic conditions and outperforms the non-MPPT device [11].

Nagaraja Kumari and Kaumudi Pravallika [12]. The goal is to use a fuzzy logic controller and an enhanced incremental conductance MPPT approach to closely manage the maximum power point (MPP). Under various heat and irradiation environments, the fuzzy controller shows its effects in preventing oscillations in output with quick answer. Under the MATLAB/Simulink software setting, a comparison of improved incremental conductance algorithm with fuzzy controller, improved incremental conductance algorithm with PI controller, perturb and observe MPPT process, and incremental conductance. MPPT methods can be observed, and the suggested method shows its popular efficacy under various atmospheric conditions [12].

Mousavi et al. (2020). To address the parameter identification issue of a WDPS system, enhanced whale optimization algorithms (EWOA) are suggested. By dividing the crowd into two subpopulations or adjusting the location of each whale depending on the strength of the best representative in its present subpopulation, the location of the best agent in the other subpopulation, and the location of the best opposing agent; the suggested EWOA involves addressing the premature convergence issue of WOA. Moreover, fractional chaotic maps are incorporated into EWOA's search method in order to improve its accuracy. The experimental results are used to classify the statistical values of WDPS for verification purpose, with various statistical analyses and results compared with other recent state-of-the-art implementations. In contrast to other equations, simulation findings show that the algorithms have less numerical simulation deviation, easy implementation, or better precision [13].

5 Proposed Methodology

The solar PV system is today one of the leading renewable energies. Solar photovoltaic (PV) is a technology which uses semiconductors to transform sunlight into direct current electricity (DC). A solar technology-based inverter that can be supplied to a commercial electric grid system or a local electrical. Network can transform the variable direct current (DC) supply of a photovoltaic solar panel into useful frequency alternation currents (AC). The boost circuit is used in the device that raises the power of the PV panel. The MPPT guarantees that the service cycle of the boost converter is governed by the impedance between the PV and the load for optimal power transfer. The photovoltaic system efficiency may be substantially enhanced with the help of maximum power point tracker (MPPT). Maximum power point tracking is primarily used by the maximum power point tracking controller in extracting the maximum PV modules with their respective solar irradiance and thermal temperature at the same moment in time. Several algorithms have been developed to effectively track the maximum performance. It has been obtained from the literature survey that in the previous work; the PI MPPT algorithm was used for tracking PV solar pump which provides the desired output. However, in a recent approach, whale optimization algorithm based on the PI controller was used for solar-based PV system. Slow convergence speed is the main problem associated with WOA. So CWOA is used to enhance the global convergence speed and to get better performance. Due to this, the requirement arises of upgrading the existing system which helps to achieve more efficient results. We will propose chaotic WOA (CWOA) methods for tuning the main parameter of WOA which helps in controlling exploration and exploitation.

Objectives:

1. To implement Chaotic-based Whale Optimization Algorithm (CWOA).
2. To update MPPT technique by using Chaotic WOA technique for PV system.
3. To carry out the performance evaluation of the proposed work and comparison with traditional approach.

6 Results

WOA is used as a direct control MPPT method in this study, i.e., duty cycle control using whale population as duty ratios to minimize steady-state oscillations. Direct control MPPT reduces voltage drop and thus increases device performance. The controller senses the corresponding voltage as well as currents for each population of whales, i.e., duty ratios and computes output power.

Maximum power point tracking (MPPT) is a useful tool for improving the performance of solar PV systems. The key component of the MPPT method is the MPPT control algorithm, which is utilized to ensure the proper functioning of the MPPT device by pushing the operating point to the maximum power point (MPP). The main variables utilized for simulating the curve seen in figures are as follows: voltage, current, voltage at max. power (VMP), current at max. power (IMP), irradiance, and temperature of electrical equivalent circuit of solar cell (Figs. 2, 3, 4, 5, 6 and 7; Table 1).

By modifying the peak power, the corresponding P–V typical curve was obtained (simulated in MATLAB). To achieve the current form, this curve is the amplitude of several closely related curves. This graph depicts the shifting of the solar PV system's maximum power point (MPP) for different solar irradiation levels. By varying the peak power, various local MPPs on P–V characteristic curves yield global MPP.

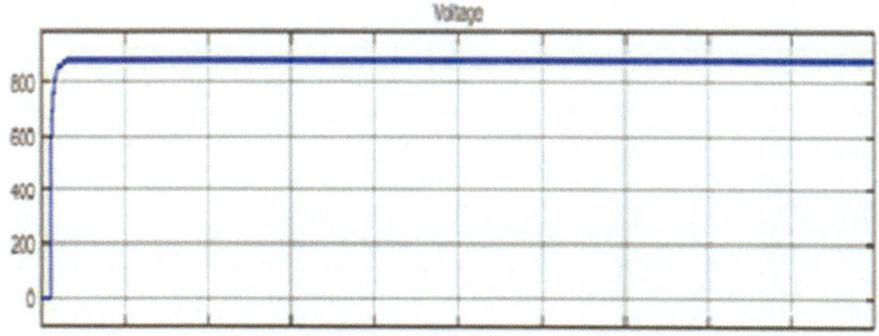

Fig. 2 PV voltage generated by proposed model

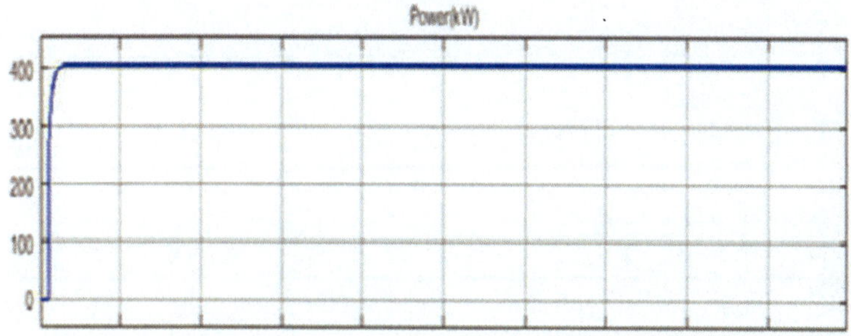

Fig. 3 PV power (kW) generated by proposed model

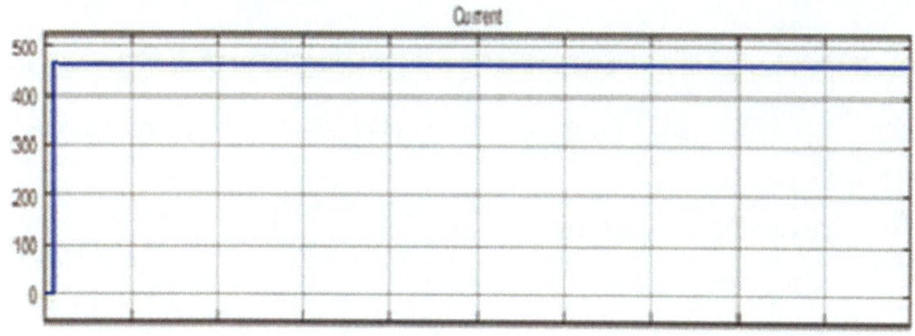

Fig. 4 PV current generated by proposed model

7 Conclusion

This paper investigates a whale optimization algorithm (WOA) for maximizing the PI processing parameters of PMSG-based WECS to obtain the maximum dynamic efficiency (PMSG). The whale optimal solution (WOA) is enhanced to the standard PI controller on machine side adapter tool to detect the optimal level benefits. The computational model for PV panel identification (I–V and P–V qualities) is displayed. Simulation results of the electric circuit utilizing the proposed technique were also

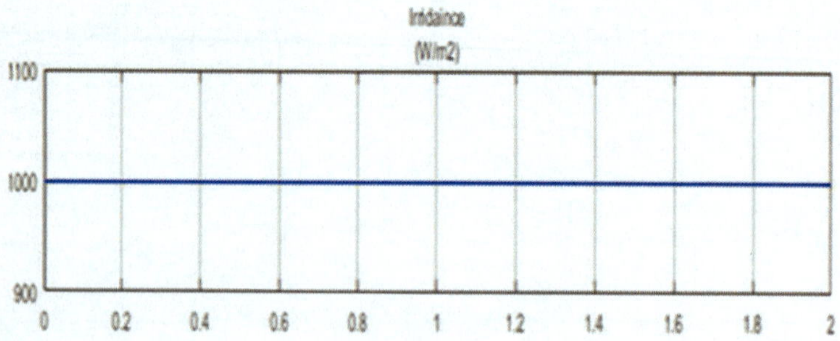

Fig. 5 PV irradiance (W/m^2) of proposed model

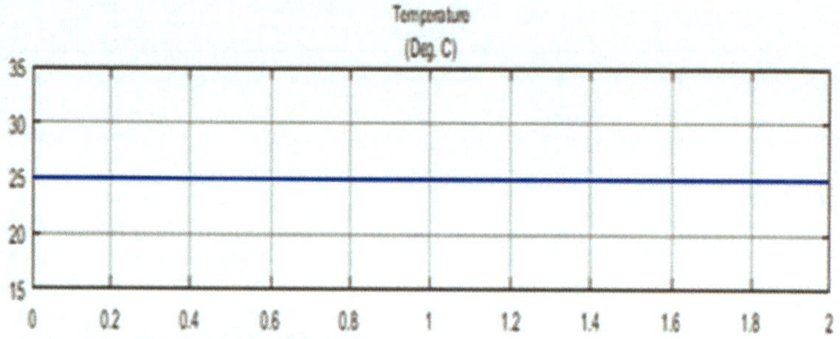

Fig. 6 PV temperature (°C) of proposed model

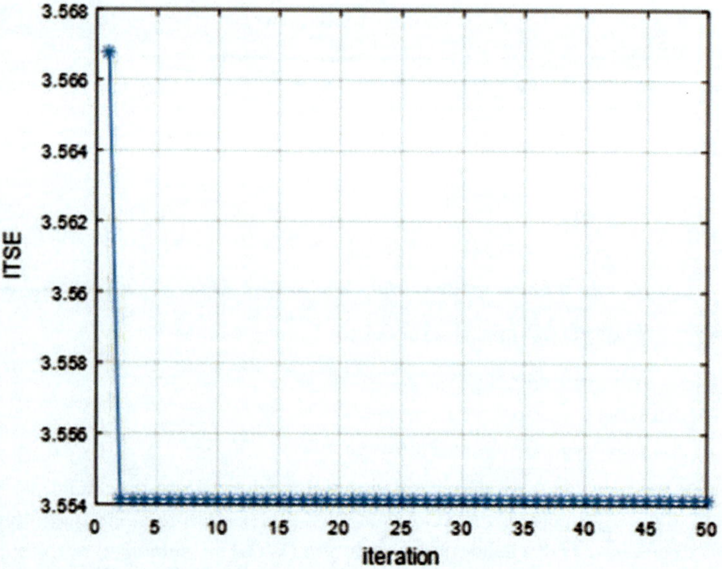

Fig. 7 Convergence curve of proposed optimization algorithm

Table 1 Comparison of proposed CWOA with conventional WOA

Parameter	IC_FOPI_WOA	Proposed CWOA
Power (kW)	401.5	405
Settling time	0.9561	0.041262
Risetime	0.7436	0.0093021
Overshoot	5.1901e−05	1.3365e−10

seen, displaying the existence of a steady voltage level for maximum constant power output from a solar PV app. As a result, the proposed MPPT officially recognize acquires a solar PV platform's-controlled power output (output voltage or current).

References

1. Amir A, Amir A, Selvaraj J, Rahim NA (2016) Study of the MPP tracking algorithms: focusing the numerical method techniques. Renew Sustain Energy Rev 62:350–371
2. Rahim NA, Chaniago K, Selvaraj J (2011) Single-phase seven-level grid-connected inverter for photovoltaic system. IEEE Trans Industr Electron 58(6):2435–2443
3. Bialasiewicz JT (2008) Renewable energy systems with photovoltaic power generators: operation and modeling. IEEE Trans Industr Electron 55(7):2752–2758
4. Nianchun W, MeiYue W, GuoSHeng S (2011) Study on characteristics of photovoltaic cells based on MATLAB simulation. In: Proceedings of the Asia-Pacific power and energy engineering conference (APPEEC), pp 1–4
5. Villalva MG, Gazoli JR, Filho ER (2009) Comprehensive approach to modeling and simulation of photovoltaic arrays. IEEE Trans Power Electron 24(5):1198–1208
6. Ye X, Liu W, Li H, Wang M, Liang G, Chen H, and Huang H (2021) Modified whale optimization algorithm for solar cell and PV module parameter identification. Hindawi
7. Parida B, Iniyan S, Goic R (2011) A review of solar photovoltaic technologies. Renew Sustain Energy Rev 15(3):1625–1636
8. Yashar M, Alireza A, Ibrahim Beklan K (2020) Enhanced fractional chaotic whale optimization algorithm for parameter identification of isolated wind-diesel power systems. IEEE 8
9. Nasri D, Mokeddem D, Bourouba B (2020) Estimation of photovoltaic cells parameters using chaos embedded Salp swarm algorithm. Int J Intell Eng Syst 13(6)
10. Maniraj B, Peer Fathima A (2020) PV output power enhancement using whale optimization algorithm under normal and shading conditions. Int J Renew Energy Res 10(3)
11. Singh A, Shimi S (2017) MATLAB/SIMULINK simulation of PV system based on MPPT in variable irradiance with EV battery as load. In: IEEE international conference on computational intelligence and computing research (ICCIC)
12. Nagaraja Kumari CH, Kaumudi Pravallika VSV (2020) Fuzzy based improved incremental conductance MPPT algorithm in PV System. IEEE
13. Borkar P, Kumar N, Choudhary D (2018) Active power rescheduling of generator for congestion management using whale optimization technique. In: International conference on smart electric drives and power system (ICSEDPS)

A Novel Cuckoo Search with Levy Distribution-Optimized Density-Based Clustering Model on MapReduce for Big Data Environment

T. Gayathri and D. Lalitha Bhaskari

Abstract In recent years, big data have been valuable in a variety of applications, including e-commerce, banking, industries, healthcare, and so on, and it has become difficult to manage the processing of enormous amounts of data. At the same time, data clustering is a significant analytics tool for data mining and is mainly employed for data analytics. Besides, a commonly employed technique to effectually handle big data is the MapReduce model. Several metaheuristic optimizations-based clustering algorithms were presented to manage big data analytics. In this view, this paper presents a novel cuckoo search with levy distribution-optimized (CSLD) density-based clustering (CSLD-ODBC) model on MapReduce for big data environment. The DBSCAN method is a popular density-based clustering model approach that can recognize random forms with changing clusters and noisy data. But, the DBSCAN algorithm encounters a significant problem of setting the optimal parameter values involved in it. To resolve this issue, the presented CSLD-ODBC model makes use of CSLD algorithm to search the parameter space and determine the optimal parameters for the DBSCAN. Besides, a fitness function is derived to decide the cluster count involved in the DBSCAN algorithm. A series of simulations were run to validate the CSLD-ODBC technique's competent clustering outcomes on MapReduce, and the results were analyzed in terms of several performance indicators.

Keywords Big data · Clustering · DBCAN algorithm · Parameter tuning · Cuckoo search

1 Introduction

The enormous development of information results in various issues in computation instead of storage and retrieving the information. Since information gathering is a costlier process, it is important to utilize the information efficiently, and thus, several effective techniques are to be designed to process big data. Big data are commonly

T. Gayathri (✉) · D. L. Bhaskari
Department of CS&SE, AUCE(A), Visakhapatnam, Andhra Pradesh, India

P. S. R. Chowdary et al. (eds.), *Evolution in Signal Processing and Telecommunication Networks*, Lecture Notes in Electrical Engineering 839,
https://doi.org/10.1007/978-981-16-8554-5_35

employed in several areas, including industry, businesses, healthcare, and also they undergo maximum generation of huge quantity of raw information. Machine learning (ML) and data mining technologies are used to solve a major challenge in data analytics. The use of a single personal computer (PC) to do data mining algorithms on a big dataset requires expensive processing. As a result, for analyzing and processing massive data, it is critical to have a highly efficient computing environment [1]. Smart data analytic approaches, such as automated classification, image processing, data fusion, and multi-temporal processing, are becoming more important as a result of big data. The parallelization approaches are established for scaling with the existing data through increasing the computation significantly.

The MapReduce method is a mixture of file system distributed offers an easy and strong situation to handle the computation of large-scale dataset on collection of systems. Under data mining concepts, this method is currently regarded compared to parallelization methods, such as message passing interface (MPI) because of its fault-tolerance system that is needed for the processes which take significant time period, also the straightforwardness of MPI. Commonly, the MapReduce methodology is performed by means of an efficient parallel programming method, named Hadoop. The MapReduce method includes map and reduces functions. The former one is utilized to filter and sort, while the latter one accomplishes a summarized process to produce the result. K-means is an elementary and widespread procedure, which is working to unfold the numerous clustering issues. Though the outcomes of k-means technique are extremely dependent upon early cluster centroids; also, the possibility of stuck into local optimum is maximum. For mitigating this problem, several metaheuristics-depended clustering techniques are developed.

To ensure that the final cluster has a greater density, Bandyopadhyay and Maulik [2] suggested using a genetic algorithm to identify an optimal centroid in the feature space. [3] presents the Bat algorithm (BA) for efficiently clustering data. For dealing with large datasets, the developed approach uses MapReduce to accomplish parallelization. The k-means approach was used to establish the center head in a gravitational search algorithm-enabled (GSA) clustering methodology published in [4]. Kumar et al. [5] developed a clustering approach that mimics gray wolf hunting behavior. [6] presents an adaptive metaheuristic searching-based approach for clustering sensor nodes and placing them in the IoT environment. Pal and Saraswat [7] presented an improved bio-geography-based data clustering approach that outperformed traditional clustering algorithms. In addition, an exponential k-best GSA is presented for determining the best threshold for multi-level picture segmentation. For grouping tweets and assessing client moods, Pandey et al. [8] presented the hybrid cuckoo search (CS) method. Because of their sequential implementation, the previously stated clustering approaches have failed to attain efficiency on big datasets in terms of storage area and computing complexity. Hadoop [9] is a publicly available platform developed by Apache to manage large datasets through distributed processing. MapReduce [10] is a parallelized computing environment that has greatly improved the performance of metaheuristic algorithms for processing large datasets. Gong et al. [11] suggested many distributed evolutionary approaches while recognizing the MapReduce framework's simplicity of use in resolving several high-computing challenges.

To cluster huge datasets, the MapReduce framework provides the hybrid K-PSO technique. To cluster the large dataset, Banharnsakun [12] proposed the MapReduce-based artificial bee colony (ABC) (MR-ABC) approach. To cope with large dispersed evolutionary problems, Tripathi et al. [13] developed the dynamic frequency-based K-BA (DFBPKBA). The bat's occurrence is adaptively modified to improve cluster precision, and the MapReduce framework is used to handle the huge dataset. To assess the efficacy of mining data from big data, Zhao et al. [14] utilized a parallelized version of the k-means method.

This paper introduces an effective cuckoo search with levy distribution-optimized (CSLD) density-based clustering (CSLD-ODBC) model on MapReduce for big data environment. The CSLD-ODBC model employs the CSLD approach to explore the parameter space and find the optimum parameters for the DBSCAN algorithm, which tackles the density-based spatial clustering of applications with noise (DBSCAN) technique's parameter setting issue. In addition, the DBSCAN algorithm uses a fitness function to calculate the number of clusters involved. For ensuring the goodness of the CSLD-ODBC technique on MapReduce, a set of experimentations were carried out, and the experimental results are analyzed with respect to several evaluation metrics.

In short, the contributions to the paper are as follows:

- Propose a new CSLD-ODBC technique for data clustering on MapReduce in big data environment and employ CSLD algorithm for parameter setting of DBSCAN algorithm
- Derive a fitness function for computing the optimum cluster count involved in the DBSCAN algorithm and validate the CSLD-ODBC technique's performance on two benchmark datasets and evaluate the results using several metrics.

The remaining part of the paper is shown as follows: The given CSLD-ODBC approach is then designed in Section 3, and the model's performance is validated in Section 4. Finally, Section 5 brings the process to a close.

2 The Proposed CSLD-ODBC Technique

The presented CSLD-ODBC technique aims to perform effective big data clustering process on MapReduce in big data environment. It makes use of DBSCAN technique to perform data clustering process where the optimal parameter setting of DBSCAN technique is carried out through the CSLD algorithm. The detailed working processes involved in the CSLD-ODBC technique are defined in the subsequent sections.

2.1 Overview of DBSCAN Technique

DBSCAN technique is used to simply determine the random cluster shapes by the detection of high-density hyperspheres and combines hyperspheres to clusters. The DBSCAN involves two important variables, namely hypersphere radius ($\in$) and minimum number of points in every hypersphere (Min_pts). The clustering efficiency of the DBSCAN mainly relies on these two variables, and the pseudocode is provided in algorithm 1. Pattern result (PTR) of a clustering process represents a list $[c_1, c_2, \ldots, c_n]$ in which every individual component c_i is a cluster identifier (0 represents the noise cluster); n is record count in the input database, and indices specify separate record numbers for all records s in S. A mark *seen* is applied for distinguishing among the records that are processed. $N_{\in}(s; S)$ indicates a function which outputs the subset of records in S which are defined by a specific cluster (hypersphere) of radius $\in$ where $s \in S$. The role of *card* ($N_{\in}(s; S)$) returns the cardinality of the set $N_{\in}(s; S)$, whereas sid(s) returns the index in PTR of s in S. But, the DBSCAN model suffers from three limitations and needs to be solved. Initially, it requires appropriate parameter setting to achieve effective outcome, and the manual tuning process is a tedious task. Next, the cluster count could not be managed by the user as DBSCAN could not offer the conception of determining cluster count at the time of initialization. Finally, it could not be applied as a supervised learning technique to carry out the classification process. For resolving these drawbacks, CSLD algorithm is incorporated into the DBSCAN technique and derives a fitness function to determine the cluster count.

2.2 Cuckoo Search with Levy Distribution

CSA algorithm is an efficient metaheuristic algorithm presented in [15]. It is based on the intriguing phenomenon of obligatory interspecific brood parasitism, which occurs in some cuckoo species. This behavioral form depends upon the fact that few species utilize an appropriate host for raising the offspring. It is noticeable that this concept is followed to escape from the parent speculation to raise the offspring and minimize the danger of losing eggs to other species. As a result, it can be accomplished by laying eggs in many nests. The intriguing breeding behavioral pattern is utilized as a model in CSA to create successful metaheuristic algorithms to solve optimization problems. The nest's eggs represent a set of possible solutions to the given optimization issue. It is also considered that every individual nest has only one egg. An intention of CSA is for utilizing the recent and significantly effective solutions related to the cuckoo eggs for replacing the present solution linked to the eggs previously placed in the nest. It is performed in an iterative manner and considerably enhance the solution quality over the iterations, and results in optimal solution.

For making the CSA appropriate for the optimization problem, three ideal rules are formulated as given below:

1. Each cuckoo only lays one egg at a time in a randomly selected nest.
2. The best nests (i.e., high-quality solutions) pass to the following round, ensuring that the best solutions are saved throughout iterations.
3. The existing host nest count remains same. The host can explore a strange egg with a possibility $p_a \in [0, 1]$. It undergoes approximation in such a way that the fraction p_a of the n accessible host, nest gets changed by new nest.

It is noted that the solution of the optimization issue is proportionate to the objective function or its contradictory, based on the minimization or maximization issue.

The steps involved in the CSLD are provided in Algorithm 2. It starts with an initialized population of N host nests, and the *k*th element of the *j*th nest has the beginning values *x* jk (0) = (up kj-low jk) + low jk, where up jk and low jk correspond to the upper and lower bounds of the kth element, respectively, and denotes a uniform arbitrary parameter on the open interval (0,1). They are managed at every individual round to make sure that the values have existed in the searching area. Every round *t*, a cuckoo egg I is picked at random, and new solutions *x i*(*t* + 1) are created. Instead of a basic random walk, it may be accomplished with ease using LFs. The LFs are a form of random walk in which the steps are measured in length and follow a predetermined probability distribution, with isotropic and variable step orientations.

$$x_i^{t+1} = x_i^t + \alpha \oplus \text{levy}(\lambda) \tag{1}$$

where t indicates the present round, $\oplus$ denotes the entry-wise multiplication, and $\alpha > 0$ signifies the step size. The step size computes how far a particle moves through random walk for a predefined iteration count.

Algorithm: Pseudo-code of CSLD algorithm
Start
Objective function $f(x)$, $x = (x\_1,\ldots,x\_d)\,\hat{}\,T$ with $d = \dim(\Omega)$
Produce initial population of N host nests $x\_i(i = 1,2,\ldots,N)$
while (t < Max_Generation) or (termination condition)
Consider a cuckoo (i) arbitrarily using LFs
Determine the fitness value *F_i*
Select a nest amongst N (say, j) arbitrarily
if (*F_i* > *F_j*)
Exchange *j* with new solution
End
A fraction (*p_a*) of worse nests are discarded, and the fresh nests are developed using LFs
Retain the optimal solutions
Sort the solutions and determine the present_best
end while
Post-processing of outcome
End

The transition probability of the LFs in Equation (1) undergoes modulation using the Lévy distribution as given below:

$$\text{levy}(\lambda) \sim g^{-\lambda}, (1 < \lambda \leq 3) \tag{2}$$

which holds infinite variance and mean. The creation of arbitrary numbers using LFs includes two stages. Primarily, an arbitrary direction based on uniform distribution is elected. Next, a series of processes following the selected Lévy distribution is produced.

Here, Mantegna's algorithm is employed for symmetric distribution, and it determines a factor as follows:

$$\hat{\phi} = \left(\frac{\Gamma\left(1 + \hat{\beta}\right). \sin\left(\frac{\pi.\widehat{\beta}}{2}\right)}{\Gamma\left(\left(\frac{1+\widehat{\beta}}{2}\right).\widehat{\beta}.2^{\frac{\widehat{\beta}-1}{2}}\right)} \right)^{\frac{1}{\widehat{\beta}}} \tag{3}$$

where Γ is the Gamma function and $\hat{\beta} = \frac{3}{2}$. It is applied to Mantegna's technique for calculating the step length ς as:

$$\varsigma = \frac{u}{|v|^{\frac{1}{\beta}}} \tag{4}$$

where u and v follow the normal distribution of zero mean and deviation σ_u^2 and σ_v^2, correspondingly. In addition, σ_u follows the Lévy distribution provided in Equation (3) and $\sigma_v = 1$. Afterward, the step size ζ can be determined by

$$\zeta = 0.01\varsigma(x - x_{\text{best}}) \tag{5}$$

where ς is attained based on Equation (4). Lastly, x is altered by $x \leftarrow x + \zeta.\ \Psi$, where Ψ is an arbitrary vector of the dimension of the solution x and follow the normal distribution $N(0, 1)$. The CSA determines the fitness value of fresh solution and compares it with the present one. When enhancement is noticed, the new solution gets replaced by the existing solution. Next, some of the poor nests are discarded and exchanged with fresh arbitrary solutions to raise the exploration of the searching area. This process gets iterated till the specified termination condition is satisfied.

2.3 Fitness Function of CSLD-ODBC Technique

In this section, a fitness function is derived for supervised/unsupervised learning by the use of CSLD technique. The primary difference is that the ground truth target class values of database entries are not employed in the CSLD-ODBC algorithm. The internal and external indices are utilized for determining the clustering efficiency. The class labels are required to calculate the external index, whereas the calculation of internal index does not necessitate ground truth values. The fitness function of the CSLD-ODBC, FF, can be represented as:

$$\mathrm{FF} = f_{\mathrm{Int}} + f_{\mathrm{NK}} \tag{6}$$

where f_{Int} denotes internal clustering index function, and f_{NK} indicates the totaling of the function for controlling the cluster count (f_K) and the noise minimization function (f_{Noise}). Here, Silhouette (SIL) and Davies Bouldin (DB) indices are utilized for f_{Int}. Provided a collection of N data points $s = (s_1, \ldots, s_N)$ allocated to K clusters $C = \{C_1, \ldots, C_i, \ldots, C_K\}$ and the centroids of all clusters m_i, $i = 1, \ldots, K$. $C_i = \left\{s_1^i, \ldots, s_j^i, \ldots, s_{n_i}^i\right\}$ is the ith cluster, where n_i indicates the data point count in C_i. The DB index can be determined by [16]:

$$f_{\mathrm{DB}} = \frac{1}{K}\sum_{i=1}^{K} \max_{i' \in \{1,\ldots,K\}, i' \neq i} \left\{ \frac{e_i + e_{i'}}{\left\|m_i - m_i'\right\|^2} \right\}, e_i = (1/n_i)\sum_{j=1}^{n_i}\left\|s_j^i - m_i\right\|^2 \tag{7}$$

where e_i and $e_{i'}$ are the metrics of scattering in the clusters C_i and C_i', correspondingly. The Silhouette statistic (SIL) can be determined by

$$f_{\mathrm{SIL}} = \frac{1}{K}\sum_{i=1}^{K}\left(\frac{1}{n_i}\sum_{j=1}^{n_i}\frac{b_j^i - a_j^i}{\max\left(a_j^i, b_j^i\right)}\right), \tag{8}$$

$$a_j^i = \frac{1}{n_i - 1}\sum_{k=1, k\neq j}^{n_i}\left\|s_j^i - s_k^i\right\| \tag{9}$$

$$b_j^i = \min_{h \in \{1,\ldots,K\}, h \neq i}\left\{\frac{1}{n_h}\sum_{k=1}^{n_h}\left\|s_j^i - s_k^h\right\|\right\} \tag{10}$$

where a_j^i denotes the average distance among data point s_j^i belong to a cluster C_i, and other data points in C_i and b_j^i are the minimal average distance among the jth data point in the cluster C_i and every data point in the remaining clusters $\{C_h : h \neq i\}$. When the value of DB index becomes low, the clustering efficiency can be improved, whereas the higher value of SIL index leads to effective clustering performance. The CSLD-ODBC algorithm aimed to minimize the fitness value, so f_{Int} is f_{DB} or $-f_{\text{SIL}}$.

2.4 *Parallelization of CSLD-ODBC on MapReduce Architecture*

For employing the presented CSLD-ODBC algorithm on massive dataset, a parallelized version of CSLD-ODBC algorithm by the use of Hadoop MapReduce-based CSLD is developed. It operates on two major stages, namely CSLD-ODBC-Map and CSLD-ODBC-Reduce. The MapReduce model will divide the massive dataset into a small set of chunks and disseminate them regularly between the Hadoop nodes. Next, all the data samples are transformed into a pair of key/values using the record reader. The CSLD-ODBC-Map stage afterward processes the input key/value pair to the cluster centroid parallelly and determines the centroid indexes of all the data objects. The ID, centroid-ID, is used as the key element in this phase, and the distance between data items and the corresponding centroid-ID is used as the value element [17]. In addition, the CSLD-ODBC reduction stage's reduce function combines the determined values with comparable keys and produces the CSLD algorithm's respect fitness function. It represents a single round of the CSLD-ODBC technique, and this procedure gets iterated till the termination criteria are satisfied.

3 Experimental Evaluation

In the proposed model, research is carried on Hadoop cluster along with one name node and seven data nodes. In addition, the results are examined against two datasets, namely PG and US climate reference network (USCRN) dataset [18, 19]. The PG dataset includes ten clusters, namely crafts, music, children, audio, education, health, comics, finance, entertainment, and markets. The USCRN, on the other hand, is a nationwide network of climate monitoring stations that is constantly updated. To assess soil condition, temperature, precipitation level, and wind speed, 114 stations are connected to high-quality equipment.

Table 1 examines the execution time analysis of the CSLD-ODBC model with other clustering techniques on the applied dataset. However, the presented CSLD-ODBC technique has necessitated only a minimum total execution time of 1447.817 s including 183.341, 1154, and 110.476 under partition time, MapReduce time, and merge time, respectively.

Table 2 depicts the speedup factor analysis of the CSLD-ODBC technique with other data clustering algorithms. The table values notified that the presented CSLD-ODBC technique has a speedup factor of 1.818, 1.274, and 1.242 s over the GRID-DBSCAN, DBSCAN-MR, and VDMR-DBSCAN models.

Table 1 Execution time of existing and proposed clustering algorithms on the applied dataset

Methods	Partition time (s)	MapReduce time (s)	Merge time (s)	Total time (s)
GRID-DBSCAN	268.936	2341	128.394	2738.330
DBSCAN-MR	183.341	1624	112.083	1919.424
VDMR-DBSCAN	183.341	1589	114.385	1886.726
MR-VDBSCAN	183.341	1196	112.159	1491.500
CSLD-ODBC	**183.341**	**1154**	**110.476**	**1447.817**

Table 2 Speedup factor results of proposed clustering algorithms on the applied dataset

Methods	Execution time (in secs)
GRID-DBSCAN/CSLD-ODBC	1.818
DBSCAN-MR/CSLD-ODBC	1.274
VDMR-DBSCAN/CSLD-ODBC	1.242

Table 3 investigates the performance analysis of the CSLD-ODBC technique with other models on the applied dataset in terms of Jaccard-Index. The presented CSLD-ODBC technique has reached to effective clustering outcome over the compared methods. The CSLD-ODBC approach achieved a maximum Jaccard-Index of 0.981 At the same time, on the applied USCRN dataset, the CSLD-ODBC method has resulted in a maximal Jaccard-Index of 0.965.

Table 4 inspects the performance analysis of the proposed CSLD-ODBC model with other techniques on the applied dataset with respect to Fowlkes_Mallows_Index of 0.974.

Table 3 Jaccard-Index of existing and proposed clustering algorithms on the applied dataset

Methods	Dataset-PG	Dataset-USCRN
GRID-DBSCAN	0.794	0.760
DBSCAN-MR	0.890	0.840
VDMR-DBSCAN	0.956	0.850
MR-VDBSCAN	0.970	0.945
CSLD-ODBC	**0.981**	**0.965**

Table 4 Fowlkes_Mallows_Index of existing and proposed clustering algorithms on the applied dataset

Methods	Dataset-PG	Dataset-USCRN
GRID-DBSCAN	0.740	0.961
DBSCAN-MR	0.675	0.785
VDMR-DBSCAN	0.943	0.847
MR-VDBSCAN	0.969	0.985
CSLD-ODBC	**0.974**	**0.989**

Table 5 Rand measure of existing and proposed clustering algorithms on the applied dataset

Methods	Dataset-PG	Dataset-USCRN
GRID-DBSCAN	0.814	0.836
DBSCAN-MR	0.783	0.795
VDMR-DBSCAN	0.890	0.846
MR-VDBSCAN	0.920	0.889
CSLD-ODBC	**0.937**	**0.913**

Similarly, on the applied USCRN dataset, the CSLD-ODBC manner has resulted in a higher Fowlkes_Mallows_Index of 0.989.

Table 5 compares the CSLD-ODBC method's performance to those of other approaches on the applicable dataset in terms of Rand measure. The projected CSLD-ODBC algorithm has achieved to effective clustering outcome over the compared techniques. On the applied PG dataset, the CSLD-ODBC method has resulted in a higher Rand measure of 0.937. Similarly, the CSLD-ODBC technique had a superior Rand-measure of 0.913 when utilizing the USCRN dataset, whereas the other algorithms VDMR-DBSCAN, DBSCAN-MR, GRID-DBSCAN and MR-VDBSCAN had Rand measures of 0.836, 0.795, 0.846, and 0.889, respectively.

4 Conclusion

This paper has presented a new CSLD-ODBC model on MapReduce to perform data clustering in big data environment. In a large data context, the proposed CSLD-ODBC approach attempts to execute an effective big data clustering procedure using MapReduce. To address the parameter setting issue of the DBSCAN algorithm, the presented CSLD-ODBC model utilizes CSLD algorithm for searching the parameter space and determine the optimal parameters for the DBSCAN. Moreover, a fitness function is derived for deciding the optimal cluster count involved in the DBSCAN algorithm. For confirming the betterment of the CSLD-ODBC technique on MapReduce, a series of simulations were performed, and the experimental results are analyzed with respect to several evaluation metrics. The collected experimental values demonstrated that the CSLD-ODBC methodology outperformed the other techniques. As a part of future work, MapReduce-based hybridization of optimization algorithms can be developed for clustering large-scale datasets.

References

1. Ulfarsson MO, Palsson F, Sigurdsson J, Sveinsson JR (2016) Classification of big data with application to imaging genetics. Proc IEEE 104(11):2137–2154

2. Tsai C-F, Lin W-C, Ke S-W (2016) Big data mining with parallel computing: a comparison of distributed and MapReduce methodologies. J Syst Softw 122:83–92
3. Maulik U, Bandyopadhyay S (2000) Genetic algorithm-based clustering technique. Pattern Recognit 33(9):1455–1465
4. Ashish T, Kapil S, Manju B (2018) Parallel bat algorithm-based clustering using mapreduce. Networking Communication and Data Knowledge Engineering, Springer, pp 73–82
5. Hatamlou A, Abdullah S, Nezamabadi-Pour H (2012) A combined approach for clustering based on k-means and gravitational search algorithms. Swarm Evol Comput 6:47–52
6. Kumar V, Chhabra JK, Kumar D (2017) Grey wolf algorithm-based clustering technique. J Intell Syst 26(1):153–168
7. Alam S, Dobbie G, Riddle P (2008) Particle swarm optimization based clustering of web usage data. Proceedings of the 2008 IEEE/WIC/ACM international conference on web intelligence and intelligent agent technology, vol 3, IEEE Computer Society, pp 451–454
8. Ebrahimi M, ShafieiBavani E, Wong RK, Fong S, Fiaidhi J (2017) An adaptive metaheuristic search for the internet of things. Future Gener Comput Syst 76:486–494
9. Pal R, Saraswat M (2017) Data clustering using enhanced biogeography-based optimization. 2017 Tenth international conference on contemporary computing, IC3, IEEE, pp. 1–6.
10. Pandey AC, Rajpoot DS, Saraswat M (2017) Twitter sentiment analysis using hybrid cuckoo search method. Inf Process Manag 53(4):764–779
11. Shvachko K, Kuang H, Radia S, Chansler R (2010) The hadoop distributed file system. 2010 IEEE 26th symposium on mass storage systems and technologies, MSST, IEEE, pp 1–10
12. Dean J, Ghemawat S (2008) Mapreduce: simplified data processing on large clusters. Commun ACM 51(1):107–113
13. Gong Y-J, Chen W-N, Zhan Z-H, Zhang J, Li Y, Zhang Q, Li J-J (2015) Distributed evolutionary algorithms and their models: a survey of the state-of-the-art. Appl Soft Comput 34:286–300
14. Banharnsakun A (2017) A mapreduce-based artificial bee colony for large-scale data clustering. Pattern Recogn Lett 93:78–84
15. Tripathi AK, Sharma K, Bala M (2017) Dynamic frequency based parallel k-bat algorithm for massive data clustering (DFBPKBA). Int J Syst Assur Eng Manag, 1–9
16. Zhao W, Ma H, He Q (2009) Parallel k-means clustering based on mapreduce. IEEE International conference on cloud computing, Springer, pp 674–679
17. Khezr SN, Navimipour NJ (2015) Mapreduce and its application in optimization algorithms: a comprehensive study. Majlesi J Multimed Process 4(3):31–33
18. Iglesias A, Gálvez A, Suárez P, Shinya M, Yoshida N, Otero C, Manchado C, Gomez-Jauregui V (2018) Cuckoo search algorithm with Lévy flights for global-support parametric surface approximation in reverse engineering. Symmetry 10(3):58
19. Guan C, Yuen KKF, Coenen F (2019) Particle swarm optimized density-based clustering and classification: supervised and unsupervised learning approaches. Swarm Evol Comput 44:876–896

Taxonomy on Breast Cancer Analysis Using Neural Networks

S. Sri Durga Kameswari and V. Vijayakumar

Abstract Breast cancer is a foremost cause of women deaths from the past decades. Early diagnosis and proper treatment can help to cause a drop in the death rates due to this malignancy. In order to provide an appropriate treatment and to decide the nature of the treatment required, the diagnosis must be carried out in a proper way. Neural networks have turn out to be the trend for classifying now-a-days, and their impact can be felt in every field. They also play a vital role in biomedical signal analysis. In this study, we would like to observe the application of neural networks in analyzing the breast cancer. Out of several types of screening tests performed, we have chosen to analyze the works done on mammographs, ultrasound, and Wisconsin dataset and Histopathology images. The performance of several types of networks employed was studied, and it is observed that the input dataset also plays a crucial role on the performance of the network.

Keywords Breast cancer · CNN · Neural networks · Mammography · Ultrasound

1 Introduction

Nearly, 15% cancer associated deaths in women are due to breast cancer, and it accounts for approximately 24.2% of new cancer diagnoses. It is the most prevalent disease diagnosed in advanced and developing nations [1]. In India, the incidence of breast cancer has steadily increased over the years, and every year so many as one lakh new patients are identified. Cancer is one of the major diseases that is widespread across the world. Early detection of such syndrome can decrease the risk of cancer

S. S. D. Kameswari (✉)
Sathyabama Institute of Science and Technology, Chennai, India
e-mail: durgakameswari.s@gmrit.edu.in

ECE Department, GMRIT, Rajam, India

V. Vijayakumar
ECE Department, Sathyabama Institute of Science and Technology, Chennai, India
e-mail: vijayakumar.ece@sathyabama.ac.in

P. S. R. Chowdary et al. (eds.), *Evolution in Signal Processing and Telecommunication Networks*, Lecture Notes in Electrical Engineering 839,
https://doi.org/10.1007/978-981-16-8554-5_36

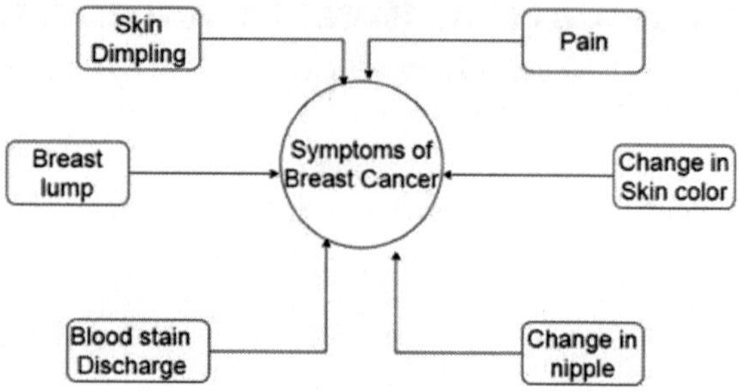

Fig. 1 Symptoms of breast cancer

deaths. Among various cancers that are prevalent, breast cancer stands as a major threat to the female population. Breast cancer usually arises as lobular carcinoma or ductal carcinoma. Lobular carcinoma is a type of cancer originating from the milk-making glands, while ductal carcinoma occurs in ducts that pass milk to the nipple. It is not completely known what causes breast cancer, which makes it hard to understand why one woman will develop breast cancer and another cannot. Some factors are unchangeable like age and family history, while some are controllable like a healthy diet and a good life style. Breast cancer can affect both men and women; however, women are more prone to this cancer. One should consult a physician or opt for a screening if the following symptoms are experienced. The various symptoms experienced by women with breast cancer are breast or chest pain, nipple retraction, fluid discharge from nipples, a lump or swell in breast, dimple on breasts' skin, a rash on nipple. The various symptoms of breast cancer are shown in Figure 1. It is to be recognized that most breast lumps are benign rather than cancerous. Non-cancerous tumors on the breast neither spread beyond the breast nor grow irregularly. Any breast lump or shift should be examined for benign or malignant (cancer) by a health professional and for any potential effect on your future risk of cancer.

They are not life-threatening, but other kinds of benign lumps can help raise a woman's risk of breast cancer. Most women are experiencing very common (benign), (non-cancerous) breast conditions. Currently, most cases of breast tumors are benign. Benign breast conditions are not critical, unlike breast cancers. However, they are significantly associated to increased risk breast cancer later. Sometimes, there may be no signs of any breast changes during a mammogram. However, they may also cause such symptoms as those of breast cancer, and the distinction on the basis of the symptoms alone can be difficult to say. If the findings of your symptoms or mammogram indicate a breast problem, the doctor will suggest a way to diagnose it. Once any symptoms are identified, the person should go for screening. The diagnosis of cancer plays a key role for providing proper treatment.

In recent days, neural networks have spread very vastly, and their impact can be felt in almost all the areas of science and technology. Neural networks can be applied for supporting a decision or to categorize the data into classes. In this paper, the convolutional neural networks (CNNs) approach for detecting and analyzing breast cancer is discussed. The types of breast cancer-related tests are as follows:

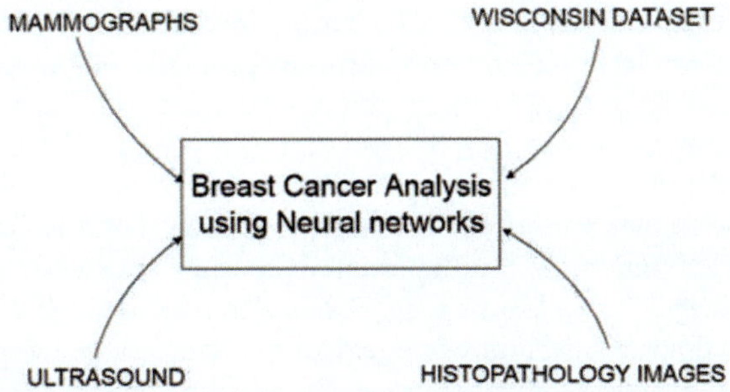

Fig. 2 Various breast cancer input types that can be applied to neural networks

Screening tests: Screening tests are performed on regular basis so that cancer (if present) can be detected early and to give proper treatment [2]. Examples are different imaging tests like mammograms and ultrasound.

Diagnostic tests: Based on the symptoms or screening test results, sometimes physicians may suggest to have diagnostic tests (eg., Biopsy) to gather additional data about the cancer disease to guide proper care and treatment [2].

Monitoring tests: When a person is diagnosed with breast cancer, there will be some tests that are performed to observe the working of therapies. These types of tests may also be used to check for any indications of recurrence [2].

A cancer is uncontrollable and anomalous body cell development. Cancers are generally associated with tumors which represent abnormal growth of body tissues. However, all tumors are not cancerous. The tumors which are non-cancerous are called 'Benign' and cancerous tumors are called 'Malignant'. A tumor can be identified by performing screening tests, but in order to decide whether it is benign or malignant, a biopsy need to be performed. In biopsy, a tissue from the tumor is acquired, and further, diagnosis is done from it. There are many cases in which biopsies are performed for malignant cases. Further, there are chances of true negative and false positive if the diagnosis is not done in a proper manner. In order to minimize the human errors, reduce the number of unwanted biopsies and to support the observer or physician, computer aided diagnosis (CAD) has been implemented. There are many techniques for detecting and analysis of tumors (Fig. 2).

In other words, if we give an input to a trained network, it could tell about its nature. In this paper, the work done in the field related to breast cancer and neural networks has been reviewed considering four different types of inputs: mammograms, ultrasound and Wisconsin dataset, and histopathology images.

2 Mammography

It is indeed a low-dose breast X-ray and one of the most relevant approaches to determine breast cancer during initial stages. Deep learning algorithms also increased the efficiency of other mainstream computer and artificial intelligence substantially.

Deep learning was used by many researchers for investigation of the medical image. Deep learning's progress depends primarily on the availability of several training examples to practice specific image mapping, which provides very better classification accuracy [3]. Saira et al. [4] implemented a CNN on mammograms-MIAS dataset containing normal and abnormal classes. In this work, the network not only classifies normal and abnormal categories, but it is indeed a low-dose breast X-ray and one of the most relevant approaches to determine breast cancer during initial stages. Deep learning algorithms also increased the efficiency of other mainstream computer and artificial intelligence substantially. Deep learning was used by many researchers for investigation of the medical image. Deep learning's progress depends primarily on the availability of several training examples to practice specific image mapping, which provides very better classification accuracy [3].

Saira et al. [4] implemented a CNN on mammograms-MIAS dataset containing normal and abnormal classes. In this work, the network not only classifies normal and abnormal categories but also could sub-classify various types in abnormal category. A model of two CNNs with similar architecture to detect and segment breast microcalcifications within mammography images has been developed by Valvano et al. [5]. Its first CNN is a detector, and the latter was a segmentator. The role of the detector realizes the region of interest of the candidate to be analyzed, while the segmentator categorizes all pixels within the given ROI. This model achieved very high accuracy. Ragab et al. [6] proposed a CAD with two segmentation approaches for classifying benign as well as malignant tumors. In their work, AlexNet is perfectly tuned for two levels, and the last fully connected (fc) layer connects to the vector support (SVM) classifier for enhanced accuracy. Xi et al. [7] trained the CNN on cropped image patches instead of training on full images through transfer learning and data augmentation. They have implemented the patch classifier to identify anomalies in full mammogram images through class activation mapping.

3 Ultrasound

Yap et al. [8] reviewed four lesion detection methods and suggested deep learning approaches to recognize breast ultrasound lesion and studied three distinct approaches: 1. LeNet, a patch, 2. U-Net, and 3. a transmission learning process with an FCN-AlexNet pretrained. Two datasets—Dataset A and Dataset B—have been used to learn deeply, and their output has been compared. Out of these two datasets, one Dataset B is made available for research purposes. For Dataset A, the transfer learning-based FCN-AlexNet has obtained the best results among the different methodologies discussed in this paper; patch-based LeNet has achieved the preeminent results for Dataset B in terms of FPs/image and F-measure. Deep learning approaches should be applied to the different features of a dataset because they are machine learning based; with each and every dataset, a particular model is created. Joo et al. [9] worked on digital ultrasonic images with database of 584 images in which are 300 benign and 284 are malignant images. After preprocessing, they extracted five features and observed the characteristic of each one for benign and malignant cases. An ANN classification system was developed to describe whether the breast tumor is benign or malignant, and the precision for this classification is 91.4%; sensitivity is 92.3%, and specificity is 90.7%. Cao et al. [10] evaluated the performance of some existing advanced detection and classification techniques for breast lesions. It was observed that single shot multi-box detector with input size 300 × 300 (SSD300) achieves best performance in terms of F1 value, average precision rate (APR), and average call rate (ARR) for the lesion regions detection. For the classification task, DenseNet is more plausible. Xu et al. [11] proposed to segment three-dimensional (3D) breast ultrasound images into four primary types of tissues by applying convolutional neural networks (CNNs). They are skin, fibro glandular tissue, mass, and fatty tissue. They designed an eight-layer CNN with input size 128 × 128 to perform pixel labeling in three ultrasound orthogonal image planes including the 1–3 convolution layers, the 1–3 pooling layers, the totally linked (FC) layer, and the softmax layer. The experimental findings have shown that the approach developed possesses the capability to differentiate functional tissues in breast ultra sound images. One more parameter termed as Jaccard similarity index (JSI) that produces an 85.1% value, outclassing their former study in which watershed approach was used with 74.54% JSI value.

4 Wisconsin Dataset

The UCI machine learning repository was used to compile both the Wisconsin diagnostic breast cancer dataset (WDBC) and the Wisconsin prognostic breast cancer (WPBC), respectively. Features are found in a digital image of an aspiration to a fine needle (FNA). The WDBC consists of 569 instances, while WPBC consists of 198 instances and 33 properties, and it has 30 real-valued input functions. Both datasets have nearly identical attributes, but WPBC has three additional features: time, tumor size, and status of the lymph node.

Abdel-Zaher et al. [12] developed a deep belief network (DBN)—neural network (NN) for testing Wisconsin breast cancer dataset (WBCD). They compared two phase and one phase, and it is observed that two phase yields higher classifier accuracy when compared to one phase. The DBN-NN achieved 99.68% of accuracy with sensitivity and specificity being 100% and 99.47%, respectively. Karabatak et al. [13] designed an automatic diagnosis system for detecting breast cancer using a combination of association rules (AR) and neural network (NN). For this analysis, AR was used to curtail the breast cancer database dimension, and NN was used for classification. The proposed system which is a combination of AR and NN performance is observed to be superior compared with NN model alone. In the test stage, to assess the proposed technique performances, threefold cross-validation method was used on the Wisconsin breast cancer database. The exact accuracy rate of the suggested approach is 95.6%. This research showed that the machine could be used to decrease the size of feature volume, and the suggested model is capable of quickly detecting other diseases. Ahmet et al. [14] used six different classifiers, namely support vector machine (SVM), k-nearest neighbor, Naïve Bayes, J48, random forest, and multilayer perceptron methods on the Wisconsin diagnostic breast cancer (WDBC) dataset. An interpretation of the results showed that random forests were the most effective solution with an accuracy of 98.77%. The 'double continuous gradient' method had the highest degree of precision at 98.41% That was evaluated using the tenfold cross-validation method Salama et al. compared multiple classification classifiers by using ROC curves and uncertainty matrixes. The analysis is implemented as a decision tree (J48), a multi-layer perception (MLP), a Naive Bayes (NB), an SMO, and an IBK on three common breast cancer datasets—Wisconsin breast cancer (WBC), Wisconsin diagnosis breast cancer (WDBC), and Wisconsin prognosis breast cancer (WPBC). Experimental research has shown that the classification of MLP and J48 fusion mixtures with the PCA is better than other dataset classifiers. The principal components found by canonical variates analysis are uniformity of cell size, mitoses, clump thickness, bare nuclei, single epithelial cell size, marginal adhesion, bland chromatin, and class. Abien et al. [15] compared six machine learning methods: linear regression, multilayer perceptron, nearest neighbor, softmax regression, and support vector machine on the Wisconsin diagnostic breast cancer (WDBC) dataset. The MLP algorithm is the top-performing algorithm, achieving test accuracy of approximately 99.04%. Şahan et al. [16] created a hybrid system of immunology and neural network for detecting breast cancer. They try to reduce the size of the training data. The data reduction was achieved using an artificial intelligence software, an artificial intelligence software algorithm. By using KNN, we were able to achieve a 99.14% accuracy rate and tenfold cross-validation.

5 Histopathology Images

It is the diagnosis and study of tissue disorders and includes the analysis of tissue and cells. The samples collected from breast biopsy were analyzed under the microscope by a specialist physician called pathologist. For each sample taken, the pathologist sends the doctor a report that gives a diagnosis. The biopsy results will convey the details about where the cancer has initially developed, the type of cancer and whether the cells are cancerous. A biopsy also lets the doctor decide the cancer's grade as histology images are color images, many pretrained neural networks like AlexNet, ResNet50, VGG-16, EfficientNet, and so on. In addition to these, researchers have developed several architectures either by modifying the existing ones or building new architectures for classification. Dabeer et al. [17] applied CNNs on the popular BreakHis database, which has been accumulated from the result of a survey by P&D Lab, Brazil during the span of January 2014 to December 2014. These histopathology images have a RGB (three channel) TrueColor (8 bits-red, 8 bits-green, 8 bits-blue) color coding scheme. This database contains a total of 7009 images of 700X460 pixel resolution. They developed a deep neural network comprising of seven layers where convolution layer, and pool layers are employed alternatively to classify the benign and malignant cases. This method achieved a training accuracy of 93.45% with a test train split of 0.2. Noorul et al. [18] devised a pretrained CNN that can perform segmentation and mitosis identification in breast histopathological images. During the training process, separate (optional) datasets from various sources are used to achieve a generic model. A pretrained CNN has been revised and organized into an FCN and then fine-tuned for mitotic nuclei segmentation. The false positives and true positives that are obtained were then referenced for classification by another hybrid CNN which uses weight transfer in the initial two convolution layers of a pretrained network with customized layers to support the task. This approach minimized the class imbalance created by the significant proportion of non-mitosis. This hybrid CNN hits an F-measure of 0.713 with an area of 76% under the PR curve. Yao et al. [19] proposed a new deep learning model to classify four major categories of cancer: normal tissues, benign lesions, in situ, and invasive. This approach utilizes two types of neural networks; one of them is CNN (DenseNet) and the other being RNN (LSTM) that uses a special perceptron attention system, which fuses the image features of the two distinct neural network types. To increase model accuracy and reduce overfitting, the new shifting normalization approach was combined with a targeted dropout regularization technique. In this model, test time augmentation (TTA) was performed during test phase and was able to obtain optimum results on three datasets: BACH2018, Bioimaging2015, and Extended Bioimaging2015. This indicates the advantage of parallel hierarchical deep neural network using CNN and RNN with attention mechanism. Physicians use automated histopathology image recognition in order to render a rapid diagnosis in a timely manner. Through the assembly of various compact CNNs, Zhu et al. [20] were able to build an inventive compact breast cancer histopathology image classification model that may be employed for assisting breast cancer diagnosis, which makes the job of pathologists simple, and improves the quality of diagnosis. A CNN is established which consists of one main branch and several sub-branches for local and global processing. Our hybrid model is able to establish good representation due to its voting structure and two-branch contact. As the proposed Squeeze-Excitation-Pruning (SEP) block is integrated into the hybrid solution, redundant channels are removed, and the overall channel importance is thus increased. With the same model size, the suggested channel pruning scheme allows for lower risk of overfitting while also maintaining the same accuracy. A variety of models can be built based on the data's partitioning and composition, and how they are assembled to improve generalization ability.

6 Experimental Results

In this work, we have implemented different pretrained networks on a popular breast histology dataset—BACH dataset that contains 400 digital histology images; 100 from each class (i) benign (ii) invasive (iii) in situ and normal. The experimentation was done for multi-class and binary classification. Multi-classification classifies four classes, and for binary class, the normal and benign images are collectively treated

Table 1 Accuracy of various networks

S.No	Name of the network	Overall accuracy (%) (Multi-class)	Overall accuracy (%) (Binary class)
1	SqueezeNet	91.02	94
2	AlexNet	83.76	89.75
3	VGG-16	95.23	96.4

as non-cancerous type, and invasive and in situ classes are combined and treated as cancerous class. We fine-tuned the networks like SqueezeNet, AlexNet, and VGG-16 to suit the dataset, and the obtained results are shown in Table 1.

7 Conclusion

In this study, we tried to observe the work done in the area of analysis of breast cancer using neural networks on different kinds of inputs like mammography, ultrasound and Wisconsin datasets, and histopathology images. It is understood that obtaining a dataset is a crucial part for the works done and also choosing the proper networks and features. Most of the research is done on the openly available datasets which makes the researchers compare their models with the existing ones. Also, in order to reduce the severity of cancer, early diagnosis is essential, and more awareness should be created particularly in rural areas of India.

References

1. Ali I, Waseem W, Saleem K (2011) Cancer scenario in india with future perspectives. Cancer Therapy 8:56–70
2. https://www.breastcancer.org/
3. Gardezi SJS, Elazab A, Lei B, Wang T (2019) Breast Cancer detection and diagnosis using mammographic data: systematic review. J Med Internet Res 21(7):e14464. https://doi.org/10.2196/14464
4. Charan S, Khan MJ, Khurshid K (2018) "Breast cancer detection in mammograms using convolutional neural network." 2018 International conference on computing, mathematics and engineering technologies (iCoMET), Sukkur, pp 1–5. https://doi.org/10.1109/ICOMET.2018.8346384
5. Valvano G, Santini G, Martini N, Ripoli A, Iacconi C, Chiappino D, Latta DL (2019) Convolutional neural networks for the segmentation of microcalcification in mammography imaging. J Healthc Eng 9:9360941
6. Ragab DA, Sharkas M, Marshall S, Ren J. 2019. Breast cancer detection using deep convolutional neural networks and support vector machines. PeerJ 7:e6201
7. Xi P et al (2018) Abnormality detection in mammography using deep convolutional neural networks. IEEE international symposium on medical measurements and applications (MeMeA). https://doi.org/10.1109/MeMeA.2018.8438639

8. Yap MH et al (2018) Automated breast ultrasound lesions detection using convolutional neural networks. IEEE J Biomed Health Inform 22(4):1218–1226
9. Joo S, Yang Y, Moon WK, Kim HC (2004) Computer-aided diagnosis of solid breast nodules: use of an artificial neural network based on multiple sonographic features. IEEE Trans Med Imaging 23:1292–1300. https://doi.org/10.1109/TMI.2004.834617
10. Cao Z, Duan L, Yang G et al (2019) An experimental study on breast lesion detection and classification from ultrasound images using deep learning architectures. BMC Med Imaging 19:51. https://doi.org/10.1186/s12880-019-0349-x
11. Xu Y, Wang Y, Yuan J, Cheng Q, Wang X, Carson PL (2019) Medical breast ultrasound image segmentation by machine learning. Ultrasonics 91:1–9
12. Abdel-Zaher AM, Eldeib AM (2016) Breast cancer classification using deep belief networks. Expert Syst Appl 46:139–144. https://doi.org/10.1016/j.eswa.2015.10.015
13. Karabatak M, Ince MC (2009) An expert system for detection of breast cancer based on association rules and neural network. Expert Syst Appl 36(2):3465–3469
14. Saygili A (2018) Classification and diagnostic prediction of breast cancers via different classifiers. Int Sci Vocational J 2(2):48–56
15. Agarap AFM (2018) "On breast cancer detection." Proceedings of the 2nd international conference on machine learning and soft computing—ICMLSC '18. n. pag. Crossref. Web
16. Şahan S, Polat K, Kodaz H, Günes S (2007) A new hybrid method based on fuzzy-artificial immune system and k-nn algorithm for breast cancer diagnosis. Comput Biol Med 37:415–423
17. Dabeer S, Khan MM, Islam S (2019) Cancer diagnosis in histopathological image: CNN based approach. Inform Med Unlocked 16:100231
18. Wahab N, Khan A (2019) Yeon Soo Lee, Transfer learning based deep CNN for segmentation and detection of mitoses in breast cancer histopathological images. Microscopy 68(3):216–233
19. Yao H, Zhang X, Zhou X, Liu S (2019) Parallel structure deep neural network using cnn and rnn with an attention mechanism for breast cancer histology image classification. Cancers 11:1901
20. Zhu C, Song F, Wang Y et al (2019) Breast cancer histopathology image classification through assembling multiple compact CNNs. BMC Med Inform Decis Mak 19:198. https://doi.org/10.1186/s12911-019-0913-x

Circle-to-circle Fractal Antenna Fed by Microstrip for Tri-band Applications

K. Yogaprasad, M. Nanda kumar, and V. R. Anitha

Abstract A circle-to-circle fractal antenna is invented for six resonance tri-band applications. The proposed antenna consists of circle-to-circle fractal antenna which is investigated up to three iterations, feeding used in this design is microstrip, and FR-4 substrate material with a thickness of 1.6 mm is used for design. The antenna prototype is designed, fabricated, tested, and validated with simulation. The antenna resonates at 4.4 GHz, 6.4 GHz, 8.27 GHz, 10.63 GHz, 12.9 GHz, 14.89 GHz and used for tri-band; those are C-band (4.4 GHz, 6.4 GHz), X-band (8.27 GHz, 10.63 GHz), ku-band (12.9 GHz, 14.89 GHz) and operated with an average gain of 5.5 dB.

Keywords Circle-circle fractal · Tri-band · Microstrip · Vector network analyzer (VNA)

1 Introduction

The multiple resonant frequencies are used to operate in wireless applications, which are mainly deployed to cover the microwave frequencies of the proposed design; those are L, S, C, X, K_u bands. The biggest task is the design of an antenna to operate in multiband applications, that is, two or more bands [1]. This is a big advantage to substitute a multiple antennas by single and decreases a cost of the design but also space constraint.

The microstrip-based antennas were used due to its plain design scenario and contain a patch which was positioned on top of the substrate, bottom of the substrate coated with the conductor which will be acted as a ground and very useful for

K. Yogaprasad (✉)
Department of ECE, Rayalaseema University, Kurnool, India

M. N. kumar
Department of Electronics and Computer Engineering, Sreenidhi Institute of Science and Technology (SNIST), Hyderabad, India

V. R. Anitha
Department of ECE, BMS Institute of Technology and Management, Bengaluru, India

P. S. R. Chowdary et al. (eds.), *Evolution in Signal Processing and Telecommunication Networks*, Lecture Notes in Electrical Engineering 839,
https://doi.org/10.1007/978-981-16-8554-5_37

wireless applications [2–4] due to its a smaller amount weight, low cost. In the last few decades, the growth of microstrip-based antennas was rapidly increasing and overcomes many restrictions as per user applications. The general shapes are used such as circular, triangular, rectangular, and it changes shapes day by day as per user perspective.

In starting days, different methods are introduced to get dual resonance, and those are slot, shorting pin, array, different size of pole, etc. [6, 7]. The size miniaturization is the major task in the present day scenario, and the best solution is to choose fractal-based antenna. A fractal antenna was a type of antenna that uses fractals which means the design that gives self-similarity property that was used for maximizing the length [8]. The fractals can be constructed much smaller in size because it was used for make operational frequency independent of its scale and some of the geometries such as Minkowski fractals, Koch curves, Sierpinski triangles [5].

In this research paper, a design of circle-to-circle fractal antennas with three iterations is introduced to C/X/Ku-band applications and discussed in Sect. 2. The results are analyzed in Sect. 3, and finally, the conclusion is stated in Sect. 4.

2 Design of an Antenna

The circle-to-circle fractal antenna with three iterations is represented in Fig. 1, FR-4 substrate material is used to print the antenna with a thickness of 1.6 mm, copper material is used in ground material with 35um thickness, and microstrip feed system is used in this design. The total occupied dimension of the antenna is 0.82 λ_o* 0.952 λ_o* 0.05 λ_o, where λ_o is a guided wavelength at a free space frequency of 9.5 GHz. Figures 2 and 3 describe parameter representation of the antenna and fabricated prototype. The coordinate system used to create a circle-to-circle monopole is mentioned in Tables 1 and 2 that represent parameters used in this design. The electromagnetic tool is used for design, development of the antenna, and results are validated with vector network analyzer, anechoic chamber.

The design steps are as follows:

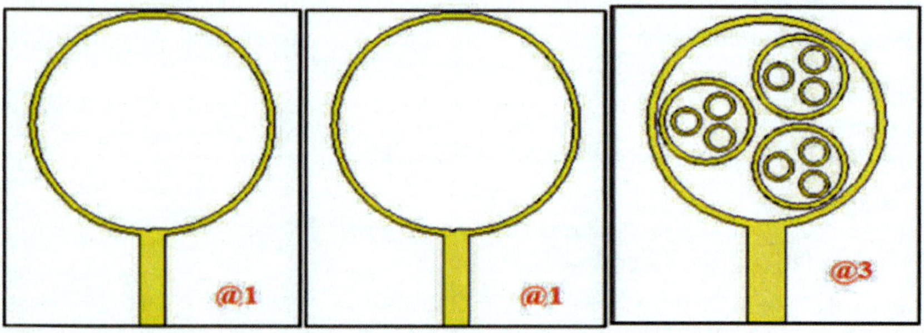

Fig. 1 Proposed antenna structure

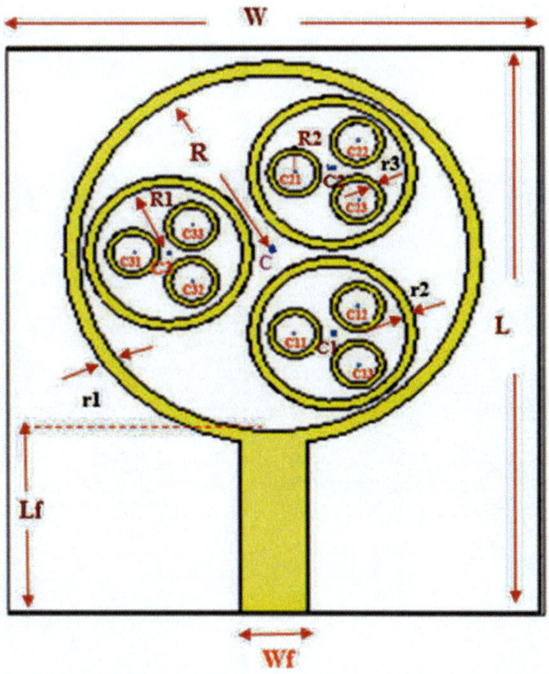

Fig. 2 Parameter representation of proposed design

Step 1: Introduce the circular patch, and radius will be calculated based on Equations 1, 2 at a frequency of 9GHz and optimized value of 9.5mm. Finally, it is converted to ring slot with a thickness of r_1 (@1).

The radius of the circle patch is represented by below equation

$$R = \frac{F}{\left(1 + \frac{2h}{\pi \varepsilon_r F}\left(\ln\left(\frac{\pi F}{2h}\right)\right) + 1.7726\right)^2} \tag{1}$$

where

$$F = \frac{8.791 * 10^9}{f_r \sqrt{\varepsilon_r}} \tag{2}$$

Step 2: With separation of 120, circle is divided into three parts, their centers are C_1, C_2, C_3, and inner radius is R_1 with a thickness of r_2 (@2).

Step 3: Same process is recurring for third iteration (@3).

Microstrip design Equations:

The width (W) and height (h) of the microstrip are [9]

Fig. 3 Fabricated prototype

Table 1 Design parameters

Parameter	Dimension (mm)
Lf	8.5
L	30
W	26
L	3.2
r1/r3/r2	0.7/0.3/0.5
R/R1/R2	9.5/4/1.2

Table 2 Circle positions

Name	Position	Name	Position
C1	(8,18)	C21	(14, 22.2)
C	(13, 18)	C31	(14, 13.8)
C3	(15.8, 13.8)	C23	(17, 20.4)
C11	(6, 18)	C2	(15.8, 22.2)
C13	(9.2, 16.4)	C32	(17, 15.2)
C33	(17, 12.4)	C12	(9.2, 19.4)
C22	(17, 23.8)		

$$\frac{W}{h} = \begin{cases} \frac{2}{\pi}\left\{ \begin{array}{c} a - 1 - \ln(2a-1) + \ldots \\ \frac{\varepsilon_r - 1}{2\varepsilon_r}\left[\ln(a - 1 + 0.39) - \frac{0.1}{\varepsilon_r}\right] \end{array} \right\} & \frac{W}{h} > 2 \\ \frac{8e^a}{e^a - 2}, & \frac{W}{h} < 2 \end{cases} \tag{3}$$

where

$$a = \frac{377\pi}{2Z_o\sqrt[2]{\varepsilon_r}} \tag{4}$$

$$b = \frac{Z_o}{60\sqrt[2]{\varepsilon_r}}\sqrt[2]{\frac{\sqrt[2]{\varepsilon_r} + 1}{2}} + \frac{\varepsilon_r - 1}{\varepsilon_r + 1}\left(0.23 + \frac{0.11}{\varepsilon_r}\right) \tag{5}$$

The length of the microstrip is [10, 11],

$$L_m = n * \lambda_g; \quad n = 1, 3, 5, 7 \ldots \tag{6}$$

3 Results and Analysis

The reflection coefficient results have been represented in Fig. 4 over a frequency range of 4 to 15 GHz. The proposed antenna is of use for C/X/k_u band applications and resonated at six frequencies; those are 4.4 GHz, 6.4 GHz, 8.27 GHz, 10.63 GHz, 12.9 GHz, 14.89 GHz, and their S_{11} values are −13.3 dB, −14.9 dB, −17.7 dB, −35.1 dB, −28.9 dB, −19.7 dB. The impedance bandwidth with respect to -10 dB reference line is 0.8 GHz at 4.4 GHz resonant frequency, 2 GHz at 6.4 GHz resonant frequency, 1.3 GHz at 8.27 GHz resonant frequency, 1.3 GHz at 10.63 GHz resonant frequency, 2 GHz at 12.9 GHz resonant frequency, and 1 GHz at 14.89 GHz resonant frequency. The 4.4 GHz frequency is used for aircraft radar applications, 6.4 GHz for C-band applications, 8.27 GHz for point-to-point wireless application, 10.63 GHz for police radar applications, and 12.9 GHz, 14.89 GHz for Ku-band applications.

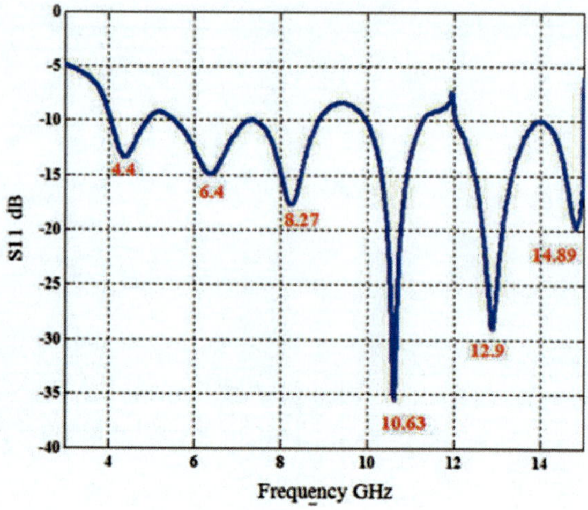

Fig. 4 Reflection coefficient

The VSWR is represented in Fig. 5 over frequency range of 4–15 GHz, and it is matched with S_{11} with respect to -10 dB.

The comparison of measured and simulation results is plotted in terms of reflection coefficient, VSWR is represented in Figs. 6 and 7 over 4–15 GHz frequency range and also observed that there is a small deviation in simulation and measured results due to small fluctuations occurred at connector as well as atmosphere. The comparison of gain in terms of simulation and measurement results is represented in Fig. 8 and observed close match between both results. The simulation gain values are 4.2dBi at 4.4 GHz, 6.2 dBi at 6.4 GHz, 6.6 dBi at 8.27 GHz, 7.1 dBi at 10.63 GHz, 5.2 dBi

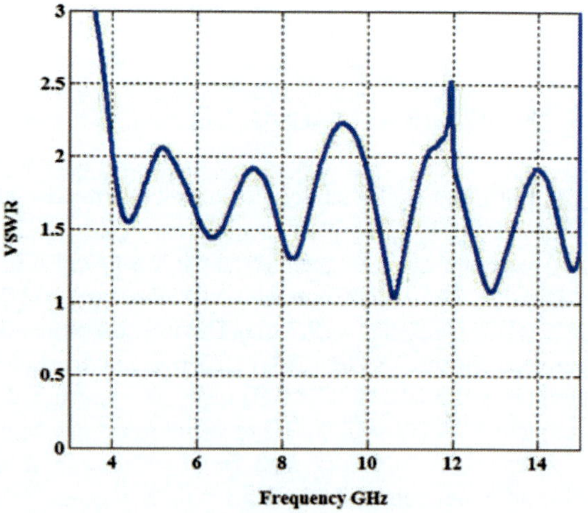

Fig. 5 VSWR

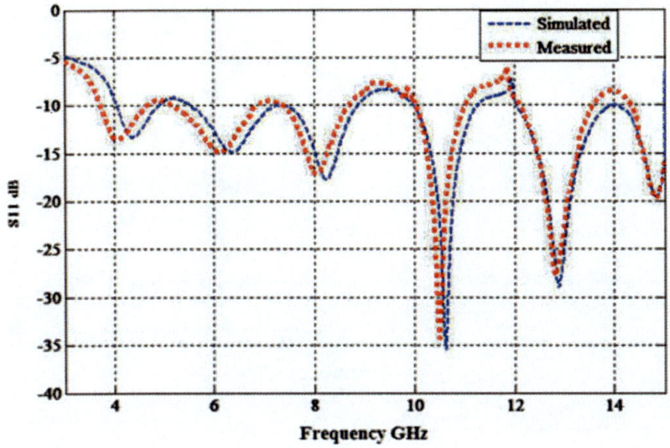

Fig. 6 Reflection coefficient for simulation and measured results

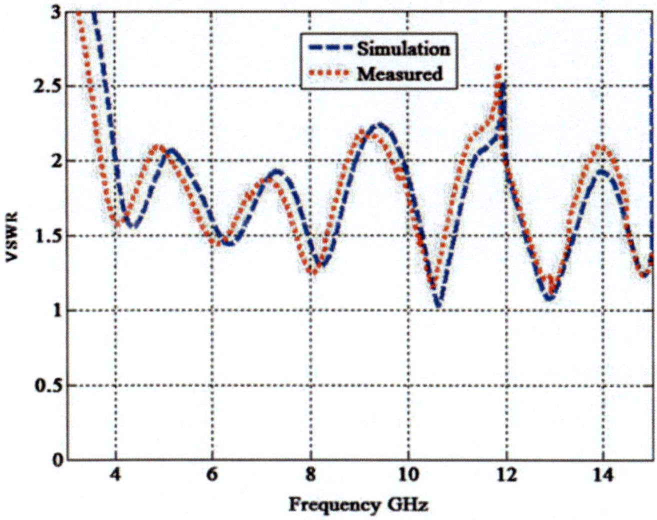

Fig. 7 VSWR for simulation and measured results

at 12.9 GHz, 4.3 dBi at 14.89 GHz, 4.9 dBi at 17.29 GHz and observed an average gain of 5.5 dB, and this gain is useful to operate all communication applications.

The far filed patterns are represented in Fig. 9 at resonant frequencies and observed bidirectional radiation pattern in both E-field as well as H-field. Table 3 describes the clear overview of simulation and measurement results in terms of S_{11}, gain, VSWR.

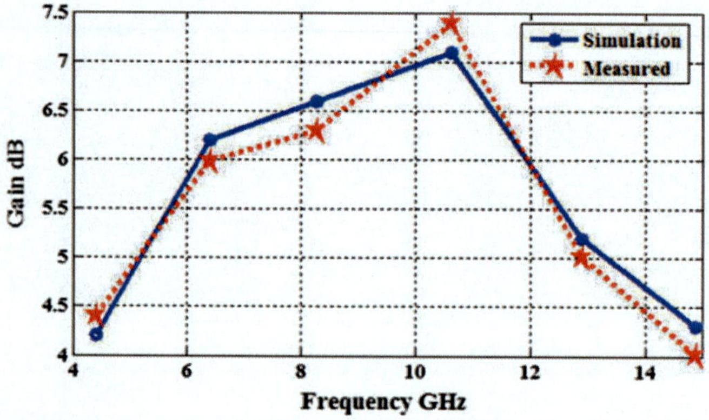

Fig. 8 Gain over frequency

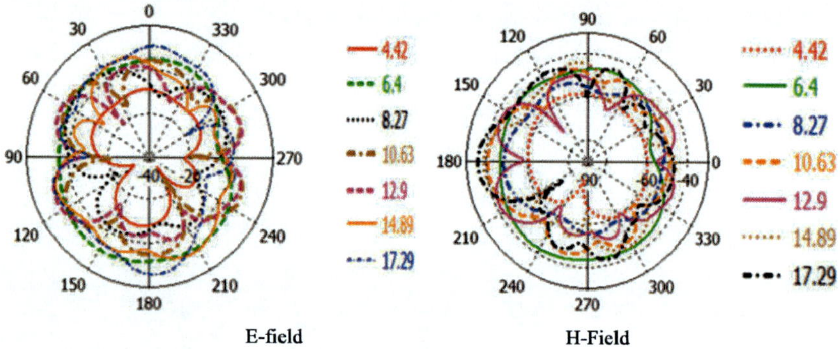

Fig. 9 Radiation patterns

Table 3 Comparison of simulation and measured results

S. no	Resonant frequency (GHz)		S11 (dB)	VSWR	Gain (dB)
1	SR	4.4	−13.3	1.563	4.2
	MR	4.03	−13.35	1.56	4.4
2	SR	6.4	−14.9	1.442	6.2
	MR	6.12	−14.65	1.45	6.02
3	SR	8.27	−17.7	1.297	6.6
	MR	8.09	−17.2	1.3	6.32
4	SR	10.63	−35.1	1.044	7.1
	MR	10.48	−34.25	1.046	7.4
5	SR	12.9	−28.9	1.074	5.2
	MR	12.84	−27.45	1.078	5
6	SR	14.89	−19.7	1.254	4.3
	MR	14.83	−19.3		4.05

4 Conclusion

In this paper, the fractal-based antenna has introduced for tri-band, six resonance applications, and size is $24 \times 30 \times 1.6\ mm^3$. The top of the substrate introduced circle-to-circle fractal with three iterations for radiation and 50Ω microstrip feed used in this antenna. The EM tool is used in this design, simulated the antenna, printed, measured, and validated the results. The measured impedance bandwidth with -10 dB reference line is 0.9 GHz at 4.03 GHz resonant frequency, 1.67 GHz at 6.12 GHz resonant frequency, 1.25 GHz at 8.09 GHz resonant frequency, 0.98 GHz at 10.48 GHz resonant frequency, 1.6 GHz at 12.84 GHz resonant frequency, and 0.7 GHz at 14.89 GHz resonant frequency. Moreover, the outstanding features of proposed antenna are like more resonance, multiband, low-cost fabrication, moderate gain, which are fitted for practical applications.

References

1. Bashar B, Elias Q (2016) Design of broadband circular patch microstrip antenna for Ku-band satellite communication applications. Int J Microw Optic Techn 11:362–368
2. Sung Y (2012) Bandwidth enhancement of a microstrip line-fed printed wide- slot antenna with a parasitic center patch. IEEE Trans Antenna Propag 60:1712–1716
3. Kumar MN, Yogaprasad K, Anitha VR (2020) A quad band Sierpenski based fractal antenna fed by CPW. Micro-wave Opti Techn Lette 2:893–898
4. Sedghi MS, Moghadasi MN, Zarraabi FB (2016) A dual band fractal slit antenna loaded by jerusalem crosses for wireless plus WiMAX communications. Prog Electro-magnet Res Lett 61:19–24
5. Nandakumar M, Shanmugnantham T (2019) Broad-band i-shaped siw slot antenna for v-band applications. Appl Comput Electroma Soc 34:1719–1724
6. Dinesh M, Kumar MN, Balachandra K (2018) Micro-strip feed reconfigurable antenna for wideband applications. J: Lecturer Notes Electric Eng 665–671
7. Nandakumar M, Shanmugnantham T (2018) Design of substrate integrated waveguide back to back π-shaped slot antenna for 60GHz applications. J: Lecturer Notes Electric Eng 215–224
8. Kumar MN, Shanmugnantham T (2019) Broad band substrate integrated waveguide venus shaped slot antenna for v-band applications. Micro-wave Opti Techn Lett 61:2342–2347
9. Nandakumar M, Shanmugnantham T (2019) Back to back Pi-shaped slot with siw cavity-backed antenna for 60GHz applications. Int J Microwave Optical Technol 14:402–408
10. Yogaprasad K, Anitha VR (Jan 2018) "Microstrip feed dumbbell shaped patch antenna for multiband applications." Microelectronics, electromagnetics and telecommunications. Lecture notes in electrical engineering, Springer, Singapore, vol 471
11. Yogaprasad K, Anitha VR (2019) "CPW fed Hexa to hexa fractal antenna for multiband applications." Lecture notes in electrical engineering, Springer, Singapore

A Comprehensive Study of Linear Antenna Arrays Using Nature-inspired Algorithms

Nagavalli Vegesna, G. Yamuna, and Terlapu Sudheer Kumar

Abstract With the increasing needs of the technologies, the role of wireless communications has become more prominent in different applications like RADAR. The ability to use RADARs can be extended by the use of antenna arrays. The efficiency of the communication systems will be increased with reduction in the side lobe levels, nulls in the desired directions, and also beam shaping like pencil beams and sector beams. Various nature-inspired algorithms are used to achieve improvisation in scan efficiency by the action of pencil and sector beams over the target range.

Keywords Linear antenna array · Optimization · Side lobe level

1 Introduction

In the increasing technological world of communications, there is great need of antennas with large directivity [1]. A single antenna itself is not enough to produce high directivity. Hence, we go for antenna arrays [2]. Because of their high collective gain, the use of array of antennas for long range communication has proven effective. Therefore, the idea of using array of antennas gained great importance. Antenna arrays are graded into linear, circular, and hexagonal based on their geometry. Array of antennas is well known broadly for their greater directivity [3]. The greater directivity for antenna arrays is generally achieved by narrow radio wave beams. Parameters like the physical position of elements and their excitations affect the radiation pattern shape. The antenna array forms radiation patterns with the main beam. It also controls the angular placement, and this is shown by the radiation patterns. Also achievable are different other features such as choosing sharpness of

N. Vegesna (✉) · G. Yamuna
Annamalai University, Chidambaram, Tamil Nadu, India
e-mail: nagavalli.v@srkrec.ac.in

T. S. Kumar
Shri Vishnu Engineering College for Women, Bhimavaram, AP, India
e-mail: profsudheer@ieee.org

P. S. R. Chowdary et al. (eds.), *Evolution in Signal Processing and Telecommunication Networks*, Lecture Notes in Electrical Engineering 839,
https://doi.org/10.1007/978-981-16-8554-5_38

the beam by considering array length, producing pencil beam patterns [4] by generating unique patterns, because of which the beam shape changes to sector beam, staircase beam, ramp beam, cosine beam, etc., [5, 6]. Based on the different application fields, such shaped beams are scrutinized with reference to their amplitude, phase, and spacing. In cellular systems, the synthesis of sector beams is performed in phased antenna arrays using the technique of cell sectorization [7, 8] and can also be accomplished by combining amplitude and phase synthesis [9]. To provide higher signal-to-noise ratio (S/N), an alternative effective idea of null and beam steering has been prominent. For null steering of the array of antennas for a variety of applications, various other algorithms are being used [10, 11]. Previously, conventional methods like Taylor's method, fast Fourier transform (FFT), Dolph—Chebyshev method [12], matrix pencil method [13], eigen value method [14] are used. The problem of using these traditional methods is that they consume more time and are mathematically complicated. Hence, various metaheuristic optimization algorithms that are inspired from the nature become as prominent alternatives for optimization of antenna array. Examples of such algorithm are firefly algorithm, flower pollination algorithm, bat algorithm, genetic algorithm, ant lion optimization algorithm, cat swarm optimization algorithm, particle swarm optimization algorithm, cuckoo search algorithm, differential evolution algorithm, etc.

2 Problem Statement

In RADAR applications to detect the targets, both sector and pencil beams are required. First sector beams are sent to avoid multiple scans; once the target is detected to know the correct position of the target, pencil beams are sent for more efficiency. So, the beam-forming techniques and nulls in desired positions are important in RADAR applications. To meet these requirements, it is of interest to investigate in beam-forming techniques and to reduce SLL using nature-inspired algorithms in solving complex nonlinear equations.

Array factor formulation of Antenna.

Uniformly spaced linear array with isotropic elements in even number is presented in Fig. 1, and the array factor (AF) can be expressed as,

$$\mathrm{AF}(\phi) = 2\sum_{n=1}^{N} I_n \cos[kx_n \cos(\phi) + \varphi_n]$$

Fig. 1. 2 N elements in a linear array with symmetricity

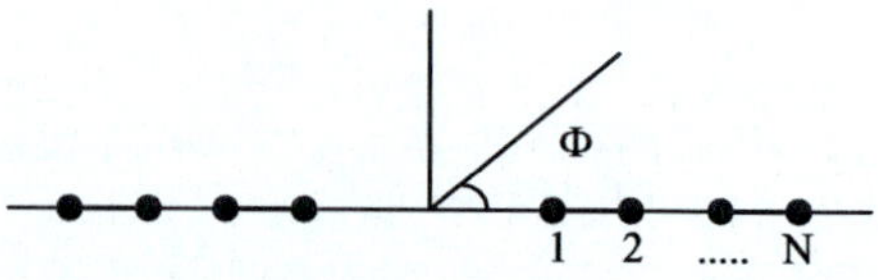

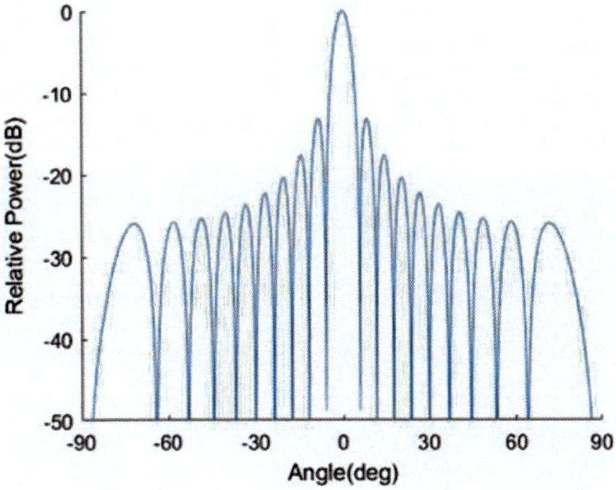

Fig. 2 Template of radiation pattern

where

k	is the wave number
φ_n	is the excitation in phase
I_n	is the excitation in amplitude
x_n	is excitation in space

In Fig. 2, the side lobes are very high which may affect the detection of targets in RADAR applications. So, the reduction of side lobes, control over main beam width with different shapes, and nulls are main objectives to be achieved. These objectives are likely to be obtained by using parameters like spacing, phase, amplitude between the elements. A few or sometimes all the parameters can be varied and controlled to obtain above objectives. From these objectives, optimization problems are differentiated based on.

1. Single objective with single variable
2. Multiple objective with single variable
3. Single objective with multiple variable
4. Multiple objective with multiple variable.

In any application like RADAR, multiple objective is mainly preferred. But, single objective function is easier to implement than multiple objective function. So, single fitness function is chosen which covers multiple objective.

By considering parameters like amplitude, spacing, phase as variables for array factor and given as input to the fitness function. Optimization algorithm takes fitness function as input and evaluates. Optimization algorithm gives the optimized result which is compared with desired pattern. If there is mismatch between best result and desired pattern, the optimization algorithm has to rerun to get no error. This process is shown in Fig. 3.

Flower pollination algorithm (FPA) is proposed by Xin-She-Yang which was popular in recent years [15]. FPA is inspired by pollination process of plants which

Fig. 3 Block diagram depicting the use of optimization algorithm for array pattern synthesis

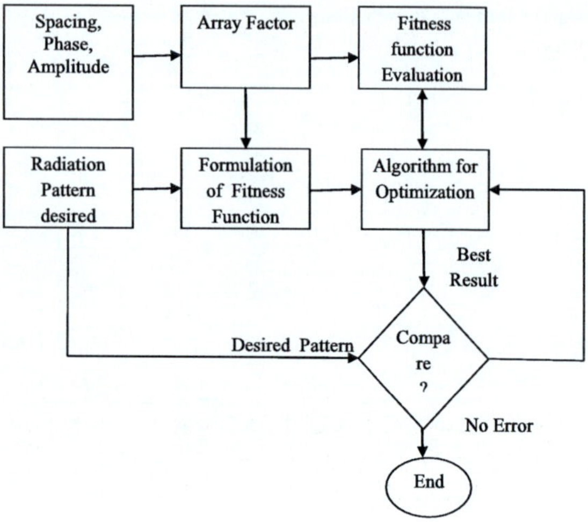

occurs either biotically or abiotically, but the majority are from biotic process. It has implemented in different fields like image compression, antennas, structural engineering [16]. In RADAR applications, there is a challenge in detecting the target due to interference caused by side lobe levels. To reduce side lobe levels, FPA is well suited when compared with GA, PSO, BBO [17, 18].

FPA has better depth of the nulls in specified direction when compared with evolutionary nature-inspired algorithms like PSO, ACO, and CSO [19]. FPA is best suited for different array patterns like circular array and has best result when compared with GA. [20]. But, FPA has certain limitations like slow convergence and local minima which are rectified by EFPA and BFPA [21, 22].

In 2010, Yang proposed bat algorithm. Bats which generally prey at night, find their prey, and avoid obstacles by using a technique called echolocation. Based on this behavior of bats, Yang proposed the bat algorithm [23]. The bat algorithm can suppress the side lobe levels and can precisely place single, multiple, and broad nulls. The beam foming is much faster in bat algorithm than compared to that of APSO and GA [24].

The genetic algorithm (GA) is well-used algorithm from many years to solve complex problems in optimization. GA is best in aperiodic, sparse linear array, but it has the problem of local minima, and to rectify this, GA is hybridized with a local search method [25]. GA can solve complex radiation pattern of planar array which is difficult with conventional methods. By using GA, any arbitrary array can be optimized without any coding and simplifies computational time. Non-uniform circular array gets good side lobe level reduction when compared with uniform case using GA [26].

By studying the flashing patterns of the fireflies, Xin-She-Yang proposed firefly algorithm. The fireflies regroup automatically by using local or global. Firefly algorithm yields better results of reduced side lobe levels and null steering for linear antenna arrays compared to PSO, GA, TS [27]. The major drawback of firefly algorithm is its slow rate of convergence which can be improved by enhanced firefly algorithm [28].

The hunting nature of the ant lions have inspired Mirjalili to propose an optimization algorithm, namely ant lion optimization algorithm [ALOA]. It makes use of diverse search spaces to solve any sort of optimization problem including the constrained and unconstrained. The advantage of ALOA is that it has very few parameters that are to be tuned which are the greatest benefit to solve problems [29]. ALOA is widely considered due to its capability in providing the LAA enhancement to a better extent compared to the synthesis results of other optimization techniques [30].

It is found to be an optimum solution in the field of engineering for solving various optimization problems such as gear train design, three-bartruss design [31]. In the year 1995, one more nature-inspired algorithm is proposed by Eberhart and Kennedy which is called the particle swarm optimization algorithm (PSOA). This algorithm is developed on the basis of the behavior of fishes during prey collection [32]. Having a simple structure and easy to implement make the PSO more useful for practical applications. Using PSO, we can achieve improvisation in side lobe level, reduction of nulls [33], and extended control of array pattern. Using this algorithm, discrete parameters can be optimized, and various side lobe heights can be created [34]. This is the reason why many design issues in the concept of linear arrays implement particle swarm optimization.

In 1986, Glover has proposed a heuristic algorithm to find the global minimum of a function for both linear and nonlinear arrays is the tabu search algorithm (TSA). This algorithm makes use of the data from the previous steps by retaining it and helps in producing new solutions [35] in the phase of searching. In linear array antennas, tabu search is very efficient as it can generate radiation patterns with a maximum reduction in the side lobe level at a fixed beam width [36].

In 2009, Yang and Deb have proposed cuckoo search algorithm (CSA) which is inspired by the process of breeding in certain species of cuckoo. These cuckoo species select the nests of different birds to lay their eggs for breeding. In recent times, this algorithm have proved to give best optimized results. They have good convergence rates compared to the conventional particle swarm and genetic algorithms [37]. Its greatest advantage is to abandon worst solutions and randomly generate new ones [38]

Differential evolution algorithm was introduced in the year 1996 by Storn and Price. It is a robust, simple, and efficient evolutionary algorithm. Synthesis of unequally spaced uniform amplitude arrays using differential algorithm is achieved by using global optimization method to reduce side lobe levels. [39]. When compared with GA and PSO, differential evolution algorithm has improved directivity and side lobe levels for scannable circular antenna arrays [40].

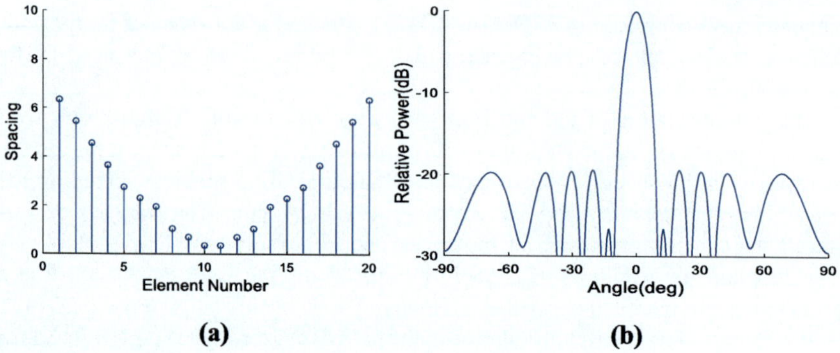

Fig. 4. a Optimized spacing, **b** Radiation pattern for 20 elements using FA

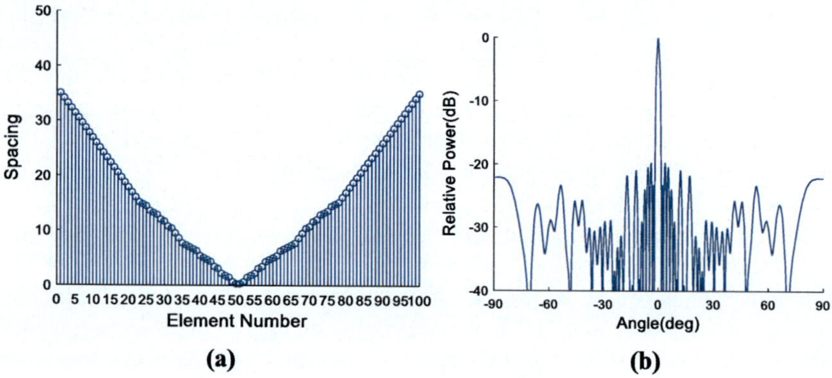

Fig. 5. a Optimized spacing, **b** Radiation pattern plot for 100 elements using FA

3 Results

The synthesis of array antennas is obtained by using firefly algorithm to get reduced side lobe levels in the radiation pattern when compared with general radiation pattern. Optimized spacing between the elements with 20 and 100 is obtained with reduced side lobe levels as shown in Figs. 4 and 5.

4 Conclusion

The side lobe levels and nulls play vital role in the mechanism of antenna arrays. Many nature-inspired algorithms like particle swarm, flower pollination, firefly, bees algorithm, cuckoo search are used for these parametric enhancements. The best possible side lobe level and nulls can be obtained by using these algorithms and also by suitable way of choosing variables and their parameters.

References

1. Manica L, Rocca P, Massa A (2008) On the synthesis of sub-arrayed planar array antennas for tracking radar applications. IEEE Antennas Wirel Propag Lett 7:599–602
2. Di Palma L et al (2016) Radiation pattern synthesis for monopulse radar applications with a reconfigurable transmitarray antenna. IEEE Trans Antennas Propagation 64(9):4148–4154
3. Goudos SK, Diamantoulakis PD, Karagiannidis GK (2018) Multi-objective optimization in 5g wireless networks with massive MIMO. IEEE Commun Lett 22(11):2346–2349
4. Vicente-Lozano M, Ares-Pena F, Moreno E (2000) Pencil-beam pattern synthesis with a uniformly excited multi-ring planar antenna. IEEE Antennas Propag Mag 42(6):70–74
5. Chatterjee A, Mahanti GK, Mahapatra PRS (2011) "Generation of phase-only pencil-beam pair from concentric ring array antenna using gravitational search algorithm." 2011 International conference on communications and signal processing. IEEE
6. Zheng L et al (2011) "Synthesis of pencil-beam patterns with time-modulated concentric circular ring antenna arrays." Progress Electromagnetics Res 373
7. Sabharwal A, Avidor D, Potter L (2000) Sector beam synthesis for cellular systems using phased antenna arrays. IEEE Trans Veh Technol 49(5):1784–1792
8. Ksienski A (1960) Maximally flat and quasi-smooth sector beams. IRE Trans Antennas Propag 8(5):476–484
9. Khzmalyan AD, Kondrat'yev AS (1996) Phase-only synthesis of antenna array amplitude pattern. Int J Electron 81(5):585–589
10. Akdagli K (2001) Null steering of linear antenna arrays using a modified tabu search algorithm. Process Electromagnetics Res 33:167–182
11. Ismail TH, Dawoud MM (1991) Null steering in phased arrays by controlling the element positions. IEEE Trans Antennas Propag 39(11):1561–1566
12. Karimkashi S, Kishk AA (2009) Focused microstrip array antenna using a Dolph-Chebyshev near-field design. IEEE Trans Antennas Propag 57(12):3813–3820
13. Liu Y, Nie Z, Liu QH (2008) Reducing the number of elements in a linear antenna array by the matrix pencil method. IEEE Trans Antennas Propag 56(9):2955–2962
14. Sahalos J, Zimourtopoulos P (1980) Simultaneous optimization of an index and determination of the polarization of arrays of nonparallel wire antennas by the eigenvalue method. IEEE Trans Antennas Propag 28(5):724–730
15. Kaur G, Singh D, Kaur M (2013) Robust and efficient 'RGB' based fractal image compression: flower pollination based optimization. Int J Comput Appl 78(10)
16. Nigdeli SM, Bekdaş G. Yang XS (2016) Application of the flower pollination algorithm in structural engineering. In Metaheuristics and optimization in civil engineering. Springer, Cham, pp 25–42
17. Singh U, Salgotra R (2018) Synthesis of linear antenna array using flower pollination algorithm. Neural Comput Appl 29(2):435–445
18. Chakravarthy VS, Terlapu SK, Chowdary PSR, Rao TV, Satapathy C (2018) On the convergence of synthesis of desired nulls from circular arrays using flower pollination algorithm. In Proceedings of 2nd international conference on micro-electronics, electromagnetics and telecommunications. Springer, Singapore, pp 475–489
19. Saxena P, Kothari A (2016) Linear antenna array optimization using flower pollination algorithm. Springerplus 5(1):306
20. Vedula VSSS, Paladuga SR, Prithvi MR (2015) Synthesis of circular array antenna for sidelobe level and aperture size control using flower pollination algorithm. Int J Antennas Propag

21. Singh U, Salgotra R (2017) Pattern synthesis of linear antenna arrays using enhanced flower pollination algorithm. Int J Antennas Propag
22. Salgotra R, Singh U (2018) A novel bat flower pollination algorithm for synthesis of linear antenna arrays. Neural Comput Appl 30(7):2269–2282
23. Yang XS (2012) Bat algorithm for multi-objective optimisation. arXiv preprint arXiv:1203.6571
24. Van Luyen T, Giang TVB (2018) Null-steering beamformer using bat algorithm. Appl Comput Electromagnetics Soc J 33(1)
25. Lommi A, Massa A, Storti E, Trucco A (2002) Sidelobe reduction in sparse linear arrays by genetic algorithms. Microw Opt Technol Lett 32(3):194–196
26. Panduro MA, Mendez AL, Dominguez R, Romero G (2006) Design of non-uniform circular antenna arrays for side lobe reduction using the method of genetic algorithms. AEU-Int J Electron Commun 60(10):713–717
27. Kaur K, Banga VK (2013) Synthesis of linear antenna array using firefly algorithm. Int J Sci Eng Res 4(8):601–606
28. Singh U, Salgotra R (2019) Synthesis of linear antenna arrays using enhanced firefly algorithm. Arab J Sci Eng 44(3):1961–1976
29. Saxena P, Kothari A (2016) Ant lion optimization algorithm to control side lobe level and null depths in linear antenna arrays. AEU-Int J Electron Commun 70(9):1339–1349
30. Talatahari S (2016) Optimum design of skeletal structures using ant lion optimizer. Iran Univ Sci Technol 6(1):13–25
31. Kennedy J (2010) Particle swarm optimization. Encyclopedia of machine learning, 760–766
32. Khodier MM, Christodoulou CG (2005) Linear array geometry synthesis with minimum sidelobe level and null control using particle swarm optimization. IEEE Trans Antennas Propag 53(8):2674–2679
33. Robinson J, Rahmat-Samii Y (2004) Particle swarm optimization in electromagnetics. IEEE Trans Antennas Propag 52(2):397–407
34. Terlapu SK, Raju GSN (2016) Synthesis of linear array for sidelobe redu-ction using particle swarm optimization. Int J Control Theory Appl 9(23):165–173
35. Glover F, Taillard E (1993) A user's guide to tabu search. Ann Oper Res 41(1):1–28
36. Merad L, Bendimerad F, Meriah S (2008) Design of linear antenna arrays for side lobe reduction using the tabu search method. Int Arab J Inf Technol (IAJIT) 5(3)
37. Yang, X. S., & Deb, S. (2009, December). Cuckoo search via Lévy flights. In 2009 World Congress on Nature & Biologically Inspired Computing (NaBIC) (pp. 210–214). IEEE.
38. Khodier M (2013) Optimisation of antenna arrays using the cuckoo search algorithm. IET Microwaves Antennas Propag 7(6):458–464
39. Yang S, Gan YB, Qing A (2002) Sideband suppression in time-modulated linear arrays by the differential evolution algorithm. IEEE Antennas Wireless Propag Lett 1:173–175
40. Panduro MA et al (2009) A comparison of genetic algorithms, particle swarm optimization and the differential evolution method for the design of scannable circular antenna arrays. Progress Electromagnetics Res 13:171–186

Automatic Modulation Recognition of Analog Modulation Signals Using Convolutional Neural Network

N. Venkateswara Rao and B. T. Krishna

Abstract In this study, different deep learning approaches are applied to automatic modulation classifier (AMC). AMC can change its characteristics based on channel conditions. It gained importance in crowded spectrum due to its numerous advantages. The primary goal of the study is to guide researchers to choose appropriate technique based on channel conditions and modulation classes from pool of available deep learning (DL) techniques. In this paper, simulations are carried on analog signals using CNN. Furthermore, merits and demerits of proposed approach are discussed.

Keywords AMC · SDR · CR

1 Introduction

Development of sophisticated information systems for civilian and military applications in congestion is a difficult thing [1]. During such imperfect situations, advanced systems are required for monitoring signal processing at regular intervals. In last decade, huge number of innovations are being done in communications [2]. AMC is one such innovation to enable higher transmission reliability and transmission rate by altering the modulation format according to channel characteristics. Implementation of AMC requires the receiver to have the intelligence of the modulation technique to demodulate it [3]. To achieve this, additional data is included in each frame, so that the receivers will know changes in modulation technique and respond according to it. But spectrum efficiency is greatly affected in this case [4].

To combat this problem, automatic modulation recognition (AMR) was proposed for recognizing the type of modulation without requiring additional data. So, AMR becomes an essential part of the receiver especially for future adaptive radio systems [5]. Figure 1 shows block diagram of adaptive modulation system (AMS).

In general, there are three groups of typical AMC approaches: 1. time–frequency analysis (TFA), 2. decision-theoretic methods and 3. pattern recognition solutions.

N. V. Rao (✉) · B. T. Krishna
Department of Electronics and Communication Engineering, JNTU Kakinada, Kakinada, Andhra Pradesh, India

P. S. R. Chowdary et al. (eds.), *Evolution in Signal Processing and Telecommunication Networks*, Lecture Notes in Electrical Engineering 839,
https://doi.org/10.1007/978-981-16-8554-5_39

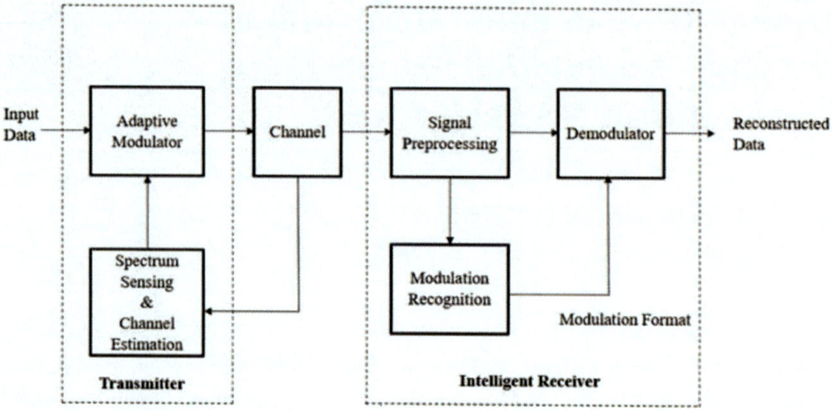

Fig. 1 Block diagram of AMS [6]

Amplitude and frequency variations of the signal can be efficiently traced by TFA, but it is unable to trace out the phase variations in the signal [7]. Decision-theoretic approach is carried on likelihood basis [8]. Likelihood-based [LB] classifiers perform hypothesis testing that leads to optimal solutions [9].

In case of pattern recognition type, classification is broadly categorized into two types: 1. feature extraction and 2. pattern recognizer [10–15]. Features are extracted from the signal samples in feature extraction. In pattern recognizer, features are processed (usually spectrograms), and they will be trained for classification purpose. This article mainly focuses on pattern recognizer techniques. Various authors made use of different modulation classes like PSK, quadrature PSK (QPSK) and QAM. They used higher-order cumulants (HOCs), quadrature (Q) and in-phase (I) to represent the modulation signal. In this paper, we considered analog modulation signals rather than digital modulation classes.

The rest of this paper is summarized as follows. Section 2 discusses the deep learning (DL) approach. Section 3 describes the CNN approach. Section 4 gives insights into simulation results. Section 5 concludes the work.

2 Deep Learning Approaches

DL is a part of machine learning. Almost all DL algorithms are regarded as deep neural networks (DNNs). It is capable of learning unstructured data. Figure 2 shows the structure of artificial neural network (ANN). In DL, all layers are interconnected, and they learn activities layer by layer. DL uses a greater number of hidden layers for efficient feature extraction [16]. Reasons behind the demand for DL in this era are extended learning capability, high performance and recent developments in machine learning (ML).

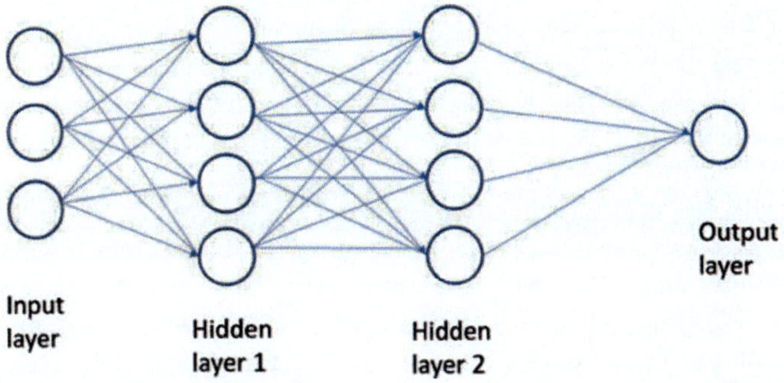

Fig. 2 Basic structure of ANN

Typical DL approaches are long short-term memory (LSTM), convolutional neural network (CNN), autoencoder (AE) and recurrent neural network (RNN). LSTM is one of the most important RNNs. LSTM cells have an internal memory to learn efficiently. Input gate (I), output gate (O) and forget gate (F) are utilized by LSTM architectures [17]. Based on input data and previous state, gate weights are updated. This gating approach helps LSTM cells to have long-term learning. LSTM cells can learn temporal dependencies efficiently. It is used in various fields like speech recognition and machine translation [18].

Recently, deep learning is applied in various fields like image classification [19], medical [20], authentication [21] and speech recognition [22]. In [23], they proposed deep CNN for image classification. In [24], they proposed 3-D input CNN model for P300 signal detection. In [25], they provided user authentication based on mouse movements. For feature extraction, they used CNN, RNN and hybrid model combining both CNN and RNN. Layer-wise relevance propagation (LRP) algorithm is proposed to calculate the relevance scores for mouse movements. In [26], they combined LSTM with RNNs for improved speech recognition. In [27], deep learning algorithms are used in various tasks, such as detection, segmentation and classification of various microscopy images.

In [24], authors proposed a new method Bhattacharyya distance-based feature selection (BDFS) algorithm for feature selection. The dissimilarity between probability density functions (PDFs) acts as a criterion for feature selection. The proposed classifiers used are RBFN, CNN and sparse autoencoder. MPSK and 16QAM are used as modulated signals. They have chosen frequency selective fading and AWGN with SNR values ranging (0–15) dB. Their feature selection model achieved an accuracy of 100% at 15 dB.

In [25], authors presented AMC based on VGGNet model. VGGNet is a CNN with a greater number of hidden layers. Initially, sampled values of signals are converted into gray images, and they will be trained on VGGNet. 4ASK, 2PSK, 4PSK, 2FSK, 4FSK and 8FSK are used as modulated signals. They observed their performance under different SNRs from (−5 dB to 10 dB). Their proposed model got an accuracy of 98% even at low SNR (−2 dB).

In [26] for enlarging the dataset, they proposed auxiliary classifier generative adversarial networks (ACGANs). AlexNet is used as classifier. 4ASK, BPSK, OQPSK, QPSK, 16QAM, 8PSK, 64QAM and 32QAM are used as modulated signals. Contour stellar image is used to train the algorithm. They compared original dataset and enlarged dataset. They found that there is a 6% rise in classification accuracy while considering ACGAN-based dataset.

In [27], authors considered carrier phase offset (PO) effect which most of them neglect while dealing with AMR. They proposed a CNN model and verified with and without PO effect. They used HOC and instantaneous values as features. They compared their proposed model with DT, random forest (RF) and DNN with and without PO. The four modulation classes chosen are BPSK, QPSK, 8PSK and 16QAM. They verified their performance under different SNRs from (−20 dB to 20 dB). Proposed model nullified the PO effect and got a highest classification accuracy 100% at 10 dB.

In [28], adversarial transfer learning architecture (ATLA) is proposed for AMC. It is a unified form of adversarial training and knowledge transfer. This model improves the performance when there is insufficient data. I and Q are used as features, and AM-DSB, 8PSK, BPSK, AM-SSB, 4PAM, GFSK, 64QAM, 16QAM, WBFM and QPSK are used as modulated signals. They observed their classification accuracy under different SNRs from (−20 dB to 18 dB). Their proposed model worked exceptionally well even when the training data is lowered.

The merits of DL approaches are high dimensionality, robust to changes in data and good classification accuracy when dataset is high. The demerits are it requires more training time and computationally expensive.

3 Convolutional Neural Network (CNN)

CNNs are similar to traditional ANNs in some aspects. CNNs self-optimize themselves through learning. The main difference between CNNs and traditional ANNs is that CNNs are mostly suited for pattern recognition. The schematic diagram of CNN architecture is shown in Fig. 3. CNNs are having three layers: 1. convolutional layers, 2. pooling layers and 3. fully connected layers. CNN architectures are varied according to application. They are formed by stacking these three layers [29, 30].

Convolutional layer plays a key role for CNNs to operate. This layer makes use of kernels and activation functions. Rectified linear unit (ReLU) performs elementary-wise activation function (preferably sigmoid function). In the next stage, pooling layer will perform downsampling, to reduce the number of parameters inside the activation function. In the last stage, it generates class scores from activation function. To improve performance, ReLU can be used between layers.

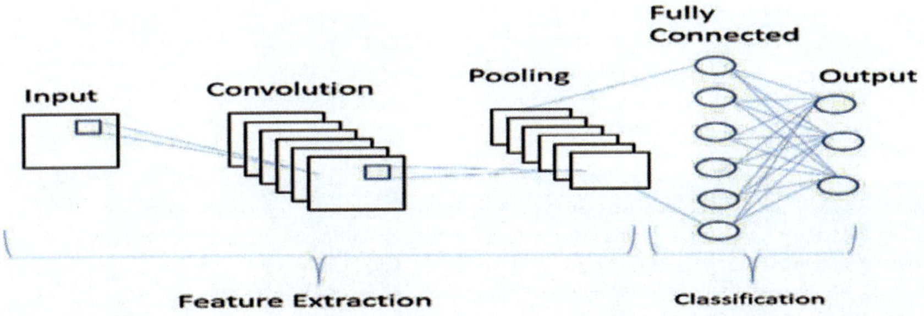

Fig. 3 Basic CNN architecture [29]

4 Simulation Results

Analog modulated signals such as Double Sideband Amplitude Modulation (DSB-AM), 4-ary PAM, Frequency Modulation (FM) and Single Sideband Amplitude Modulation (SSB-AM) are considered for classification. These signals are trained, tested and validated at different scenarios [31, 32]. For each modulation class, 10,000 frames are generated. Signal frames for modulation classes are shown in Fig. 4. CNN classifier is trained under imperfect conditions. Spectrograms of modulation classes are plotted utilizing STFT. Figure 5 shows the spectrograms of modulation types.

Figure 4 interprets the confusion matrix of CNN at 30 dB SNR. The test accuracies observed at 80%, 70% and 60% are 100%, 99.5% and 99.5%, respectively.

Table 1 provides insights to the accuracies of modulation class at different SNR values. Table 2 summarizes the test accuracies for AMS with different training rates and SNR.

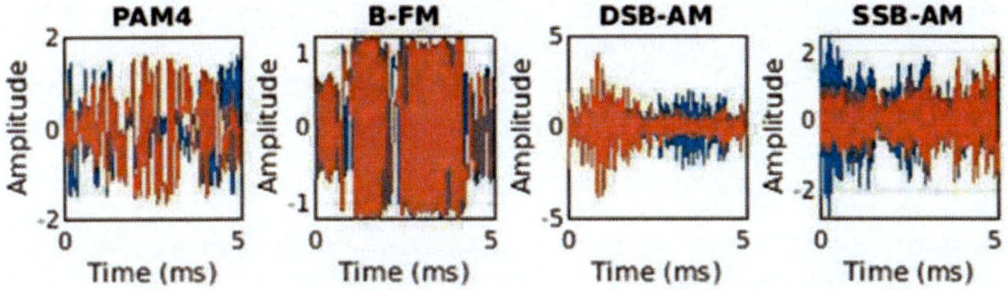

Fig. 4 Signal frames for modulation classes

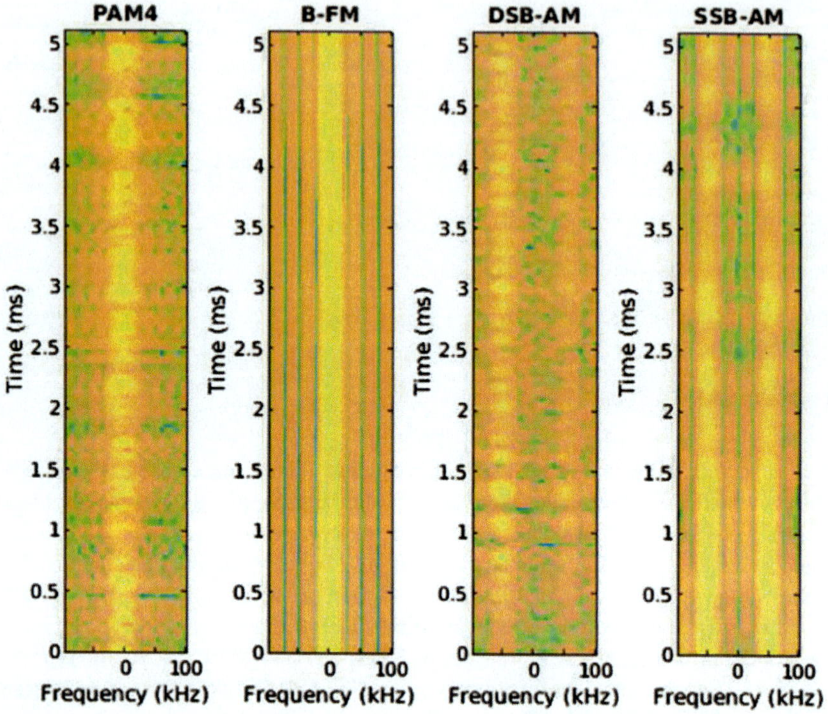

Fig. 5 Spectrograms of four modulation classes

5 Conclusion

This study initially presents a comprehensive review of AMR using DL approaches. In the latter part, simulations are carried out for analog modulation signals using CNN approach. From simulation results, it is clear that the proposed approach achieved 100% test accuracy at SNR 30 dB for 80% training rate.

Table 1 Accuracies of modulation class at different SNR

Class/Training rate	Accuracy (%)															
	SNR 30 dB				SNR 20 dB				SNR 10 dB				SNR 0 dB			
	80	70	60	50	80	70	60	50	80	70	60	50	80	70	60	50
1	100	100	100	100	100	100	100	100	100	100	100	100	100	100	96.1	97.1
2	100	98	98	98	96	95	94	94.5	96	95	94	94.5	96	95	90.7	87.5
3	100	100	100	100	100	99	99.3	99.5	100	99	99.3	99.5	100	99	48.2	50.3
4	100	100	100	100	100	100	100	99.5	100	100	100	99.5	100	100	72.7	74.5

Table 2 Test accuracy summary for different training rates

SNR (dB)	Accuracy (%) at training rate of			
	80	70	60	50
30	100	99.5	99.5	99.5
20	99	98.5	98.33	98.375
10	99	98.5	98.33	98.375
0	99	98.5	72.5	72.875

References

1. Ikuma T, Naraghi-Pour M (2008) A comparison of three classes of spectrum sensing techniques. IEEE GLOBECOM 2008—2008 IEEE global telecommunications conference, pp 179–182
2. Celebi H, Guvenc I, Gezici S, Arslan H (2010) Cognitive-radio systems for spectrum, location, and environmental awareness. IEEE Antennas Propag Mag 52(4):41–61
3. Wu TW, Lin YE, Hsieh HY (2008) Modeling and comparison of primary user detection techniques in cognitive radio networks. IEEE GLOBECOM 2008—2008 IEEE global telecommunications conference, pp.1–5
4. Lee C, Wolf W (2007) Multiple access-inspired cooperative spectrum sensing for cognitive radio. MILCOM 2007—IEEE military communications conference, pp.1–6
5. Yuanzeng C, Hailong Z, Yu W (2011) Research on modulation recognition of the communication signal based on statistical model. 2011 Third international conference on measuring technology and mechatronics, vol 3, pp 46–50
6. Venkata Subbarao M, Samundiswary P (2020) Performance analysis of modulation recognition in multipath fading channels using pattern recognition classifiers. Wireless Pers Commun 115:129–151. https://doi.org/10.1007/s11277-020-07564-z
7. Subbarao MV, Samundiswary P (2019) K-nearest neighbors based automatic modulation classifier for next generation adaptive radio systems. Int J Security Appl 13:41–50
8. Wu HC, Saquib M, Yun Z (2008) Novel automatic modulation classification using cumulant features for communications via multipath channels. IEEE Trans Wireless Commun 7(8):3098–3105
9. Wei W, Mendel JM (2000) Maximum-likelihood classification for digital amplitude-phase modulations. IEEE Trans Commun 48(2):189–193
10. Wang S, Zhang L, Zuo W, Zhang B (2019) Class-specific reconstruction transfer learning for visual recognition across domains. IEEE Trans Image Process 29:2424–2438
11. Subbarao MV, Samundiswary P (2021) Automatic modulation classification using cumulants and ensemble classifiers. In: Kalya S, Kulkarni M, Shivaprakasha KS (eds) Advances in VLSI, signal processing, power electronics, IoT, communication and embedded systems. Lecture notes in electrical engineering, vol 752. Springer, Singapore. https://doi.org/10.1007/978-981-16-0443-0_9
12. Subbarao MV, Samundiswary P (2018) Spectrum sensing in cognitive radio networks using time–frequency analysis and modulation recognition. In: Anguera J, Satapathy S, Bhateja V, Sunitha K (eds) Microelectronics, electromagnetics and telecommunications. Lecture notes in electrical engineering, vol 471. Springer, Singapore. https://doi.org/10.1007/978-981-10-7329-8_85
13. Dobre OA, Bar-Ness Y, Wei S (2003) Higher-order cyclic cumulants for high order modulation classification. IEEE Military Commun Conf 1:112–117
14. Soliman SS, Hsue S-Z (1992) Signal classification using statistical moments. IEEE Trans Commun 40(5):908–916
15. S. Kadambe, QJ (2004) Classification of modulation of signals of interest. 3rd IEEE Signal processing education workshop. 2004 IEEE 11th digital signal processing workshop, pp 226–230

16. Zhang L, Wang S, Liu B (2018) Deep learning for sentiment analysis: A survey. Wiley Interdiscipl Rev: Data Min Knowl Discovery 8(4):e1253
17. Rajendran S, Meert W, Giustiniano D, Lenders V, Pollin S (2018) Deep learning models for wireless signal classification with distributed low-cost spectrum sensors. IEEE Trans Cognitive Commun Netw 4(1):433–445
18. Skansi S (2018) Introduction to deep learning. Undergraduate topics in computer science, pp 1–191
19. Krizhevsky A, Sutskever I, Hinton GE (2017) ImageNet classification with deep convolutional neural networks. Commun ACM 60(6):84–90
20. Oralhan Z (2020) 3D input convolutional neural networks for P300 signal detection. IEEE Access 8:19521–19529
21. Chong P, Elovici Y, Binder A (2019) User authentication based on mouse dynamics using deep neural networks: a comprehensive study. IEEE Trans Inf Forensics Secur 15:1086–1101
22. Graves A, Mohamed A, Hinton G (2013) Speech recognition with deep recurrent neural networks. 2013 IEEE International Conference on Acoustics, Speech and Signal Processing, 6645–6649
23. Xing F, Xie Y, Su H, Liu F, Yang L (2017) Deep learning in microscopy image analysis: a survey. IEEE Trans Neural Netw Learn Syst 29:4550–4568
24. Shah MH, Dang X (2019) Novel feature selection method using bhattacharyya distance for neural networks based automatic modulation classification. IEEE Signal Process Lett 27:106–110
25. Sun D, Chen Y, Liu J, Li Y, Ma R (2019) Digital signal modulation recognition algorithm based on vggnet model. 2019 IEEE 5th international conference on computer and communications (ICCC), pp 1575–1579
26. Tang B, Tu Y, Zhang Z, Lin Y (2018) Digital signal modulation classification with data augmentation using generative adversarial nets in cognitive radio networks. IEEE Access 6:15713–15722
27. Shi J, Hong S, Cai C, Wang Y, Huang H, Gui G (2020) Deep learning-based automatic modulation recognition method in the presence of phase offset. IEEE Access 8:42841–42847
28. Bu K, He Y, Jing X, Han J (2020) Adversarial transfer learning for deep learning based automatic modulation classification. IEEE Signal Process Lett 27:880–884
29. Phung VH, Rhee EJ (2018) A deep learning approach for classification of cloud image patches on small datasets. J Inf Commun Convergence Eng 173–178
30. O'Shea K, Nash R (2015) An introduction to convolutional neural networks. ArXiv e-prints, pp 1–11
31. Subbarao MV, Samundiswary P (2016) An intelligent cognitive radio receiver for future trend wireless applications. Int J Comput Sci Inform Sec 14:7–16
32. Subbarao MV, Samundiswary P (2018) Automatic modulation recognition in cognitive radio receivers using multi-order cumulants and decision trees. Int J Rec Technol Eng (IJRTE) 7:61–69

Design of Wearable Microstrip Patch Antenna for Biomedical Application with a Metamaterial

D. Ramesh Varma, M. Murali, and M. Vamshi Krishna

Abstract Numerous therapeutic and light-based diagnostic devices are habitually used in the health centers. These devices are bulk in size, and they appear as items mounted on the wall. Recently, lot of research is undergoing to detect the diseases like cancer at its early stage. The above devices serve the purpose, but they seem to be large in size and bulky with wires. There is a need of device which is compact and simple. Wearable antennas serve this purpose. These antennas are capable of serving the medical needs. These wearable antennas are very useful to track the patient's health and remotely monitor their condition. The user comfort for wearing this antenna is a crucial task. This paper reviews various wearable antennas used for medical fields and for commercial purpose. This includes textile antennas, non-textile antennas and patch antennas which work at a frequency of ISM band. Each of them having advantages and disadvantages which are highlighted in the review of the paper. Also keeping all these review essentials into note, a new antenna is designed which will resonate at a 2.45 GHz frequency along with metamaterial to reduce the surface currents which will affect when placed on a human body. The return loss is −15 dB, the gain is 2 dBi, and Voltage standing wave ratio is 1 as the resonant frequency is about 2.43 GHz.

Keywords Textile antenna · Wearable antenna · Non-textile antenna · ISM band

D. R. Varma (✉)
Centurion University of Technology and Management, Bhubaneswar, Odisha, India

Shri Vishnu Engineering College for Women (A), Kovvada, India

M. Murali
Department of ECE, Centurion University of Technology and Management, Visakhapatnam, Andhra Pradesh, India

M. V. Krishna
Department of ECE, Dhanekula Institute of Technology and Management, Vijayawada, Andhra Pradesh, India

P. S. R. Chowdary et al. (eds.), *Evolution in Signal Processing and Telecommunication Networks*, Lecture Notes in Electrical Engineering 839,
https://doi.org/10.1007/978-981-16-8554-5_40

1 Introduction

Wearable devices are censorious in the era of information technology and communication technology for off-body and on-body applications. These can be deployed for wireless sensor networks (WSN) and Internet of Things (IoT). The main goal is to enhance and improve the functionality of wearable devices by combining with electronics and fabrics [1, 2]. These devices have become part in our life. These wearable devices should be reliable, small and compact in size, easy to install and its reliability. These devices should be inexpensive, low maintenance, robust and flexible enough to operate in all the environments.

Present research is going on these wearable electronics which serves the future needs. Wearable antennas are used to detect the diseases like cancer, heart diseases and lung diseases. According to the survey conducted by Mason, the revenue of wearable electronics will reach 22 billion dollars in 2020 from 3 billion dollars in 2014 [3]. A wearable antenna is placed on the body, and it is affected by motion of the body. The parameters of the antenna will change due to these motions. There will be gain variations, radiation pattern, impedance mismatch and frequency detuning happen when antenna is operating on or near the human body [4]. According to FCC and ERC, 402.5–405.5 MHz is for biotelemetry applications [5]. In this paper, design of current wearable antennas in accordance to medical and health purpose is studied, and its challenges related to it are also.

2 Review of Wearable Antenna

In 2017, Thangaselvi proposed an innovative approach for wearable antennas. In [6], a textile antenna is consisting of conductive patch and ground and also substrate. The inset feed is provided to the patch. This textile antenna is connected to the recorder, and this recorder will receive the signals coming from different parts of the body. These recorded signals are subjected to the scattering matrix. This antenna can be placed on back side of the body for convenience. There is a coupling between transmitter and a receiver (Fig. 1).

In [7], the wearable antenna is developed for the medical applications. This antenna is capable of detecting the breast cancer. It helps in cancer diagnoses and

Fig. 1 Front view and back view of nickel copper textile antennas, designed by E. Thangaselvi, 2018, Springer Science + Business Media, LLC, part of Springer Nature

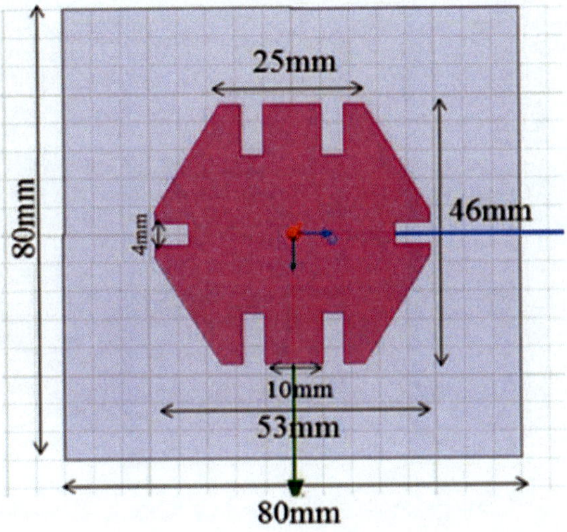

Fig. 2 Textile wearable antenna designed by Dhamodharan Srinivasan, 2019, Journal of Medical Systems

reduces the risk of death. Jeans is used as a substrate material whose dielectric constant ∈ is 1.7, copper (Cu) is used as a conducting material for both radiating patch and a ground plane, and the slot is made on the patch as shown in Fig. 2. The jean material is placed on a breast. Co-axial feeding technique is preferred which will improve the performance of an antenna. Antenna radiation pattern is 145.33 dB which is perfect for the cancer diagnosis (Fig. 2).

In 2018, Sajith. K proposed a new technique for health monitoring applications. In [8], wearable antenna is designed to monitor the patient status. In this study, split ring resonator loaded conductor which is backed with the coplanar wave guide (CPW) is used for monitoring the health. This wearable antenna is made up of two similar ground planes, and at the middle, one metal strip conductor is lying in the same plane. As it is clear from Fig. 3, this antenna is designed with Teflon substrate whose conductivity is 0.91 s/m. Since it is a split ring resonant, it is having high-gain flexible frequency tuning (FFT) (Fig. 3).

In [9], the split ring resonator (SRR) is loaded with the radiating patch which is a diamond-shaped one, which is backed with the coplanar waveguide. This antenna is operated at 2.4, 3.5 and 3.65 GHz, which are WLAN and Wi-MAX frequency bands (Fig. 4). The wearable antenna is multi-layered antenna; therefore, it is having higher gain and higher efficiency. The proposed antenna has Teflon as a substrate. The encircled, Complementary split ring resonator and split ring resonator are sandwiched between patch and skin layers. Therefore, gain is high, and there is less effect of back radiation.

In [10], a wearable antenna is designed which is used to measure the abnormalities in the brain. It is small, and it is able to provide better results when compared to the EEG. In this, wearable antenna is made up of textile materials which are cheap, compact and easily available. The wearable antenna is composed of conducting patch,

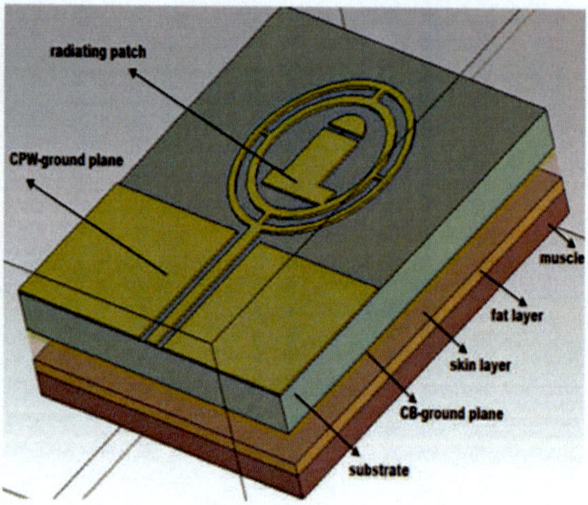

Fig. 3 SRR loaded antenna designed by Sajith. K, J. Gandhimohan, 2018, IEEE International Conference on Current Trends toward Converging Technologies

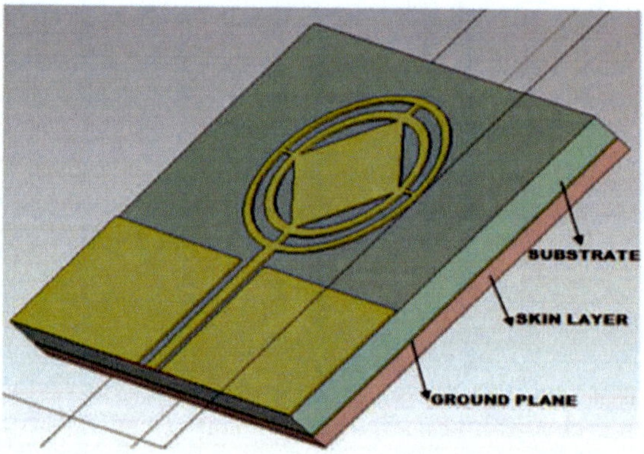

Fig. 4 CSRR loaded antenna designed by Sajith.K, J.Gandhimohan, 2017, IEEE International Conference on Circuits and Systems

Fig. 5 Fabricated textile antenna designed by Imran Saied, 2018, IEEE 29th Annual International Symposium on Personal, Indoor, and Mobile Radio Communications

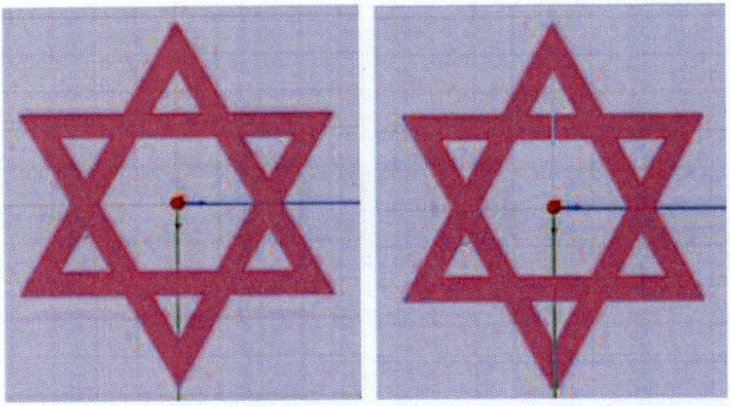

Fig. 6 Antenna without splits and with splits on X-axis designed by Neha, 2016, International Conference on Next Generation Computing Technologies

felt like substrate. The microstrip transmission line is used to feed the sensing antenna (monopole structure). The range of this wearable antenna is from 1 to 4.2 GHz when placed in human phantom (head phantom) (Fig. 5).

In [11], an antenna is designed to detect the early stage of skin cancer. This antenna is made up of metamaterial with the help of SRR. In this work, a wearable antenna operates at X-band, i.e., 8–12 GHz. It will sense variations of electrical properties. Star-shaped ring metamaterial antenna is used in this design. Initially, antenna is simulated without any cuttings or splits, and it is observed that it is operated for only three bands with 185, 390 and 732 MHz bandwidths. For cancer detection, it is not sufficient. Therefore, the above structure is modified with splits or cuttings on the X-axis. Now, bandwidth is about 1.34 GHz (Fig. 6).

In [12], a concept of defected ground structure is introduced. Microstrip patch antenna is designed with the defected ground structure (DFS). An antenna is made up of felt and Teflon as substrate materials, whose dielectric constants are 1.3 and 2.1, respectively. DGS with toy shape is also simulated which got more accurate results. Fabricated antenna is shown in Fig. 7.

In [13], two antennas are designed: a textile antenna which is a normal one and other is annular slot textile antenna. In this study, two papers are studied, and also performance of the two antennas is compared. First, the textile antenna is made up of polyester and rubber as a substrate as they are compact and easily wearable. Second, a slot is made in the textile antenna which is an annular slot whose performance is measured. By looking at the results, it is clear that annular slot antenna performs better than the normal textile antenna (Fig. 8).

In [14], a wearable patch antenna is designed with an electromagnetic band gap ground planes under bended and normal conditions. Its operating frequency is about 2.4 GHz which is an unlicensed ISM band. In this, electromagnetic bandgap surfaces are used to design a wearable antenna. The overall dimension of this wearable antenna is $113 \times 96 \times 3$ cubic meters. Substrate material is wash cotton. Zelt is used for the

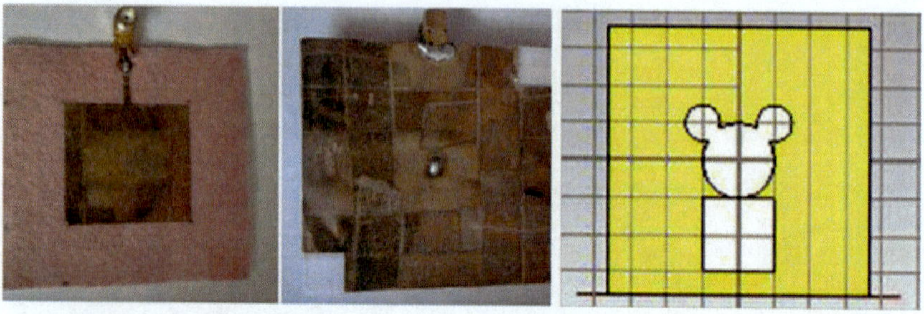

Fig. 7 Fabricated antenna with DGS designed by Ameena Banu Mustafa, 2019, Journal of Medical Systems

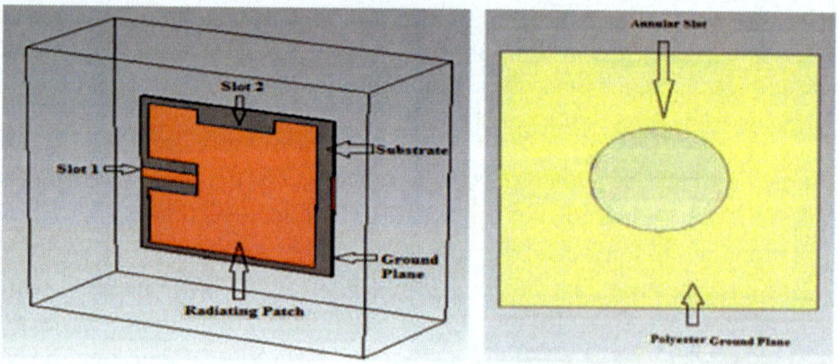

Fig. 8 Textile antenna without slot and with slot designed by Jaspreet Singh, 2018, International Journal of Innovative Science, Engineering & Technology

radiating patch. The dielectric constant is about 1.5, and loss tangent of the substrate is 0.02. The back radiation is very harmful for the biological tissues. So, it should be eliminated with the ground planes as shown in Fig. 9. The metamaterials are used to shield the body from the surface currents and the electromagnetic radiation. The specific absorption rate (SAR) is a performance metric to know the antenna's performance. The phantom model consists of fat, skin and muscle. The obtained SAR is about 1.77 W/kg which is less than 2 W/kg which is a predefined value. The efficiency is improved from 52 to 75% when compared with the conventional parts. The return loss is about −30 dB and gain of 6 dB for non-bend condition (Fig. 9).

In 2017, Sajith. K proposed a new technique for health monitoring applications. In [15], wearable antenna is designed to monitor the patient status. In this study, split ring resonator loaded conductor backed with the coplanar wave guide (CPW) as shown in Fig. 10 is used for monitoring the health. This antenna is capable of fine tuning at the center frequency's 2.45 GHz. This wearable antenna is made up of two similar ground planes, and at the middle, one metal strip copper conductor is placed on the same plane. The antenna overall dimension is 33 × 35x1.6 cubic meters. This antenna is designed with Rogers RT6002 substrate whose conductivity is 0.95 s/m. Since it is a split ring resonant, it is having high-gain flexible frequency tuning (FFT).

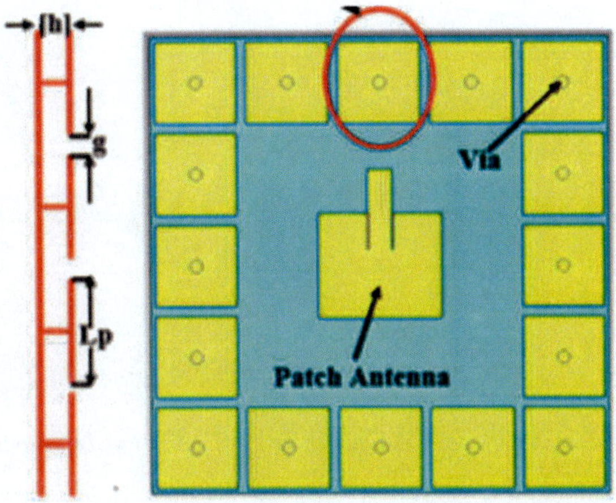

Fig. 9 Metamaterial patch antenna designed by Usman Ali, 2017, Journal of Electrical Engineering & Technology

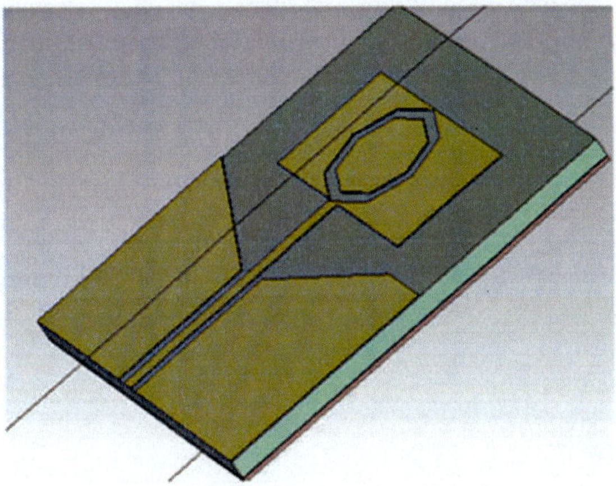

Fig. 10 SRR loaded antenna designed by Sajith. K, 2017, IEEE International Conference on Circuits and Systems

For non-SRR loaded antenna, the reflection coefficient s11 is −15 dB for operating frequency 2.26 GHz. For SRR loaded antenna, −25.4 dB for operating frequency 2.405 GHz, −36.52 dB for operating frequency 2.452 GHz and −22.491 dB for operating frequency 2.432 GHz (Fig. 10).

In [16], from Fig. 11, a compact and small wearable slot antenna is designed which is operated at a 2.4 GHz operating frequency which is an unlicensed ISM band. The wearable antenna is composed of substrate patch and a ground plane. In this electronic band gap with defected ground structures, the usage of EBG is used for shielding the body from the back radiations from the patch. The defected ground structure (DGS) is used to widen the bandwidth. This enhances the bandwidth and leads to broadening. It is very useful for loading and deformation on the human body.

Fig. 11 Fabricated antenna with EBG-DGS designed by Adel Y, Ashyap, 2020, IEEE Access

The bandwidth is about 32.08%, and a gain of 6dBi, efficiency 72.3% and front-to-back ratio 15.8 dB. For free space, −40 dB is return loss, and for on body, it was −29 dB. The SAR is less than 1.67 W/kg standard FCC (Fig. 11).

In [17], a wearable ultra-wide band antenna is designed for detecting the skin cancer at the early stage. The concept of metamaterials is used in this work. This wearable antenna is integrated with the magnetic conductor which is an artificial one. This antenna is designed with two sections. Hexagonal-shaped patch was fed with the CPW feed. The felt material is used for a substrate whose dielectric constant $\in$ is 1.2 and 0.0162 loss tangent. Conducting material, Zelt, is used for the microstrip patch. Zelt has many advantages like washable, flexible and tear resistant. This antenna is designed at X-band (8–12 GHz). Since these are high frequency, it cannot penetrate into deeper of the skin, so it is shielded from the back radiations from the microstrip patch antenna. Here, the artificial magnetic conductor (AMC) is used to shield the body from back radiations. The return loss is about −27 dB for free space, −31 dB for cancer tissue and −32 dB for normal skin. Figure 12 shows the UWB antenna for cancer detection (Fig. 12).

In [18], a wearable antenna is designed which is capable of providing solution to the medical monitoring systems. This designed antenna is capable of detecting tumor cells of the cancer. This antenna is able to detect the healthy tissues and unhealthy tissues, and it is able to differentiate among those two. The coplanar waveguide is used for feeding the antenna. Artificial magnetic conductor is used over the substrate to hold the back radiations from the antenna. This will provide better SAR when compared to the conventional ground planes (Fig. 13).

In [19], Sakthi Abirami proposed Yagi-Uda antenna which is backed with EBG structure. This wearable antenna is made up of copper radiating patch and a polyester substrate material. From the results, it is observed that SAR is reduced from 8.56 W/kg to 0.075 W/kg when backed with EBG. The simulations are performed in various conditions like Yagi in free space, Yagi on body, Yagi + EBG in free space and Yagi + EBG on human body (Fig. 14).

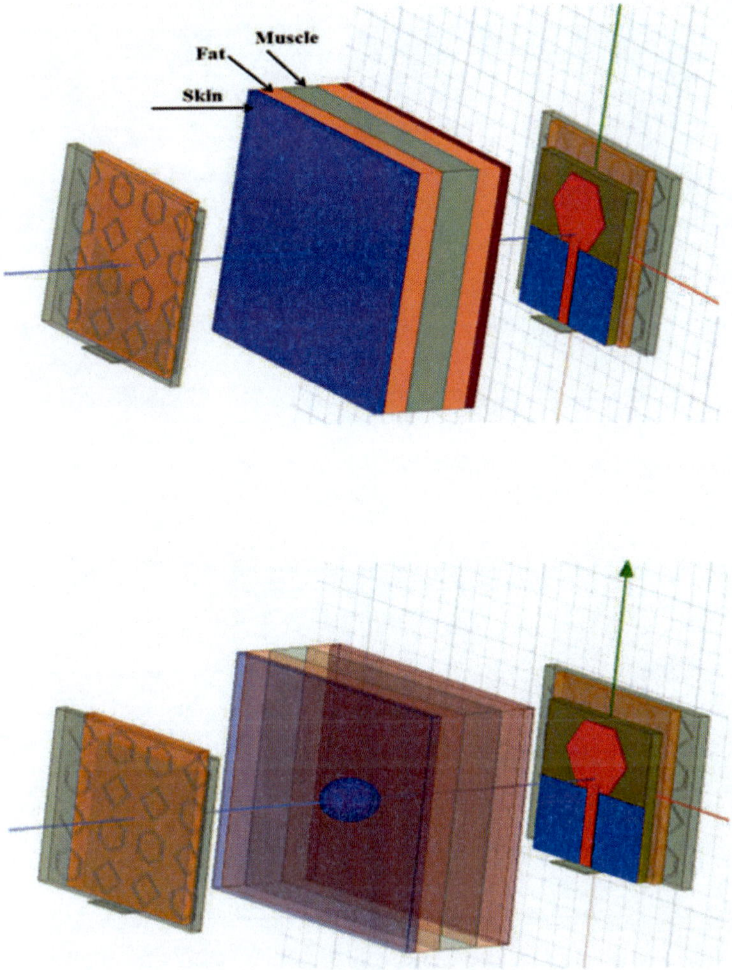

Fig. 12 Setup for normal skin and tumor designed by Anamika Mani Tripathi, 2020, International Conference on Electrical and Electronics Engineering

3 Proposed Antenna Design

The dimensions of the proposed antenna is of rectangular shape with the radiating patch, substrate and ground plane. The substrate dimensions are 86 × 86 × 1 mm. The ground plane is also having similar to that of substrate, i.e., 86 × 86 × 0.035 mm (Fig. 15). The antenna is designed for the 2.43 GHz resonant frequency. The patch dimensions are 43 × 40 × 0.035 mm. To reduce the surface currents, metamaterial is also designed (Fig. 16).

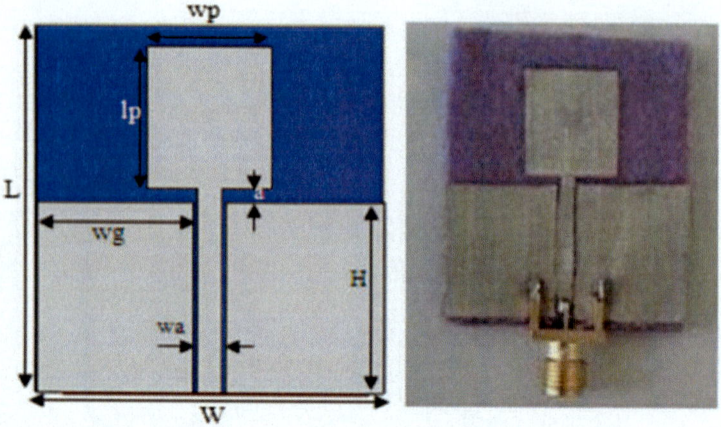

Fig. 13 Fabricated patch antenna designed by Ameni Mersani, 2019, Progress In Electromagnetics Research M

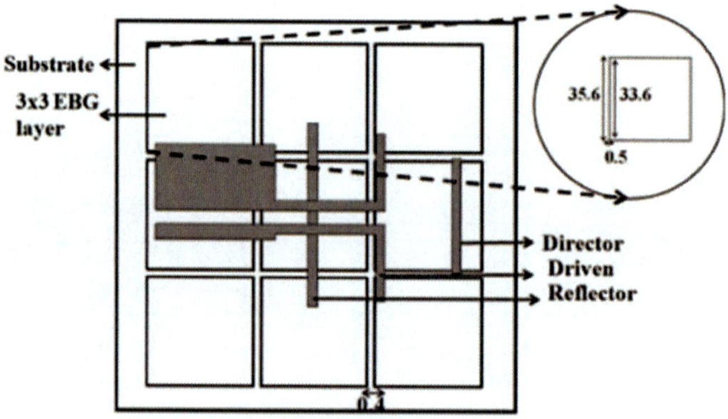

Fig. 14 Yagi-Uda antenna with EBG designed by B. Sakthi Abirami, 2017, IEEE TRANSACTIONS ON ANTENNAS AND PROPAGATION

The metamaterial is made up of three planes: patch, substrate and ground planes. The substrate dimensions of an array are 86 × 86 × 1 mm. The patch is having dimensions of 23 × 29 × 0.035 (Fig. 16).

4 Results and Discussion

From the below figures, it is clear that the antenna is resonating at the 2.43 GHz with the bandwidth and the voltage standing wave ratio less than 2 at the resonant

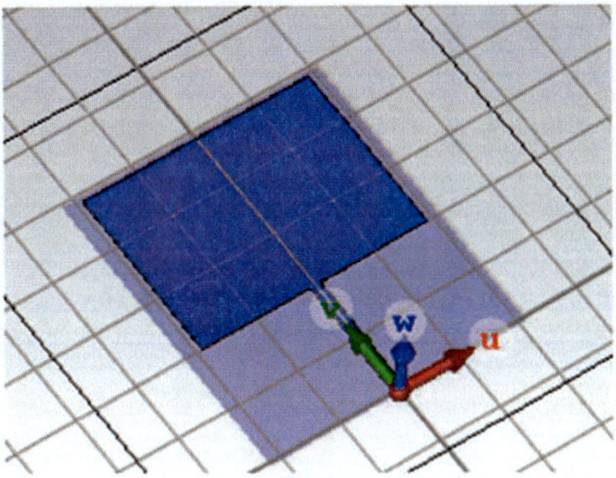

Fig. 15 Proposed antenna design

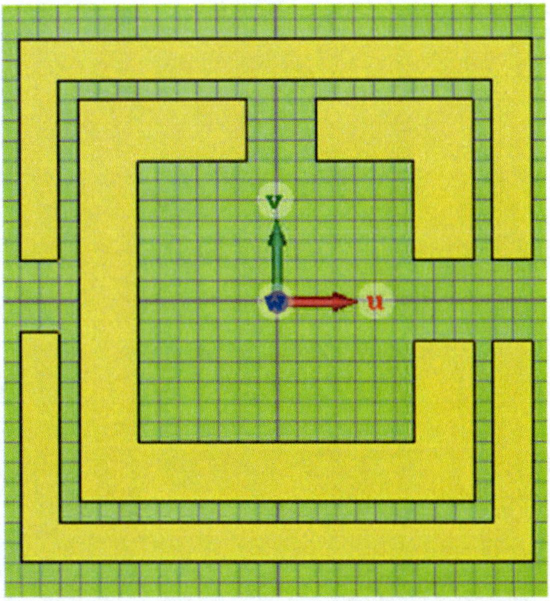

Fig. 16 Metamaterial unit cell

frequency. The gain, directivity and the E-pattern are drawn from the results, and it is clear that it is well suited for the wearable biomedical applications.

From Fig. 17, it is clear that the antenna is resonating at the required frequency range with the return loss of about −15 dB and VSWR of about 1 which conveys that it is showing the desired characteristics for a wearable antenna.

From Fig. 18, the value is about 2dBi which is acceptable for a wearable biomedical application.

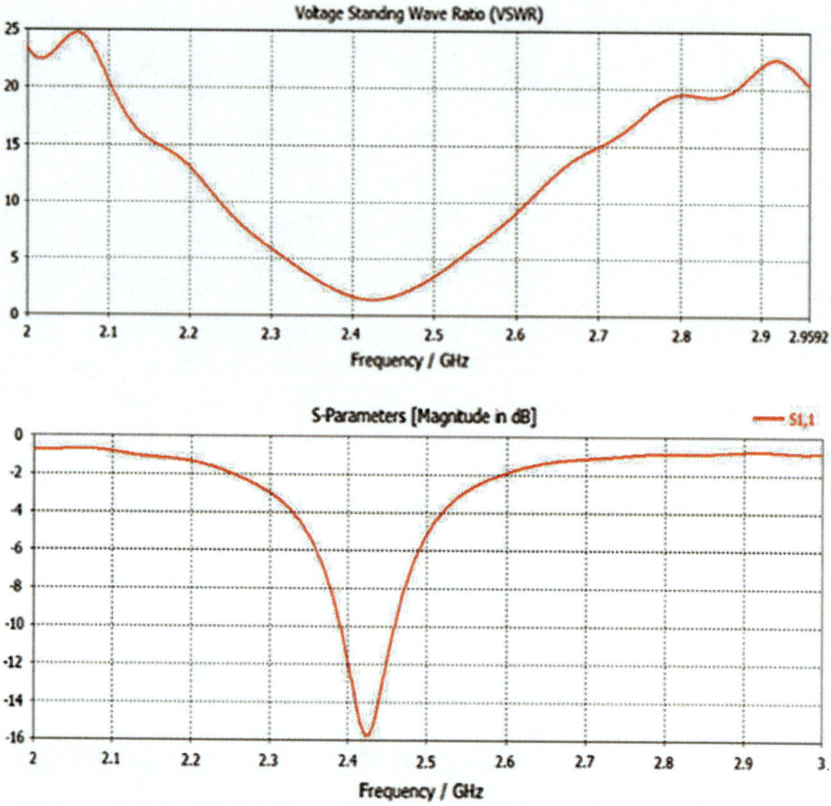

Fig. 17 **a** Return loss and **b** voltage standing wave ratio

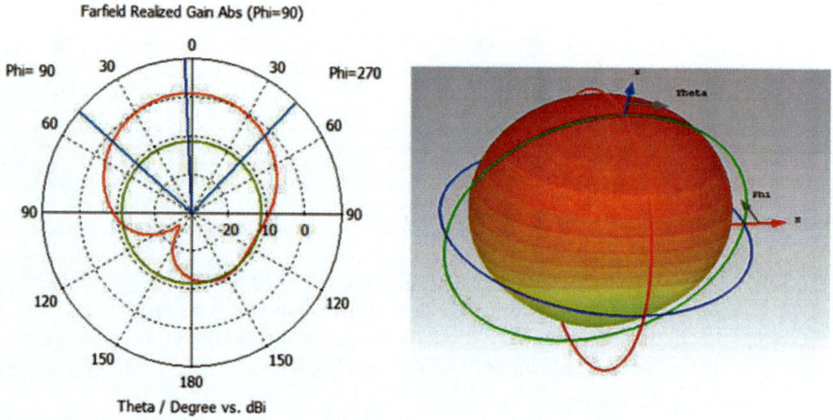

Fig. 18 Polar and 3D plot of gain of proposed antenna

5 Challenges

In medical field, there should be continuous monitoring of the patient's health. The transmitted data from the monitoring system should be reliable and accurate. Few challenges are identified to achieve it: 1. power efficiency, 2. radiation degradation due to human tissues, 3. environmental factors such as temporal physical obstructions and 4. antenna form factor and user comfort.

Design of the patch antenna using fractal geometry can be utilized for the ultra-wide applications which is discussed in [20–23].

6 Conclusion

In medical applications, designed antenna should have limited SAR value when placed on the body. The antenna structure maybe patch, non-textile and textile; the basic priority is to achieve optimum performance during its operation. During the design, bandwidth, power consumption and throughput and safety should be considered. The review is performed on the current design of the wearable antennas based on different purposes and objectives, and it is observed that the EBG and defected ground structures are needed to get the optimum performances. It is used to reduce the SAR values as per IEEE standards. The above-designed antenna is resonating at a 2.43 GHz with a return loss of −15 dB, VSWR of 1 and gain of 2dBi.

References

1. Ashyap AYI, Zainal Abidin Z, Dahlan SH, Majid HA, Kamarudin MR, Alomainy A, Abd-Alhameed RA, Kosha JS, Noras JM (2018) Highly efficient wearable CPW antenna enabled by EBGFSS structure for medical body area network applications. IEEE Access 6:77529–77541
2. Ashyap AY, Dahlan SH, Abidin ZZ, Majid HA, Seman FC, Ngu X, Cholan NA (Dec 2018) "Flexible antenna with HIS based on PDMS substrate for WBAN applications," in Proc IEEE Int RF Microw Conf (RFM), pp 69–72
3. Velasco-castillo E (2014) The smart wearables market will be worth USD22.9 billion worldwide by 2020. Analysys Mason Limited
4. Ur-Rehman M, Abbasi QH, Akram M, Parini C (2014) Design of band-notched ultra-wideband antenna for indoor and wearable wireless communications. IET Microwaves Antennas Propag 9(3):243–251
5. Wu Z, Liu X, Di Y (2015) A novel stacked inverted-f antenna for implanted bio-devices in the mics band. IEEE Int Conf Comput Electromagn 355–357
6. Thangaselvi E, Meena alias Jeyanthi1 K (2018) "Implementation of flexible denim nickel copper rip stop textile antenna for medical application." Springer Science+Business Media, LLC, part of Springer Nature
7. Srinivasan D, Gopalakrishnan M (2019) Breast cancer detection using adaptable textile antenna design. J Med Syst 43(177):2019

8. Sajith K, Gandhimohan J, Shanmuganantham T "Multilayered and dual-band CB-CPW fed wearable antenna for healthcare monitoring applications". Proceeding of 2018 IEEE international conference on current trends toward converging technologies, Coimbatore, India
9. Sajith K, Gandhimohan J, Shanmuganantham T (2017) "Design of SRR loaded CB-CPW fed diamond shaped patch on-body antenna for ecg monitoring applications". Proceedings of 2017 IEEE international conference on circuits and systems (ICCS)
10. Saied I, Arslan T "Wideband textile antenna for monitoring neurodegenerative diseases". 2018 IEEE 29th annual international symposium on personal, indoor, and mobile radio communications (PIMRC) 356–360
11. Neha, Kaur A (14–16 Oct 2016) "Wearable antenna for skin cancer detection", 2016 2nd international conference on next generation computing technologies (NGCT-2016) Dehradun, India
12. Mustafa AB, Rajendran T (2019) An effective design of wearable antenna with double flexible substrates and defected ground structure for healthcare monitoring system. J Med Syst 43:186
13. Lee CS, Wu CY, Kuo YL "Wearable belt resonators model for wrist detection", 2016 IEEE 5th Asia-Pacific conference on antennas and propagation (APCAP) 35–36
14. Ali U, Ullah S, Khan J, Shafi M, Kamal B, Basir A, Flint JA, Seager RD (2017) Design and sar analysis of wearable antenna on various parts of human body, using conventional and artificial ground planes. J Electric Eng Technol 12(1):317–328
15. Sajith K, Gandhimohan J, Shanmuganantham T (2017) " Proceedings of 2017 IEEE international conference on circuits and systems (ICCS)" 49–53
16. Ashyap AY, Dahlan SHB, Abidin ZZ, Dahri MH, Majid HA, Kamarudin MR, Yee SK, Jamaluddin MH, Alomainy A, Abbasi QH (2020) Robust and efficient integrated antenna with ebg-dgs enabled wide bandwidth for wearable medical device applications. IEEE Access 8:56346–56358
17. Tripathi AM, Rao PK, Mishra R (2020) "An amc inspired wearable uwb antenna for skin cancer detection". 2020 International Conference on Electrical and Electronics Engineering (ICE3)
18. Mersani A, Osman L, Ribero JM (2019) Flexible uwb amc antenna for early stage skin cancer identification. Progress Electromagn Res M 80:71–81
19. Abirami BS, Sundarsingh EF (Jul 2017) "EBG-backed flexible printed yagi–uda antenna for on-body communication". IEEE Trans Antennas Propag 65(7)
20. Terlapu SK, Jaya C, Raju GS (2017) On the notch band characteristics of koch fractal antenna for uwb applications. Int J Control Theory Appl 10(06):701–707
21. Terlapu SK, Chowdary PSR, Jaya C, Sameer Chakravarthy VVSS, Satpathy SC (2018) On the design of fractal uwb wide-slot antenna with notch band characteristics. In: Anguera J, Satapathy S, Bhateja V, Sunitha K (eds) Microelectronics, electromagnetics and telecommunications. Lecture notes in electrical engineering, vol 471. Springer, Singapore
22. Jagadam K, Terlapu SK, Raju G (Jan 2017) "Design and analysis of slot coupled cylindrical dielectric resonator antenna array". Springer J: "Lecture Notes in Electrical Engineering"
23. Terlapu SK, Subba Rao MV, Chowdary PSR, Satapaty SC (2021) Design and analysis of koch fractal slots for ultra-wideband applications. In: Anguera J, Satapathy S, Bhateja V, Sunitha K (eds) Microelectronics, electromagnetics and telecommunications. Lecture notes in electrical engineering, vol 471. Springer, Singapore

Emotion Recognition from Speech Audio Signals Using Convolution Neural Network Model Architectures

M. Nanda Kumar, Thokala Eswar Chand, Mettu Joseph Rithvik Reddy, and Narra Pranay Reddy

Abstract Speech is one of the traditional ways of expressing oneself. We rely so much on it that we accept its value while using other means of communication, such as emails and text messages, where we use emojis regularly to express feelings sent to messages. Because emotions are so important in communication, discovering and exploring them is essential in today's digital age of remote communication. Because emotions are subdued, seeing them is hard work. There is no universal agreement on how to measure or combine them. The program has a set of methods for processing and strengthening speech signals to detect the emotions embedded in them. A program like this can be used for a variety of purposes, such as voice-based assistants or agent chat analysis. By examining the acoustic structures of audio recording data, the study created hopes of seeing the basic emotions in recorded speech. People-to-computer communication, collaborative teaching, recreation, security and other fields have all benefited from the research. The visual acuity of speech is directly affected by the level of feature release in the system. The whole emotional phrase was commonly used as a unit output element, and the output content was four elements of emotional expression, incorporating various acoustic structures such as time constructions, size and quantity. The developed model can even classify the emotions based on the vocabulary it is used without losing any context which differentiates from other existing models.

Keywords Mel frequency cepstral coefficients · Chroma · Mel spectrogram · Convolution neural networks · Dimensionality reduction

1 Introduction

Speech recognition is a computer-based process that removes emotional traits from speech signals and comparisons and analyzes the parameters of acquired traits and emotional changes. Speech and emotional law were eventually abolished, and

M. Nanda Kumar (✉) · T. E. Chand · M. J. R. Reddy · N. P. Reddy
Department of Electronics and Computer Engineering, Sreenidhi Institute of Science and Technology, Hyderabad, India

P. S. R. Chowdary et al. (eds.), *Evolution in Signal Processing and Telecommunication Networks*, Lecture Notes in Electrical Engineering 839,
https://doi.org/10.1007/978-981-16-8554-5_41

emotional states were determined by the law [1, 2]. Speech recognition has now become a popular area of research in signal processing and pattern recognition, as well as the evolving branch of artificial intelligence and artificial psychology [3]. People-to-computer communication, collaborative teaching, recreation, security and other fields have all benefited from the research. The visual acuity of speech is directly affected by the level of feature release in the system. The whole emotional phrase was commonly used as a unit output element, and the output content was four elements of emotional expression, incorporating various acoustic structures such as time constructions, size and quantity [4].

Speech is one of the traditional ways of expressing oneself. We rely so much on it that we accept its value while using other means of communication, such as emails and text messages, where we use emojis regularly to express feelings sent to messages. Because emotions are so important in communication, discovering and exploring them is essential in today's digital age of remote communication [5, 6]. Because emotions are subdued, seeing them is hard work. There is no universal agreement on how to measure or combine them [10]. The SER program is defined as a set of methods for processing and strengthening speech signals to detect the emotions embedded in them. A program like this can be used for a variety of purposes, such as voice-based assistants or agent chat analysis. By examining the acoustic structures of audio recording data, the study created hopes of seeing the basic emotions in recorded speech [7].

2 Solution Overview

One or more of these features can be used to solve the problem of voice emotion recognition [8]. If one wants to estimate emotions from audio in real time, following the vocabulary features would necessitate a speech transcript, which would necessitate a text from speech. Furthermore, going forward with features which are visible like facial expressions analysis would necessitate access to the conversations where video can be on, which may not be suitable in all the cases, whereas acoustic feature analysis can be implemented in real time while the communication is in progress because the study implemented only needs the audio data to complete our work. As a result, the study decided to focus more on the acoustic aspects [9].

Furthermore, there are two methods for representing emotions:

Discrete Classification is the process of categorizing emotions into discrete labels such as anger, happiness, boredom and so on.

Dimensional Representation is the process of changing the representation of dimensions from low to high, active to passive and negative to positive scale.

The above-mentioned approaches have their own advantages and disadvantages. For a prediction, the model provides more priority to context which makes the dimensional procedure more complex. As the process is more complex, it is somewhat difficult to execute and also lack in annotated audio data. The discrete classification is a basic process which is easier to apply as compared to dimensional representation. By implementing discrete classification, there are chances of losing the context while prediction. As the data is not available in the public domain, the study implemented discrete classification strategy in this work.

3 Data Sources

To differentiate the emotions behind the audio clip, it is necessary to train the model with enough data. The data must be well organized without having any wrong labels. If the labels assigned to the model are mismatched, then there are bright chances that our model may detect some unpleasant patterns which leads to lower accuracy and greater loss which is not suggestable. Before selecting the correct dataset, there are some constraints which need to match. The criteria which are to be noticed prior selecting the dataset are

a. The audio files present in the dataset must contain less noise. If there is lot of noise in the background, there are chances that our model can detect some unnecessary disturbances as main features. This drastically plummets the model's accuracy.
b. The distribution of the emotions comprised in the dataset must be of equal in number. If there are more number of audio files related to one certain emotion and very less number of audio files related to some other emotions, then model cannot understand the exact characteristics of certain emotions which have very less number of audio files. This uneven distribution of emotions in dataset also damages the reputation of the model.
c. The audio available in the dataset must be audible enough. The base of the audio must be higher. If there is very low sound, then it is very difficult in identifying the features which are needed to classify. We can also visualize the audio waveforms by satisfying this constraint.
d. The audio files in dataset must show some difference in pitch, frequency and base; else model might understand them as similar audio types that fall under same emotion class.
e. The pronunciation of some words will be different from country to country, place to place. The dataset must contain the audio files which consists of the recordings of audio from speakers who are from different places and having different ages.

The study used these five datasets as they are satisfying the constraints. They are Toronto Emotional Speech Set (TESS), Survey Audio-Visual Expressed Emotion (SAVEE), RAVDESS, CREMA-D and Berlin. The results obtained from the study are based on combined usage of these above five datasets.

4 Features

To accurately detect the emotions underlying in the audio, there are some important features which are to be considered. In this study, key features like Mel frequency cepstral coefficients (MFCCs), Mel spectrogram and Chroma were extracted from the audio files. After loading the necessary datasets into the Python environment, the above-mentioned features were extracted to perform exploratory data analysis followed by modeling. To extract these features, there is package in Python environment named Librosa which is used throughout the study to deal with audio files.

Any machine learning model can only understand numerical values and only take numbers as input. Here, we have audio files in the dataset which is impossible for the model to detect the underlying patterns between different emotions. So, here by using the package Librosa, we are taking out the important features which are necessary in classifying emotions. For every audio, the values of the features will be varied. With the help of the values of the features extracted from the audio files, we train the model. We give these feature values as input to the model. As the feature values are numerical, model can understand and take those values as input. This is the reason why extracting these features is very important especially when we are dealing with audio files.

The characteristics of each of the features are mentioned below in detailed manner.

1. **Mel Frequency Cepstral Coefficients (MFCCs)**
   Analyzing any kind of time signals or any type of periodic components such as echoes ends up showing up a sharp peak in the respective frequency spectrum. The frequency spectrum may be Fourier spectrum which is obtained by imposing a Fourier transform on given time signal. By imposing the Fourier transform on a spectrogram, the feature cepstrum is achieved. The scale which relates the perceived frequency to measured actual frequency is called Mel scale. MFCC is taken on a Mel scale which is a special characteristic. This scales the frequency of the audio file loaded from the dataset such that humans can hear. Vocal tract and MFCC together represents the envelope of the speech signal.
2. **Mel Spectrogram**
   The spectrogram is generated by imposing fast Fourier transform on overlapping windowed audio signal. The representation of amplitude in the produced spectrogram is mapped on a Mel scale.
3. **Chroma**
   The feature Chroma has a twelve element feature vector which indicates the energy usage of pitch present in the audio signal. Here the feature values are based on a standard chromatic scale.

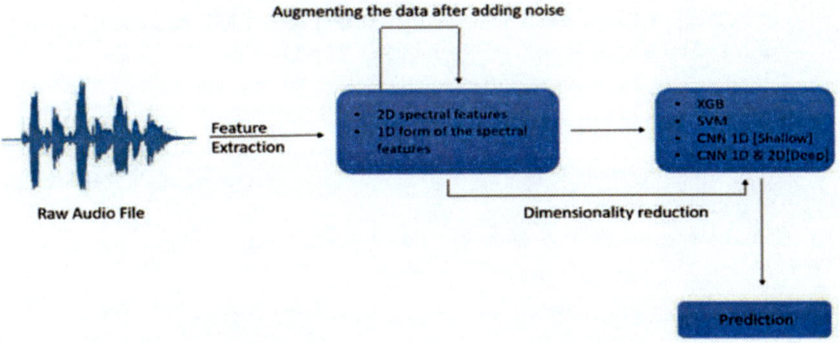

Fig. 1 Model pipelining schematic diagram

5 Modeling

5.1 Model Pipeline

The solution pipeline for this study is shown in Fig. 1 as a plan. Input is a green signal, processed as shown. 2D elements were first found in the database and then converted into 1D form using linear methods. In our four databases, the sound is presented in crude audio (excluding CREMA-D as others record studio and clean). Features were then found in audio files and included in our database. Following the release of the feature, we used SVM, XGB, CNN-1D (Shallow) and CNN-1D in our 1D data frame and CNN-2D in our 2D tensor to use several machine learning techniques. The study used size reduction to see if it was appropriate and re-trained models as well because some models had transferred data and looked at a number of factors (181 in 1D).

5.2 CNN Model Architecture Implemented

- **Convolution Neural Networks—1D (Shallow):**
  The model implemented in the study included a 64-channel convolution layer with the same padding, a dense layer and an output layer.
- **Convolution Neural Networks—2D (Deep):**
  The model built in the study is the same way as VGG-16, with the exception that the last two blocks of three convolution layers were deleted to reduce complexity. The architectural complexity of this CNN model was as follows:
  - Two layers of 64 channels with a kernel size of 33% and parallel padding, followed by a multi-layer compound size 22 and stride 22.

- Two layers of 128 channel mixers with a kernel size of 33% and equal padding, followed by a multi-layer composite layer 22 and stride 22.
- Three mixing layers with 256 channels, 33 kernel size, and the same padding, followed by a multi-layer with 22 and stride 22.
- The 'relu' activation function was applied to each convolution layer.
- After softening, two dense layers of 512 units were applied each, followed by 0.1 and 0.2 drop layers.
- Finally, a 'softmax' activation function was introduced to the output layer.

6 Results

After training the model using a series of convolution layers, the accuracy obtained by con2d model is 98%. 98% accuracy obtained is training accuracy. Validation accuracy obtained is just 72% which is very less compared to training accuracy. Training accuracy is the accuracy obtained by using the training data used to train the model. Validation accuracy is the accuracy obtained by using the test data which is new to the model. The less accuracy obtained by the model using test dataset and very high accuracy using training dataset indicate that the model is over fitted as shown in Fig. 2.

In order to tackle this situation, the study used MFCC augmentation to reduce the testing loss. In MFCC augmentation, it prepares the data using the function prepare_data(). As we are using augmentation, we change the parameter aug to 1. If augmentation is not needed, the parameter aug is set to 0. As the parameter aug is set to 1, the model prepares the data using augmentation. In augmentation, the noise will be removed from the audio so that the model can only detect the original voice without any disturbances. If there is noise, there will be deviation in the accuracy. The pitch will also be changed according to the emotion, so that the model cannot decide the emotion based on pitch.

After training the model with MFCC augmentation using a series of convolution layers, the accuracy obtained by con2d model is 88%. 88% accuracy obtained is

Fig. 2 Loss versus epoch using convolution networks

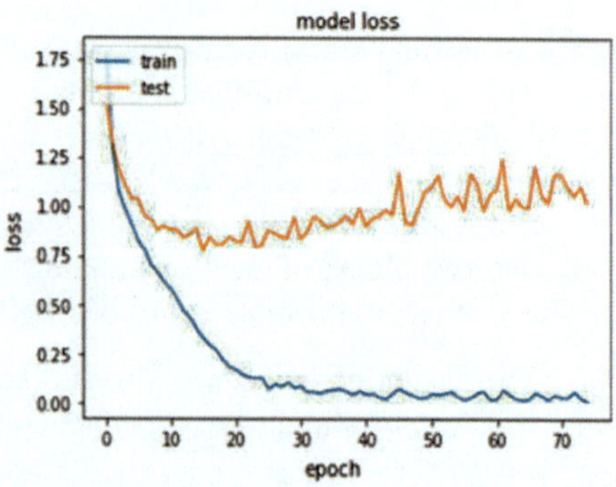

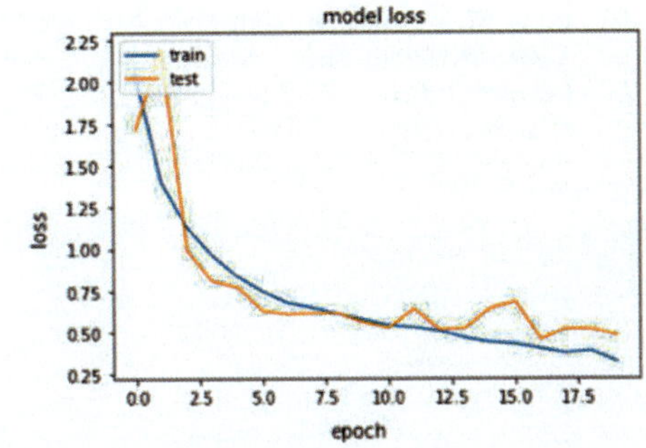

Fig. 3 Loss versus epoch using convolution networks by performing MFCC augmentation

training accuracy. Validation accuracy obtained is just 82% which is better compared to validation accuracy without performing any augmentation. Training accuracy is the accuracy obtained by using the training data used to train the model. Validation accuracy is the accuracy obtained by using the test data which is new to the model. After performing augmentation, the model over fitting reduced and improved test accuracy drastically as shown in Fig. 3.

7 Conclusion

Research has shown how to use machine learning to extract subconsciousness from spoken audio data, as well as some insights into the expression of human emotions by voice, in this study. This method can be applied to a variety of settings, including customer service or marketing call centers, voice-based assistants or bots chat, language research and so on. The developed model is well trained, and it is capable enough to classify the emotions. The additional unique part is it can also detects the gender. When the audio file is given as input, the model detects the gender and also emotion that underlied in the audio file.

References

1. https://www.analyticsinsight.net/speech-emotion-recognition-ser-through-machine-learning/
2. https://medium.com/prathena/the-dummys-guide-to-mfcc-aceab2450fd
3. http://practicalcryptography.com/miscellaneous/machine-learning/guide-mel-frequency-cepstral-coefficients-mfccs/
4. https://towardsdatascience.com/getting-to-know-the-mel-spectrogram-31bca3e2d9d0
5. https://tspace.library.utoronto.ca/handle/1807/24487
6. https://zenodo.org/record/1188976#.XvbvZudS_IU
7. https://github.com/CheyneyComputerScience/CREMA-D
8. Al-Talabani A, Sellahewa H, Jassim SA (2015) Emotion recognition from speech: tools and challenges. In: Mobile multimedia/image processing, security, and applications, pp 9497–9505

9. Sezgin MC, Gunsel B, Kurt GK (2012) Perceptual audio features for emotion detection. EURASIP J Audio Speech Music Process 1:1–6
10. Maddileti T, Chand T, Reddy VSK, Anurag KS (2020) Analyzing the impact of covid-19 on pollution and forecasting the pollutants levels in India. Turkish J Physiotherapy Rehab 32:1–3

Performance Analysis of Automatic Modulation Recognition Using Convolutional Neural Network

N. Venkateswara Rao and B. T. Krishna

Abstract In recent trends, automatic modulation recognition (AMR) has become a primary candidate in cognitive radio (CR). Modulation recognition using deep learning (DL) drawn attention of users in recent times due to its capability of performing in blind scenarios. In this article, convolutional neural network (CNN) model has been preferred for modulation classification purpose. This model learns from the features of modulation schemes present in the training data. CNN classifier operates in all adverse channel conditions. MPSK and MQAM modulation schemes are chosen in this paper. Total 7 modulation classes is trained in CNN with different training rates at different SNR. In this paper, proposed approach is compared with traditional models. Simulation results shows that proposed model performs well at different SNR.

Keywords AMR · CNN · DL · CR

1 Introduction

Future wireless applications demand continuous growth in certain aspects like mobility and quality of service. In telecommunication field, several challenges still need to be covered like providing encrypted wireless communication, accurate channel estimation and upgrading the QoS of the wireless methodologies used [1]. AMR can be an ultimate solution to the challenges which are stated above. AMR is used as a primary module in cognitive radio, which identifies the modulation class in presence of noise [2]. AMR is having a large research scope in wireless communication. It is used for spectrum management and interference rejection in civilian applications. It is used for jamming and demodulation of the signal in battlefield scenarios. It has been extended to various fields like radar [3, 4] and medical. Al-

N. V. Rao (✉) · B. T. Krishna
Department of ECE, JNTU Kakinada, Kakinada, A.P, India

P. S. R. Chowdary et al. (eds.), *Evolution in Signal Processing and Telecommunication Networks*, Lecture Notes in Electrical Engineering 839,
https://doi.org/10.1007/978-981-16-8554-5_42

Qatab et al. proposed a framework that incorporates feature selection and acoustic features that will be helpful in speech classification, depending on the levels of deterioration [5].

In the last few years, various algorithms are proposed for AMR in the literature [6]. In general, signal parameter detection methods are usually classified into likelihood ratio-based (LB) approach and feature-based (FB) approaches [7]. LB approach is formulated as a hypothesis problem [8]. It is computed by setting a specific threshold value; later, it will compare the likelihood ratio with that threshold value [9]. Zhu et al. selected the modulation type based on probability distribution that increases the loglikelihood function [10]. In LB approach, some prior probability information is required, and it is selected at the expense of high computational complexity. FB approaches have dominance in practical implementations due to their low complexity. FB approach makes decisions based on extracted features. For feature extraction characteristics like cyclic statistics [11], wavelet transform [12] and high-order cumulants [13, 14] are used. ML algorithms like ANN [11], KNN [15] and SVM [16] have been widely used for classification. But these methods were developed such that it can be operated in limited environments. Feature extraction and classification both are independent processes. Due to their mismatch, the performance will be degraded for the above approaches.

With the recent advances in wireless technologies, a new paradigm based on DL is adopted for AMR. In [17–20], high performance is achieved, which decreases human participation. DL mainly focuses on the network models, which includes CNN and recurrent neural network (RNN) for AMR. DL has numerous applications in natural language understanding [21], speech recognition and image recognition. DL has been considered as important tool in many areas such as IoT detection [22, 23], MIMO detection [24, 25], radio control [26–28] and channel estimation [29, 30].

Recently, CNNs have created a big impact in computer vision [31–33]. With this success, it is now playing a lead role in AMR [34]. Initial design model of CNN was developed by O'Shea et al. for classification [35], which gave better performances compared to previous approaches, and it is more robust for different kind of signals. Authors at [36] developed a model cascading two CNN in their work. Out of these two, first CNN utilized in-phase and quadrature (IQ) components and the second one made use of the constellation diagrams. Long-term dependency is better handled by RNN. Karpathy et al. [37] combined RNN and CNN for image classification problem, in which CNN is used for feature extraction, and RNN is used for modeling temporal dependencies.

This paper is presented as follows. Section 2 provides a brief introduction of proposed model. Section 3 gives an overview of experimental setup. Section 4 shows the results and comparison tables with traditional approaches. Concluding remarks are presented in Sect. 5.

2 Proposed Model

CNN model is preferred in this article for modulation recognition. Almost all CNN models comprise various layers like convolutional layer, batch normalization layer, rectification layer (ReLU), max pooling layer, fully connected layer, Softmax layer and output layer. Convolution layer is utilized for extracting features from input data, and these are processed to next stage. Excluding input layer, remaining layers needs an activation function. It is provided by ReLU and parametric ReLU. Pooling layer is mainly used to improve the performance. It is also used to reduce the dimensions of feature to overcome overfitting. In CNN model, fully connected layer will be used as a classifier. For better classification, fully connected layer makes use of results from previous layers. Softmax layer utilizes Softmax activation function for mapping the output data. It is preferred for multi-classification problem.

The proposed CNN model comprises image input layer, convolution 2D layer, batch normalization layer, ReLU layer and max pooling 2D layer. The following layers are repeated in 6 stages. In the last stage, average pooling 2D layer is used rather than max pooling 2D layer. Finally, fully connected layer is utilized followed by Softmax layer and classification output layer. The proposed CNN model is shown in Fig. 1.

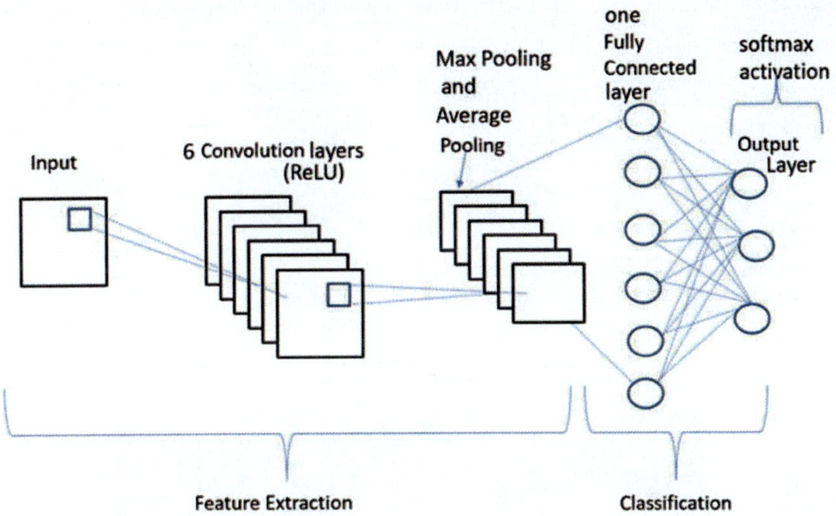

Fig. 1 Proposed CNN model

3 Experimental Setup

For modulation recognition, 7 modulation classes are considered, i.e., 64QAM, 16QAM, CPFSK, GFSK, QPSK, BPSK and 8PSK. For each modulation class, several frames are generated for effective training, where 80% of the features are used for training, and 10% of them is utilized for testing phase. In the network training stage, 10% of features are used for validation. Parameters chosen for waveform generation are samples per frame is 1024 samples, samples per symbol is 8, sample rate (f_s) is 200 kHz and center frequency is chosen as 902MHzfor digital signal modulation types and 100 MHz foe analog signal modulation types. Figure 2 represents the real and imaginary parts of different modulation classes considered for simulation.

Spectrograms of modulated signals are generated with the help of short-time Fourier transform (STFT).

Figure 3 represents the spectrograms of different modulation classes at 20 dB SNR. Each frame in the channel passes through different channel conditions like AWGN and Rician multipath fading.

CNN model makes use of 1024 samples for training. Based on training, it accurately finds out the type of modulation. CNN model is given training with different training rates and different timings. Parameters chosen for training are batch size, epochs and learning rate. The learning rate and epochs are varied based on training.

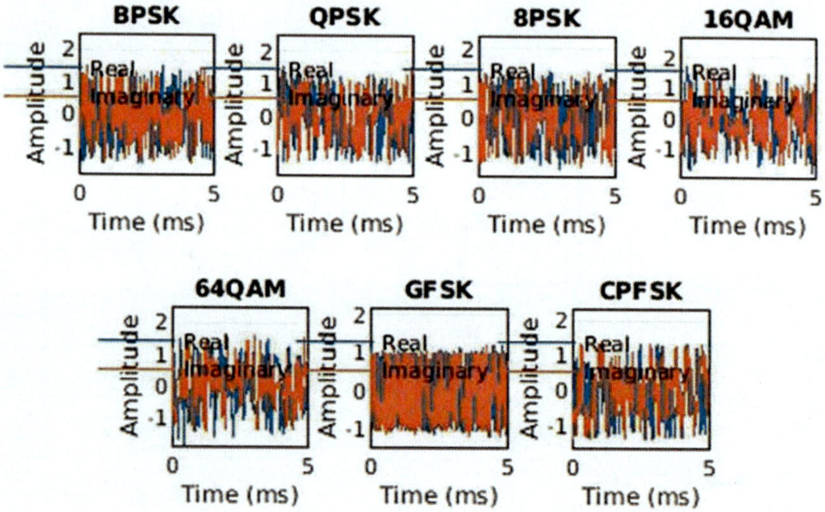

Fig. 2 Real and imaginary parts of the signal frames

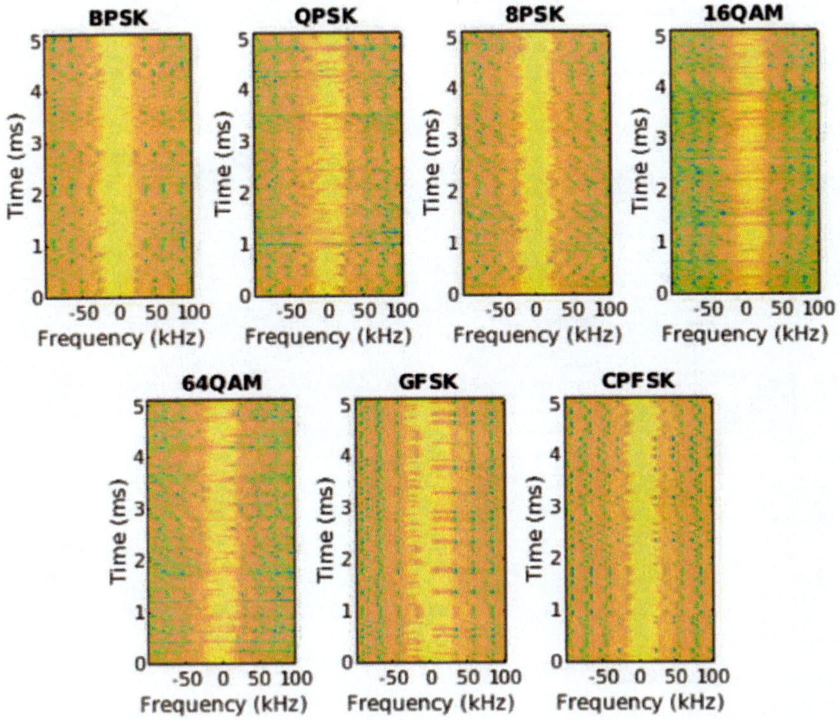

Fig. 3 Spectrograms of seven modulation classes

4 Simulation Results

QPSK, BPSK, 8PSK, CPFSK, GFSK, 16QAM and 64QAM signals are considered for training to check the performance of CNN at non-ideal channel conditions with SNR = (0,10,20 and 30) dB. CNN model is trained with different training rates at (80, 70, 60 and 50) and different testing times at (10, 20, 30 and 40). The confusion matrices of proposed CNN in terms of true class and prediction class at 20 dB SNR are shown in Fig. 4. For 80% training, performance reached is 93.14%. The performance levels at 70, 60 and 50% are 92.71, 92.71 and 92.43%.

Table 1 summarizes the accuracy of each modulation class at different training rates and SNR values. And Table 2 summarizes the test accuracies of CNN model for AMR with different SNR and different training rates. Table 3 shows the performance comparison of the proposed approach with other existing approaches.

From the simulations, it is clear that event at 50% training rate, the proposed classifier performance better in classification of modulations for SNR 10 dB and above.

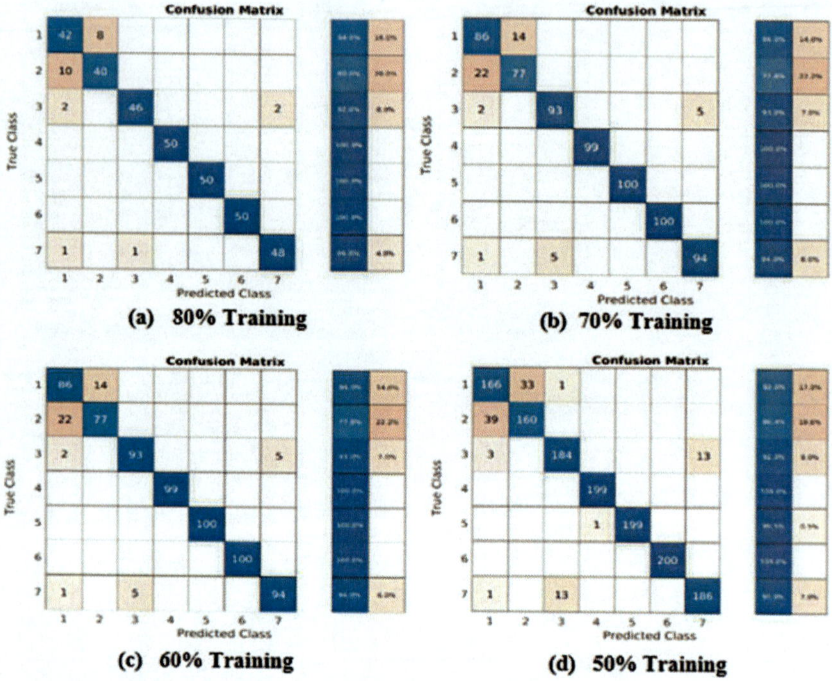

Fig. 4 Confusion matrix at 20 dB SNR

5 Conclusion

This paper presented CNN model approach for AMR of digitally modulated signals. From the simulation results, it is proven that CNN works in non-ideal channel conditions. It achieved an accuracy of 94% at 30 dB SNR. The proposed classifier outperformed all existing approaches. In future, this model can be applied to different signal classification problems in all adverse conditions.

Table 1 Accuracy of each class at different training rates and SNR

Class/training rate	Accuracy (%)															
	SNR 30 dB				SNR 20 dB				SNR 10 dB				SNR 0 dB			
	80	70	60	50	80	70	60	50	80	70	60	50	80	70	60	50
1	78	81	79.3	80	84	86	86	83	89.8	90.9	85.9	84.8	48.1	40.7	41.8	42.2
2	88	85.9	85.9	86.9	80	77.8	77.8	80.4	57.4	48.9	48.6	49.5	20	15.4	14.9	19
3	96	93	90.7	92	92	93	93	92	89.8	87.9	82.6	82.4	43.8	38.7	40	39.2
4	100	100	100	100	100	100	100	100	100	100	100	99.5	92.1	89.2	88.5	88.6
5	100	100	99.3	99.5	100	100	100	99.5	100	100	99.3	99.5	87.5	90.3	89.5	90.2
6	100	100	100	100	100	100	100	100	100	100	100	100	93.9	93.9	91.9	92.9
7	96	93	91.3	92	96	94	94	93	76	80	75.2	76.3	63.3	63.2	59.8	58

Table 2 Summary of test accuracies for different SNR/training rates

SNR (dB)	**Accuracy (%) at training rate of**			
	80	**70**	**60**	**50**
30	94	93.14	92.29	92.86
20	93.14	92.71	92.71	92.43
10	86.57	85.86	83.62	83.79
0	50	46.43	45.91	46.36

Table 3 Comparison of proposed approach with existing approach

Reference no	Approach	Features	Modulation classes	Accuracy
[38]	MTL-CNN	Signal samples	9	86.97
[39]	LSTM	Time domain Amplitude and Phase	11	90
[40]	CNN and LSTM	I, Q and HOC	11	88
[41]	RNN	Signal samples	4	92
Proposed model	CNN	Spectrograms	7	94%

References

1. Almohamad TA, Salleh MFM, Mahmud MN, Karaş İR, Shah NSM, Al-Gailani SA (2021) Dual-Determination of modulation types and signal-to-noise ratios using 2D-ASIQH features for next generation of wireless communication systems. IEEE Access 9:25843–25857. https://doi.org/10.1109/ACCESS.2021.3057242
2. Venkata Subbarao M, Samundiswary P (2020) Performance analysis of modulation recognition in multipath fading channels using pattern recognition classifiers. Wireless Pers Commun 115:129–151. https://doi.org/10.1007/s11277-020-07564-z
3. Wan T, Jiang K, Tang Y, Xiong Y, Tang B (2020) Automatic LPI radar signal sensing method using visibility graphs. IEEE Access 8:159650–159660. https://doi.org/10.1109/ACCESS.2020.3020336
4. Wei S, Qu Q, Wang M, Wu Y, Shi J (2020) Automatic Modulation Recognition for Radar Signals via Multi-Branch ACSE Networks. IEEE Access 8:94923–94935. https://doi.org/10.1109/ACCESS.2020.2995203
5. Al-Qatab BA, Mustafa MB (2021) Classification of dysarthric speech according to the severity of impairment: an analysis of acoustic features. IEEE Access 9:18183–18194. https://doi.org/10.1109/ACCESS.2021.3053335
6. Subbarao MV, Samundiswary P (2021) Automatic modulation classification using cumulants and ensemble classifiers. In: Kalya S, Kulkarni M, Shivaprakasha KS (eds) Advances in VLSI, signal processing, power electronics, IoT, Communication and embedded systems. Lecture notes in electrical engineering, vol 752. Springer, Singapore. https://doi.org/10.1007/978-981-16-0443-0_9
7. Subbarao MV, Samundiswary P (2018) Spectrum sensing in cognitive radio networks using time–frequency analysis and modulation recognition. In: Anguera J, Satapathy S, Bhateja V, Sunitha K (eds) Microelectronics, electromagnetics and telecommunications. Lecture notes in electrical engineering, vol 471. Springer, Singapore. https://doi.org/10.1007/978-981-10-7329-8_85

8. Zhang T, Shuai C, Zhou Y (2020) Deep learning for robust automatic modulation recognition method for iot applications. IEEE Access 8:117689–117697. https://doi.org/10.1109/ACCESS.2020.2981130
9. Chen S, Zhang Y, He Z, Nie J, Zhang W (2020) A Novel attention cooperative framework for automatic modulation recognition. IEEE Access 8:15673–15686. https://doi.org/10.1109/ACCESS.2020.2966777
10. Zhu D, Mathews VJ, Detienne DH (2018) A likelihood-based algorithm for blind identification of qam and psk signals. IEEE Trans Wireless Commun 17(5):3417–3430
11. Ramkumar B (2009) 'Automatic modulation classification for cognitive radios using cyclic feature detection.' IEEE Circuits Syst Mag 9(2):27–45
12. Kumar S, Bohara VA, Darak SJ (Mar 2017) "Automatic modulation classification by exploiting cyclostationary features in wavelet domain," in Proc 23rd Nat Conf Commun (NCC), pp 1–6
13. Subbarao MV, Samundiswary P (2018) Automatic modulation recognition in cognitive radio receivers using multi-order cumulants and decision trees. Int J Recent Technol Eng (IJRTE) 7(4):61–69
14. Swami A, Sadler BM (2000) Hierarchical digital modulation classification using cumulants. IEEE Trans Commun 48(3):429–461
15. Subbarao MV, Samundiswary P (2019) K nearest neighbors based automatic modulation classifier for next generation adaptive radio systems. Int J Security Appl 13:41–50
16. Xie L, Wan Q (2018) Automatic modulation recognition for phase shift keying signals with compressive measurements. IEEE Wireless Commun. Lett. 7(2):194–197
17. O'Shea TJ, Corgan J, Charles Clancy T (2016) "Convolutional Radio Modulation Recognition Networks"
18. Zeng Y, Zhang M, Han F, Gong Y, Zhang J (2019) Spectrum analysis and convolutional neural network for automatic modulation recognition. IEEE Wireless Commun Lett 8(3):929–932
19. Bu K, He Y, Jing X, Han J (2020) Adversarial transfer learning for deep learning based automatic modulation classification. IEEE Signal Process Lett 27:880–884
20. Wang Y, Yang J, Liu M, Gui G (2020) LightAMC: lightweight automatic modulation classification via deep learning and compressive sensing. IEEE Trans Veh Technol 69(3):3491–3495
21. LeCun Y, Bengio Y, Hinton G (2015) Deep learning. Nature 521(7553):436
22. Diro AA, Chilamkurti N (2018) Distributed attack detection scheme using deep learning approach for Internet of Things. Future Gener Comput Syst 82:761–768
23. Lopez-Martin M, Carro B, Sanchez-Esguevillas A, Lloret J (2017) Conditional variational autoencoder for prediction and feature recovery applied to intrusion detection in IoT. Sensors 17(9):1967
24. Samuel N, Diskin T, Wiesel A (Jul 2017) "Deep MIMO detection," in Proc IEEE 18th Int Workshop Signal Process Adv Wireless Commun (SPAWC), pp 1–5
25. Yan X, Long F, Wang J, Fu N, Ou W, Liu B (May 2017) "Signal detection of MIMO-OFDM system based on auto encoder and extreme learning machine." In Proc Int Joint Conf Neural Netw (IJCNN), pp 1602–1606
26. Wijaya MA, Fukawa K, Suzuki H (Sep 2015) "Intercell-interference cancellation and neural network transmit power optimization for MIMO channels," in Proc. IEEE 82nd Veh Technol Conf (VTC-Fall), pp 1–5
27. Naparstek O, Cohen K (Dec 2017) "Deep multi-user reinforcement learning for dynamic spectrum access in multichannel wireless networks," in Proc. GLOBECOM—IEEE Global Commun Conf, pp 1–7
28. Wijaya MA, Fukawa K, Suzuki H (2016) 'Neural network based transmit power control and interference cancellation for MIMO small cell networks.' IEICE Trans Commun E99.B(5):1157–1169
29. D. Neumann, T. Wiese, and W. Utschick, "Deep channel estimation," in Proc. WSA 21th Int. ITG Workshop Smart Antennas, 2017, pp. 1–6.
30. Huang H, Yang J, Huang H, Song Y, Gui G (2018) Deep learning for super-resolution channel estimation and DOA estimation based massive MIMO system. IEEE Trans Veh Technol 67(9):8549–8560

31. Simonyan K, Zisserman A (2015) “Deep convolutional networks for largescale image recognition,” in Proc ICLR, pp 1–14
32. Huang G, Liu Z, Van Der Maaten L, Weinberger KQ (Jul 2017) “Densely connected convolutional networks,” in Proc IEEE Conf Comput Vis Pattern Recognit (CVPR), pp 4700–4708
33. He K, Zhang X, Ren S, Sun J (Jun 2016) “Deep residual learning for image recognition,” in Proc IEEE Conf Comput Vis Pattern Recognit (CVPR), pp 770–778
34. Lin R, Ren W, Sun X, Yang Z, Fu K (2020) A hybrid neural network for fast automatic modulation classification. IEEE Access 8:130314–130322. https://doi.org/10.1109/ACCESS.2020.3009471
35. O’Shea TJ, Corgan J, Clancy TC (2016) “Convolutional radio modulation recognition networks,” in Proc Int Conf Eng Appl Neural Netw, Cham, Switzerland, Springer, pp 213–226
36. Wang Y, Liu M, Yang J, Gui G (2019) Data-driven deep learning for automatic modulation recognition in cognitive radios. IEEE Trans Veh Technol 68(4):4074–4077
37. Karpathy A, Fei-Fei L (Apr 2017) “Deep visual-semantic alignments for generating image descriptions,” IEEE Trans. Pattern Anal. Mach. Intell 39(4):664–676 Ettefagh Y, Moghaddam MH, Eghbalian S (2017) An adaptive neural network approach for automatic modulation recognition. 2017 51st Annual Conf Inf Sci Syst 1–5
38. Mossad OS, ElNainay M, Torki M (2019) Deep convolutional neural network with multi-task learning scheme for modulations recognition. 2019 15th International Wireless Communications & Mobile Computing Conference (IWCMC), pp 1644–1649
39. Rajendran S, Meert W, Giustiniano D, Lenders V, Pollin S (2018) Deep learning models for wireless signal classification with distributed low-cost spectrum sensors. IEEE Trans Cognitive Commun Netw 4:433–445
40. Zhang M, Zeng Y, Han Z, Gong Y (2018) Automatic modulation recognition using deep learning architectures. 2018 IEEE 19th International Workshop on Signal Processing Advances in Wireless Communications (SPAWC), pp 1–5
41. Hu S, Pei Y, Liang PP, Liang Y (2018) Robust modulation classification under uncertain noise condition using recurrent neural network. IEEE Trans Vehicular Technol 69:564–577

Real-Time Image Enhancement Using DCT Techniques for Video Surveillance

D. Girish Kumar, G. Challa Ram, and M. Venkata Subbarao

Abstract Video surveillance at night time is very difficult task to identify the objects under different environmental conditions. The surveillance task plays a vital role for easily identifying the abnormal scenarios. This tracking data is quite helpful for armed services and civilians. The proposed algorithm enhances the intensity of real-time dark images. This can be achieved by gathering the daytime image and night time image of same place, and the noise can be removed by background subtraction after that image is fused with same place daytime image by discrete cosine transformation (DCT) technique. This algorithm provides best results in terms of reducing noise, and it justifies through quality metrics like peak signal-to-noise ratio (PSNR) and mean squared error (MSE).

Keywords Discrete cosine transform (DCT) · Background subtraction · Night vision images · Image fusion · Image enhancement

1 Introduction

Twilight vision image enhancement is necessary for monitoring the sensitive areas in night times to prevent the unwanted activities. Night-time-captured images are not clear due to poor illumination; identifying the entities is too difficult in images so image enhancement is necessary, and it is done by traditionally several methods like slicing, contrast stretching and histogram equalization; but these methods are failed to remove the noise and problem for maintaining the required threshold level due to some troubleshooting factors like bleedings, dull illumination, color shift [1], etc.

The proposed method strengthens the intensity levels present in night-captured images. Initially, subtracting the background details from the required exact night time image by using thresholding technique. The resultant subtracted image is fused with daytime image by DCT transform technique [2] which yields good results and

D. Girish Kumar (✉) · G. Challa Ram · M. Venkata Subbarao
Department of ECE, Shri Vishnu Engineering College for Women, Bhimavaram, India
e-mail: girishkumar.d@svecw.edu.in

P. S. R. Chowdary et al. (eds.), *Evolution in Signal Processing and Telecommunication Networks*, Lecture Notes in Electrical Engineering 839,
https://doi.org/10.1007/978-981-16-8554-5_43

recollects more details for changing different threshold values and helps to sustain the accurate results in peak-to-signal noise ratio (PSNR) and mean squared error (MSE).

This paper is ordered as follows; initially, image enhancement existing techniques are discussed in Sect. 2, the suggested algorithm implemented for low-light condition images is discussed in Sect. 3, the real-time results are shown in Sect. 4 and at last closure note in Sect. 5.

2 Existing Fusion Techniques for Image Enhancement

The different fusion techniques are used for enhance the night time image for the purpose of surveillance. Anwaar-ul-Haq et al. [3] implemented a semi-automatic colorization and reliable image fusion (Scarf) algorithm which helps to recollect the features in dull-illuminated images from different places to compliance the human visual system by pixel-level image fusion. Naji and Aghagolzadeh [4] enhance the images by interconnecting the coefficients using discrete cosine transform (DCT) techniques in multi-focus images which offer low computation time but applicable to only related standard set images present in the jpeg format.

Asmare et al. [2] increase the image internal features by using the multi-resolution approach useful to monitor the adverse environmental condition images. Contourlet decomposition technique achieves accurate results in dull-light conditions but fails to eliminate noise in input images. Fan et al. [1] enhanced the visibility by considering infrared night time images among different resolutions by multi-scale approach, and preprocessing is done by wavelet transform techniques which is helpful to assist the surveillance systems in real time but requires more computation time. Histogram equalization method extract more details in input image but preserve the hue and saturation components [5].

Liu et al. [6] assessed the images by comparative trails on different multi-resolution fusion algorithms based on quality metrics like correlation coefficients and power spectrum for objection evaluation but slightly fails to avoid blurring effects in images. Ancuti et al. [7] researched on dehazy images especially in night time by multi-scale fusion approach and by decomposition techniques which used to get finest details from low light condition images by estimating hazy pixels brings more efficiency in computation compared to other techniques.

Choi et al. [8] develop multi-spectral dataset useful for image enhancement for the purpose of self-driving cars. This technique is used to extract more details in low-light illumination images and to guide toward detection of objects easily; further, this dataset helps researchers to study their specific applications. Waxman et al. [9] proposed that fused images bring more qualitative information in night scenes when infrared image fused with night time image, but at same time, the computational performance is lack behind due to this technique. Gabriel Zahi and Shigang Yue [10] reduce the blurring in motion images by using temporal summation procedure and enhance the quality in texture details present in dark illumination-captured images during movement of car cameras or surveillance cameras, but it fails to reduce noise in some scenarios particularly during movement of objects.

3 Proposed Algorithm for Real-Time Images

The proposed algorithm requires preprocessing, low-light image enhancement and image fusion. The input image obtained from surveillance video converted in to frames after the filtering of noise for identification of objects [8]. The steps for convert the video to frames are represented in Fig. 1.

- Collect the surveillance video file in .avi format.
- Transform the night time video into frames.

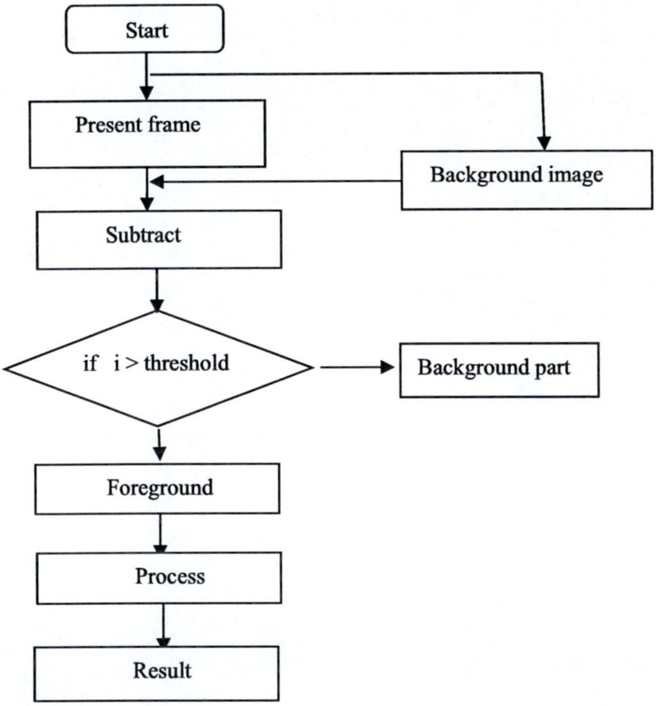

Fig. 1 Background subtraction

- Interpret the number of frames shown in command window.
- Designate the numbers to frames and add .jpeg or .jpg extension.

The main concern for image enhancement is to increase the internal details of image-like object boundaries, contrast, brightness, etc. useful for analysis and to avoid illegal activities; here, image enhancement is done by background subtraction, and the flowchart is shown below.

The steps involved are represented in Fig. 2

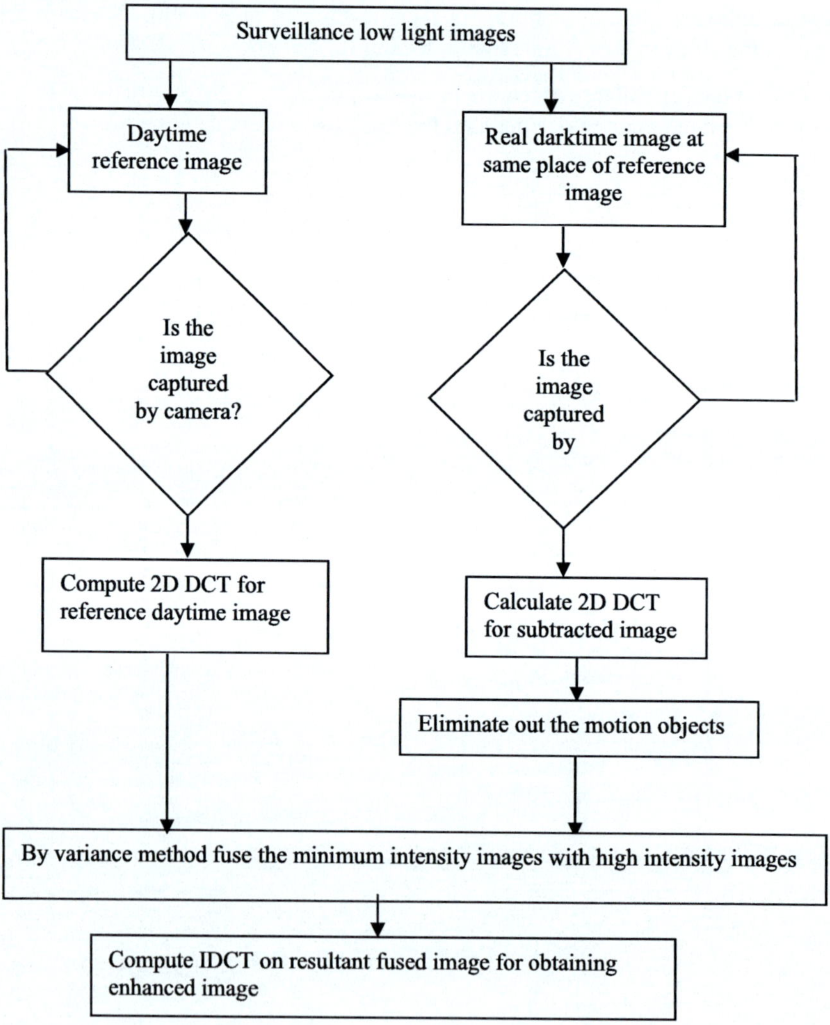

Fig. 2 Image fusion and enhancement

- Gather the actual daytime image and reference night time image.
- Transform both images as grayscale images and subtract reference image from actual image.
- Enhance the subtracted image by using a threshold value.
- Transform subtracted image to color image by referring actual image.

Figure 2 shows image fusion is done by DCT, and the series of steps are as follows.

- Collect the two images to be fused and segregate the image in to 8X8 size.
- Compute 2D DCT every block; it is given by

$$G(x, y) = \frac{1}{\sqrt{2n}} h(x) f(y) \sum_{x=0}^{n-1} \sum_{y=0}^{n-1} g(x, y) \cos \frac{(2x+1)u\pi}{2n} \cos \frac{(2y+1)v\pi}{2n} \quad (1)$$

where $h(x) = \begin{cases} \frac{1}{\sqrt{2}} & u = 0 \\ 1 & u > 0 \end{cases}$ $f(y) = \begin{cases} \frac{1}{\sqrt{2}} & v = 0 \\ 1 & v > 0 \end{cases}$

- Choose maximum variance value sub-block.
- Compute inverse discrete cosine transform (IDCT) for all sub-blocks for obtaining the enhanced fused image.

$$g(z, s) = \frac{1}{\sqrt{2n}} \sum_{u=0}^{n-1} \sum_{v=0}^{n-1} h(x) f(y) G(x, y) \cos \frac{(2x+1)u\pi}{2n} \cos \frac{(2y+1)v\pi}{2n} \quad (2)$$

4 Analysis of Proposed Algorithm Results

The proposed algorithm was implemented in MATLAB R2020. The selected real-time dark source image for identifying the human details after preprocessing is shown in Fig. 3.

The darktime background reference of selected target image is shown in Fig. 4. In this work, we consider the background in daytime and night time. The actual daylight reference image of selected target is shown in Fig. 5.

The background subtracted image is shown in Fig. 6. The resultant fused image after applying DCT transformation function and also removing the motion objects shown in Fig. 7.

$$\mathrm{PSNR} = 20 \log_{10}\left(\frac{\mathrm{MAX}_f}{\sqrt{\mathrm{MSE}}}\right) \quad (3)$$

where the mean squared error (MSE) is:

Fig. 3 Darktime image with required target

Fig. 4 Darktime image for background reference

$$\mathrm{MSE} = \frac{1}{mn}\sum_{0}^{m-1}\sum_{0}^{n-1}||f(i,j) - g(i,j)||^2 \tag{4}$$

The proposed algorithm accuracy efficiency is increased compared to traditional algorithms by metrics PSNR and MSE. The values of PSNR and MSE of actual image are 25.43 and 187.51, respectively, and for resultant image 25.31 and 193.18, respectively.

Fig. 5 Daylight image for reference

Fig. 6 Resultant image after background subtraction

5 Conclusion

Image fusion by DCT transform enhances the intensity of dark images and justifies through results. Further, this algorithm helps to gather more details from clumsy images like dull-light condition due to atmospheric changes, smoky images and motion images. Improve PSNR and reduce the MSE that represent robustness of algorithm for tracking video surveillance data. In future, this algorithm is quite helpful

Fig. 7 Image fused with daytime reference image and background subtracted image

to enhance and to serve the purpose of armed forces, wildlife conservation and assisted ETC applications and also dealing with remote-sensing images.

References

1. Fan Z, Yan L, Xia Y, Fu M, Xiao B (2018) Fusion of multi-resolution visible image and infrared images based on guided filter. In: Proceedings of the 37th Chinese control conference, July 25–27, Wuhan. IEEE, pp 4449–4454. https://doi.org/10.23919/ChiCC.2018.8483810
2. Asmare MH, Asirvadam VS, Iznita L (2009) Multi-sensor image enhancement and fusion for vision clarity using Contourlet transform. In: 2009 international conference on information management and engineering. IEEE, pp 352–356. https://doi.org/10.1109/ICIME.2009.112
3. Anwaar-ul-Haq, Gondal I, Murshed. M (2010) Scarf: semi-automatic colorization and reliable image fusion. IEEE, pp435–440. https://doi.org/10.1109/DICTA.2010.80
4. Naji MA, Aghagolzadeh A (2018) Multi-focus image fusion using singular value decomposition in DCT domain. In: 10th Iranian conference on machine vision and image processing (MVIP). IEEE, pp 2166–6784. https://doi.org/10.1109/IranianMVIP.2017.8342367
5. Lee S-L, Tseng C-C (2017) Color image enhancement using histogram equalization method without changing hue and saturation. In: IEEE international conference on consumer electronics—Taiwan (ICCE-TW). IEEE, pp 305–306. https://doi.org/10.1109/ICCE-China.2017.7991117
6. Liu Z, Blasch E, Xue Z, Zhao J, Laganie R, Wu W (2012) Objective assessment of multiresolution image fusion algorithms for context enhancement in night vision: a comparative study. IEEE Trans Pattern Anal Mach Intell 34(1):94–109. https://doi.org/10.1109/TPAMI.2011.109
7. Ancuti C, Ancuti CO, De Vleeschouwer C, Bovik AC (2016) Night time Dehazing by fusion. In: ICP 2016. IEEE, pp 2256–2260
8. Choi Y, Kim N, Hwang S, Park K, Yoon JS (2018) KAIST multi-spectral day/night dataset for autonomous and assisted driving. IEEE Trans Intell Transp Syst 19(3):934–948. https://doi.org/10.1109/TITS.2018.2791533

9. Waxman A, Fay D, Ilardi P, Savoye D (2017) Sensor fused night vision: assessing image quality in the lab and in the field. In: 2006 9th international conference on information fusion. IEEE. https://doi.org/10.1109/ICIF.2006.301767
10. Zahi D, Yue S (2014) Reducing motion blurring associated with temporal summation in low light scenes for image quality enhancement. In: 2014 International conference on multisensor fusion and information integration for intelligent systems (MFI). IEEE. https://doi.org/10.1109/MFI.2014.6997725

A Novel Semi-blind Digital Image Watermarking Using Fire Fly Algorithm

Ch. Ravi Kumar and P. Rajesh Kumar

Abstract The proposed firefly optimization technique is more robust and imperceptible, while compared with particle swam optimization. Firefly is able to optimize the system fast for the proposed innovative scheme of watermarking. The insertion of watermark, discrete cosine transform (DCT) applied to the sub-bands with the coefficients to generate the watermarked image. Embedding and extraction have been done with parameters and FF algorithm to optimize the values of peak signal-to-noise ratio (PSNR), normalize cross correlation. Firefly algorithm is an objective function to get more robustness and imperceptibility. The results obtained offer a superior approach to traditional methods and are more effectively resistant to various attacks.

Keywords Image watermarking · Firefly algorithm · MSFs

1 Introduction

Recently, digital watermarking [1, 2] is the innovative approach used to authenticate content and secure copyrights. The process of inserting the watermark into original image changes the content. The robustness, imperceptibility, and fragility are the main features of watermarking technique. Imperceptibility refers to PSNR and robustness refers to (NC). The watermark may semi-fragile [3] over different attacks like Gaussian noise, median filtering, and cropping rotation attacks. The watermark brief description of various watermarking schemes explained in [4, 5]. Embedding done by using DCT-SVD transforms. To optimize the embedding constant their FIREFLY (FF) adapted. The embedding of watermark constant gone through different transforms like NBP-IWT-DCT-SVD and FF algorithm, and extracting will be done inversely. Proposed method has been done on the image and experiments held for different attacks. Section 2 gives the related work. Section 3 gives preliminaries, Sect. 4 demonstrates the details of proposed approach, and Sect. 5 describes about performance evaluation and ends by Sect. 6.

Ch. Ravi Kumar (✉) · P. Rajesh Kumar
Department of ECE, AUCE, Visakhapatnam, India

P. S. R. Chowdary et al. (eds.), *Evolution in Signal Processing and Telecommunication Networks*, Lecture Notes in Electrical Engineering 839,
https://doi.org/10.1007/978-981-16-8554-5_44

2 Related Work

Imperceptibility and robustness of image watermarking is the most effective thing for content hiding and copyright protection of the data. Various watermarking schemes, i.e., spatial and frequency domain [6–8] represents the transformation techniques. The image is directly changed by pixel values in the spatial domain. This technique improves robustness without significantly degrading the image. In SVD [9, 10], the singular values of the entire image enhanced and combined to host image. Embedding the watermarking pixel values into the high-frequency region of the DCT coefficients. The optimization is carried through FF for an embedding constant [11–13]. Various algorithms carried to optimize the system parameters. It gives better results than (GA) genetic algorithm [14–16] and PSO. FF algorithm explains about the fireflies that produce light has no infrared or ultraviolet frequency attacks mates or prey.

3 Preliminaries

In this section, different transforms which are used in this work are presented.

A. *Integer Wavelet Transform (IWT)*

The integer wavelet transform involves three factors, i.e., split, predicting, and updating. Loss of information is the main problem of IWT.

(1) *Split*: splitting of two subsets into odd and even.

$$\begin{cases} x_e = x(2n) \\ x_o = x(2n+1) \end{cases} \tag{1}$$

Split is carried until odd and even sets are closely correlated. Nearby samples correlated than the far.

(2) *Prediction*: A predictor for one set with other set needed.x_o is odd index samples, x_e is even index samples and p is predictor:

$$\widetilde{x_o} = P(x_e) \tag{2}$$

Coefficient of signal *x*(*n*) is considered as the difference 'd'

$$d = x_o - \widetilde{x_o} = x_o - P(x_e) \tag{3}$$

$VV^T = I_n$; The U columns are orthonormal AAT own vectors, the V columns are orthonormal ATA vectors, and the S, a diagonal matrix which holds in descending order the roots of the U or V proper values. If r in which condition is a ($r \leq n$) is a rank of matrix A, then the diagonal matrix S elements satisfy (4) and the A can written as (5):

$$\lambda_1 \geq \lambda_2 \geq \cdots \geq \lambda_r > \lambda_{r+1} = \lambda_{r+2} \cdots = \lambda_n = 0, \tag{4}$$

$$A = \sum_{k=1}^{r} \lambda_k u_k v_k^T \tag{5}$$

where λ_k is kth singular value, u_k and V_k are the kth eigenvectors of U and V.

B. *Discrete Cosine Transform (DCT)*

The signal is transformed into frequency domain from the spatial domain using DCT, which has excellent energy compaction properties. It is a famous transformation technique. The small high-frequency components are eliminated with the use of DCT.

C. *Particle Swarm Optimization (PSO)*

The PSO search for the population represents a velocity, location.
(g best) (p best) by the equations below:

$$v_i(n+1) = w_i v_i + c_1 \text{rand}_1(p\text{best} - x_i(n)) + c_2 \text{rand}_2(g\text{best} - x_i(n)) \tag{6}$$

inertia weight:

$$w_i = w_{\max} - \frac{w_{\max} - w_{\min}}{\text{iter}_{\max}} \cdot \text{iter} \tag{7}$$

4 Proposed Watermarking Scheme

Yang (2008) introduced firefly algorithm (FA). Firefly characteristics depend on light flashing patterns and brightness. FA is based on the following idealized firefly flashing behavior. Properties:

(1) Every firefly is unisexual so the firefly gets attracted to another firefly regardless of its sex. (2) The attraction is directly proportional to its intensity of light, and the lower is brighter. (3) The objective function's landscape is used to determine the brightness of the firefly (Fig. 1).

Each firefly relates to the problem in the firefly algorithm, and its objective function is calculated using fitness/shine (I). Firefly with a higher fitness is given a lighter firefly. Lighter fireflies switch to the brighter firefly by changing the values with a lower fitness score. Attractiveness (β) varies exponentially between brighter firefly I and brighter firefly j at a Cartesian distance (r_{ij}) (Figs. 2 and 3).

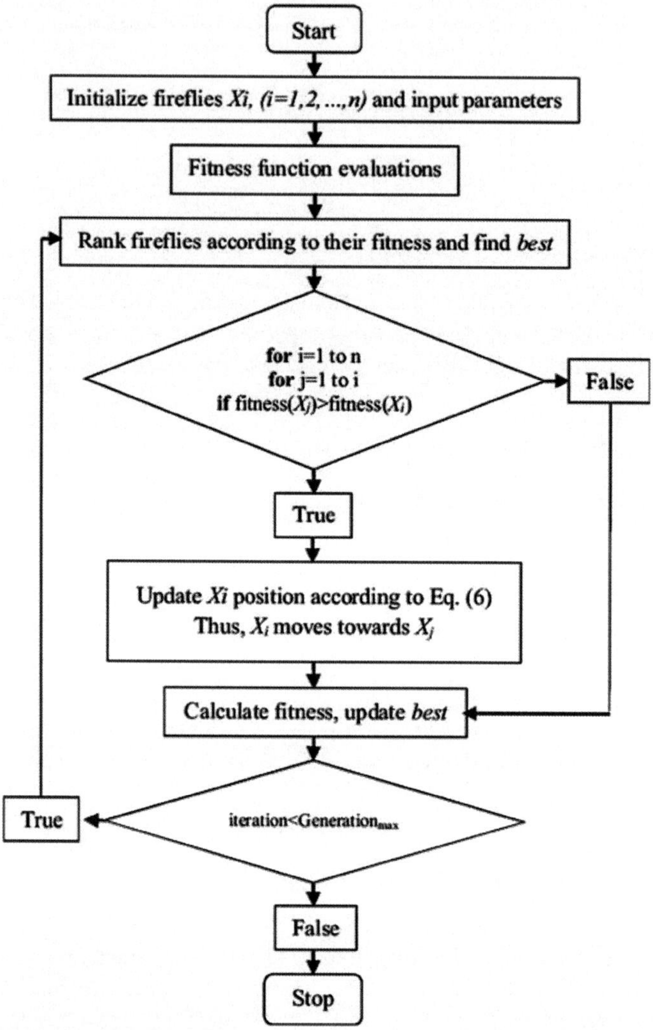

Fig. 1 Flowchart of firefly algorithm

A. ***Embedding Pattern***

Step 1: Perform w and DCT

$$\{W_{LL}, W_{LH}, W_{HL}, W_{HH}\} = lwt2(W) \tag{8}$$

Step 2: Apply SVD

$$\left[u_{LL}^{w} s_{LL}^{w} v_{LL}^{w}\right] = svd(W_{LL}),$$

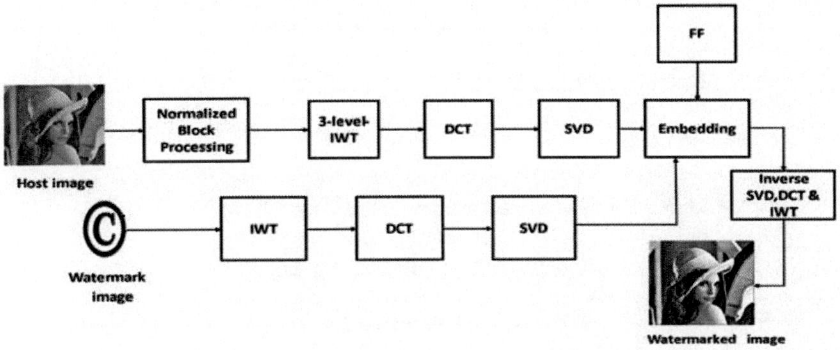

Fig. 2 Embedding diagram

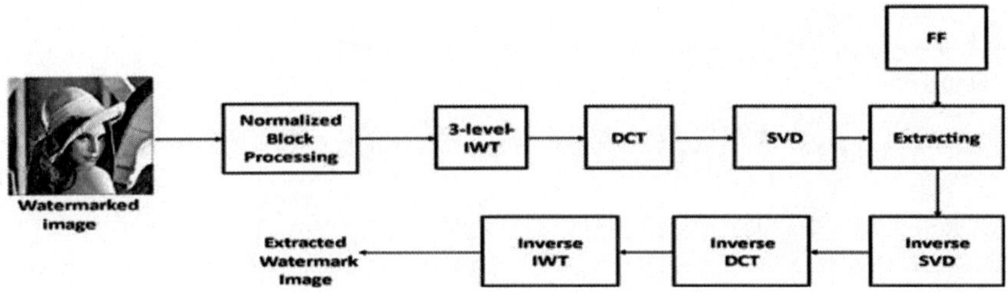

Fig. 3 Extracting diagram

$$\left[u^w_{LH} s^w_{LH} v^w_{LH}\right] = svd(W_{LH}),$$
$$\left[u^w_{HL} s^w_{HL} v^w_{HL}\right] = svd(W_{HL}),$$
$$\left[u^w_{HH} s^w_{HH} v^w_{HH}\right] = svd(W_{HH}) \quad (9)$$

Step 3: Multiply matrices u^w_i and s^w_i

$$W^{us}_i = u^w_i * s^w_i \quad (10)$$

Step 4: Perform SVD

$$\left\{H^1_{LL}, H^1_{LH}, H^1_{HL}, H^1_{HH}\right\} = lwt2(H), \left\{H^2_{LL}, H^2_{LH}, H^2_{HL}, H^2_{HH}\right\} = lwt2\left(H^1_{LL}\right),$$
$$\left\{H^3_{LL}, H^3_{LH}, H^3_{HL}, H^3_{HH}\right\} = lwt2\left(H^2_{LL}\right) \quad (11)$$

Step 5: perform 3-level IWT and DCT

$$\left[u^H_{LL} s^H_{LL} v^H_{LL}\right] = svd\left(H^3_{LL}\right),$$
$$\left[u^H_{LH} s^H_{LH} v^H_{LH}\right] = svd\left(H^3_{LH}\right),$$
$$\left[u^H_{HL} s^H_{HL} v^H_{HL}\right] = svd\left(H^3_{HL}\right),$$

$$\left[u_{HH}^{H} s_{HH}^{H} v_{HH}^{H}\right] = svd\left(H_{HH}^{3}\right) \tag{12}$$

Step 6: embedding watermark

$$s_i^* = s_i^H + \alpha * W_i^{us} \tag{13}$$

Step 7: Modified bands obtained are represented as follows

$$\begin{aligned} &WM_{LL}^3, WM_{LH}^3, WM_{HL}^3, WM_{HH}^3 \\ &WM_i^3 = u_i^H * s_i^* * \left(v_i^H\right)^T \end{aligned} \tag{14}$$

Step 8: Inverse IWT is applied.
Step 9:

$$WM_{LL}^2 = ilwt2(WM_{LL}^3, WM_{LH}^3, WM_{HL}^3, WM_{HH}^3) \tag{15}$$

$$WM_{LL}^1 = ilwt2(WM_{LL}^2, H_{LH}^2, H_{HL}^2, H_{HH}^2) \tag{16}$$

$$WM = ilwt2\left(WM_{LL}^1, H_{LH}^1, H_{HL}^1, H_{HH}^1\right) \tag{17}$$

Step 10: Various attacks applied.

B. ***Extraction pattern***

Step 1: On distorted watermarked image apply NBP (WM*).
Step 2: DCT is followed by N_{WM}^*

$$\left\{(N_{WM}^*)_{LL}^1, (N_{WM}^*)_{LH}^1, (N_{WM}^*)_{HL}^1, (N_{WM}^*)_{HH}^1\right\} = lwt2\left(N_{WM}^*\right) \tag{18}$$

$$\left\{(N_{WM}^*)_{LL}^2, (N_{WM}^*)_{LH}^2, (N_{WM}^*)_{HL}^2, (N_{WM}^*)_{HH}^2\right\} = lwt2\left((N_{WM}^*)_{LL}^1\right) \tag{19}$$

$$\left\{(N_{WM}^*)_{LL}^3, (N_{WM}^*)_{LH}^3, (N_{WM}^*)_{HL}^3, (N_{WM}^*)_{HH}^3\right\} = lwt2\left((N_{WM}^*)_{LL}^2\right) \tag{20}$$

Step 3: Apply SVD

$$\begin{aligned} &(N_{WM}^*)_{LL}^3, (N_{WM}^*)_{LH}^3, (N_{WM}^*)_{HL}^3, (N_{WM}^*)_{HH}^3 \\ &\left[u_{LL}^{wm} s_{LL}^{wm} v_{LL}^{wm}\right] = svd\left((N_{WM}^*)_{LL}^3\right), \left[u_{LH}^{wm} s_{LH}^{wm} v_{LH}^{wm}\right] = svd\left((N_{WM}^*)_{LH}^3\right) \\ &\left[u_{HL}^{wm} s_{HL}^{wm} v_{HL}^{wm}\right] = svd((N_{WM}^*)_{HL}^3, \left[u_{HH}^{wm} s_{HH}^{wm} v_{HH}^{wm}\right] = svd\left((N_{WM}^*)_{HH}^3\right) \end{aligned} \tag{21}$$

Step 4: extracting the distorted components

$$W_i^{us^*} = \frac{\left(s_i^{wm} - s_i^H\right)}{\alpha},$$

Step 5:

$$W_i^* = W_i^{us^*} * \left(v_i^w\right)^T \tag{22}$$

Step 6: Apply inverse DCT and IWT

$$W^* = ilwt2\left(W_{LL}^*, W_{LH}^*, W_{HL}^*, W_{HH}^*\right) \tag{23}$$

5 Results and Discussions

The evaluation of the experimental results of the proposed watermark scheme with the existing method is explained in this section. In this, two figures are considered in which Fig. 4a is a host image and Fig. 4b is a watermark image. The watermarked picture is subjected to nine forms of attack to test the validity and robustness of the proposed solution. As a similarity measure between the actual and the obtained watermark image, the normalized correlation (NC) coefficient [17–20] is utilized. The reports are compared to the previous method in order to illustrate the performance of the proposed scheme.

The results of the proposed method is evaluated for the test cases of Lena as a host image and copyright as a watermark image [21–26].

Table 1 shows how the proposed method is compared to the other literature methods in terms of performance, i.e., the new attacks on FF method contrasted with the approach introduced previously PSO method. The proposed methods performance metrics are shown in Table 2. The proposed method attacks are contrasted to the previously proposed method.

(a) (b)

Fig. 4 **a** Lena host image, **b** copyright watermark image

Table 1 Evaluating the results of proposed approach with existing approach

Attacks	Existing method (PSO)		Proposed method (FF)	
	PSNR	NC	PSNR	NC
NA	46.89	0.9692	68.4575	0.9900
GNA (0.01)	43.67	0.9590	66.2655	0.8891
SPA (0.05)	42.85	0.9482	64.9340	0.9671
MFA (3 × 3)	41.02	0.9403	66.0829	0.9649
HEA	41.26	0.9465	38.1137	0.8902
CEA	38.79	0.9612	30.1469	0.8876
RA (50%)	41.38	0.9368	41.4870	0.9382
RSA (50%)	42.78	0.9390	26.3934	0.7303
CA (35%)	34.35	0.9025	50.5859	0.9808

6 Conclusion

In the previous paper (PSO), the robustness and imperceptibility is less compared with proposed approach (firefly). This paper describes the details of the proposed NBP-IWT-DCT-SVD-FF. This approach used FF as an optimization algorithm to achieve an optimal watermarking constant, PSNR and NCC. Here, the FF optimizes the watermarking constant twice, one for approximation band and another for remaining bands. Since, the approximation band carries more important information. The appropriate watermarking constant for the LL band and the remaining bands is 0.05 and 0.005, respectively, from the simulation results. Various simulations are carried out over the proposed approach by varying watermark and host images. Through this simulation, it can be assumed that the new approach exceeds the previously proposed method. The proposed approach provides an increase of 8.8306 dB in the PSNR over the existing approach.

Table 2 Results of the proposed NBP-IWT-DCT-SVD-FF

Attack	Watermarked image	Extracted watermark
NA	 PSNR:68.4575	 NC:0.9900
GNA	 PSNR:66.2655	 NC:0.8891
SPA	 PSNR:64.9340	 NC:0.9671
MFA	 PSNR:66.0829	 NC:0.9649
HEA	 PSNR:38.1137	 NC:0.8902
CEA	 PSNR:30.1469	 NC:0.8876

(continued)

Table 2 (continued)

Attack	Watermarked image	Extracted watermark
RA	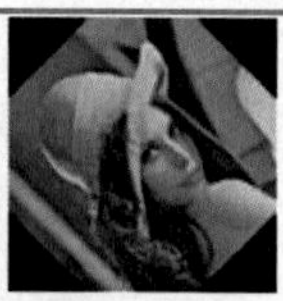 PSNR:41.4870	 NC:0.9382
RSA	 PSNR:26.3934	 NC:0.7303
CA	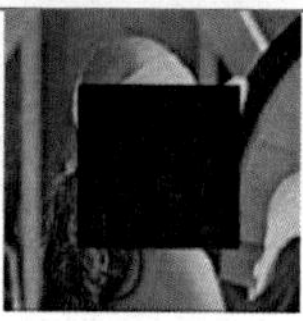 PSNR:50.5859	 NC:9808

References

1. Cox IJ, Kilian J, Leighton FT, Shamon T (1997) Secure spread spectrum watermarking for multimedia. IEEE Trans Image Process 6(12):1673–1687
2. Liu R, Tan T (2002) An SVD-based watermarking scheme for protecting rightful ownership. IEEE Trans Multimedia 4(1):121–128
3. Miller ML, Cox IJ, Linnertz J-PMG, Kelkar T (1999) A review of watermarking principles and practices. In: Digital signal processing in multimedia systems, pp 461–485
4. Hussein E, Belal MA (2012) Digital watermarking techniques, applications and attacks applied to digital media: a survey. Int J Eng Res Technol 1(7):1–8
5. Surya Prakasarao R, Rajesh Kumar P (2016) An efficient genetic algorithm based grayscale digital image watermarking for improving the robustness and imperceptibility. In: Proceedings of international conference on electrical, electronics, and optimization techniques (ICEEOT)
6. Vahedi E, Lucas C, Zoroofi RA, Shiva M (2007) A new approach for image watermarking by using particle swarm optimization. In: Proceedings of IEEE ICSPC, pp 1383–1386
7. Ravi Kumar CH, Surya Prakasa Rao R, Rajesh Kumar P (2018) GA based lossless and robust image watermarking using NBP-IWT-DCT-SVD transforms. JARDCS
8. Ravi Kumar Ch, Rajesh Kumar P, NBP-IWT-DCT-SVD-PSO based secure image watermarking. IJEAT. ISBN:978-93-89107-11-1
9. Khan A, Siddiqa A, Munib S, Malik SA (2014) Recent survey of reversible watermarking techniques. Inf Sci 279:251–272
10. Li L-D, Guo B-L (2009) Localized image watermarking in spatial domain resistant to geometric attacks. AEU: Int J Electron Commun 63(2), 123–131
11. Su Q, Niu Y, Wang Q, Sheng G (2013) A blind color image watermarking based on DC component in the spatial domain. Optik 124(23):6255–6260
12. Civicioglu P (2012) Transforming geocentric Cartesian coordinates to geodetic coordinates by using differential search algorithm. Comput Geosci 46:229–247

13. Li H, Sun Z, Tan T (2012) Robust iris segmentation based on learned boundary detectors. In: 2012 5th IAPR international conference on biometrics (ICB), New Delhi, India, pp 317–322 (2012)
14. Yang C-H, Weng C-Y, Wang S-J, Sun H-M (2008) Adaptive data hiding in edge areas of images with spatial LSB domain systems. IEEE Trans Inf Forensics Secur 3(3):488–497
15. Jamal SS, Shah T, Hussain I (2013) An efficient scheme for digital watermarking using chaotic map. Nonlinear Dyn 73(3):1469–1474
16. Lin SD, Shie S-C, Guo JY (2010) Improving the robustness of DCT-based image watermarking against JPEG compression. Comput Standards Interf 32(1–2):54–60
17. Venkata Subbarao M, Sayedu Khasim N, Thati J, Sastry MHH (2013) Hybrid image compression using DWT and neural networks. Int J Adv Sci Technol 53:31–41
18. Lang, Zhang Z-G (2014) Blind digital watermarking method in the fractional Fourier transform domain. Opt Lasers Eng 53:112–121
19. Li L, Xu H-H, Chang C-C, Ma Y-Y (2011) A novel image watermarking in redistributed invariant wavelet domain. J Syst Softw 84(6):923–929
20. Liu N, Li H, Dai H, Guo D, Chen D (2015) Robust blind image watermarking based on chaotic mixtures. Nonlinear Dyn 80(3):1329–1355
21. Fan M-Q, Wang H-X, Li S-K (2008) Restudy on SVD-based watermarking scheme. Appl Math Comput 203(2):926–930
22. Mohammad AA, Alhaj A, Shaltaf S (2008) An improved SVD based watermarking scheme for protecting rightful ownership. Signal Process 88(9):2158–2180
23. Makbol NM, Khoo BE (2013) Robust blind image watermarking scheme based on redundant discrete wavelet transform and singular value decomposition. AEU—Int J Electron Commun 67(2):102–112
24. Makbol NM, Khoo BE (2014) A new robust and secure digital image watermarking scheme based on the integer wavelet transform and singular value decomposition. Digital Signal Process 33:134–147
25. Mishra A, Agarwal C, Sharma A, Bedi P (2014) Optimized gray-scale image watermarking using DWT-SVD and firefly algorithm. Expert Syst Appl 41(17):7858–7867
26. Ali M, Ahn CW, Pant M (2014) A robust image watermarking technique using SVD and differential evolution in DCT domain. Optik 125(1):428–434

Antenna Array Synthesis of Shaped Beam Using Deterministic Method

R. Krishna Chaitanya, P. Mallikarjuna Rao, and K. V. Satya Narayana Raju

Abstract Array antennas have broad application range in telecommunication systems. Practical antennas like dipole, wave guide, and patch used as array element in array antennas have significant effect on performance of array antenna to generate shaped beams. Deterministic method like quasi-Newton method (QNM) is applied to synthesize stair step pattern using linear antenna arrays (LAAs). QNM is used to find excitation coefficients to synthesize the shaped beam and presented in U domain. The convergence plots are generated with respect to number of iterations for different array configurations of LAA.

Keywords Quasi-Newton method · Array synthesis · Stair step pattern

1 Introduction

Shaped beams are generated with different types of linear array antennas using optimization methods and deterministic methods. Optimization methods like particle swarm optimization algorithm (PSO) are used to synthesize planar fractal sub-array with dipole as array element to generate shaped beam. Koch fractal dipole antenna [1] is used to generate multi-beam and narrow beam using random optimization method. PSO is used to determine amplitude distribution and phase distribution with the effect of mutual coupling to generate the shaped beam. An antenna-independent design [2] for the synthesis of sum and difference pattern is presented. Different types of antennas like symmetric and unsymmetrical antennas are used. Liang et al. [3] presented a new synthesis method for the synthesis of linear array and planar array. The new method used for shaped beam synthesis is done without specifying the mask function. A function based on maximum side lobe level to minimum main

R. Krishna Chaitanya (✉) · K. V. Satya Narayana Raju
Department of ECE, S.R.K.R. Engineering College, Bhimavaram, India
e-mail: rkchaitanya@srkrec.ac.in

P. Mallikarjuna Rao
Department of ECE, Andhra University, Visakhapatnam, India

P. S. R. Chowdary et al. (eds.), *Evolution in Signal Processing and Telecommunication Networks*, Lecture Notes in Electrical Engineering 839,
https://doi.org/10.1007/978-981-16-8554-5_45

lobe level is considered for optimization. Sinha et al. [4] presented a new algorithm for the synthesis of linear array antennas. A soft algorithm, hybrid WDO/IWO, is used to generate radiation pattern with low side lobes, narrow beam width, and nulls in desired locations to reduce interference coming in particular direction. The new algorithm is compared with other optimization algorithms like PSO and IWO algorithms. Hybrid WDO/IWO is applied to determine both amplitude and phase distributions for generation of radiation pattern with low side lobes and nulls in desired location. Sudheer Kumar Terlapu [5] presented the synthesis of linear array antenna. The design involves optimizing antenna parameters of linear array. PSO [6] is applied to determine the amplitude excitation and phase excitation to obtain the desired pattern. Improved teaching learning-based optimization (ITLBO) [7], teaching learning-based optimization (TLBO) [8], and flower pollination algorithm [9] are used in previous research studies. In [10], Yazhong used deterministic method to solve the nonlinear equations. Li [3] used QNM in dynamic economic dispatch to solve Lagrangian relaxation problem in multipliers.

In this paper, deterministic method like QNM is used for the synthesis of linear array antennas with practical antennas like dipole, waveguide, and patch antenna.

In this paper, Sect. 2 discusses the expression to represent array factor and fitness function. Section 3 presents the deterministic method (QNM). Section 4 describes the results obtained using QNM. Section 5 gives the overall conclusions obtained from the results.

2 Field Pattern of Arrays

Array factor equation [11] for linear array for N elements is shown in Eq. (1)

$$\left.\begin{aligned} AF_f(\theta) &= \sum_{n_1=-N}^{-1} a_{n_1} e^{j(kd_{n_1}x+\phi_{n_1})} + \sum_{n_2=1}^{N} a_{n_2} e^{j(kd_{n_2}x+\phi_{n_2})} \\ d_{n_1} &= \frac{(2n+1)}{2} \times \frac{\lambda}{2} \text{ for } -N \le n_1 \le -1 \\ d_{n_2} &= \frac{(2n-1)}{2} \times \frac{\lambda}{2} \text{ for } 1 \le n_2 \le N \\ x &= \cos\theta \\ k &= 2\pi/\lambda \end{aligned}\right\} \tag{1}$$

a_{n_1}, a_{n_2} elements' amplitude weights.
d_{n_1}, d_{n_2} elements' distance weights.
N Total number of elements.
λ in terms wavelength.

Objective Function

Objective function is to generate shaped beam by minimizing the mean square error value.

$$\text{Fitness function} = \frac{1}{N_x} \sum |V_i(n)|^2 \tag{2}$$

$$V(n) = X_1(n) - X_2(n) \tag{3}$$

$X_1(n)$ desired pattern.
$X_2(n)$ obtained pattern.

3 Quasi-Newton Method

Quasi-Newton method is a deterministic method used for nonlinear programming. QNM is used when full Newton method is time-consuming or difficult to use. QNM is used to find double differential functions. QNM is computationally cheap and no need for second derivative. QNM does not require solving linear system of equations and requiring more convergence steps. QNM has less precise convergence path.

Procedure

(1) Choose the function $F(x)$ min or max.
(2) Choose the initialization point.
(3) Estimate the H.
(4) New value of X

$$X^{r+1} = X^r - [H^{(-1)}]_r \times \text{Grad}(X^r) \tag{4}$$

(5) Update the value of X.
(6) Find if the method has converged using convergence criteria.

If not optimum, repeat from Step 2.

4 Results

In this paper, shaped beam like stair step pattern is synthesized utilizing QNM. Array antenna with dipole, waveguide, and patch as array elements are utilized to generate the shaped beam. Different array configurations are utilized to synthesize the shaped beam, i.e., stair step pattern. The shaped beam generated for $n = 20$, 60, and 80 is presented in Figs. 1, 2, and 3. The error minimum values along with number of iterations are presented for different array configurations in Table 1. When the

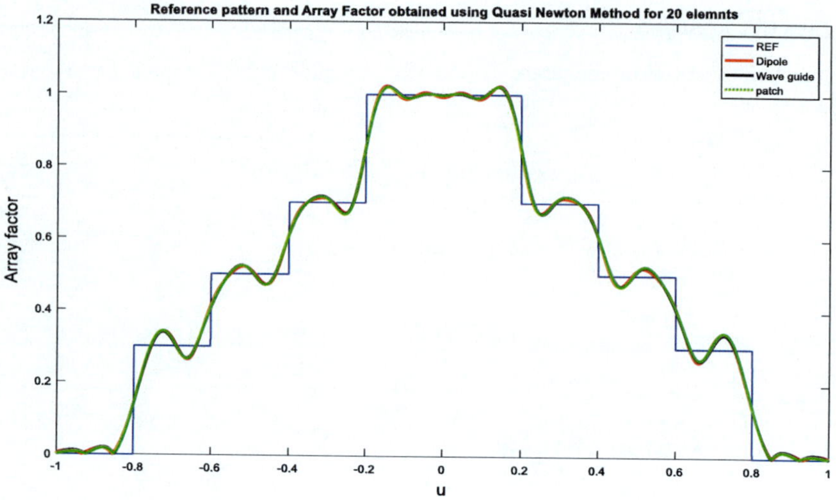

Fig. 1 Shaped beam pattern using QNM for LAA with practical antenna elements of dimension $n = 20$

number of elements of the array is 20, the error value is minimized to 0.002 in 33, 40, and 29 iterations, for LAA with dipole, waveguide, and patch as elements of the array. The convergence plots are shown in Fig. 4. When the number of elements is 40, the error value is minimized to 0.002 in 49, 17, and 45 iterations, for LAA with dipole, waveguide, and patch as elements of the array. When the number of elements is 60, the error value is minimized to 0.002 in 49, 17, and 40 iterations for LAA with dipole, waveguide, and patch as element of the array. The convergence plots are shown in Fig. 5. When the number of elements is 80, the error value is minimized to 0.002 in 51, 20, and 50 iterations for LAA with dipole, waveguide, and patch as element of the array. The convergence plots are shown in Fig. 6.

5 Conclusion

It is observed from the results that QNM converges faster than optimization methods. It is able to generate the shaped beam in very less number of iterations for 20, 40, 60 ,and 80 elements LAA. For less than 50 iterations, QNM is able to minimize the error and synthesize the shaped beam. Ripples generated in stair step pattern for different antenna array configurations and reduced significantly as the number of elements increased from 20, 40, 60, and 80 elements.

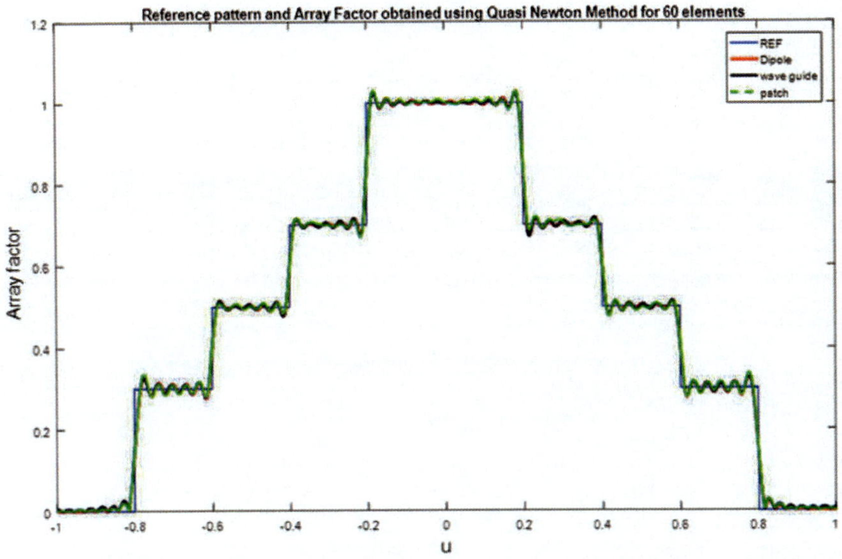

Fig. 2 Shaped beam pattern using QNM for LAA with practical antenna elements of dimension $n = 60$

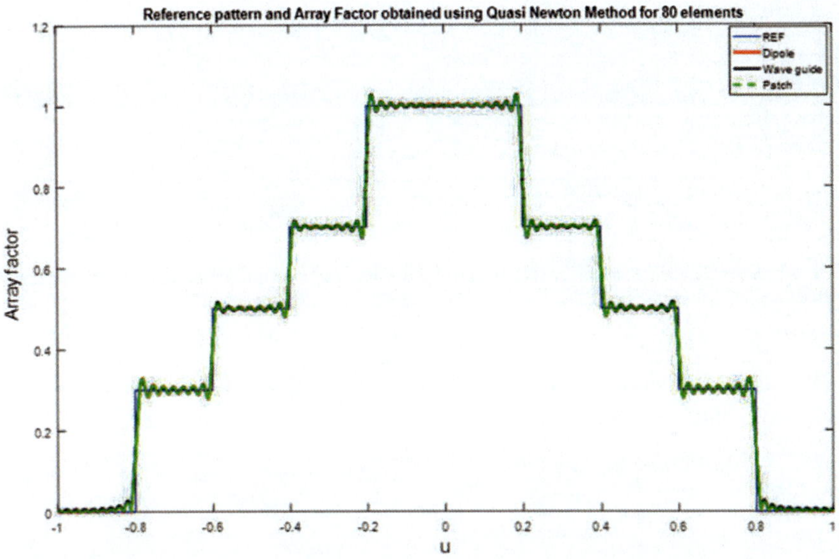

Fig. 3 Shaped beam pattern using QNM for LAA with practical antenna elements of dimension $n = 80$

Table 1 Total iterations taken for error minimization to generate stair step pattern using quasi-Newton method

No. of elements	Dipole antenna		Waveguide		Patch antenna	
	Error value	No. of iterations	Error value	No. of iterations	Error value	No. of iterations
20	0.002	33	0.002	40	0.002	29
40	0.002	49	0.002	17	0.002	45
60	0.002	49	0.002	17	0.002	40
80	0.002	51	0.002	20	0.001	50

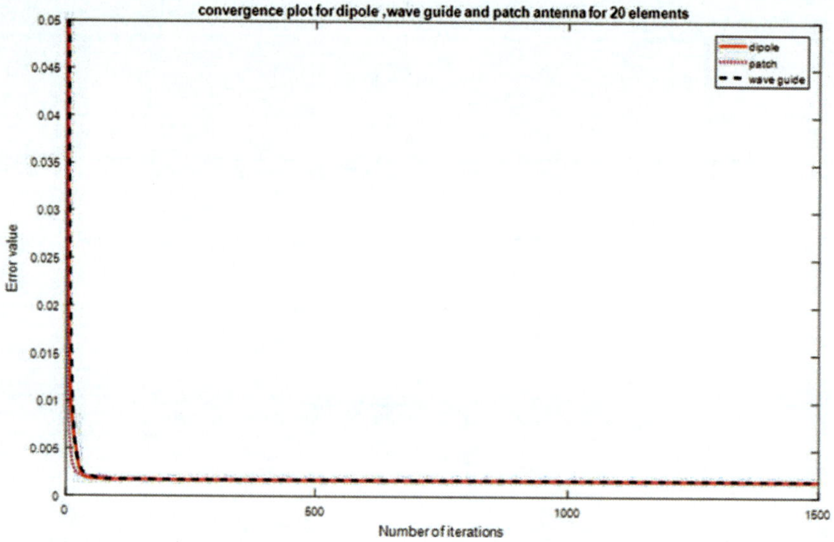

Fig. 4 Plot presenting the convergence of error value to number of iterations for 20 elements array with practical antenna elements

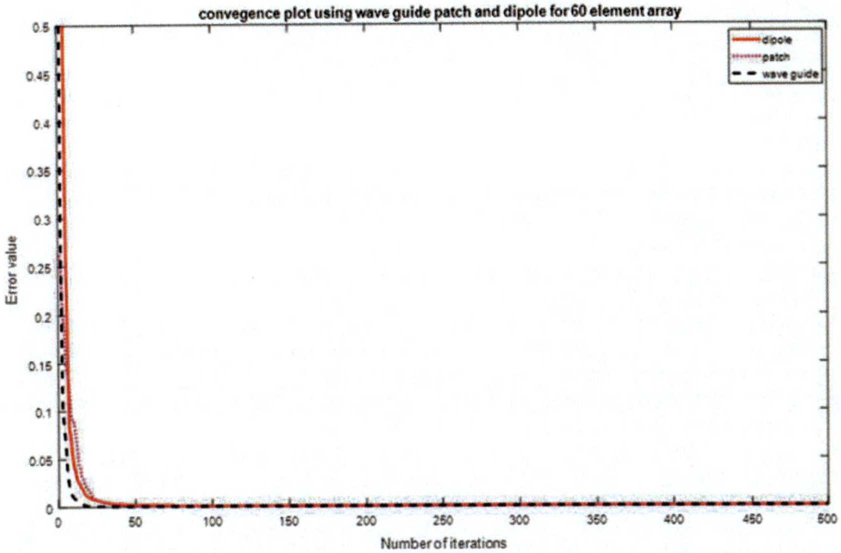

Fig. 5 Plot presenting the convergence of error value to number of iterations for 40 elements array with practical antenna elements

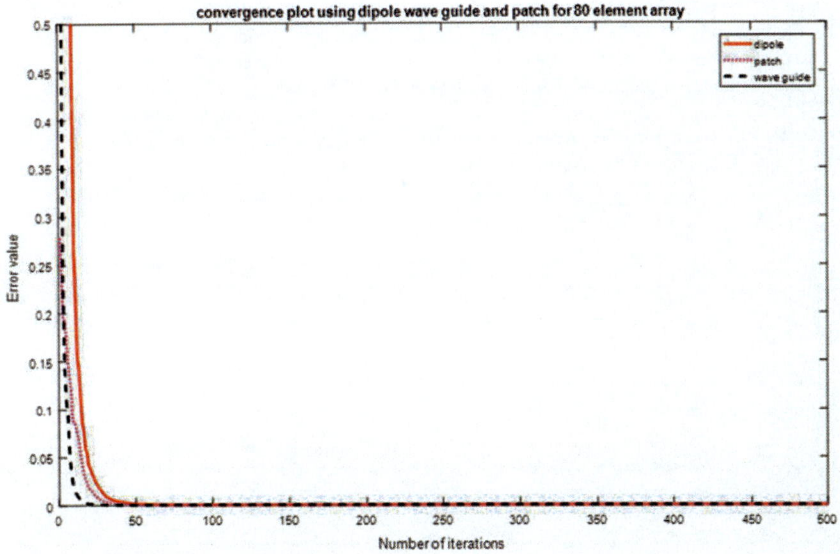

Fig. 6 Plot presenting the convergence of error value to number of iterations for 80 elements array with practical antenna elements

References

1. Azzam DM, Zainud-Deen SH, Malhat HA (2019) Shaped beam synthesis of antenna arrays formed of fractal sub-array lattices. In: 2019 7th international Japan-Africa conference on electronics, communications, and computations (JAC-ECC). IEEE, pp 192–195
2. Fan X, Liang J, Zhao X, Zhang Y, So HC (2020) Optimal synthesis of sum and difference beam patterns with a common weight vector for symmetric and asymmetric antenna arrays. IEEE Trans Antennas Propag 68(10):6982–6996
3. Li Z, Wu W, Zhang B, Sun H, Guo Q (2013) Dynamic economic dispatch using Lagrangian relaxation with multiplier updates based on a quasi-Newton method. IEEE Trans Power Syst 28(4):4516–4527
4. Sinha R, Choubey A, Mahto SK, Ranjan P, Barde C (2019) Synthesis of linear array antenna using hybrid IWO/WDO Algorithm. In: 2019 PhotonIcs & electromagnetics research symposium-Spring (PIERS-Spring). IEEE, pp 4144–4151
5. Terlapu SK, Raju GSN (2016) Synthesis of linear array for sidelobe reduction using particle swarm optimization. Int J Control Theory Appl 9(23):165–173
6. Terlapu SK, Raju GSN (2014) Design of array antenna for symmetrical sum patterns to reduce close-in Sidelobes using Particle Swarm Optimization. J Adv Model Simul Techn Entrep (AMSE) 87(3):44–56
7. Satapathy S, Naik A (2013) Improved teaching learning based optimization for global function optimization. Decision Sci Lett 2(1):23–34
8. Chakravarthy VVSSS, Babu KN, Suresh S, Devi PC, Rao PM (2015) Linear array optimization using teaching learning based optimization. In: Emerging ICT for bridging the future-proceedings of the 49th annual convention of the computer society of India CSI, vol 2. Springer, pp 183–190
9. Chakravarthy VVSSS, Chowdary PSR, Panda G, Anguera J, Andújar A, Majhi B (2018) On the linear antenna array synthesis techniques for sum and difference patterns using flower pollination algorithm. Arab J Sci Eng 43(8):3965–3977
10. Luo YZ, Tang GJ, Zhou LN (2008) Hybrid approach for solving systems of nonlinear equations using chaos optimization and quasi-Newton method. Appl Soft Comput 8(2):1068–1073
11. Balanis CA (2015) Antenna theory: analysis and design. Wiley

Miniaturized Edge Slotted Dual-Band PIFA Antenna for Wearable Medical Applications at 2–9 GHz

T. V. S. Divakar, P. Krishna Rao, and A. Sudhakar

Abstract This paper recommends a lesser size planar inverted F-antenna (PIFA) for two bands operation. The proposed antenna contains a stepped rectangular radiating component near to the edge of the PIFA besides having a ground plane with complete copper coating. The dimensions of this antenna are $27.7 \times 16 \times 1$ mm^3 and feeding technique used is microstrip line. Here, simulated S11 values, voltage standing wave ratio (VSWR) and gain are fairly related to each other. The proposed antenna structure operates at 2.24 and 8.8 GHz frequency range below −15 dB. The mentioned antenna structure is suitable for wearable medical applications.

Keywords Microstrip feed · PIFA · Stepped rectangular slot and wearable antenna

1 Introduction

In recent years, wearable antennas are showing predominant thrust in antenna designs due to many applications in well-known areas, like health monitoring, physical training and tracking. Since this type of antenna is close to the human body and moreover, it has many bends, the antenna performance must take into account the losses because of the human body and also bending losses, besides having the optimum performance. So this type of antenna systems must have low radiation effect which can be measured by specific absorption rate (SAR). Also, the designed antenna should have a flexibility to attach to the human bodies to increase comfort while humans wear it. Since fabric materials are light in weight and easy to wear, they can be

T. V. S. Divakar (✉) · A. Sudhakar
Department of Electronics and Communication Engineering, GMR Institute of Technology, Rajam, AP, India
e-mail: divakar.tvs@gmrit.edu.in

A. Sudhakar
e-mail: sudhakar.a@gmrit.edu.in

P. Krishna Rao
Aiditya Institute of Technology and Management, Tekkali, AP, India

P. S. R. Chowdary et al. (eds.), *Evolution in Signal Processing and Telecommunication Networks*, Lecture Notes in Electrical Engineering 839,
https://doi.org/10.1007/978-981-16-8554-5_46

used to incorporate this type of antennas. Besides the advantages of PIFA antennas like size and cost, there are some disadvantages like "poor impedance matching, poor proficiency and excitation of surface waves that could bring down the radiation effectiveness." A small-sized and low-profile fabric electromagnetic bandgap (EBG)-based antenna for wearable health applications which operates at 2.1 GHz is explored in [1], with help of EBG structure reduction in the back radiation and improvement in FBR was achieved, SAR of 0.0368 w/kg was achieved with dimensions 46 × 46 × 2.4 mm^3. Investigation on textile antenna with artificial magnetic conductor is carried out in [2, 3]. A dual resonant mode wearable textile PIFA for 5 GHz WLAN applications is designed in [4]. In this design, using hollow copper rivets and wool felt SAR of 0.9307 was achieved. A small-sized dual-band fabric PIFA for 432 MHz/2.4 GHz ISM bands was proposed which is made of 6 mm thick felt and 0.17 mm thick conductive textile sheet at the back of the PIFA. [5–7]. Supple fractal EBG for mm-Wave wearable antennas was projected in [8].

This paper explains a two band miniaturized PIFA antenna with an edge slot at the end of the PIFA when viewed from feed element to gain good impedance match and to have double-band action at 2.2 GHz and 8.8 GHz. The projected antenna is small in size when compared with other antennas in the literature [9].

The organization of this manuscript is like this. Section 2 describes the Antenna Design with all dimensions of the antenna and Sect. 3 describes the Results and discussion followed by conclusions.

2 Antenna Design

The complete size of proposed design is 27.7 × 16 × 1 mm^3 as shown in Fig. 1. It comprises of PIFA structure which can be divided as two vertical patches of 13.57 × 2.72 mm^2 and 14.10 × 2.72 mm^2 when viewed from microstrip feeding end. Two more rectangular patches of 11.57 × 1.38 mm^2 and 12.10 × 1.38 mm^2 were placed to fulfill PIFA structure. One square slot of 1 × 1 mm^2 placed near to the two patches to decrease the return loss and to get perfect impedance matching for the structure. Ground plane is completely shielded with copper coating. Simulations were carried out using HFSS software, and all the measurements were augmented by trial and error process besides maintaining primary values constant to get preferred features. The thickness, dielectric constant of the dielectric material and loss tangent of FR4 substrate are 1.6 mm, 4.4 and 0.02, respectively. Resonating frequencies are twofold at 2.24 GHz and 8.8 GHz. Sizes of the recommended antenna are accessible from Table 1.

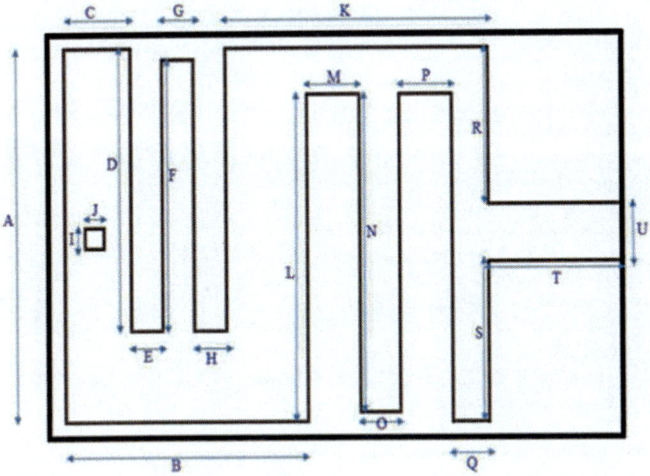

Fig. 1 Structure of the antenna—top view

Table 1 Dimension of the proposed antenna

Dimension name	Value (mm)	Dimension name	Value (mm)
A	16	L	14.10
B	12	M	2.72
C	3.65	N	13.57
D	12.10	O	1.96
E	1.38	P	2.72
F	11.57	Q	1.7
G	1.44	R	6.77
H	1.38	S	6.77
I	1	T	6.6
J	1	U	2.45
K	13.25		

3 Results and Discussion

The characteristics of reflection coefficient were displayed in Fig. 2 and 4 and we can observe that the antenna is resonating at the higher frequency with good resonance compared to the lower frequency with edge slot compared to without slot. Figures 3 and 5 show the corresponding VSWR characteristics. We can observe that there is a good agreement with the reflection coefficient and VSWR at the desired frequencies. Recommended antenna works in the double band of frequencies at 2.24 GHz and 8.8 GHz. Because of the edge slot, there is a notable change in the reflection

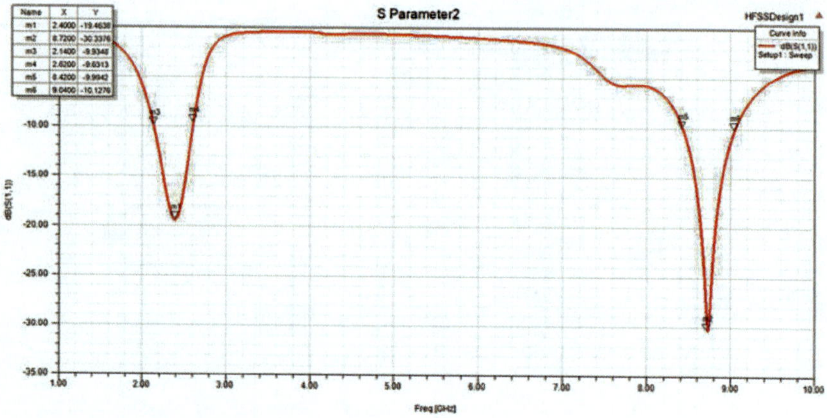

Fig. 2 Reflection coefficient versus frequency with edge slot

coefficient, and it was observed that if the measurements of air box and operation frequency was changed, there are minor variations in the operating frequency.

Figure 6 shows the fabricated antenna with edge slot. SMA connector is used to feed the antenna for power.

Figure 7 displays the E plane pattern for maximum phi values at 2.24 GHz, and Fig. 8 displays the E plane pattern for maximum phi values at 8.8 GHz. It can be observed that a bidirectional pattern at 2.24 GHz associated to E plane pattern at 8.8 GHz. The E plane pattern at 8.8 GHz is a better unidirectional pattern.

Figure 9 displays the H plane pattern for maximum theta values at 2.24 GHz, and Fig. 10 displays the H plane pattern for maximum theta values at 8.8 GHz. It can

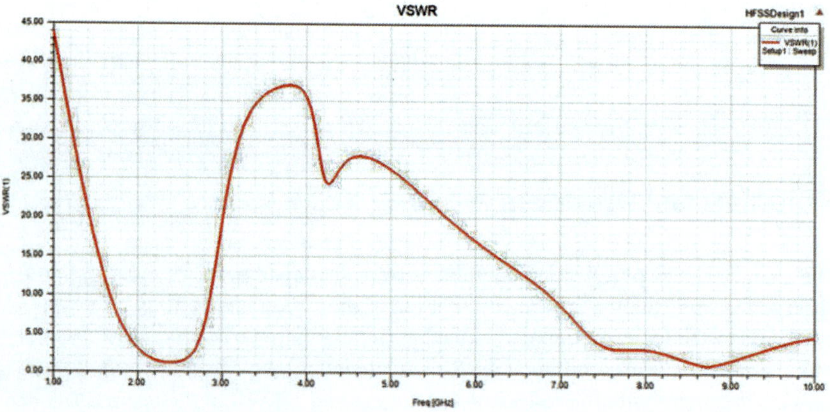

Fig. 3 VSWR versus frequency with edge slot

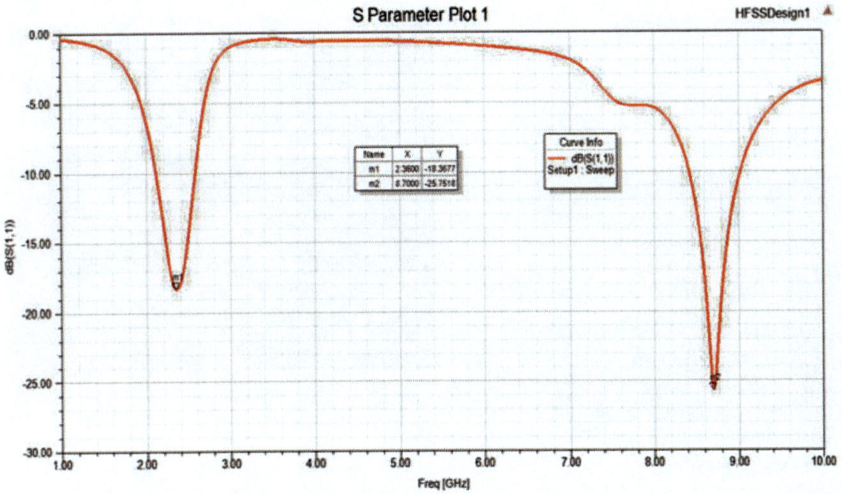

Fig. 4 Reflection coefficient versus frequency without edge slot

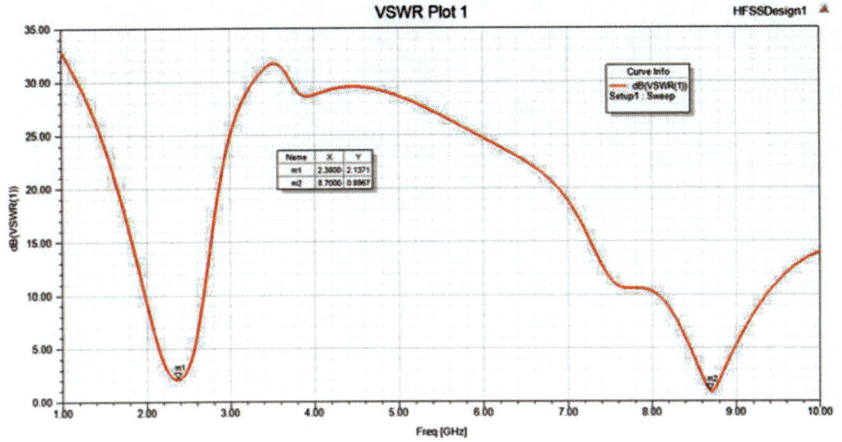

Fig. 5 VSWR versus frequency without edge slot

be observed that a better H plane pattern at lesser frequency which is at 2.24 GHz compared to upper frequency of 8.8 GHz.

4 Conclusion

A PIFA antenna with edge slot of overall size $27.7 \times 16 \times 1$ mm^3 with fully covered ground structure with copper has been simulated and fabricated. Microstrip line

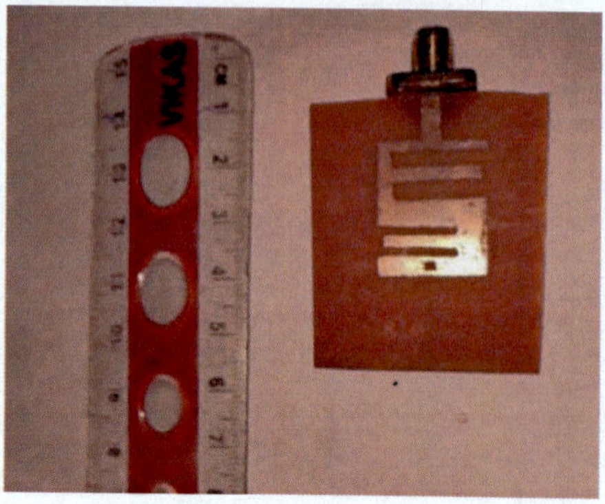

Fig. 6 Fabricated antenna

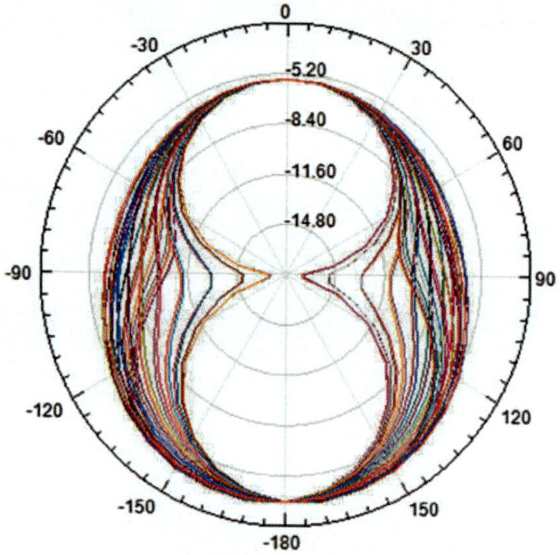

Fig. 7 E plane pattern at all phi values at 2.24 GHz

feed is used to feed the structure. Because of small size, it is useful for medical applications and wearable applications. This antenna meets requirements of 2.24 GHz and 8.8 GHz applications. Determined gain of the PIFA antenna is found to be 3.5 dB. Simulated reflection coefficient and VSWR properties are fairly in good agreement.

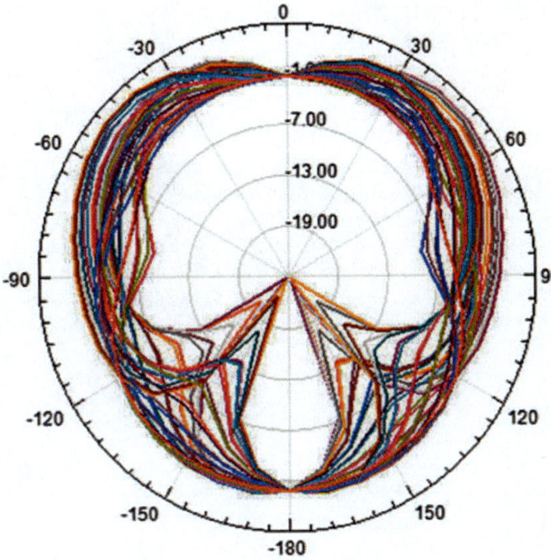

Fig. 8 E plane pattern at all phi values at 8.8 GHz

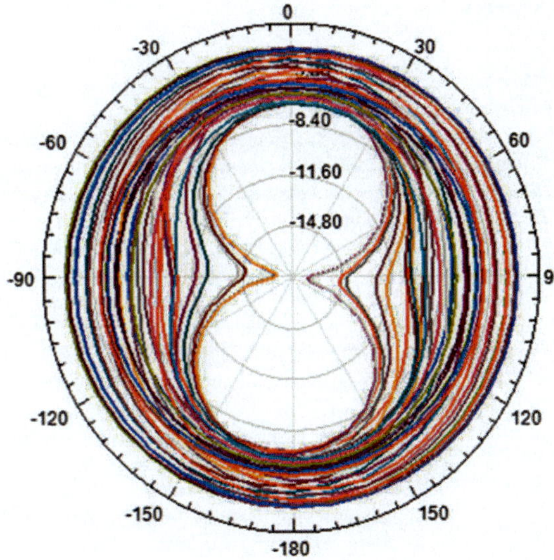

Fig. 9 H plane pattern at all Theta values at 2.24 GHz

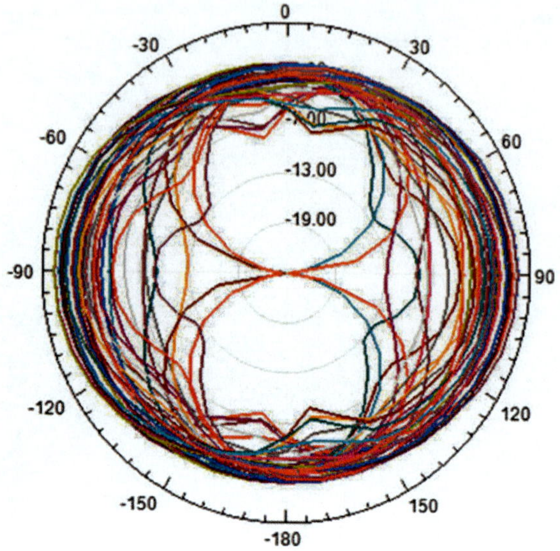

Fig. 10 H plane pattern at all Theta values at 8.8 GHz

References

1. Kaschel H et al (2018) Design of rectangular microstrip patch antenna at 2.4 GHz for WBAN applications. In: ICA-ACCA, pp 17–19
2. Ashyap AYI et al (2017) Compact and low-profile textile EBG-based antenna for wearable medical applications. IEEE Antennas Wirel Propag Lett 16:2550–2553
3. Gao G-P et al (2019) A wide bandwidth wearable all-textile PIFA with dual resonance modes for 5-GHz WLAN applications. IEEE Trans Antennas Propag 67(6)
4. Alonso-González L et al (2019) On the development of a novel mixed embroidered-woven slot antenna for wireless applications. IEEE Access 7:9476–9489
5. Ahmad MS et al (2014) Multi shorting pins PIFA design for multiband communications. Int J Antennas Propag 6
6. Viswanadh Raviteja G et al (2019) Multiband planar inverted-F antenna employing rectangular SRR for UMTS and WiMax/WiFi applications. Int J Comput Appl 182(36)
7. Ray JA et al (2011) A review of PIFA technology. In: 2nd international conferences on information technology, information systems and electrical engineering (ICITISEE). https://doi.org/10.1109/IndianAW.2011.6264946
8. Rama Krishna K et al (2015) Design and simulation of dual band planar inverted F Antenna (PIFA) for mobile handset applications. Int J Antennas (JANT) 1(1)
9. Ruaro A et al (2016) Wearable shell antenna for 2.4 GHz hearing instruments. IEEE Trans Antennas Propag 64(6):2127–2135

Nearest Neighbor Classification of Remote Sensing Images with the Statistical Features and PCA-Based Features

Telagarapu Prabhakar, Padmavathi Kora, and A. Sudhakar

Abstract The technique of assigning land cover classifications to pixels is known as image classification. It may divide pictures into categories such as forest, urban, agricultural, and others. A huge picture dataset of 21 land use types is used to test techniques in this research. In addition to conventional methods, there are comparisons to be made. The texture characteristics are extracted from remote sensing pictures using DWT at two degrees of decomposition. The UC-Mercedes dataset is used to explain the results. At the second level of decomposition, 72 features are retrieved using DWT and different feature extraction techniques at the approximation sub-band. The PCA method is used to find the best features. The two most popular methods are unsupervised and supervised image categorization. The k-NN supervised method is used to classify the data, and the best accuracy results are determined.

Keywords Texture feature extraction · Optimal features · Classification · Remote sensing images

1 Introduction

Remote sensing is a way of gathering information about various items all around the world without directly interacting with them. It is a method of gathering and analyzing non-instant geographic data to extract information about features and objects on the Earth's surface, oceans, and atmosphere. Existing Earth monitoring technologies (e.g., multi/hyper spectral [1], synthetic aperture radar, and so on) generate an

T. Prabhakar (✉) · A. Sudhakar
GMR Institute of Technology, Rajam, Srikakulam, Andhra Pradesh, India
e-mail: prabhakar.t@gmrit.edu.in

A. Sudhakar
e-mail: sudhakar.a@gmrit.edu.in

P. Kora
GRIET, Hyderabad, India

P. S. R. Chowdary et al. (eds.), *Evolution in Signal Processing and Telecommunication Networks*, Lecture Notes in Electrical Engineering 839,
https://doi.org/10.1007/978-981-16-8554-5_47

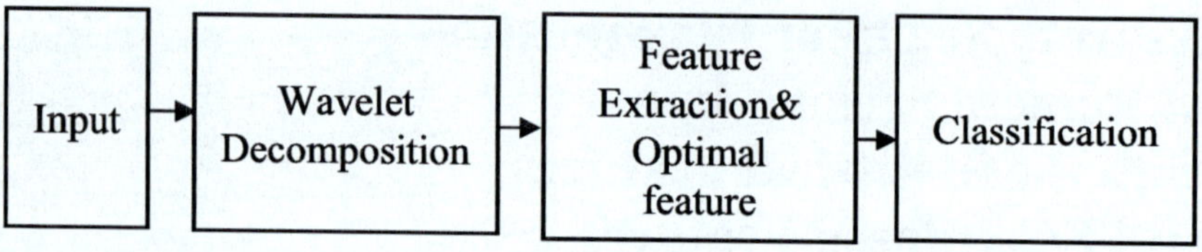

Fig. 1 Classification of remote sensing. Images with the GLCM and PCA-based features

increasing number of various types of aerial or satellite pictures with varying resolutions (spatial resolution, spectral resolution, and temporal resolution). This demands the use of intelligent earth observation via remote sensing pictures, which enables intelligent detection and categorization of land use and land cover (LULC) conditions from airborne or space platforms. The goal of remote sensing image scene classification, a prominent issue in aerial and satellite image analysis, is to categorize scene pictures into a discrete set of significant LULC classes depends on their contents. Due to the importance of remote sensing image scene classification in a wide range of applications, such as natural hazards detection [2, 3], LULC determination, geospatial object detection, geographic image retrieval, vegetation mapping, environmental monitoring, and urban planning, significant efforts have been made in developing various methods for the task over the past decades. Low-level features or hand-crafted features were utilized in early scene categorization algorithms, with the goal of creating distinguishing human-engineering attributes locally or globally, such as shape, color, texture, and spatial information. In scene classification, representative features such as the scale invariant feature transform (SIFT), color histogram (CH), local binary pattern (LBP), Gabor filters, gray level co-occurrence matrix (GLCM), and histogram of oriented gradients (HOG) or their mixes are frequently employed. While these low-level characteristics work well on photos with uniform texture or spatial arrangements, they are restricted in their capacity to differentiate images with more difficult and complicated situations. This is since human engagement in feature design has a substantial impact on the efficacy of scene picture representation capability. Mid-level feature techniques, in contrast to low-level feature-based approaches, aim to calculate a holistic picture representation based on local visual characteristics such as SIFT, color histogram, or LBP of local image patches. The standard approach for generating mid-level features in remote sensing pictures is to first extract local characteristics of image patches and then encode them. Because of its simplicity and efficiency, the well-known bag-of-visual-words model is the most common mid-level approach and has been widely used for remote sensing picture scene categorization [4–6].

2 Methodology

Figure 1 shows the block diagram of classification of remote sensing images with the GLCM and PCA-based features. The images acquired from UC Merced Land Use Dataset. This is a 21-class land use image dataset meant for research purposes. There are 100 images for each class measures 256 × 256 pixels of the following 21 classes: (1) Agricultural, (2) airplane, (3) baseball, diamond, (4) beach, (5) buildings, (6) chaparral, (7) dense residential, (8) forest, (9) freeway, (10) golf course, (11) harbor, (12) intersection, (13) medium residential, (14) mobile home park, (15) overpass, (16) parking lot, (17) river, (18) runway, (19) sparse residential, (20) storage tanks, and (21) tennis court. For this study, used remote sensed photos from the public source UC MERCEDES, with each class including 100 images. Retrieved 72 texture characteristics in all directions after applying 2D Haar to the images, and computed mean and range for each of them. The top 10 qualities were then chosen using principal component analysis [7]. k-NN classification is utilized for the images on which wavelet transform and optimal feature selection are performed [8].

2.1 Wavelet Decomposition

Figure 2 shows two-level wavelet decomposition features. To denoise the pictures, a 2D Haar wavelet is used to each one. Mother wavelet is another name for Haar Wavelet. Wavelet is used for a limited amount of time. Wavelet is becoming more widely used in image processing and feature extraction. Wavelet transforms provide the advantages of multi-resolution and space-frequency localization. Because of the wavelet decomposition, layers of approximated and detailed coefficients are produced. A wavelet transform is a method for multi-resolution analysis. Analyze the

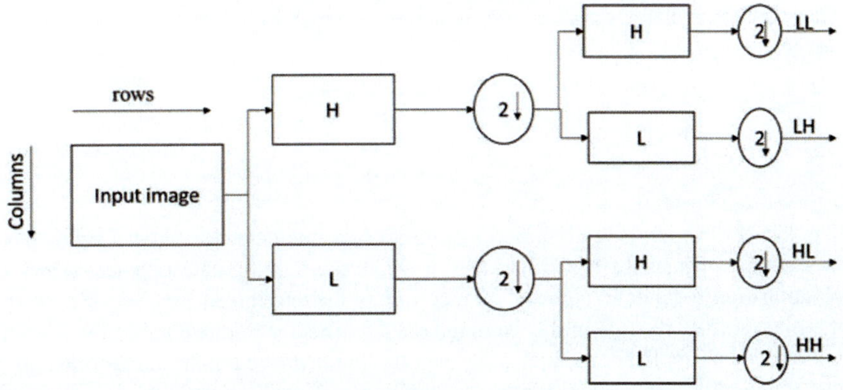

Fig. 2 Two-level wavelet decomposition features

signal in various frequency bands using this multi-resolution technique. The time–frequency representation is provided by the wavelet transform. It is not uncommon for a certain spectral component to be of special interest at any one time. The wavelet and resolution levels were used to apply the 2D-DWT to the remote sensed pictures, producing detail coefficient matrices and one approximation coefficient matrix. The high-pass filter at each level provides fine details, whereas the low-pass filter linked with the scaling function produces crude approximations. The high-pass filter at each level provides fine details, whereas the low-pass filter linked with the scaling function produces crude approximations [9].

2.2 Texture Feature Extraction and Feature Selection

This technique extracts the GLCM features and feeds them to the classifier. (1) Statistical features are among the GLCM Features [9]. (SF), (2) Matrices of spatial gray level dependence (SGLDM), (3) Statistics on gray level differences (GLDS), (4) Matrix of gray tone differences in the neighborhood (NGTDM), (5) Matrix of statistical features (SFM), (6) Texture energy measurement laws (LTEM), (7) Fractal texture dimension analysis (FDTA), (8) Fourier power spectrum (Fourier power spectra) (FPS), (9) Parameters for the shape, (10) Gray-level run length matrix (GRLLM). After extracting GLCM texture features, must decrease the feature set using principal component analysis to get optimum features. As a result, the performance of the optimum features classifier may improve. PCA is defined as "an orthogonal linear transformation that transfers data to a new coordinate system such that the greatest variance by any projection of the data lies on the first coordinate (first principal component), the second greatest variance lies on the second coordinate (second principal component), and so on" (second principal component). PCA is a technique for reducing data dimensionality without sacrificing too much information. Used in signal processing and picture compression, as well as machine learning. These optimum characteristics serve as an input vector for classifying remote sensing images.

2.3 Classification

The k-NN unsupervised classifier was utilized in this case. k-nearest neighbor is an example of instance-based learning in which the training dataset is saved and a classification for a new unclassified record may be determined by comparing it to the most similar records in the training set [10]. This technique has been used to a wide range of applications, including data mining, statistical pattern identification, and image processing. Handwriting recognition and satellite pictures are examples of successful uses.

3 Results and Discussion

This article examines the texture feature extraction and classification method, which was trained on 50 remote sensing pictures per class and evaluated on 50 remote sensing images per class. Remote sensed pictures are used to extract texture characteristics. Used remote sensed pictures from the public source UC MERCEDES for this research, with each class including 100 images. After applying 2D Haar to the pictures, extracted 72 texture features in all directions and calculated mean and range for all of them. Next, utilized principal component analysis to choose the top ten characteristics. Based on k-NN classification, categorized different classes using these optimum characteristics. For the pictures on which wavelet transform and optimum feature selection are conducted, k-NN classification is used. Figure 3 shows database of various classes. Figure 4 shows Wavelet Decomposition and Fig. 5 shows Denoised Images.

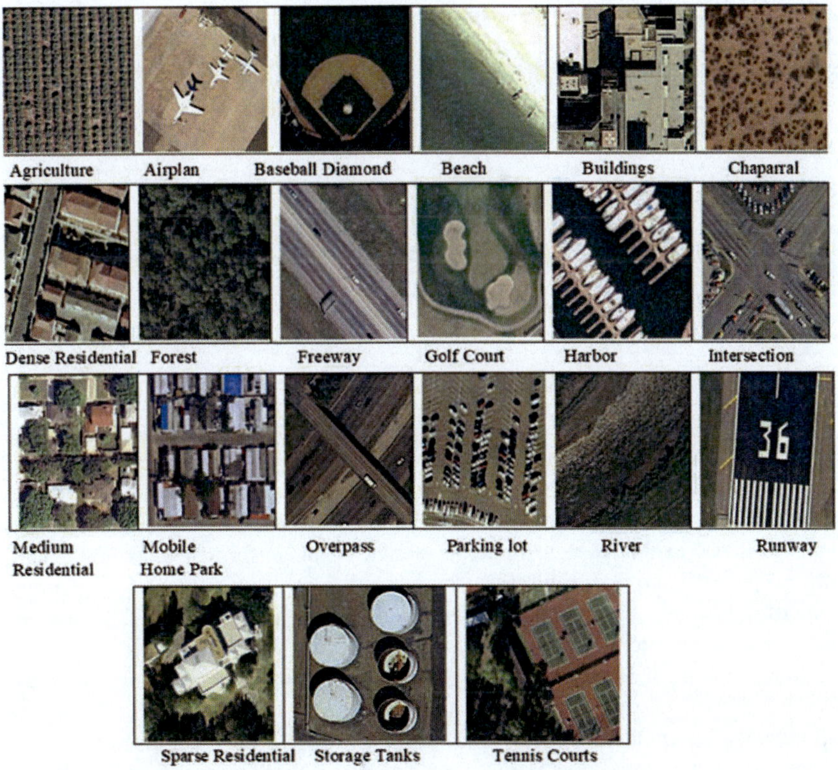

Fig. 3 Database of various classes

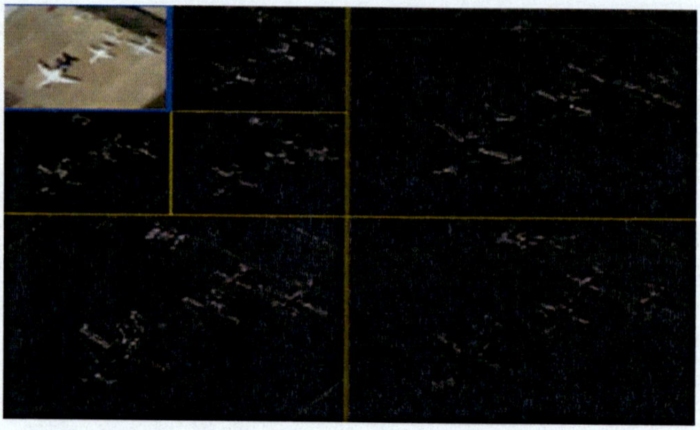

Fig. 4 Wavelet decomposition (tree level)

3.1 2D Wavelet Decomposition

See Fig. 4.

3.2 Preprocessing

See Fig. 5.

3.3 Texture Feature Extraction

Figure 4 shows wavelet decomposition (tree level) and Fig. 5 shows denoised images. The GLCM features extracted from with and without wavelet. Following feature extraction, these features were provided to the best feature selection technique, which utilized in this case to improve classification performance. From agriculture images, it is observed that there is a variation in mean values of M_entropy, R-entropy, low gray-level run emphasis, and standard deviation of R_ASM, M-difference, and R_entropy for before and after applying wavelet to the images. From Airplane images, it is observed that there is a variation in R-entropy, low gray-level run emphasis, and standard deviation of R_ASM, R_SUM, and R_entropy for before and after applying wavelet to the images. From Baseball Diamond images, it is observed that there is a variation in M_contrast, M_difference, R_ASM, R-sum variance, R_entropy, R_information, short run low gray-level emphasis, and standard deviation of M_contrast, M_difference for before and after applying wavelet to the images. Observed almost no change in some features like complexity, area,

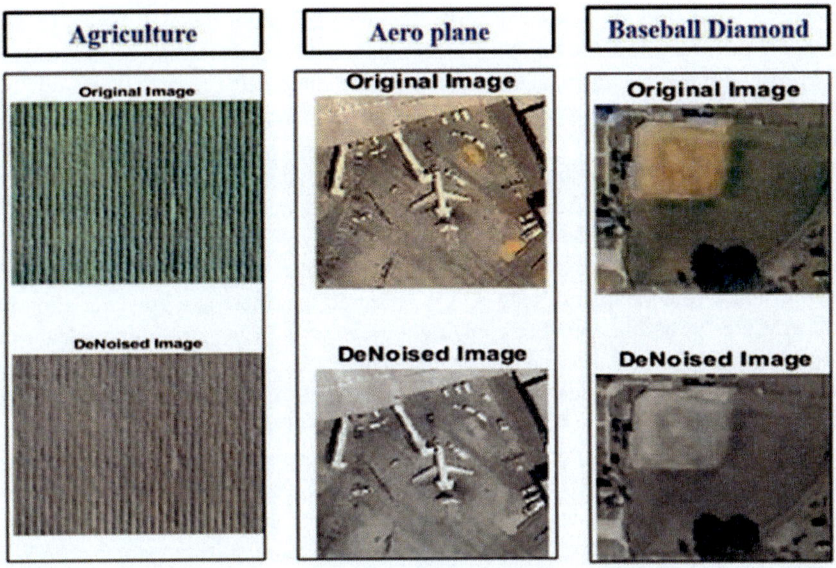

Fig. 5 Denoised images

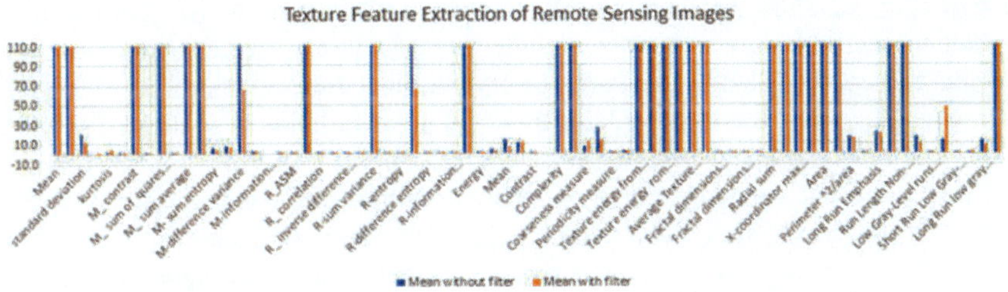

Fig. 6 GLCM-based texture feature extraction of mean values of all features

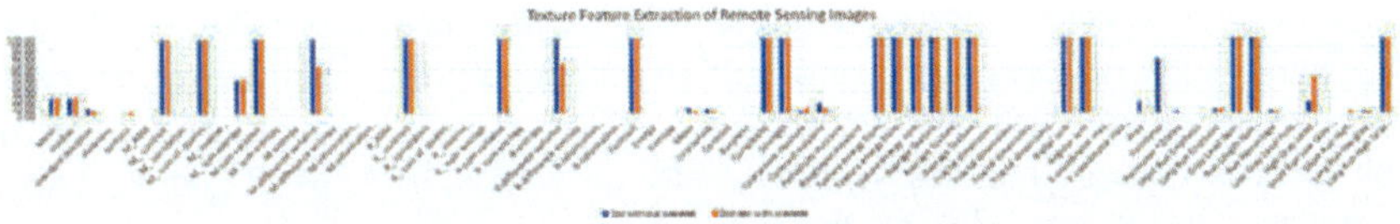

Fig. 7 GLCM-based texture feature extraction of std. deviation values of all features

run length features radial sum etc. From beach images, it is observed that there is a variation in mean values of M_contrast, M_difference, R_ASM, R-sum variance, R_entropy, R_information, short run low gray-level emphasis, and standard deviation of M_contrast, M_difference, R-sum variance, R_ASM, and R_information for before and after applying wavelet to the images. Figures 6 and 7 show GLCM-based texture feature extraction of mean and standard deviation values of all features. Observed almost no change in some features like complexity, mean, texture-related features, radial sum etc. From buildings images, it is observed that there is a variation in mean values of M_contrast, M_difference, R_ASM, R-sum variance, R_entropy, R_information, short run low gray-level emphasis, and it is observed that there is a great variation in standard deviation values of before and after applying wavelet to the building's images in the feature perimeter. Observed almost no change in almost all features like complexity, M_contrast, R_ASM, area, texture kernel features, and radial sum. From Chaparral images, it is observed that there is a great variation in mean values of before and after applying wavelet to the Chaparral images features R_entropy. Observed almost no change in remaining features. It is observed that there is a great variation in standard deviation values of before and after applying wavelet to the Chaparral images in the features R_ASM, R_SUM, and R_entropy. Observed almost no change in some features like complexity, strength, radial sum etc. From dense residential images, it is observed that there is a great variation in mean values of before and after applying wavelet to the dense residential images features low gray-level run emphasis. Observed almost no change in remaining features. It is observed that that there is a great variation in standard deviation values of before and after applying wavelet to the dense residential image's features area, perimeter, R_entropy. Observed almost no change in some features like complexity, texture kernel features, radial sum etc. From forest images, it is observed that there is great variation in mean values of before and after applying wavelet to the forest images features M_difference, R_ASM, R-sum variance, R_entropy. Observed almost no change in remaining features. It is observed that there is a great variation in standard deviation values of before and after applying wavelet to the forest images features M_contrast, M-difference, and run length features. Observed almost no change in some features like complexity, strength, radial sum etc. From the freeway images, it is observed that there is a great variation in mean values of before and after applying wavelet to the freeway images features low gray-level run emphasis. Observed almost no change in remaining features. It is observed that there is a great variation in standard deviation values of before and after applying wavelet to the freeway images features M_contrast, perimeter. Observed almost no change in some features like complexity, M_sum of squares, area, radial sum etc. From the golf court images, it is observed that there is a great variation in mean values of before and after applying wavelet to the golf court images features M_contrast, M_difference, R_ASM, R-sum variance, R_entropy, and R_information. Observed almost no change in remaining features. It is observed that there is a great variation in standard deviation values of before and after applying wavelet to the golf court images features M_contrast, M-difference, short run low gray-level emphasis. Observed almost no change in some features like complexity, strength, area, radial sum etc. From the harbor images, it is observed that there is a great variation in mean values of before and after applying wavelet to the harbor images features coarseness, low gray-level run emphasis, perimeter, short run low-gray level emphasis. Observed almost no change in remaining features. It is observed that there is a great variation in standard deviation values of before and after applying wavelet to the Harbor images features low gray level run emphasis, short run gray level emphasis. Observed almost no change in some features like complexity, M_contrast, M_sum of squares, radial sum etc. From the intersection images, it is observed that there is a less variation in Mean values of before and after applying wavelet to the Intersection images features coarseness, low gray-level run emphasis, and perimeter. Observed almost no change in remaining features. From the results of GLCM texture feature extraction, it is observed that there is a great variation in standard deviation values of before and after applying wavelet to the Intersection images features low gray-level run emphasis. Observed almost no change in some features like M_contrast, M_sum of squares, texture kernel features etc. From the medium residential images, it is observed that there is a great variation in mean values of before and after applying wavelet to the medium residential images features low gray-level run emphasis. Observed almost no change in remaining features. From the results of GLCM texture feature extraction, it is observed that there is a great variation in standard deviation values of before and after applying wavelet to the medium residential images features low gray-level run emphasis. Observed almost no change in almost all features. From the mobile home park images, it is observed that there is a little variation in mean values of before and after applying wavelet to the mobile home park images features coarseness, and low gray-level run emphasis. Observed almost no change in remaining features. From the results of GLCM texture feature extraction, it is observed that there is a great variation in standard deviation values of before and after applying wavelet to the mobile home park images features R_sum of squares (variance). Observed almost no change in most of the features in this class. From the overpass images, it is observed that there is a great variation in mean values of before and after applying wavelet to the overpass images features low gray-level run emphasis. Observed almost no change in remaining features. From the results of GLCM texture feature extraction, it is observed that there is a great variation in standard deviation values of before and after applying wavelet to the overpass images features low gray-level run emphasis, area. Observed almost no change in most of the features. From the parking lot images, it is observed that there are fewer variations in Mean values of before and after applying wavelet to the parking lot images features standard deviation, coarseness, perimeter^2/area, low gray-level run emphasis. Observed almost no change in remaining features. From the results of GLCM texture feature extraction, it is observed that there is a great variation in standard deviation values of before and after applying wavelet to the parking lot images features low gray-level run emphasis, area. Observed almost no change in most of the features. From the River images it is observed that there is a great variation in mean values of before and after applying wavelet to the river images features M_difference, R_ASM, R-sum variance, R_entropy, low gray-level run emphasis. Observed almost no change in remaining features. From the results of GLCM texture feature extraction, it is observed that there is a great variation in standard deviation values of before and after applying wavelet to the River images features M_contrast, M-difference, R_ASM, R_sum variance, R-entropy, R-information. Observed almost no change in some features like m_sum of squares, textural kernel-related features, complexity etc. From the runway images, it is observed that there are fewer variations in mean values of before and after applying wavelet to the runway images features short run low gray level emphasis. Observed almost no change in remaining features. From the results of GLCM texture feature extraction, it is observed that there is a great variation in standard deviation values of before and after applying wavelet to the Runway images features short run low gray level emphasis, area. Observed almost no change in most of the features. From the sparse residential images, it is observed that there are fewer variations in mean values of before and after applying wavelet to the sparse residential image's features R_entropy, low gray-level run emphasis. Observed almost no change in remaining features. From the results of GLCM texture feature extraction, it is observed that there is a great variation in standard deviation values of before and after applying wavelet to the sparse residential images features M_contrast, M-difference, R_ASM, R_sum variance, R-entropy, R-information. Observed almost no change in some features like m_sum of squares, textural kernel-related features, complexity etc. From the storage tanks images, it is observed that there are fewer variations in mean values of before and after applying wavelet to the storage tanks images features perimeter^2/area, low gray-level run emphasis. Observed almost no change in remaining features. From the results of GLCM texture feature extraction it is observed that there is a great variation in standard deviation values of before and after applying wavelet to the storage tanks images features low gray-level run emphasis, short run low gray-level emphasis. Observed almost no change in most of the features. From the tennis courts images, it is observed that there is fewer variations in mean values of before and after applying wavelet to the tennis courts images features R_entropy, low gray-level run emphasis. Observed almost no change in remaining features. From the results of GLCM texture feature extraction, it is observed that there is a great variation in standard deviation values of before and after applying wavelet to the tennis court images features M-difference, R_ASM, R_sum variance, R-entropy, low gray-level run emphasis. Observed almost no change in some features like M_contrast, radial sum, m_sum of squares, textural kernel related features, complexity etc.

3.4 Optimal Features

From the texture feature extraction method, extracted 72 statistical features and from these features obtained 10 optimal features to increase the classification performance. The optimal features are listed in Table 1.

3.5 Classification

Optimal features of with filter, without filter, with feature selection, and without feature selection are given to k-NN classifier and calculated performance metrics

Table 1 Optimal features obtained using principal component analysis (PCA) algorithm

S. No.	Optimal features
1	Average texture energy from LS and SL kernels
2	Texture energy from EE kernels
3	Angular sum
4	Perimeter
5	Average texture energy from ES and E kernels
6	Texture energy rom SS kernels
7	M_sum average
8	X-coordinator max length
9	Y-coordinator max length
10	Long run high gray level emphasis

Table 2 Performance of k-nearest neighbor (k-NN) classification

	Without wavelet and without feature selection (%)	Without wavelet and feature selection (%)	With wavelet and without feature selection (%)	With wavelet and with feature selection (%)
Accuracy	83.52	87.24	89.05	91.33
Sensitivity	83.55	88.28	89.80	91.51
Specificity	99.17	99.35	99.45	99.56
Mathew correlation coefficient (MCC)	0.85	0.89	0.90	0.92

like accuracy, sensitivity, specificity and Mathew correlation coefficient (MCC). In Table 2 listed the Performance metrics of k-nearest neighbor (k-NN) classification.

$$\text{Accuracy} = \frac{\text{TP} + \text{TN}}{\text{TP} + \text{TN} + \text{FP} + \text{FN}} \tag{1}$$

$$\text{Sensitivity} = \frac{\text{TP}}{\text{TP} + \text{FN}} \tag{2}$$

$$\text{Specificity} = \frac{\text{TN}}{\text{TN} + \text{FP}} \tag{3}$$

$$\text{MCC} = \frac{(\text{TP} * \text{TN}) - (\text{FN} * \text{FP})}{\sqrt{(\text{TP} + \text{FN}) * (\text{TN} + \text{FP}) * (\text{TP} + \text{FP}) * (\text{TN} + \text{FN})}} \tag{4}$$

4 Conclusion

The database for this paper includes a truth image collection of 21 land use classifications. The texture characteristics of the multiple class pictures are extracted using textural feature analysis, and the best features are utilized in the k-NN classifier to describe the multiple classes. Wavelet decomposition with optimum feature selection and k-NN classification is the technique presented here. On the 2100 pictures, the suggested system is used. Other techniques are outperformed by the suggested method. The suggested method's accuracy, sensitivity, specificity, and MCC are 91.33%, 91.51%, 99.56%, and 0.923, respectively.

References

1. Cheng G, Han J, Lu X (2017) Remote sensing image scene classification: benchmark and state of the art. Proc IEEE 105(10):1865–1883
2. Nogueira K, Penatti OAB, Dos Santos JA (2017) Towards better exploiting convolutional neural networks for remote sensing scene classification. Pattern Recogn 61:539–556
3. Nijhawan R et al (2019) A futuristic deep learning framework approach for land use-land cover classification using remote sensing imagery. In: Advanced Computing and Communication Technologies. Springer, Singapore, pp 87–96
4. Tong X-Y et al (2018) Learning transferable deep models for land-use classification with high-resolution remote sensing images. arXiv preprint arXiv:1807.05713
5. Carranza-García M, García-Gutiérrez J, Riquelme JC (2019) A framework for evaluating land use and land cover classification using convolutional neural networks. Remote Sens 11(3):274
6. Xu L et al (2020) Multi-structure joint decision-making approach for land use classification of high-resolution remote sensing images based on CNNs. IEEE Access 8:42848–42863
7. Huang H, Xu K (2019) Combing triple-part features of convolutional neural networks for scene classification in remote sensing. Remote Sens 11(14):1687
8. Jin B et al (2019) Object-oriented method combined with deep convolutional neural networks for land-use-type classification of remote sensing images. J Indian Soc Remote Sens 47(6):951–965
9. Haralick RM, Shanmugam K, Dinstein I (1973) Textural features for image classification. IEEE Trans Syst Man Cybern 3(6):610–621
10. Telagarapu P (2011) Closed planar shape classification using nonlinear alignment. In: 2011 IEEE recent advances in intelligent computational systems. IEEE

A Nanoplasmonic Ultra-wideband Antenna for Wireless Communications

Kavitha Rani Balmuri, Srinivas Konda, Kola Thirupathaiah, Voruganti Naresh Kumar, and Jonnadula Narasimharao

Abstract This article demonstrates the design and analysis of a new plasmonic monopole antenna is designed based on circular ring resonator (CRR) using conductor-backed coplanar waveguide (CBCPW) for ultra-wideband (UWB) applications. This antenna operating at optical frequency range in between 1625 and 1675 nm (U-band). The CRR is integrated into the CBCPW feeding line for attaining the notched characteristics for desired frequency range in the UWB monopole antenna. The notch band of the antenna can be controlled by the dimensions of the CRR. For easy understanding of nanoplasmonic UWB antenna, the functionality and characteristics are investigated for its operation at optical bands. Finally, it allows us to make nanoscale wireless links such as photonic integrated circuits (PICs).

Keywords CPW · CBCPW · Monopole · PICs · UWB

1 Introduction

Surface plasmons polaritons (SPPs) are the electromagnetic waves, propagates at the interface of the metal–insulator sandwich and manipulate the light at the subwavelength range to overcome the diffraction limit of light [1–3]. The light and matter interaction leads the SPP waves with the highest momentum than light with similar frequencies. The main focus of the nanoplasmonic devices is on optical and near-infrared frequencies. Metals are the lossy conductors or perfect conductors at microwave frequencies, whereas lossy plasmas are at optical frequencies. Various nanoplasmonic devices have been designed and proposed, such as splitters [4], nanowires [5], SPPs with dielectrics [6, 7], and metal–insulator-metal (MIM) waveguides [8, 9]. Within these designs, MIM waveguides are most important to

K. R. Balmuri (✉) · S. Konda · V. N. Kumar · J. Narasimharao
Department of Computer Science & Engineering, CMR Technical Campus, Kandlakoya, Hyderabad, India

K. Thirupathaiah
Department of Electronics and Communication Engineering, MLRITM, Hyderabad, India

P. S. R. Chowdary et al. (eds.), *Evolution in Signal Processing and Telecommunication Networks*, Lecture Notes in Electrical Engineering 839,
https://doi.org/10.1007/978-981-16-8554-5_48

handle the SPPs can strongly confine the light at the interface of the metal–dielectric sandwich with an appropriate propagation length without bending loss. Recently, nanoplasmonic MIM coplanar waveguide (CPW) plays an important role because of their compact sizes such as surface plasmon coplanar waveguides [10] and nanoantennas [11]. The coplanar waveguide has widely used to realize the antennas, such as coplanar waveguide-based UWB antenna [12]. The CBCPW-based nanoplasmonic UWB antennas can play an important role in the silicon photonic integrated circuits (SPICs) [13].

Plasmonic UWB antennas are the most essential element of the UWB systems offering low down power dissipation and high bandwidth have the attractive focus of the researchers. Along with, all types of monopole UWB antennas, plasmonic antennas are best candidate for UWB applications. Plasmonic UWB antennas with notch characteristics can play an important role to avoid the undesired signals in the existing narrowband systems and UWB antennas. Recently, several UWB notch band antennas have been reported with different design techniques [14].

The traditional techniques mainly focused on cutting slots of many designs in the radiating patch [15], feeding line [16], and ground plane [17] of the structure. Additionally, several UWB antennas using notch functionality already designed for UWB communication systems [18]. This article presents a new plasmonic UWB antenna with reconfigurable band notch functionality. The designed antenna can be designed using a CRR on the front side of the dielectric substrate with a fixed circular slot upon the ground plane. The proposed antenna has been designed and numerically analyzed to explain its feasibility. Simulated results of the antenna have been carried out with FDTD technique by using perfect matched layer (PML) boundary conditions by utilizing the full wave simulation CAD software tool CST microwave studio suite.

2 The Geometry of Plasmonic Antenna

The schematic of the designed plasmonic antenna with notch band functionality is shown in Fig. 1. A CBCPW-based plasmonic antenna has been used to investigate the UWB functionality with an operational frequency range from 179.624 THz to 187.958 THz. The suggested antenna is designed by using a 50 Ω CBCPW-based feeding line, a CRR and two identical parasitic slots have shown in Fig. 1. The CRR offers significant controllability in between optical frequencies 179.624 THz and 187.958 THz.

Figure 1 demonstrates the schematic of the proposed plasmonic antenna. The antenna is positioned along XY-plane and the direction is parallel to the z-axis. The proposed CRR-based UWB antenna is designed by silica as an insulator with dielectric permittivity ($\varepsilon_i = 2.50$) sandwiched in between two silver metallic layers. The proposed antenna is designed by using a circular section of radius R and radiating with a 50 Ω fed line of slot width S, kept on a dielectric substrate with a gap width "h" between disc and ground plane and width "L" backed by a ground plane height "h_1" illustrated in Fig. 1.

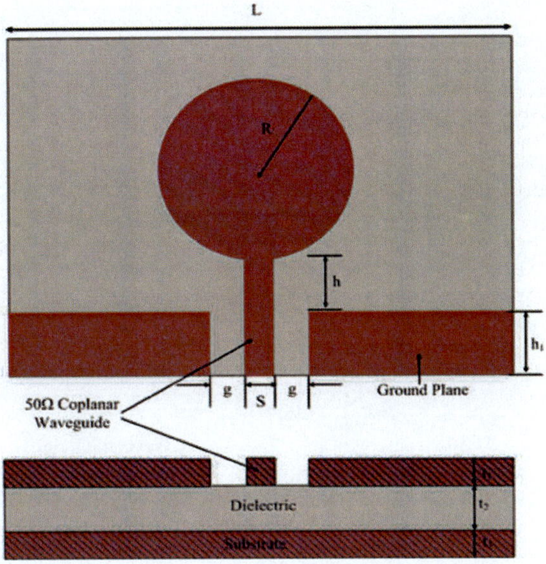

Fig. 1 The geometry of the CBCPW-fed CRR monopole antenna

The corresponding dimensions of the UWB antenna are chosen to be $R = 400$ nm, $L = 1080$ nm, $h = 150$ nm, $h_1 = 400$ nm, $t_1 = t_3 = 20$ nm, and $t_2 = 50$ nm. In order to apply the notch band functionality, CRR can be inserted below the radiating patch demonstrated in Fig. 1.

3 Parametric Study of the Antenna

The proposed plasmonic antenna is investigated by using commercial electromagnetic (EM) simulation software tool CST microwave studio suite. Figure 2 shows the simulated results of the return loss of the proposed plasmonic UWB antenna with CRR. One can observe that the notch band of the antenna around 187.958 THz (1601-nm) was obtained by using CRR shown in Fig. 2. The designed antenna with CRR covers the bandwidth from 179.624 THz (1601-nm) to 187.958 THz (1669-nm) with a band rejection, which can confirm the antenna can perfectly suitable for UWB wireless communication networks. It is observed that by changing the parameter "*h*" the return loss can be slightly shifting to the higher wavelength. To obtain the monopole UWB characteristics, the resonant modes require the optimization of the physical dimensions of the antenna.

Figure 3 demonstrates the variation of the simulated return loss of the antenna as a function of "*S*." One can be seen that the frequency of the first notch decreases from 187.37 to 179.646 THz with an increase of slot width *S* from 30 to 60 nm and not affected by the change in length on the second notch band of 181.429 THz frequency with an increase of 1 dB return loss.

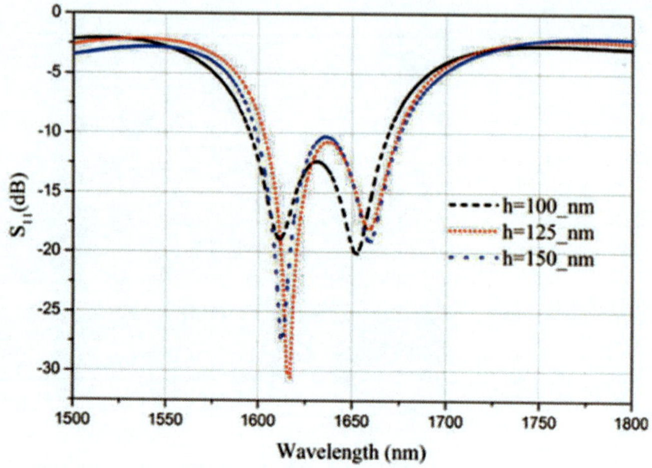

Fig. 2 Return loss as function of "h" with $S = 60$ nm, $R = 400$ nm, $L = 1080$ nm, $h_1 = 400$ nm, $t_1 = t_3 = 20$ nm and $t_2 = 50$ nm at frequencies 179.624 THz and 187.958 THz

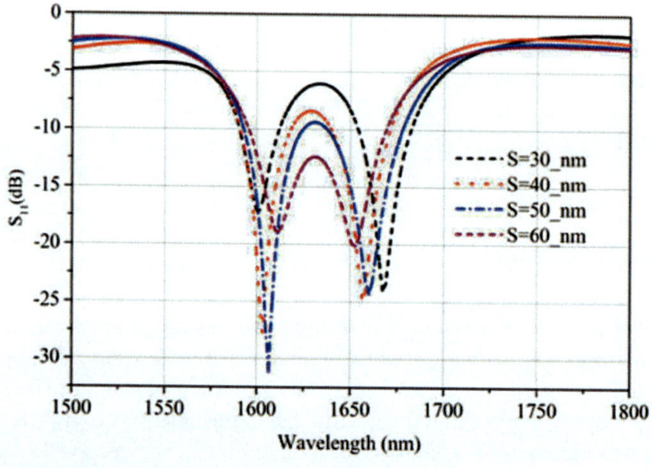

Fig. 3 Return loss as function of "S" with $R = 400$ nm, $L = 1080$ nm, $h = 150$ nm, $h_1 = 400$ nm, $t_1 = t_3 = 20$ nm and $t_2 = 50$ nm at frequencies 179.624 THz and 187.958 THz

The simulation results of the nanoplasmonic monopole antenna as a function of "R" with the physical dimensions $S = 60$ nm, $L = 1080$ nm, $h = 150$ nm, $h_1 = 400$ nm, $t_1 = t_3 = 20$ nm, and $t_2 = 50$ nm are shown in Fig. 4. The monopole UWB characteristics have been found within the frequency range of 187.958 THz (1601-nm) to 179.624-THz (1669-nm). The return loss of the monopole UWB antenna is for changing the radius "R" with the remaining fixed parameters. It is observed that by changing the parameter g, the return loss can be slightly shifting to the higher

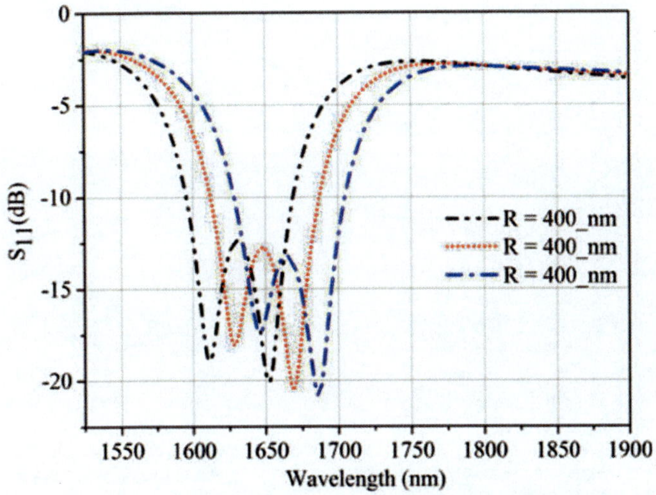

Fig. 4 Return loss as function of "*R*" with $S = 60$ nm, $L = 1080$ nm, $h = 150$ nm, $h_1 = 400$ nm, $t_1 = t_3 = 20$ nm and $t_2 = 50$ nm at frequencies 179.624 THz and 187.958 THz

wavelength side. To obtain the monopole UWB characteristics, the resonant modes require the optimization of the physical dimensions of the antenna.

Figure 5 shows the radiation patterns of the UWB antenna with the co-polarization and cross-polarization at a frequency of 187.958 THz (1601-nm) for $\theta = 90°$ and $\theta = 180°$. It is demonstrated that the impedance change occurs at these frequencies due to the band notch properties of the proposed structure at the frequency range 187.958 THz (1601-nm) to 179.624-THz (1669-nm) shown in Fig. 3a–d, respectively. Figures 6 and 7 show the electric field distribution of the plasmonic UWB antenna at frequencies 187.958 THz (1601-nm) and 179.624-THz (1669-nm).

4 Conclusion

Finally, the design and analysis of plasmonic CBCPW-based CRR monopole antenna for the UWB applications have been demonstrated. The new CBCPW-based nanoplasmonic UWB antenna with the feeding line by 50 Ω CPW on a silica layer can easily generate an ultra-wide 10 dB return loss bandwidth by reasonable radiation pattern. The antenna operating at optical frequency range in between 1625 and 1675 nm (U-band). The CRR is integrated into the CBCPW feeding line for attaining the notched characteristics for desired frequency range in the UWB antenna. For easy understanding of nanoplasmonic UWB antenna, the functionality and characteristics are investigated for its operation at optical bands. Ultimately, it allows us to make nanoscale wireless links such as photonic integrated circuits (PICs).

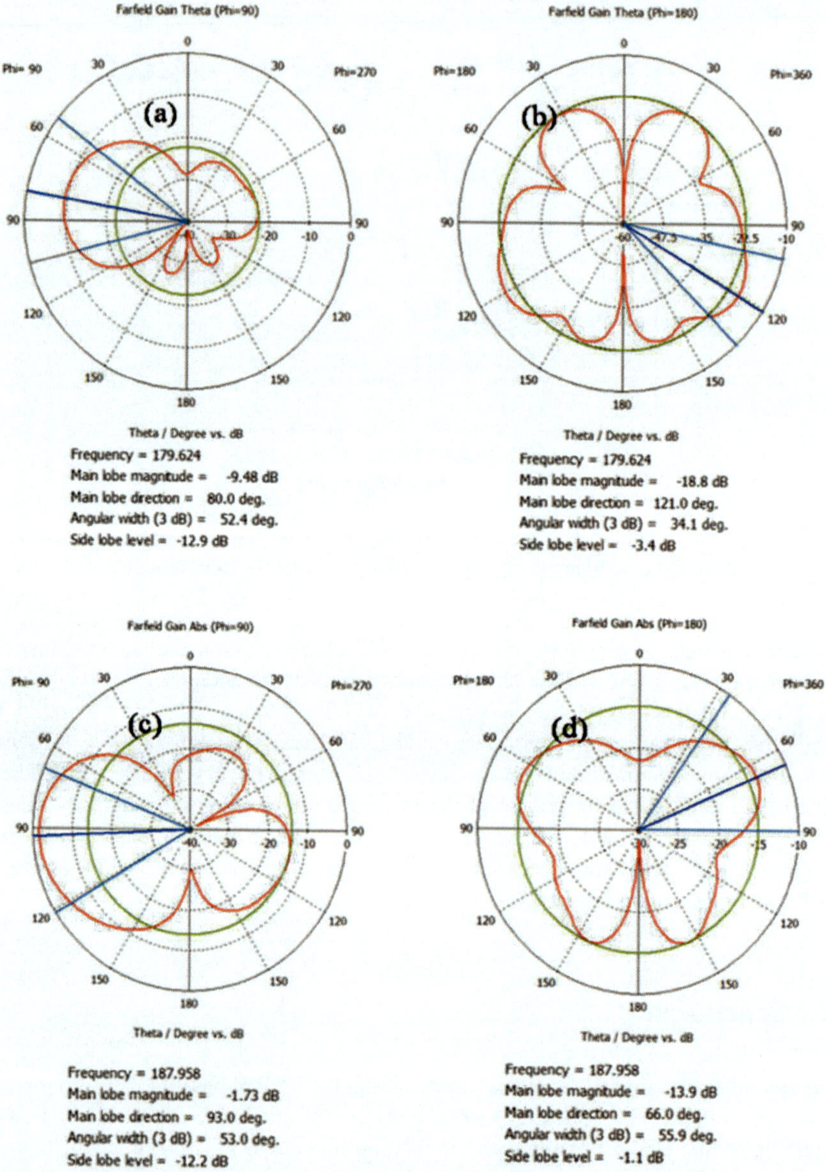

Fig. 5 Radiation patterns with $S = 60$ nm, $R = 400$ nm, $L = 1080$ nm, $h_1 = 400$ nm, $h = 150$ nm, $t_1 = t_3 = 20$ nm and $t_2 = 50$ nm at **a** $\theta = 90°$, **b** $\theta = 180°$, **c** $\theta = 90°$, **d** $\theta = 180°$ at frequencies 179.624 THz and 187.958 THz

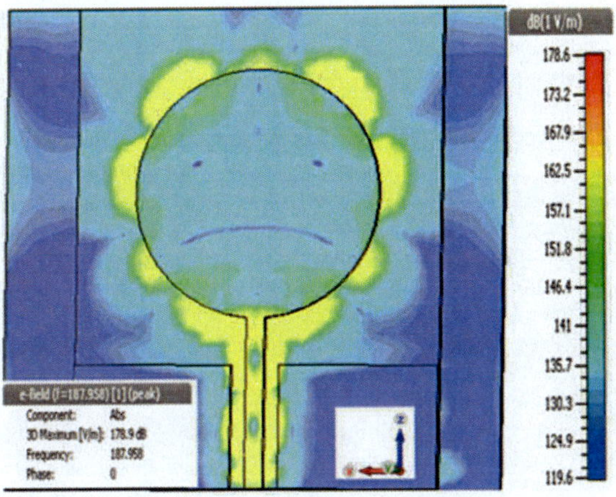

Fig. 6 Electric field distribution at 187.958 THz frequency

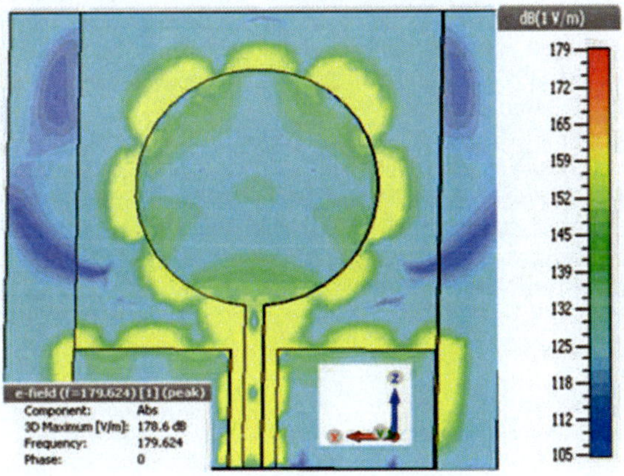

Fig. 7 Electric field distribution at 179.924 THz frequency

References

1. Luo X, Zhai X, Wang L, Lin Q, Liu J (2017) Theoretical analysis of plasmon-induced transparency in MIM waveguide bragg grating coupled with a single sub radiant resonator. IEEE Photonics J 9(5):480108 (1–9)
2. Yin L et al (2005) Subwavelength focusing and guiding of surface plasmons. Nano Lett 5(7):1399–1402
3. Wen KH et al (2012) Spectral characteristics of plasmonic metal-insulator-metal waveguides with a tilted groove. IEEE Photonics J 4(5):1794–1800
4. Wen K (2016) Plasmonic splitter based on subwavelength dual-slits waveguide structure. In: WOCC 2016—25th wireless and optical communication conference: jointly held with photonics forum of Chiao-Tung Universities
5. Kottmann JP, Martin OJF (2001) Plasmon resonant coupling in metallic nanowires. Opt Express 8(12):655–663
6. Feng NN, Brongersma ML, Dal Negro L (2007) Metal-dielectric slot-waveguide structures for the propagation of surface plasmon polaritons at 1.55 μ/m. IEEE J Quantum Electron 43(6):479–485

7. Liu Y, Yan J, Han G (2015) The transmission characteristic of metal-dielectric-metal slot waveguide-based nanodisk cavity with gain medium. IEEE Photonics J 7(2):4500608 (1–9)
8. Huang YX et al (2014) A plasmonic refractive index sensor based on a MIM waveguide with a side-coupled nanodisk resonator. In: RTCSA 2014—20th IEEE international conference on embedded and real-time computing systems and applications
9. Wang Y et al (2010) Manipulating surface plasmon polaritons in a 2-D T-shaped metal insulator metal plasmonic waveguide with a joint cavity. IEEE Photonics Technol Lett 22(17):1309–1311
10. Jung KY, Teixeira FL, Reano RM (2009) Surface plasmon coplanar waveguides: mode characteristics and mode conversion losses. IEEE Photonics Technol Lett 21(10):630–632
11. Adato R, Yanik AA, Altug H (2011) On chip plasmonic monopole nano-antennas and circuits. Nano Lett 11(12):5219–5226
12. Zhang Y, Hong W, Yu C, Kuai Z, Don Y, Zhou J (2008) Planar ultra wideband antennas with multiple notched bands based on etched slots on the patch and/or split ring resonators on the feed line. Antenna 56(9):3063–3068
13. Hosseininejad SE, Komjani N (2016) Waveguide-fed tunable terahertz antenna based on hybrid graphene-metal structure. IEEE Trans Antennas Propag 64(9):3787–3793
14. Kim K-H, Cho Y-J, Hwang S-P, Park S-O (2005) Band-notched UWB planar monopole antenna with two parasitic patches. Electron Lett 41(14):783–785
15. Yoon I-J, Kim H, Yoon YJ (2006) UWB RF receiver front-end with band-notch characteristic of 5 GHz WLAN. In: Proceedings of the IEEE international antennas and propagation symposium digest, pp 1303–1306
16. Kim J, Cho CS, Lee JW (2006) 5.2 GHz notched ultra-wideband antenna using slot-type SRR. Electron Lett 42(6):315–316
17. Kim K-H, Park S-O (2006) Analysis of the small band-rejected antenna with the parasitic strip for UWB. IEEE Trans Antennas Propag 24(6):1688–1692
18. Nikolaou S, Kingsley et al (2009) UWB elliptical monopoles with a reconfigurable band notch using MEMS switches actuated without bias lines. IEEE Trans. Antennas Propag 57(8):2242–2251

Intelligent Noise Detection and Correction with Kriging on Fundus Images of Diabetic Retinopathy

K. Nirmala, K. Saruladha, and K. Srujan Raju

Abstract Digital imaging collections are widely used in image analysis particularly in medical and healthcare systems. Medical images collected are prone to some type of inaccuracies called noise or incompleteness. Images with inaccuracy even are important in medical image analyses as they contain image components that form rare models. To overcome the problem of inaccuracies and put through the process of crucial processes of image analyses, it is very essential to correct images as required for the models of classification. Kriging interpolation technique is implemented on medical images for the removal of speckle noise. A technique of approximation known as kriging may be used to anticipate values that have not yet been seen based on observations of the value at adjacent places. In this paper, the application of kriging interpolation depending on a distance metric between the suggested speckle noise images and non-noisy images filtered with ground truth mask is experimented. The resulting images have significant improvement in the performance of the method.

Keywords DR images · Speckle noise · Kriging interpolation

1 Introduction

Medical image analysis is a most common and essential activity in the new world of health and medical sciences with advent of many digital appliances. Deep study of anatomical structures, textures, features and comparative analyses, workflow of radiological activities, etc., needs image analyses. Furthermore classification is prime job in the medical image analysis.

K. Nirmala · K. Saruladha
Pondicherry Technological University, Puducherri, India
e-mail: avnirmalas@pec.edu

K. Saruladha
e-mail: charuladha@pec.edu

K. Srujan Raju (✉)
CMR Technical Campus, Hyderabad, Telangana, India

P. S. R. Chowdary et al. (eds.), *Evolution in Signal Processing and Telecommunication Networks*, Lecture Notes in Electrical Engineering 839,
https://doi.org/10.1007/978-981-16-8554-5_49

The convolutional neural networks play very important roles in image analysis, for whatsoever the complexity is the image. CNNs can contribute much in the core of computer vision in image detection and recognition. CNNs need many layers to work on huge sizes of image data, to challenge and outperform the traditional methods based on convex optimizations. However, due to limited storage capabilities of computers, CNNs were first recommended on tiny datasets. Much research consensus is available in the collection of image datasets with different types, formats, and resolutions.

Meticulous care would be taken in the health care and treatments if effective classification is employed. Collecting medical images is very hard and time consuming with the concerns of privacy and labeling. Often, as the research initiatives are directed toward quick methods of analysis for diagnoses, they fail to target on the quality of images. The first hurdle that talks about the quality of images is noise. Conditions of light, hardware, photographic-trails, and albedo affect the images declining the quality. Speckle noise dominates in the images in spite of high definition imaging methods and hinders the analyses in extracting features, quantitative measurements, and component recognition. Noise is misunderstood as an additive with zero-mean, Gaussian, or Poisson.

Kriging emphasizes the sheer phenomenal significance in geo-statistics. Kriging is the method of interpolation based on Gaussian processes governed by prior covariance. With best suitable expectations, this method can yield best linear unbiased predictions, particularly at unsampled areas of images. However, smoothness is a challenging task during spatial computations.

Noise is seen in the form of grains in the image, which is an otherwise variation of intensities in the image. Disturbances due to natural characteristics of light and/or temperature causing brightness and warmth inside image capturing apparatus develop during the image capture. Even, noise is described as pixel intensities grabbed wrongly by the devices.

Image reconstruction is an important task in the fields of medical, geo-spatial, criminal, and other important studies of biology. Kriging is considered a best alternative method used in image restoration and reconstruction using interpolation methods. Producing new predictions of unobserved values from the observations in the neighborhood and supply weights for the new predictions based on the distances is the most used of operations.

2 Related Work

Neural Network Based Filters

If we consider denoising to be a learning issue, a feed forward conventional neural network may be used to eradicate the salt-and-pepper noise from the signal. Trimmed medial numbers are used to substitute noisy pixels values such as 0 and 255 in the chosen window with values that are less noisy [1]. If the values of all the elements in the chosen window are between 0 and 255, then noisy pixels are substituted by a mean value derived from all of the elements in the window.

In an 8-bit gray-level picture, a three weights technique combined with a different-sized window that contains pixels with values other than zero and one hundred fifty-five is used for the recovery of a central pixel containing extreme values, that is, either zero or one hundred fifty-five. In a local window, the fundamental three-values are regarded to be the high, the lowest, and the middle pixel values, respectively. In the beginning, a different-sized local window is used to assess each pixels of noise, which contains either, the highest or the lowest value of the pixel at the center of the window, depending on the case. It is necessary to extend the window size until all of the pixels with values other than 0 or 255 are in it if pixels values of local window's are zero or two hundred and fifty-five.

When the value of a pixel is very high or extremely low, a pixel may be considered noisy. During the consideration of rebuilding the pixel at center, the noisy pixel is eliminated from consideration. Based on the distance between the pixels having values other than 0 and 255, the pixels are divided into two groups: maximum and minimum groups.

As a result, the number of pixels is divided in each of these two categories by the number of weighting factors in each group. Using the weighing factors and highest pixel value and the lowest pixel value, a medium value may be produced by multiplying them together. A combination of the weighting factors from these three categories is used to weigh the non-extreme pixels with respect to the high, middle, and lowest pixel values, respectively. The weighted value is substituted for the extreme value in the central pixel, allowing the noise-corrupted pixels to be recovered [2].

An unbiased weighted mean filter performs noise detection, spatial bias reduction, and noise reduction in a sequential successive process by convolving a window over a contaminated picture [3], which is accomplished by convolving a window over a contaminated image. To replace the noisy pixels depending on the present processing window values, a mean of neighbor four pixels, an unsymmetrical trimmed median, an unsymmetrical trimmed midpoint, or a global trimmed mean are calculated depending on the content of the data handling window values [4].

A decision-based asymmetrical trimmed modified Winsorized mean filter (DATMWMF) first identifies noisy pixels based on their pixel values, which may be either 0 or 255, and then replaces them with the asymmetrically trimmed median value calculated from the pixel values. Pixels with values other than 0 or 255 are regarded to be non-noisy pixels since they do not have any noise. When applied to individual frames of video, the DATMWMF algorithm brings back the original picture [5], which is then combined.

Manhattan Distance Method

Comparisons are made between the performances of different distance measures on interpolation algorithms implemented to natural pictures affected by speckle noise. Based on the findings generated by the method utilizing Manhattan distance ($L1$ norm), it was discovered that the technique produced extremely excellent results when compared to other Pythagorean measurements. When interpolating data inside a restricted neighborhood, the method requires pixels three in minimum number to be used. The rationale for utilizing these pixels is primarily to avoid the blur at the edges as much as possible [2].

Adaptive Weighted Mean Filter (AMWF)
In the detection of noise, the method of adaptive weighted mean filter technique is applied to replace the pixel with noise with the weighted mean value of the current window, while the pixel free with noise is kept unaltered. There are two main reasons why we chose the weighted mean over the other options. The comparisons reveal that the PSNR of the picture in which noise pixels are replaced by weighted mean values is greater on average than the PSNR of the image in which noise pixels are not replaced by weighted mean values [3].

Adaptive Read Median Filter (ARMF)
An adaptive read median filter (ARMF) is a new adaptive method for salt-and-pepper noise removal. In this method, a noise image matrices is considered to obtain a binary matrix. After that, *t*-symmetric pad matrices of these matrices are constructed. If an entry of the binary matrix is equal to zero and its 1-approximate matrix differs from zero matrices, then strictly increasing regular entry matrix (SIREM) of the 1-approximate matrix is obtained, and the right median is evaluated. Afterward, this right median is overwritten to the entry. If an entry of the binary matrix is equal to zero and its 1-approximate matrix is zero matrices, but the 2-approximate matrix of this entry differs from zero matrices, then SIREM of 2-approximate matrix is obtained, and the right median is evaluated. Afterward, this right median is overwritten to the entry. Similarly, if an entry of the binary matrix is equal to zero and its (1)-approximate matrix is zero matrix, but the k-approximate matrix of this entry differs from zero matrix, then SIREM of the -approximate matrix is obtained, and the right median is evaluated. Afterward, this right median is overwritten to the entry [2].

3 Kriging Interpolation

The kriging interpolation is a geo-statistical method that generates and predicts unknown values from a known set of points. It is an interpolation method based on statistical relationships and the distance among measured points. Kriging uses the variogram and covariance to estimate the statistical relationships between measured points. Interpolation using kriging for the removal of salt-and-pepper noise from images isolates pixels that are not affected by noise. Only noisy pixels are processed by this algorithm. Damaged pixels are replaced with a value that is interpolated using weights computed based on the semi-variance in between corrupted pixels and non-noisy pixels in a restricted neighborhood of the corrupted pixels. It is necessary to have at least three non-noisy pixels in the current processing window.

4 Proposed Method

Identifying noise and eliminating noisy areas by correcting the noise is one of the best mechanisms to treat the images containing broken components. Though, high definition devices grab the qualitative images, there are opportunities that noise prevails in the images due to light and hardware reasons. On images that are grabbed at random, the procedure of purification and correction shall be applied. Particularly in the process of exudates detection in the diabetic retinopathy images, the cotton-wooly areas are covered as the exudates component.

In the process of applying the kriging on diabetic retinopathy images, the source images are filtered through an exudates image mask and are recognized as having abnormalities. The cloudy or cotton-wooly areas recognized as exudates are not perfectly shaped or as the gradients do not cover the exact boundaries. The images with such abnormalities of exudates, which are not clearly described with the presence of exudates, are treated with kriging, so that gaps or pixels with disconnected representation of exudates are interpolated. A kriging interpolation filter does the filtering of the 3×3 window of each image.

The DR images are collected from diaretb1, calibration 1 (Fig. 1).

The ground truth and masks of DR images collected from diaretb1, calibration 1 for exudates detection (Fig. 2).

The proposed filter is designed by varying the distance between non-noisy pixels existing adaptive kriging interpolation filter based on $L1$ norm or $L2$ norm. A noisy image is read through a 3×3 window and tested for the pixel, whether corrupted. The corrupted pixels are retained for reconstruction and restoration. If a pixel is relevant in the image, then it is stored in the array Z. Manhattan distance is computed between the non-noisy pixels in the array of the selected window in order to generate h-matrix. The values of the h-matrix are summed row and column wise. A weight is generated denoted by $W(j)$, from the said sum is divided by $2 \times L$, where L is the number of non-noisy pixels in the current window. The sum of product of non-noisy

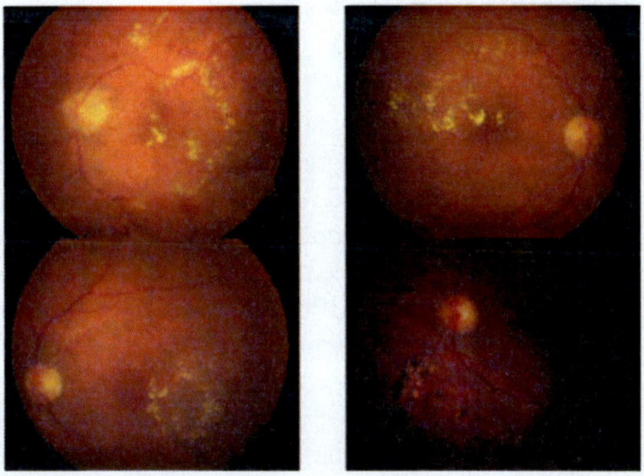

Fig. 1 Images with symptoms of various levels of exudates

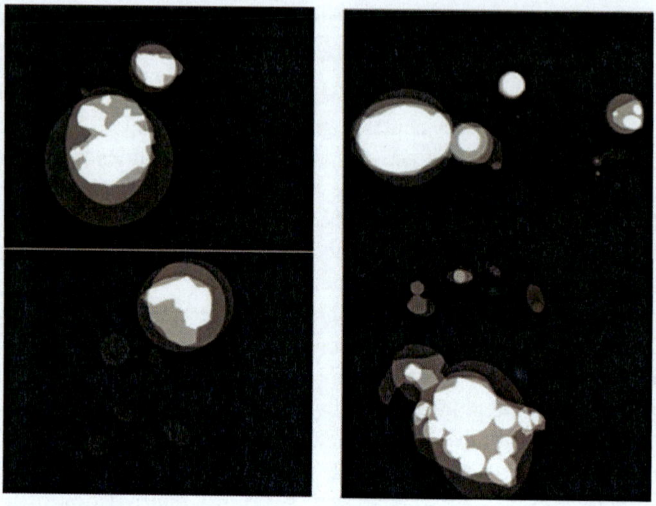

Fig. 2 Ground truth images as masks with symptoms of various levels of exudates

pixels with weights divided by the sum of weights will derive the new interpolated value. The new interpolated value is treated as good pixel and substituted in the place of noisy pixel. An assertion for restoration and reconstruction will be performed, if necessary the process is iterated for the needy areas of the image containing noisy pixels (Fig. 3).

On identifying the images containing the exudates components in the images, the interpolation method of kriging is applied based on the ground truth filters on the selected images. The diabetic retinopathy images with kriging interpolation representing enhancements of exudates are shown in Fig. 4.

5 Results and Discussion

The results were plotted for noise densities ranging from 10 to 90% in terms of mean square error (MSE), structural similarity index metric (SSIM), and cumulative probability blur detection metric (CPBD). The value of the mean square error indicates that the amount of deviation the distance measure offers while applying the proposed algorithm. The structural similarity index metric indicates the edge preservation capability of the proposed algorithm using different distance-based measures. The no reference metric CPBD indicates the amount of perceivable blur caused by the low pass filter by different distance measures. Table 1 indicates the error obtained when different distance measures are applied on adaptive kriging interpolation algorithm on a fundus images of diabetic retinopathy corrupted by speckle noise. Table 2 indicates the structural similarity index metric of different distance measures which

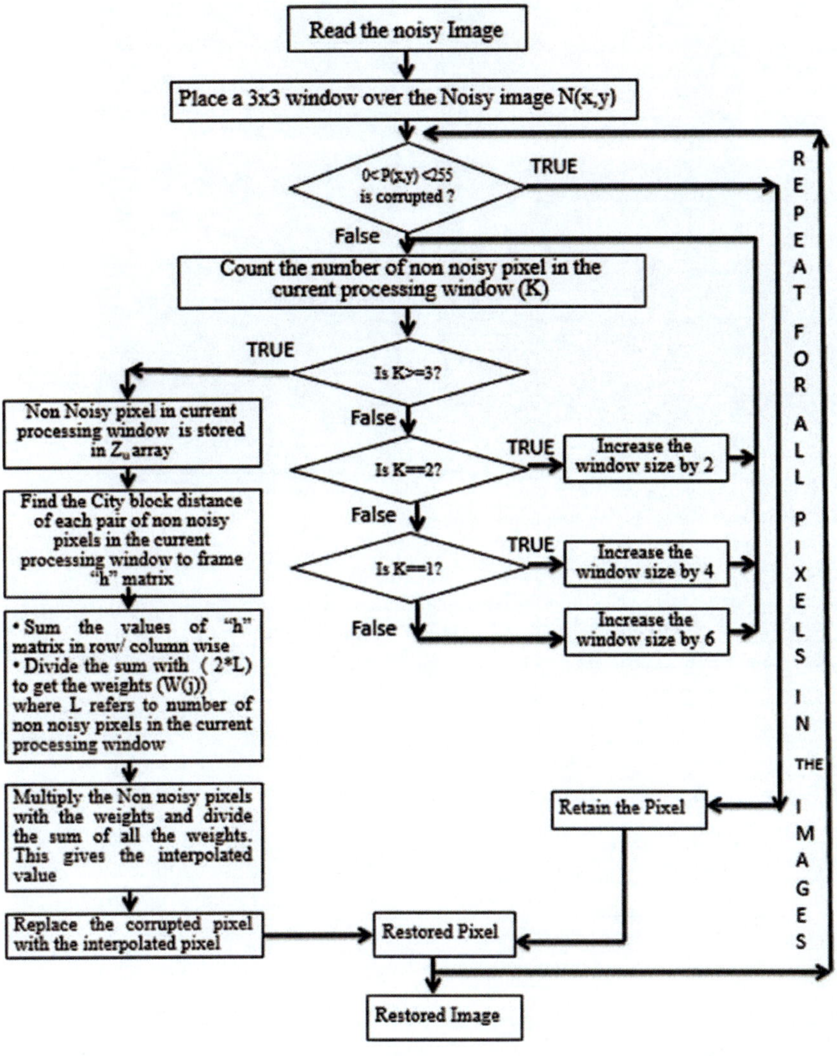

Fig. 3 Flowchart of the adaptive kriging interpolation algorithm using Manhattan distance

is applied on adaptive kriging interpolation algorithm on a fundus images of diabetic retinopathy corrupted by speckle noise. Table 3 indicates the cumulative probability blur detection metric (CPBD) of different distance measures which is applied on the adaptive kriging interpolation Algorithm on a fundus images of diabetic retinopathy corrupted by speckle noise. Figure 2 indicates the error obtained on different distance measures on adaptive kriging interpolation on fundus images of diabetic retinopathy corrupted by speckle noise.

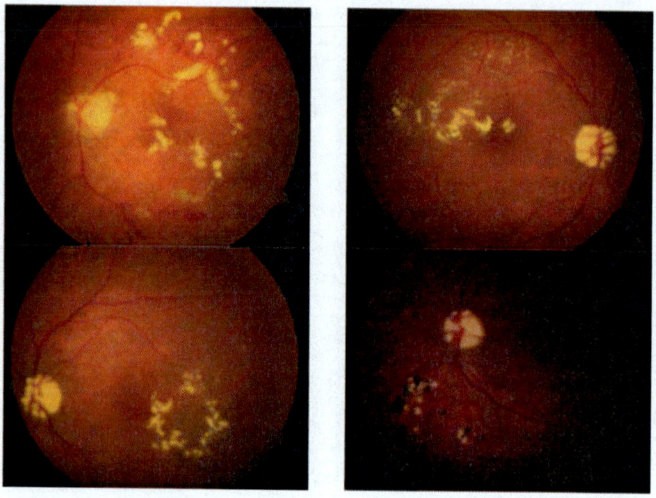

Fig. 4 Images after applying kriging interpolation

Table 1 Mean square value of adaptive kriging interpolation algorithm using various distance measures on diabetic retinopathy images corrupted by speckle noise

ND	Squared Euclidian distance	Chess board distance	Manhattan distance	Canberra distance
0.1	21.7897	21.9566	20.9497	21.0532
0.2	44.3298	43.5364	43.2465	43.676
0.3	67.7314	69.4442	68.7202	69.9579
0.4	98.2536	99.4556	99.1289	101.905
0.5	134.7089	137.9789	134.7488	138.3903
0.6	180.7331	191.0931	181.8307	193.3193
0.7	248.037	259.615	250.193	268.8124
0.8	368.0865	387.0095	365.0073	390.3282
0.9	629.9875	651.0517	639.3513	669.6858

In the adaptive kriging interpolation, experiment on fundus images of diabetic retinopathy has obtained various distance measures. As the percentage of noise detected increases from 10 to 100%, the distances between the potential pixels with noisy pixels are increasing exponentially. The Manhattan distance at the 10% level of noise is 20.9497 units and has subsequently increased to 99.1289 units at 40% level of noise. Further, at the 90% level of noise, there is an increase up to 629.9875, 651.0517, 639.3513, and 669.6858 for squared Euclidean, chess board, Manhattan distance, and Canberra distance, respectively. The following graph illustrates the curve gradually increasing and reaching the points between 600 and 700 indicatively representing the distance measures.

Table 2 Structural similarity index metric (SSIM) values of adaptive kriging interpolation algorithm using various distance measures on diabetic retinopathy images corrupted by speckle noise

ND	Squared Euclidian distance	Chess board distance	Manhattan distance	Canberra distance
0.1	0.9785	0.9792	0.9794	0.98
0.2	0.9555	0.9567	0.957	0.9575
0.3	0.9301	0.9294	0.9308	0.9308
0.4	0.8987	0.8983	0.8994	0.898
0.5	0.859	0.857	0.862	0.859
0.6	0.811	0.808	0.8122	0.8055
0.7	0.7469	0.7354	0.7429	0.7315
0.8	0.6427	0.6339	0.6429	0.6305
0.9	0.4865	0.4714	0.4805	0.4661

Table 3 Cumulative probability blur detection metric (CPBD) values of adaptive kriging interpolation algorithm using various distance measures on diabetic retinopathy images corrupted by speckle noise

ND	Squared Euclidian distance	Chess board distance	Manhattan distance	Canberra distance
0.1	0.6859	0.6864	0.6854	0.6855
0.2	0.6689	0.6689	0.6701	0.6664
0.3	0.6485	0.6514	0.6467	0.6438
0.4	0.6252	0.6208	0.6218	0.6234
0.5	0.5881	0.5907	0.5878	0.5978
0.6	0.5471	0.5526	0.5474	0.5596
0.7	0.4965	0.5016	0.499	0.5147
0.8	0.4332	0.4454	0.4403	0.4576
0.9	0.3263	0.3504	0.3422	0.3836

The structured similarity index metric (SSIM) has developed almost floating values when the adaptive kriging interpolation is applied, experimented on fundus images of diabetic retinopathy corrupted by speckle noise. As the percentage of noise detected increases from 10 to 100%, the distances between the potential pixels with noisy pixels is moderately decreasing. The Manhattan distance at the 10% level of noise is 0.9794 units and has subsequently decreased to 0.8994 units at 40% level of noise. Further, at the 90% level of noise, a drastic decrease of distance measures is noticed as said following 0.4865, 0.4714, 0.4805, and 0.4661 for squared Euclidean, chess board, Manhattan distance, and Canberra distance, respectively. The following graph illustrates the moderately decline in the curve indicatively representing the distance measures.

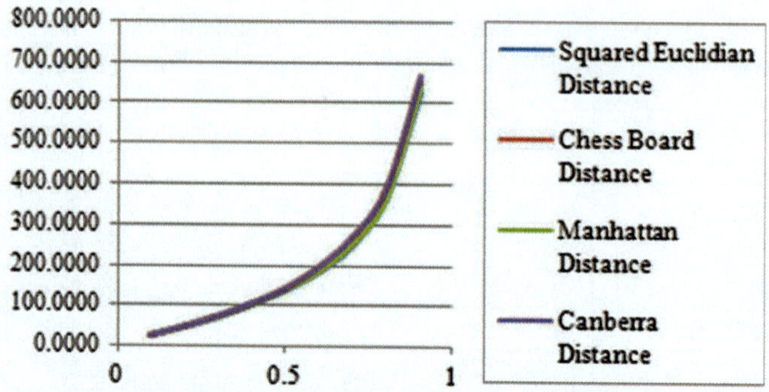

Fig. 5 Mean square value of adaptive kriging interpolation algorithm using various distance measures on diabetic retinopathy images corrupted by speckle noise

The cumulative probability blur detection metric (CPBD) has also developed certain floating values when the adaptive kriging interpolation is applied, experimented on fundus images of diabetic retinopathy corrupted by speckle noise. As the percentage of noise detected increases from 10 to 100%, the distances between the potential pixels with noisy pixels has more substantially decreased than in the experiments of SSIM. The Manhattan distance at the 10% level of noise is 0.6854 units and has subsequently decreased to 0.6218 units at 40% level of noise. Further, at the 90% level of noise, a substantial decrease of distance measures is noticed as said following0.3263, 0.3504, 0.3422, and 0.3836 for squared Euclidean, chess board, Manhattan distance, and Canberra distance, respectively. The following graph illustrates the substantial decline in the curve indicatively representing the distance measures.

It was observed from Table 1 and Fig. 5 that Manhattan distance offers very less error on fundus images of diabetic retinopathy when compared to other distance measure applied on adaptive kriging interpolation. Table 2 and Fig. 6 show Manhattan distance gives higher values of SSIM after 70% on natural images.

The Manhattan distance shows a high degree of structural preservation at very high noise densities when compared to other distance measures. Table 3 and Fig. 7 indicate the squared Euclidean distance offers less blurring at high noise densities when compared to another distance measure.

It is vivid that the proposed algorithm with the Manhattan distance fairs well in all the images of the database. The Manhattan distance act as a good replacement of Euclidean distance measure on natural images as the latter amplifies the noise due to the squaring nature of the measure and the former has reduced effect. But the degree of blurring is less in Euclidean distance. So in a natural image due to pixel values which occur at closer proximity, we rather use first-order distance rather than second-order distance. This reduces the complexity with enhanced noise elimination capability with good information preservation.

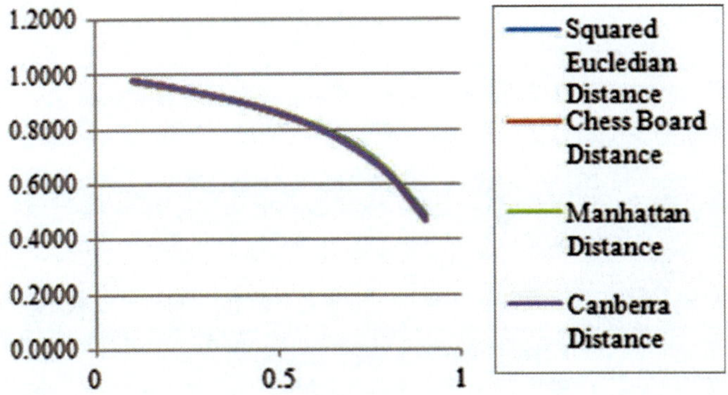

Fig. 6 Structural similarity index metric (SSIM) values of adaptive Kriging interpolation algorithm using various distance measures on diabetic retinopathy images corrupted by speckle noise

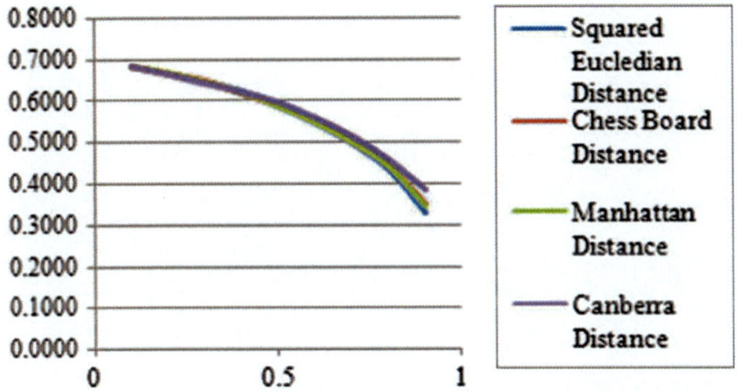

Fig. 7 Cumulative probability blur detection metric (CPBD) values of adaptive Kriging interpolation algorithm using various distance measures on diabetic retinopathy images corrupted by speckle noise

6 Conclusion

An adaptive kriging interpolation algorithm using Manhattan distance is proposed. The use of first-order measures offers reduced error and preserves the structure of the image. The use of Manhattan distance reduces the complexity of the algorithm while hardware implementation. The algorithm eliminates noise at high noise densities with excellent information capability in the presence of speckle noise.

References

1. Lu CT, Chen YY, Wang LL, Chang CF (2016) Removal of salt-and-pepper noise in the corrupted image using a three-values-weighted approach with variable-size window. Pattern Recogn Lett 80:188–199
2. Qistina A, Taha Md, Haidi I, Ina Qistina Md, Taha, Ibrahim H, Springer Nature Singapore Pte Ltd. (2020) In: Jamaluddin Z, Ali Mokhtar MN (eds) Reduction Of Sapn from digital gray scale image by using recursive switching adaptive median filter. Sympo SIMM 2019. LNME, pp 32–47, 2020.
3. Kandemir C, Kalyoncu C, Toygar Ö, Gusa G (2015) A-weighted mean filter with spatial bias elimination for impulse noise removal. Digital Signal Process 46(2015):164–174. https://doi.org/10.1016/j.dsp.2015.08.012 (1051-2004/© 2015 Elsevier)
4. Subaja Christo M, Vasanth K, Varatharajan R (2019) A decision based asymmetrically trimmed modified winsorized median filter for the removal of salt and pepper noise in images and videos. In: Multimedia tools and applications. https://doi.org/10.1007/s11042-019-08124-9
5. Zhang P, Li F (2014) A new adaptive weighted mean filter for removing salt and pepper noise. IEEE Signal Process Lett 21(10):1280–1283

A Secure and Optimal Path Hybrid Ant-Based Routing Protocol with Hope Count Minimization for Wireless Sensor Networks

Voruganti Naresh Kumar and Ganpat Joshi

Abstract In wireless sensor networks (WSNs), particular objectives are achieved by different routing protocols which are efficient for specific situations. But, some limitations are owned by WSN systems, and these are security threats or low data rates. So, the designing process of WSN routing protocols is difficult, and WSN network lifetime is effected by the factor of topology change. Wireless sensor networks (WSNs) routing protocols designing are having the key factors as easy maintenance, high reliability, and low cost. In this paper, a secure and optimal path hybrid ant-based routing protocol with hope count minimization for WSN is analyzed. An adaptive secure hybrid routing protocol is also analyzed in this paper which is based on bio-inspired mechanism that integrates ant colony optimization (ACO). Two optimal paths are selected in the scheme of minimum hop count in addition with route security. Each node balanced energy consumption, total energy consumption which is less and optimal routing path can be calculated by the proposed algorithm. Superiority is maintained by the proposed algorithm while searching the optimal path, network topology maintenance, and network load balancing. According to simulation result, a better performance is achieved by proposed routing protocol than other WSN routing algorithms.

Keywords Routing protocol · Wireless sensor networks (WSNs) · Ant colony optimization (ACO) · Minimum hop count scheme

V. N. Kumar (✉) · G. Joshi
Research Scholar, Department of Computer Science and Engineering, Madhav University, Rajasthan, India

V. N. Kumar
Department of Computer Science and Engineering, CMR Technical Campus, Hyderabad, India

P. S. R. Chowdary et al. (eds.), *Evolution in Signal Processing and Telecommunication Networks*, Lecture Notes in Electrical Engineering 839,
https://doi.org/10.1007/978-981-16-8554-5_50

1 Introduction

Low cost and small sensor manufacturing is becoming a trendy and economically feasible in present days because of improvement in technology. Ambient conditions can be measured by sensing elements in the sensor surrounding areas and converts them in the form of electrical signals. The sensor surrounding area happening events and properties of sensor are revealed by the processing electrical signals. Some unattended operations are required by different applications which significantly uses more and more disposal sensors. Too many, like thousands of sensor nodes are contained with wireless sensor network (WSN) [1]. External base-station (BS) communication with the each and every node can be done by these sensors. Greater accuracy of the sensors is obtained in geographical regions.

According to certain attributes like addressing based on attributes, sensing elements will analysis the data with data-centric networks as WSNs [2]. Pair query of attribute-value set forms the addressing based on attributes. WSNs architecture and application requirements along with features can be considered in these routing mechanisms [3]. A nontrivial maintenance and finding routes are existed in the WSNs because node changes which results unpredictable and frequent topological changes, and energy restrictions. Well-known routing tactics are employed by WSNs and tactics special to WSNs which are proposed in literature sections with routing techniques in order to minimize the energy consumption [4]. Data-centric methods, different node role assignment, clustering, data aggregation, and in-network processing are employed in tactics special to WSNs. Based on network structures like hierarchical, flatness the routing protocols are to be classified [5].

Because of inherent scalable features of bio-inspired mechanisms, the routing techniques are stronger than other methods, and interesting solutions are provided [6]. An adaptive secure hybrid routing protocol is also analyzed in this paper which is based on bio-inspired mechanism that integrates ant colony optimization (ACO). WSN ant colony optimization routing algorithms advantages and as well as disadvantages are also analyzed in this paper with WSN minimum hop count-based routing algorithms [7]. So, two algorithms are integrated as to form a hybrid WSN routing algorithm, and different critical factors which are specified for WSN are considered along with sensor nodes energy constraint, dynamic network topology, and load balancing network. A routing algorithm of ant colony optimization with hop count minimization for WSN is presented in this paper. The proposed protocol provides adaptive security mechanism. Data transmission uses two paths which are selected by routing mechanism. Both of these two paths provided the security from proposed mechanism. The optimal paths are selected cording to security, robustness, and adaptability with the proposed algorithm.

2 Security Issues and Threats in Wireless Sensor Networks

2.1 Security Services or Requirements

Confidentiality: Information reading by unauthorized persons is prevented by confidentiality. The main aim of the term confidentiality is to maintain the secrecy while data being transfer process. The technique of systematic encryption is achieved by this process [8].

Integrity: Unauthorized writing can be eliminated by the integrity of the information. Intrusion detection tools, cryptography, and firewalls are tools which help in modification of data. The data are protected by integrity on these conditions also. The data will remain same for source end to destination end in transmission process [9].

Availability: Information security is having the fundamental unit as data availability. Accepted business continuity plan-based procedure of sound disaster recovery planning is involved by data availability.

2.2 Security Issues in Wireless Sensor Networks

Large storage requirements and memory with high processing capability are required by the traditional security mechanisms. But, wireless sensor networks are not having such large memory resources. So, effective security mechanisms designing process is too difficult than wired network. Some of these restrictions are as follows:

(a) **Small memory**: Storage capacity is very small with limited memory is present in Wireless sensor node. So, the designing process of security mechanism with wireless sensor networks is having the limitations whereas wired network not.
(b) **Reduced energy levels**: Reduced energy levels must be considered in the designing of security mechanisms with wireless sensor networks which are implicit with sensor nodes. Battery is the energy source when in the sensor node deployment. Quick exhausted battery life prevention is done by security feature designing which is critical. The power consumption of security features will consume more power than operation [10].
(c) **Communication problems**: Wireless communication contains the inherent problem in which the information is lost, intercepted, and attacked. In wireless sensor networks, data packets are damaged, and in between the transmitter and receiver ends, data will be lost because of heavy traffic [11].
(d) **Physical security**: A node is efficiently damaged and captured by attacker because sensor nodes are having less strength.

2.3 *Threats and Issues in Wireless Sensor Networks*

Different security threats acquire the wireless networks.

1. **Denial of Service (DoS) attack**: Malicious action or nodes unintentional failure causes the DoS attack. The available resources are reduced by the simplest DoS attack with unnecessary packets sending. From accessing resources or services will prevent the users of legitimate network [12]. In different layers of wireless sensor networks, several DoS attacks are performed. DoS attacks are tampering and jumping at physical layer; unfairness, exhaustion, collision are at link layer; black holes, misdirection, homing, greed, and neglect are at network layer, and de-synchronization and malicious flooding is performs this attack at transport layer [13].
2. **Wormhole Attack**: Network one part is received attacker tunnels messages with low latency link, and different part may response. This is called as wormhole attack. Two distant nodes which are neighbors are convinced with these wormhole attacks [14].
3. **Sybil Attack Idea**: Network different parts are existed in the prediction of single node. Other nodes multiple identities are simply presented by a single node. Fault-tolerant schemes effectiveness is reduced by the Sybil attack and these schemes as disparity and multipath routing, distributed storage, storage partitions, and topology maintenance replicas. Disjoint nodes are used for believing the routes, and these are using multiple identities with single identity presentation. In Sybil attack, number of legitimate nodes can be identified by the imitation of sensor node. A Sybil attack can suspect all peer-to-peer networks. But, Sybil nodes detection is difficult [15].

3 'ACOMHC' Routing Scheme

Number of applications and enormous usage, the gaining of WSN is more popular nowadays. WSN important factor is security consideration even in hostile and harsh environment. So, while transmitting information from source to destination, most important parameter is secure routing. Hybrid secure routing is proposed in this paper based on ant colony optimization (ACO) method along with minimum hope count mechanism (MHC). WSNs lifetime is maximized with proposed hybrid routing algorithm with ACOMHC. Best solutions searching uses the effective routing algorithm which must meet quickly and having the high success rate for the optimal path, maintenance, route update, and fewer complications in finding routes. One of the planar network topology is hop count-classification which is used in presenting ACOMHC algorithm. Route discovery, route selection, route security, and data forwarding are the four main stages in routing protocol. Node information location based proposed method is designed, so the maintenance is easy. Therefore, many advantages are integrated with this simple routing protocol structure. Initialization, network topology generation, route discovery, route selection, route security, and data forwarding are the six major steps in ACOMHC algorithm implementation procedure. The framework of this algorithm is shown in Fig. 1.

3.1 Network Topology Generation-MHC

The combination of ant colony optimization with minimum hop count can develop the ACOMHC routing algorithm for WSN systems. Each node location information can be examined in the network by the algorithm before the ants' starts and uses the hop count-classification idea as a reference for marking all nodes according to their hop count values. The sink node hop count defines the node hop count value as 0, and other nodes hop count value is considered as their neighbor nodes smallest hop count value with addition of 1. After marking the each node hop count value, network topology of hop count-classification is formed with total network nodes. Network center square node is called as sink node, and remaining nodes are called as ordinary sensor nodes. Hop count can be represented as the number on the node. The data cannot transmit for any h-hop ($h \geq 2$) to the hop counts are $h \pm 2$ with nodes without starting power amplifier. Therefore, the h-hop ($h \geq 1$) node best next hop nodes must be $h - 1$ hop nodes without considering the node energy. While choosing the next hop node, too many alternative nodes are efficiently reduced by the topology of hop count-classification. Then, algorithm time complexity is reduced, and convergence speed is accelerated. Network topology of hop count-classification maintenance is dependent on the topology maintenance. If there is a change in network topology, then topology maintenance process starts. The changes are because of wireless link breakage or failure in sensor nodes. The values of hop count are also changed accordingly with topology changes. So, updating of hop count values is must with route.

3.2 Route discovery—ACO Technique

The shortest distance or path can be detected by ant from source to destination, i.e., home to food while moving in the paths ants can leave pheromone named chemical substance. The pheromone concentration or intensity is more in shortest path than other paths. Forward and backward ants are two important types in ants. Route reply and route request mechanisms utilize the above ant concept. Hello packets can discover the surrounding neighbors, and this simple mechanism used the proposed routing protocol before the process of route discovery. According to surrounding neighbors, received response routing table is updated. Source node broadcasts a fixed number of ants (route request packets (RREQ)) in the discovery process of route to random neighbors. RREQ ant's lifetime interval is fixed by source node, and these ants are called as forward ants. These forward ants can determine the destination or food with finding intermediate nodes. Therefore, through the intermediate nodes which are random, many RREQ ants are received at the destination side.

Source node sending RREQ ants are equal or less number of RREQ ants at destination side because some ants are may confused in finding the actual path, due to congestion and some may lose their life. Back route reply (RREP) ants are sending by the destination node by suing forward ants used paths. So, backward ants are represented with RREP ants. Forward and backward ants packet formats are almost same, but field path security is maintained by backward ant in place of reputation value. 0 or 1 is two values which are accepted by path security field. In these values, insecure path is denoted by 0 and needs further security mechanism, and secured path is indicated by 1 and no need of further security mechanism.

Two main features are owned by route discovery process as:

(i) Reputation value increment is done by fixed numbers of forward ant's propagation along multi-hop path till destination is clear.
(ii) Too many forward ants are received by destination node and respond through the propagation of backward ant toward source node. Path security field is enabled by the destination node for identifying security need of source node with reputation values observations.

3.3 Route Selection—ACO Technique

The procedure of route selection is triggered after receiving more backward ants RREPs packets by source node. WSN characteristics must to be considered during the optimal routing path searching. For path selection in ant colony algorithm, a new strategy of pheromone update is introduced in place of heuristic factor in the state transition formula of probabilistic node, and it is also be revised. According to security features, two paths are selected by the source node after received backward ants examination in route selection process. ACO inspires the proposed mechanism route selection and route discovery process.

(i) R1, R2, R3, R4 are rotes selected by ants A1, A2, A3, A4, respectively.
(ii) From destination side or food (F), the ants are A1, A2, A3, and A4 are reached randomly.
(iii) For example, fewer difficulties are faced by ant A3; then, it leaves pheromone with high concentration. Therefore, R3 route is having the high trust value than other routes.
(iv) All paths trust values are compared at destination side.
(v) For reaching the starting point, the ants should follow the same path of its own from destination.
(vi) Based on security values, two paths are selected at starting point.
(vii) Then, all ants are traveled on selected two paths which are having high security.

3.4 Routing Security—Watchdog Mechanism

Described routing protocol security mechanism follows the watchdog mechanism. At each sensor node, this watchdog security scheme is implemented. All surrounding neighbors are monitored by sensor nodes within the radio range of its own by using watchdog mechanism. Cooperative or non-cooperative neighbors are classified by node based on mutual monitoring mechanisms. Surrounding nodes data forwarding behavior observing, data collection, and neighbors monitoring are some actions of watchdog mechanism. All neighbors are assigned with reputation value in proposed method, and the next neighbor reputation value is incremented when there is arrived of forward ant. In the backward ant propagation, decisions are made with enabling 0 or 1 by destination node (sink) according to requirement of transferring encrypted data or not for source node. If backward ant with 0 of path security field is received by source node, then encrypted data ensured the security. The used encryption mechanism is very light in this process. RC4 generates the permutated key in this mechanism. RC4 key encryption changes dynamically. Different keys are dynamical generation with every data packet.

3.5 Data Forwarding

Each node balanced energy consumption and total minimum energy consumption-based mutation strategy with dynamic energy threshold strategy are described in this paper during the optimal path searching. For path selection, heuristic factor is revised especially, and the formulae of probabilistic node state transition new pheromone update strategy is replaced. Low maintenance, routing reliability, convergence speed, load balancing, and wide ranging of energy efficiency are maintained in the new algorithm. Energy efficiency is calculated during the data transmitting process (Kbit/J) as the ratio between sink node delivered total packets, and network's sensor nodes consumed total energy in data transmission path. The sensor nodes total energy consumption minimization is expected in this method on routing path. Node energy consumption is reduced with data delivery, and route discovery involvement with redundant transmissions reduction is also expected in this method.

4 Results

On C+ +, WSN environment is simulated for evaluating the ACOMHC algorithm performances. 100 nodes with 100 m $\times$ 100 m area are randomly generated in the simulation. The whole area center is assumed as sink node which is represented in Fig. 1. The best combination of the parameters described in the framework of algorithm is determined by experiments. All protocols data forwarding efficiency

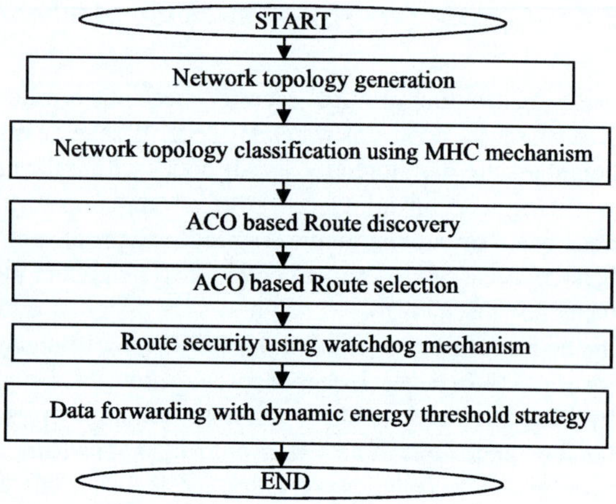

Fig. 1 Framework of 'ACOMHC' routing algorithm

is represented in Fig. 2. Because of maintenance of two paths, the data are more efficiently and quickly transfer the data from source to destination in the ACOMHC routing scheme.

The routing protocols packet loss ratio is observed by the introduction of few malicious nodes which are represented in below Fig. 3. The traffic can be misdirected

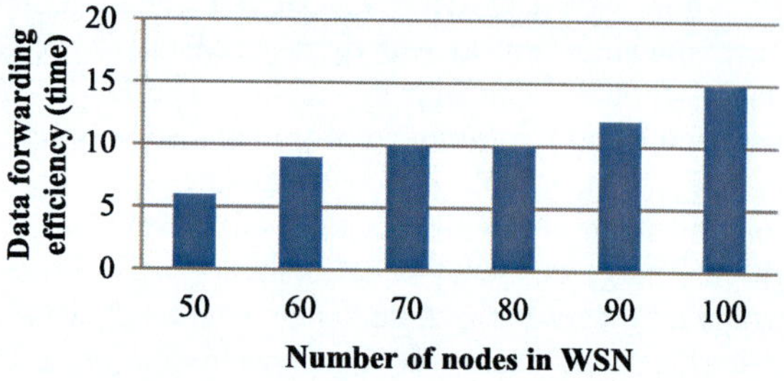

Fig. 2 Data forwarding efficiency

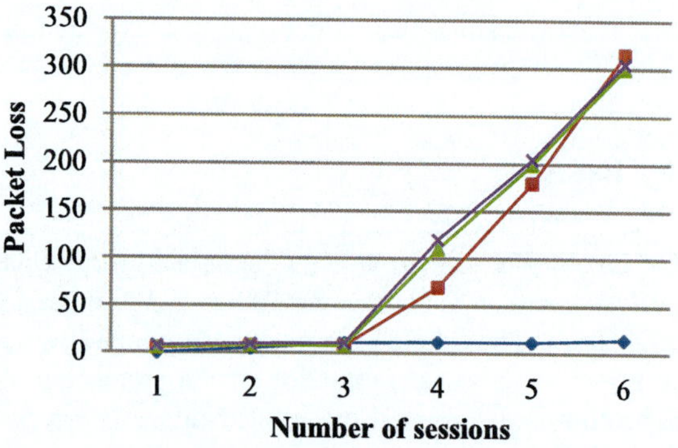

Fig. 3 Packet loss in the presence of malicious nodes

or dropped with the malicious nodes from source and destination. Therefore, the comparative analysis of packet loss of the ant colony optimization (ACO), life time aware *routing* algorithm for wireless sensor networks (LTAWSN), genetic algorithm (ACO-GA), and ACOMHC algorithm in the experiments is as follows. It is clearly to see that the packet loss of ACOMHC is the low compared to the others.

The network lifetime is defined as the duration beginning when the network is activated until when one node's battery goes flat. The networks which consist of 50, 60, 70, 80, 90, and 100 sensor nodes are generated, respectively, in the simulation area, and the nodes' routing information is produced by running five algorithms, respectively. Figure 4 is the comparison of the network lifetime of our proposed one against other four algorithms. It is clearly to see that the network lifetime of ACOMHC is the longest.

The success rate of searching for the optimal solution in 100 iterations of the four algorithms is recorded in Fig. 5.

The success rates of different algorithms are represented in Fig. 5, and it is clear that proposed ACOMHC algorithm is more superior to others. 100 iterations are considered for testing the algorithm optimal solution success rate in each test. It can be shows that ACOHCM has a strong global optimal solution searching ability, and its success rate is higher than that of the other algorithms. The above simulation

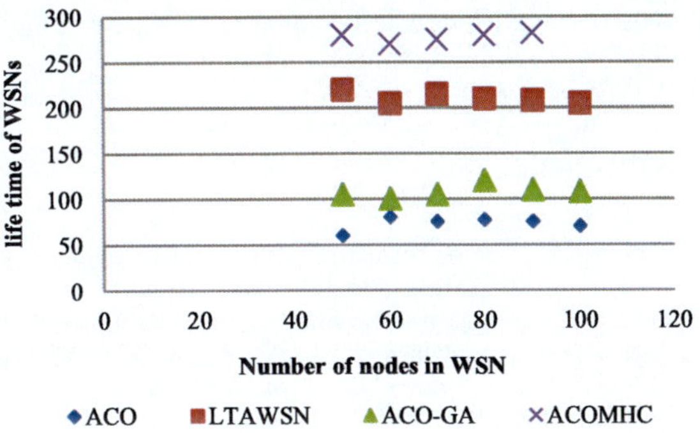

Fig. 4 Comparison of the network lifetime

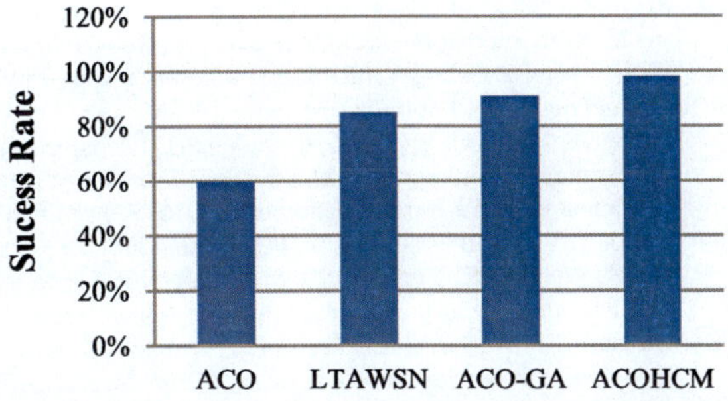

Fig. 5 The success rate of searching for the optimal solution of algorithms

results have shown that ACOHCM can greatly improves the routing computation efficiency and global optimal solution searching success rate, and verify that ACOMHC algorithm is reliable and effective.

5 Conclusion

This paper presents a secure hybrid WSN routing algorithm named ACOMHC. Our proposed ACOMHC takes into comprehensive account energy efficiency, routing reliability, and secure routing. Data security while transmitting it between the sources to destination is increased with WSN secure routing protocol which is analyzed in this paper according to real-world applications. Overall performance and efficiency are greatly improved with the combination of minimum hop count routing strategy and ant colony optimization to form a hybrid algorithm of ACOMHC. Reputation values are collected and incremented along the path by forward ants in proposed mechanism. Route security of destination node intrusion and information is carried by using destination node backward ants. Overall network efficiency improvement and node failure problem are overcome by data forwarding two paths of proposed mechanism. Mutation strategy and threshold strategy of dynamic energy are also analyzed in this method. ACOMHC routing scheme shows better performance and found to be efficient when compared to other routing protocols.

References

1. Niya JM, Azarhava H (2020) Energy efficient resource allocation in wireless energy harvesting sensor networks. IEEE Wirel Commun Lett 9(7)
2. Chen H, Feng J (2019) Repairing confident information coverage holes for big data collection in large-scale heterogeneous wireless sensor networks. IEEE Access 7
3. Abdulazeem Y, Badawy M, Farsi M, Ali HA, Moustafa M (2019) A congestion-aware clustering and routing (CCR) protocol for mitigating congestion in WSN. IEEE Access 7
4. Mazinani SM, Mohajer PAA, Daneshvar SMMH (2019) Energy-efficient routing in WSN: a centralized cluster-based approach via grey wolf optimizer. IEEE Access 7
5. Singh R, Kathuria K, Sagar AK (2018) Secure routing protocols for wireless sensor networks. In: 2018 4th international conference on computing communication and automation (ICCCA)
6. Ni Q, Sheng Z, Wang Y (2018) A microbial inspired routing protocol for VANETs. IEEE Internet of Things J 5(4)
7. Alongi A, Vitello G, Vitabile S, Conti V (2017) A bio-inspired cognitive agent for autonomous urban vehicles routing optimization. IEEE Trans Cogn Dev Syst 9(1)
8. Rebaudengo M, Ferrero R, Gandino F (2016) A key distribution scheme for mobile wireless sensor networks: q–s–composite. IEEE Trans Inf Forensics Secur 12(1)
9. Ota K, Wu J, Li C, Dong M (2016) A hierarchical security framework for defending against sophisticated attacks on wireless sensor networks in smart cities. IEEE Access 4
10. Ouyang Y-C, Wen C-Y, Hsueh C-T (2015) A secure scheme against power exhausting attacks in hierarchical wireless sensor networks. IEEE Sens J 15(6)
11. Pegatoquet A, Castagnetti A, Auguin M, Le TN (2014) A joint duty-cycle and transmission power management for energy harvesting WSN. IEEE Trans Ind Inf 10(2)

12. Luo X, Tang Y, Chang RKC, Hui Q (2014) Modeling the vulnerability of feedback-control based internet services to low-rate DoS attacks. IEEE Trans Inf Forensics Secur 9(3)
13. Krishnamurthi I, Marimuthu M (2013) Enhanced OLSR for defense against DOS attack in ad hoc networks. J Commun Netw 15(1)
14. Shi Z, Sun R, Lu R, Qiao J, Chen J, Shen X (2013) A Wormhole attack resistant neighbor discovery scheme with RDMA protocol for 60 GHz directional network. In: IEEE Trans Emerg Top Comput 1(2):341–352. https://doi.org/10.1109/TETC.2013.2273220
15. Xu F, Wei W, Li Q, Tan CC (2013) SybilDefender: a defense mechanism for sybil attacks in large social networks. IEEE Trans Parallel Distrib Syst 24(12)

Author Index

P. S. R. Chowdary et al. (eds.), *Evolution in Signal Processing and Telecommunication Networks*, Lecture Notes in Electrical Engineering 839,
https://doi.org/10.1007/978-981-16-8554-5